Multiple Time Scales

This is Volume 3 in

COMPUTATIONAL TECHNIQUES

Edited by BERNI J. ALDER and SIDNEY FERNBACH

Multiple Time Scales

Edited by

JEREMIAH U. BRACKBILL
Los Alamos National Laboratory
University of California
Los Alamos, New Mexico

BRUCE I. COHEN
Lawrence Livermore National Laboratory
University of California
Livermore, California

1985

ACADEMIC PRESS, INC.
(Harcourt Brace Jovanovich, Publishers)
Orlando   San Diego   New York   London
Toronto   Montreal   Sydney   Tokyo

COPYRIGHT © 1985, BY ACADEMIC PRESS, INC.
ALL RIGHTS RESERVED.
NO PART OF THIS PUBLICATION MAY BE REPRODUCED OR
TRANSMITTED IN ANY FORM OR BY ANY MEANS, ELECTRONIC
OR MECHANICAL, INCLUDING PHOTOCOPY, RECORDING, OR ANY
INFORMATION STORAGE AND RETRIEVAL SYSTEM, WITHOUT
PERMISSION IN WRITING FROM THE PUBLISHER.

ACADEMIC PRESS, INC.
Orlando, Florida 32887

United Kingdom Edition published by
ACADEMIC PRESS, INC. (LONDON) LTD.
24/28 Oval Road, London NW1 7DX

Library of Congress Cataloging in Publication Data

Main entry under title:

Multiple time scales.

Includes indexes.
1. Differential equations, Partial--Numerical
solutions--Addresses, essays, lectures. 2. Monte Carlo
methods--Addresses, essays, lectures. 3. Nonlinear
theories--Addresses, essays, lectures. I. Brackbill,
Jeremiah U. II. Cohen, Bruce I. (Bruce Ira)
QA377.M946 1985   515.3'5   84-16833
ISBN 0-12-123420-7 (alk. paper)

PRINTED IN THE UNITED STATES OF AMERICA

85 86 87 88   9 8 7 6 5 4 3 2 1

Contents

Contributors ix

Preface xi

1. Considerations on Solving Problems with Multiple Scales
   R. C. Y. Chin, G. W. Hedstrom, and F. A. Howes

      I. Introduction   1
     II. Examples of Problems with Multiple Scales   3
    III. Numerical Methods for Multiple-Scale Problems   11
    IV. Summary and Perspectives   23
        References   24

2. Problems with Different Time Scales
   Heinz-Otto Kreiss

       I. Introduction   29
      II. Systems of Ordinary Differential Equations   32
     III. Numerical Methods for Ordinary Differential Equations   36
     IV. Partial Differential Equations   40
      V. Shallow Water Equations   44
     VI. Atmospheric Motions   49
    VII. Plasma Physics   55
        References   56

3. Nonlinear Normal-Mode Initialization of Numerical Weather Prediction Models
   C. E. Leith

     I. Introduction   59
    II. Normal-Mode Analysis   62

III. Nonlinear Analysis   65
   IV. Analysis of Observations   68
   V. Remaining Problems   70
      References   71

4. **The Diffusion-Synthetic Acceleration of Transport Iterations, with Application to a Radiation Hydrodynamics Problem**
   Raymond E. Alcouffe, Bradley A. Clark, and Edward W. Larsen

   I. Introduction   74
   II. Transport Iteration Methods   75
   III. A Problem in Radiation Hydrodynamics   84
   IV. Time-Dependent Example Calculations   104
      References   111

5. **Implicit Methods in Combustion and Chemical Kinetics Modeling**
   Robert J. Kee, Linda R. Petzold, Mitchell D. Smooke, and Joseph F. Grcar

   I. Introduction   113
   II. Stiffness and Implicit Methods   115
   III. The Method of Lines   122
   IV. Adaptive Meshing   127
   V. Solution of the Nonlinear Equations 138
      References   142

6. **Implicit Adaptive-Grid Radiation Hydrodynamics**
   Karl-Heinz A. Winkler, Michael L. Norman, and Dimitri Mihalas

   I. Introduction   146
   II. Physical Equations   148
   III. Adaptive-Mesh Equations   153
   IV. Numerical Equations   155
   V. The Adaptive Mesh   157
   VI. Numerical Techniques   160
   VII. Ordinary Gas Dynamics: Shock Tubes   169
   VIII. Radiation Hydrodynamics: A Supercritical Shock   173
   IX. A "Hilbert Program" for Nonlinear Radiation Hydrodynamics 179
      References   183

7. **Multiple Time-Scale Methods in Tokamak Magnetohydrodynamics**
   Stephen C. Jardin

   I. Introduction   186
   II. Ideal Time-Scale MHD Simulations 193

III. Resistive Time-Scale MHD Simulations 209
IV. Discussion 229
References 231

8. Hybrid and Collisional Implicit Plasma Simulation Models
Rodney J. Mason

I. Introduction 233
II. Basic Moment Method 236
III. Collisional-Hybrid Extensions 249
IV. Applications 264
V. Conclusion 267
References 269

9. Simulation of Low-Frequency Electromagnetic Phenomena in Plasmas
J. U. Brackbill and D. W. Forslund

I. Introduction 272
II. Implicit Plasma Simulation 274
III. Implicit Formulation of the Dynamic Equations 280
IV. The Algorithm for the Implicit Moment Method 293
V. Properties of the Implicit Moment Method 295
VI. Computational Examples 300
VII. Conclusions 308
References 309

10. Orbit Averaging and Subcycling in Particle Simulation of Plasmas
Bruce I. Cohen

I. Introduction 311
II. Electron Subcycling 316
III. Orbit Averaging 320
IV. Discussion 329
References 332

11. Direct Implicit Plasma Simulation
A. Bruce Langdon and D. C. Barnes

I. Introduction 336
II. Direct Method with Electrostatic Fields 341
III. Gyroaveraged Particle Simulation 361
IV. Electromagnetic Direct Implicit Method 366
V. Concluding Remarks 373
References 374

12. Direct Methods for N-Body Simulations
Sverre J. Aarseth

    I. Introduction   378
    II. Basic Formulation   381
    III. Ahmad–Cohen Scheme   385
    IV. Comoving Coordinates   389
    V. Planetary Perturbations and Collisions   393
    VI. Two-Body Regularization   396
    VII. Three-Body Regularization   409
    VIII. Star-Cluster Simulations   413
    References   417

13. Molecular Dynamics and Monte Carlo Simulation of Rare Events
Bruce J. Berne

    I. Introduction   419
    II. Activated Barrier Crossing Theory and Methodology   421
    III. Some Methods for Accelerating Simulations   430
    IV. Summary   435
    References   435

Subject Index   437

Contributors

Numbers in parentheses indicate the pages on which the authors' contributions begin.

SVERRE J. AARSETH (377), Institute of Astronomy, University of Cambridge, Cambridge, England

RAYMOND E. ALCOUFFE (73), Los Alamos National Laboratory, University of California, Los Alamos, New Mexico 87545

D. C. BARNES* (335), Institute for Fusion Studies, The University of Texas at Austin, Austin, Texas 78712

BRUCE J. BERNE (419), Department of Chemistry, Columbia University, New York, New York 10027

JEREMIAH U. BRACKBILL (271), Los Alamos National Laboratory, University of California, Los Alamos, New Mexico 87545

R. C. Y. Chin (1), Lawrence Livermore National Laboratory, University of California, Livermore, California 94550

BRADLEY A. CLARK (73), Los Alamos National Laboratory, University of California, Los Alamos, New Mexico 87545

BRUCE I. COHEN (311), Lawrence Livermore National Laboratory, University of California, Livermore, California 94550

D. W. FORSLUND (271), Los Alamos National Laboratory, University of California, Los Alamos, New Mexico 87545

JOSEPH F. GRCAR (113), Applied Mathematics Division, Sandia National Laboratories, Livermore, California 94550

G. W. HEDSTROM (1), Lawrence Livermore National Laboratory, University of California, Livermore, California 94550

F. A. HOWES (1), Department of Mathematics, University of California, Davis, Davis, California 95616

STEPHEN C. JARDIN (185), Plasma Physics Laboratory, Princeton University, Princeton, New Jersey 08544

---

*Present address: Science Applications, Inc., Austin, Texas 78746

ROBERT J. KEE (113), Applied Mathematics Division, Sandia National Laboratories, Livermore, California 94550

HEINZ-OTTO KREISS (29), Department of Applied Mathematics, California Institute of Technology, Pasadena, California 91125

A. BRUCE LANGDON (335), Lawrence Livermore National Laboratory, University of California, Livermore, California 94550

EDWARD W. LARSON (73), Los Alamos National Laboratory, University of California, Los Alamos, New Mexico 87545

C. E. LEITH (59), Lawrence Livermore National Laboratory, University of California, Livermore, California 94550

RODNEY J. MASON (233), Los Alamos National Laboratory, University of California, Los Alamos, New Mexico 87545

DIMITRI MIHALAS* (145), Max-Planck-Institut für Astrophysik, D-8046 Garching bei München, Federal Republic of Germany

MICHAEL L. NORMAN† (145), Max-Planck-Institut für Astrophysik, D-8046 Garching bei München, Federal Republic of Germany

LINDA R. PETZOLD (113), Applied Mathematics Division, Sandia National Laboratories, Livermore, California 94550

MITCHELL D. SMOOKE (113), Applied Mathematics Division, Sandia National Laboratories, Livermore, California 94550

KARL-HEINZ A. WINKLER† (145), Max-Planck-Institut für Astrophysik, D-8046 Garching bei München, Federal Republic of Germany

---

*Present address: High Altitude Observatory, National Center for Atmospheric Research, Boulder, Colorado 80307

†Present address: Group X-DOT, Los Alamos National Laboratory, Los Alamos, New Mexico 87544

Preface

There has been rapid progress in using large computers to solve problems possessing wide ranges of time scales. The variety of applications is bewildering, among them multiple-time-scale problems in chemical kinetics, statistical mechanics, weather prediction, astrophysics, radiation hydrodynamics, magnetohydrodynamics, and particle simulation of plasmas. Here, a number of the more successful numerical methods in each application are described. In collecting these methods, we hope to present the reader not only with a cookbook of recipes for modeling particular multiple-time-scale problems, but also with examples of inventiveness and focus that should lead to success in other, similar problems as well.

When the range of time scales in a problem is large and one is interested in the slower ones, numerical calculations can become unbearably expensive. The cost can be reduced by using mathematical analysis to eliminate the fast time scales. If one is interested in the behavior of a dynamical system after oscillations and transients have died away, for example, it is certainly less expensive to solve equilibrium rather than dynamical equations. In many cases of interest, however, simply eliminating the fast time scales results in a model that does not agree with physical observations.

Fortunately, the cost of solving multiple-time-scale problems can also be reduced using clever numerical methods that allow one to model slowly evolving phenomena embedded in rapid oscillations or transients of less physical interest. Some of the methods depend heavily on analysis in their development. Others have been developed in the heuristic, empirical way that has long characterized numerical modeling. Most of the methods have been applied to the solution of complex systems of nonlinear partial differential equations in several dimensions. All of them have been useful in modeling physical problems.

The first chapters are more analytic and general, and the later chapters are more heuristic and specific. In the only chapter that does not describe a numerical method, Chin, Hedstrom, and Howes begin the volume with a review of multiple-time-scale analysis, and place the succeeding chapters in context. They also caution that many methods are wrong when fast and slow time scales interact, an example of which is given by Jardin in a later chapter. Next, Kreiss develops the bounded derivative principle, which is then incorporated into methods using backward Euler implicit differencing and problem initialization to annihilate rapidly varying modes. Leith then describes how weather data are filtered by linear and nonlinear initialization methods so that rapid, spurious oscillations are not excited in weather prediction models.

In the next four chapters, more complex multiple-time-scale problems are treated using implicit methods. Alcouffe, Clark, and Larsen use complementary asymptotic limits of the radiation transport equation to devise an iterative method that requires no more work for long time steps than for short ones. Kee, Petzold, Smooke, and Grcar review their use of implicit differencing and adaptive gridding to model combustion and chemical kinetics. Winkler, Norman, and Mihalis describe their very successful adaptive, implicit methods for radiation transport in astrophysical applications. (In astrophysics, evidently, one must accommodate very large differences in the properties of the medium and associated time scales in closely adjacent regions of the domain.) Finally, Jardin describes a semi-implicit method for problems in magnetohydrodynamics in which fast and slow time scales are separated geometrically. He also compares the properties of analytically reduced equations with those of full equations in which the range of time scales is artificially compressed.

In the next four chapters, multiple-time-scale methods are applied to the modeling of kinetic phenomena in plasmas. In two chapters, implicitly differenced moment equations are used to estimate the long-range interactions between charged particles: Mason describes a hybrid model combining fluid and particle representations; and Brackbill and Forslund describe a model for electromagnetic phenomena in plasmas. These allow the modeling of embedded time-scale problems economically in computer time, but at the cost of damping or discarding unresolved phenomena. Cohen, by contrast, describes methods using either orbit averaging or subcycling to resolve fast kinetic phenomena occurring in plasmas whose average properties are more slowly evolving. Finally, Langdon and Barnes describe an implicit method that expands about explicitly calculated orbits.

In the final chapters, methods for multiple time scales are reviewed that depart from the more conventional ones described above. First, Aarseth describes a model for the dynamics of stars interacting through gravity, in which the equation of motion for each star is advanced at its own rate with only occasional reference to the motion of the others. (Binary pairs of stars, which interact very strongly, are identified and treated in a special way.) Finally, Berne describes a model for reactions that occur slowly because of a high energy barrier between states.

We gratefully acknowledge the guidance and assistance of B. Alder and the enthusiastic support of C. K. Birdsall who first suggested the compilation of this volume.

J. U. BRACKBILL
B. I. COHEN

1

Considerations on Solving Problems with Multiple Scales

R. C. Y. CHIN and G. W. HEDSTROM
Lawrence Livermore National Laboratory
University of California
Livermore, California

F. A. HOWES
Department of Mathematics
University of California, Davis
Davis, California

I. Introduction	1
II. Examples of Problems with Multiple Scales	3
III. Numerical Methods for Problems with Multiple Time Scales	11
IV. Summary and Perspectives	23
References	24

I. INTRODUCTION

In this chapter we give a quick overview of some of the considerations that must be taken into account in the computation of solutions of problems involving several scales. Such problems arise when several physical mechanisms are active, each associated with a particular scale. When the equations describing such phenomena are written in nondimensional form, the resulting equations contain small nondimensional parameters—ratios of scales from the original problem. Frequently, in these problems we are interested only in the long-time or large-scale behavior of the solution, and details of the local behavior are unimportant. Furthermore, it may be true that the large-scale motion is not much influenced by the small-scale motions. For example, in the computation of the lift on an airfoil in steady flight, we do not need to know the detailed motion of the turbulent wake. In such cases we should either solve some sort of reduced equations from which the small-scale effects have been removed or use numerical algorithms that automatically smooth out the small-scale effects. The papers in this volume generally take one of these two approaches.

It should be realized, however, that there exist situations in which the large-scale effects may be significantly modified by small-scale effects. Thus, the speed of a detonation front is determined by the rapid chemical reactions taking place in the front. In such cases details of the fast-scale motions are needed in order to compute the slow-scale motion correctly. There is currently no well-developed theory giving criteria for distinguishing situations with strong coupling between scales from those without. In individual disciplines, such as aerodynamics and plasma physics, however, there is a lot of experience to guide the scientist, and some of this experience is reported in this volume.

There exist several books on analytic methods for multiple-scale problems. For example, the books by Bender and Orszag (1978), Kevorkian and Cole (1981), Nayfeh (1973), and van Dyke (1975) are excellent references for analytic techniques. For a discussion of the theory, see Chang and Howes (1984), Eckhaus (1979), and Olver (1974). Finally, the books by Lin and Segel (1974) and Meyer and Parter (1980) contain discussions both of the techniques and the theory behind them.

In problems with multiple scales, the scales may be widely separated, or they may be rather evenly distributed over a broad range. When the scales are widely separated, we may hope to be able to eliminate the rapid motions and solve for the slow motions alone. For this reason the problems with widely separated scales have a natural division into two classes, depending on whether or not the motions on the fast scales influence the motions on the slow scales. We refer to the problems in which the fast and slow motions are independent as problems of Class I. The problems of Class II are those in which the rapid motions modify the motions on the slow scales. A similar classification of multiple-scale problems is given in van Kampen (1983), but his classification is based on physical properties of the solution, whereas ours is based on numerical considerations. The two classifications are related, of course, but the point of view is different.

In Section II we present examples from each of these classes. A chain of point masses joined by springs displays an evenly distributed, broad range of time scales, and the observed motions reflect an accumulation of the motions on many scales. This example displays some of the features of the prob-

1. Solving Problems with Multiple Scales

lems discussed in this book in Chapter 12 on the gravitational interaction of many particles, as well as in Chapters 7–11 on plasma physics. On a short time scale the chain exhibits propagating waves, but after a long time the motion becomes almost periodic, and we see only modes. For this reason it is natural to introduce the concept of subspace projection into the discussion of the motion of the chain. We make use of this notion repeatedly, and it plays an important role in Chapters 2–4.

In Section II we also examine a steady-state convection–diffusion equation, which is a problem of Class I. In this problem the different scales manifest themselves by the presence of boundary layers and internal layers. We consider only a time-independent problem for simplicity. For evolution problems with such layers, see Chapters 5–7 and 13 in this book. In these cases the large-scale phenomenon is that in most of the domain the solution is smooth, and small-scale effects appear in boundary layers and internal layers. The primary difficulty in such problems is in identifying and locating the layers.

The final example in Section II is from Class II, and here the multiple scales appear as oscillations at greatly differing frequencies. In this case we have ordinary differential equations in which the solution contains a rapid oscillation superimposed on a slow drift. Here, the object is to obtain equations for the drift alone, incorporating the influence of the rapid motion. This may be done by the methods of averaging or multiple scales. These ideas form the basis of Chapter 2 in this volume, and they are also behind the work of Cohen (Chapter 10).

Common to all of these examples is the need to identify the essential characteristics of the problem. In particular, we wish to emphasize the importance of identifying the small parameters in the problem and, hence, of understanding the competing physical mechanisms. We may use this information to construct a subspace of functions reflecting these characteristics, and we may then use projection onto this subspace to simplify the task of computing approximate solutions. Naturally, for complicated problems we may want to introduce several subspaces. In Section III we discuss numerical methods commonly used to solve problems such as those of Section II, and it turns out that the effective methods are based on subspace projection. The other chapters in this volume are discussed within this framework.

II. EXAMPLES OF PROBLEMS WITH MULTIPLE SCALES

In this section we examine the multiple-scale aspects of three different problems which illustrate the classification mentioned in the introduction. For more examples and a discussion of the theory, the reader is referred to Chin *et al.* (1984).

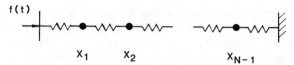

Fig. 1. Springs and point masses.

The first example is a lattice of point masses connected by springs to make a linear chain. It is possible to write the solution in several different representations, which emphasize different features. Thus, a representation in terms of propagating waves brings out the short-time behavior, a representation in terms of modes displays the long-time behavior, and a representation in terms of an inverse Laplace transform is convenient for mathematical analysis. It is also possible to combine these features. Thus, a representation using both rays and modes displays both short- and long-time behavior.

The problem may be posed as follows. We are given a linear lattice of $N - 1$ point masses, each of mass m, connected by Hookean springs with spring constant μ. The first point mass is connected to a driving function $f(t)$ by a spring, and the $(N - 1)$st point mass is connected through a spring to a rigid wall. See Fig. 1. Thus, the displacements w_n satisfy the linear system of differential equations

$$m \, d^2 w_n/dt^2 = \mu(w_{n+1} - 2w_n + w_{n-1}), \quad n = 1, \ldots, N - 1, \qquad (1)$$

with boundary conditions

$$w_0(t) = f(t), \quad w_N(t) = 0.$$

Let us introduce a scaled time,

$$\tau = 2t\sqrt{\mu/m},$$

and a dimensionless driving force

$$F(\tau) = \alpha f[t(\tau)].$$

This gives the equivalent system

$$4 \, d^2 u/d\tau^2 = Au + b, \qquad (2)$$

where A is an $(N - 1)$ by $(N - 1)$ tridiagonal matrix with -2 on the main diagonal and 1 on each adjacent diagonal, u a vector with components αw_n, and b a vector with $F(\tau)$ as its first entry and the rest zero.

In order to display the time scales in (2), we represent its solution in terms of modes. The eigenvalues of A are $-4\lambda_s^2$, where

$$\lambda_s = \sin[s\pi/(2N)], \quad s = 1, 2, \ldots, N - 1, \qquad (3)$$

and the corresponding eigenvector has components,

1. Solving Problems with Multiple Scales

$$v_{s,n} = \sin(sn\pi/N), \quad n = 1, 2, \ldots, N-1. \tag{4}$$

Thus, the modal representation of the solution of Eq. (2) is

$$u_n(\tau) = \frac{1}{N} \sum_{s=1}^{N-1} \sin\left(\frac{sn\pi}{N}\right) \cos\left(\frac{n\pi}{2N}\right) \int_0^\tau F(\xi) \sin[\lambda_s(\tau - \xi)] \, d\xi. \tag{5}$$

It is clear from (3) that the vibrational frequencies of the chain range from λ_1 to λ_{N-1}. Hence, the time scales as measured by τ range from $1/\lambda_{N-1}$ to $1/\lambda_1$. From (4) we see that the fundamental mode, $s = 1$, is associated with the collective motion of the entire lattice, whereas for the fastest mode, $s = N - 1$, each particle has opposite phase from its neighbors.

The representation (5) leads naturally to the concept of subspace projection, which is used extensively in the subsequent chapters in this book. If the sum in (5) is truncated at a certain value, say M, then the solution u_n is projected onto the subspace generated by the first M modes, containing the time scales ranging from $1/\lambda_1$ to $1/\lambda_M$. Such subspace projection is one of the popular ways of removing the fast time scales from a problem. Let us turn to projections onto other subspaces selected to emphasize other features of the motion.

Because the chain of springs may be regarded as a discrete approximation to a string and because (2) is a semidiscrete approximation to the wave equation, it is natural to expect to find traveling-wave solutions to (2). (See Section III.) Consequently, we ignore the boundary conditions for the moment and postulate solutions of the form (Lighthill, 1978; Whitham, 1974)

$$u_n(\tau) = B(k) \exp[i(kn - \omega\tau)]. \tag{6}$$

We find that k and ω must be related by

$$\omega = \pm \sin(k/2), \tag{7}$$

which is known as the dispersion relation because it shows that waves of differing wave number k have different phase velocity ω/k. For long waves, $k \approx 0$, the plus sign in (7) is associated with waves moving to the right, $\omega \approx k/2$, whereas the minus sign is associated with waves moving to the left.

By using the identity

$$J_n(z) = \frac{1}{2\pi} \int_{-\pi}^{\pi} \exp(iz \sin \zeta - in\zeta) \, d\zeta,$$

it is possible to find the combination of waves of the form (6) that satisfies the required boundary conditions,

$$u_n(\tau) = \sum_{k=0}^{\infty} (g[\tau - (2kN + n), 2kN + n] \\ - g\{\tau - [2(k+1)N - n], 2(k+1)N - n\}), \tag{8}$$

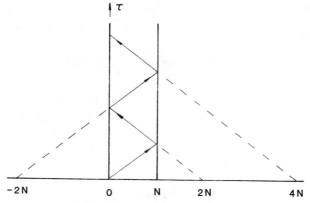

Fig. 2. Generalized ray representation.

where $1 \le n \le N - 1$,

$$g(\tau, n) = 2n \int_0^\tau F(\tau - \xi) J_{2n}(\xi) \, (d\xi/\xi),$$

and J_{2n} is the Bessel function of order $2n$ (Chin 1975). Equation (8) may be interpreted as a sum of dispersive waves emanating from sources located at the points $(2Nk, 0)$ with $k = 0, \pm 1, \pm 2, \ldots$. See Fig. 2. At any given time that

$$2(k - 1)N < \tau \le 2kN,$$

the sum (8) contains only $2k$ nonzero terms. Thus, this representation is convenient for studying the short-time behavior of the solution. Note that truncation of the sum in (8) produces another subspace projection.

Representations (5) and (8) are closely related through the representation as an inverse Laplace transform:

$$u_n(\tau) = \frac{1}{2\pi i} \int_\Gamma \tilde{F}(s) e^{s\tau} \frac{\sinh(N - n)\psi(s)}{\sinh[N\psi(s)]} \, ds, \qquad (9)$$

where

$$\psi(s) = \cosh^{-1}(2s^2 + 1),$$

Γ is the Bromwich path, and $\tilde{F}$ the Laplace transform of F. We do not give the details because the argument is similar to that used by Ahluwalia and Keller (1977) in their discussion of the ray and mode representations of the solution of the acoustic-wave equation in a stratified medium. As in acoustics, we may alternatively use a combined ray–mode representation to construct an effective approximation to the solution (Felsen and Kamel 1981). In fact, using the formula for the sum of a finite geometric series, Eq. (9) may be

1. Solving Problems with Multiple Scales

transformed to

$$u_n(\tau) = \sum_{k=0}^{M} \frac{1}{2\pi i} \int_\Gamma \tilde{F}(s) e^{s\tau}(\exp[-(2kN+n)\psi(s)]$$
$$- \exp\{-[2(k+1)N-n]\psi(s)\})\,ds$$
$$+ \frac{1}{2\pi i} \int_\Gamma \tilde{F}(s) \exp[s\tau - 2N(M+1)\psi(s)] \frac{\sinh[(N-n)\psi(s)]}{\sinh[N\psi(s)]}\,ds. \quad (10)$$

In the finite sum of integrals in (10) the contour may be deformed to a path along the branch cut for $\psi(s)$, yielding $2M$ propagating waves. If the remaining integral in (10) is evaluated by residue theory, a representation in terms of modes is obtained. In contrast with (5), however, the modes in (10) are driven by a modified forcing function,

$$G(\tau) = 2N(M+1) \int_0^\tau F(\tau-\xi) J_{2N(M+1)}(\xi) \frac{d\xi}{\xi}.$$

Thus, the forcing function is filtered by a Bessel function, illustrating the dispersive nature of the lattice chain. Note that we could evaluate a finite number of the residues in (10) and estimate the remainder, giving us a modified ray–mode representation of the solution. Such a representation contains both high- and low-frequency information about the solution, but the intermediate frequencies are absorbed in the remainder. Still, a computation based on (10) will efficiently give information about both the fast and slow time scales in the problem. In fact, this is yet another subspace projection. We choose the subspace according to the properties of the solution that we wish to emphasize. Chapter 4 in this volume makes use of a similar projection onto a subspace generated by two smaller subspaces.

The connections between the different representations of the solution are summarized in Fig. 3. A similar graph illustrating the connections between the forms of the solutions of the equations of acoustics was given by Ahluwalia and Keller (1977). Starting with the integral representation [Eq. (9)], the modal expansion [Eq. (5)] is obtained by evaluating the residues. An infinite binomial expansion of the function $[1 - \exp(-2N\psi)]^{-1}$ in (9) yields the ray representation [Eq. (8)], whereas a finite expansion with remainder yields the ray–mode representation [Eq. (10)]. The ray [Eq. (8)] and mode [Eq. (5)] expansions, in turn, are directly related through the Poisson summation formula (Courant and Hilbert 1953).

As an example of a problem of Class I, let us discuss a simple convection–diffusion equation which has "parabolic" boundary layers along characteristic boundaries, in addition to an "ordinary" boundary layer along a

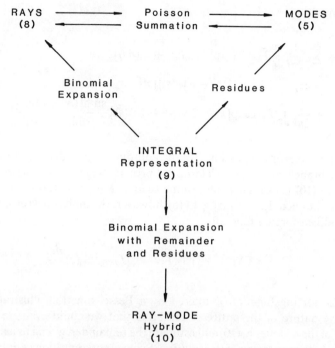

Fig. 3. Relations between the representations.

noncharacteristic boundary. The problem is

$$\varepsilon \nabla^2 u = u_x, \quad 0 < x < 1, \quad 0 < y < 1,$$
$$u(x, 0) = u(x, 1) = u(1, y) = 0, \quad u(0, y) = f(y), \quad (11)$$

where f is a twice continuously differentiable function for $0 \leq y \leq 1$. If we set $\varepsilon = 0$ in (11), we obtain the appropriate reduced equation, $U_x = 0$. It is clear that with solutions of the reduced problem we cannot satisfy all the boundary conditions of (11). In fact, the theory of Levinson (1950) shows that the boundary condition for the reduced problem is to be $U(0, y) = f(y)$, so that we have $U(x, y) = f(y)$. In fact, Levinson (1950) shows that the solution $u(x, y; \varepsilon)$ of (11) satisfies

$$\lim_{\varepsilon \to 0^+} u(x, y; \varepsilon) = f(y)$$

for all (x, y) in any fundamental rectangle $0 \leq x \leq 1 - \delta, \delta \leq y \leq 1 - \delta$ with $0 < \delta < 1$. There is an ordinary boundary layer of width $O(\varepsilon)$ along the outflow boundary $x = 1, \delta \leq y \leq 1 - \delta$, in which (11) is approximated by

1. Solving Problems with Multiple Scales

the ordinary differential equation $u_x = \varepsilon u_{xx}$, with $u(0, y) = f(y)$ and $u(1, y) = 0$. In addition, along each of the characteristic boundaries $y = 0$ and $y = 1$, we expect (Kamin 1955) to have a parabolic boundary layer of width $O(\varepsilon^{1/2})$, in which the solution to (11) is asymptotic to a solution of the parabolic equation $u_x = \varepsilon u_{yy}$ as $\varepsilon \to 0$. We also anticipate some sort of nonuniform behavior in neighborhoods of the corners $(0,0)$ and $(0,1)$. In fact, it is not hard to show by an examination of the terms in a modal decomposition,

$$u(x, y; \varepsilon) = \sum_{n=1}^{\infty} X_n(x; \varepsilon) \sin(n\pi y),$$

that the solution does, indeed, have the behavior indicated above. It should be noted that this modal expansion may be used as the basis for projection onto subspaces that bring out the different behavior of the solution u to (11) (Chin et al. 1984). It is easy to show that the boundary layer along the edge $x = 1$ is reflected in the terms with $\varepsilon n^2 \pi^2 < 1$. The parabolic boundary layers along $y = 0$ and $y = 1$ and any internal layers along a line $y = $ const are displayed by the terms with $1 < \varepsilon n^2 \pi^2 < 1/\varepsilon$. Finally, the behavior of u in the regions near the edge $x = 0$ of birth of a boundary or internal layer is given primarily by the terms with $\varepsilon n\pi > 1$.

Our last example is one in which the fast-scale variations are periodic or almost periodic, involving oscillations about a quasi-equilibrium state. For such problems the rapid motion has a significant cumulative effect on the slowly varying average solution, so that we have a problem of Class II. Thus, even though we are interested only in computing this slowly varying solution, we have to find some way to incorporate the effects of the fast variations. We could do this by computing the fast-scale motions, but the cost is likely to be prohibitive. For this example we accumulate the fast-scale motions by the method of averaging. Actually, the method of averaging has several variants, and we use the generalized method of averaging (Krylov and Bogolyubov, 1934). Alternatively, the averaged equations could be obtained by the method of multiple-scale expansions or the standard method of averaging (Bogolyubov and Mitropolsky, 1962).

As our example, we examine the motion of a charged particle in a slowly varying magnetic field $B(x, y)$. It is assumed that the direction of the field is constant (in the z direction), but the magnitude may vary. Thus, the equations of motion are

$$dx/dt = u, \quad dy/dt = v,$$
$$m\, du/dt = -evB(x, y)/c, \quad m\, dv/dt = euB(x, y)/c, \qquad (12)$$
$$x(0) = x_0, \quad y(0) = y_0, \quad u(0) = u_0, \quad v(0) = v_0.$$

Here, u and v denote, respectively, the components of the velocity in the x and y directions; m is the mass of the particle, and c the speed of light. Note that it follows immediately from (12) that the particle speed $V = (u^2 + v^2)^{1/2}$ is a constant of the motion. If B is constant (a homogeneous magnetic field), then the motion is circular with frequency $\Omega = eB/(mc)$, and the radius of gyration is $\rho = V/\Omega$, the Larmor radius.

Suppose that the magnetic field $B(x, y)$ is spatially inhomogeneous. We begin by determining a domain D that will contain the path of the electron. Then we introduce a scale for the field strength,

$$B_{\max} = \max_D B(x, y),$$

and a length scale for the variation of $B(x, y)$,

$$L_B = \max_D [B(x, y)/|\nabla B(x, y)|].$$

We make the assumption that L_B is large compared with a representative Larmor radius, $\rho_0 = V/\Omega_0$, where

$$\Omega_0 = eB_{\max}/mc.$$

Then the motion of the charged particle consists of a rapid gyration about the magnetic-field lines, coupled with a slow drift caused by the inhomogeneities in the magnetic field. Let us introduce the following dimensionless variables,

$$x = L_B \xi, \quad y = L_B \eta, \quad u = V_p, \quad v = V_q, \quad B = H B_{\max}, \quad t = \tau/\Omega_0.$$

Then with the small parameter $\varepsilon = \rho_0/L_B$, we obtain the system of equations,

$$d\xi/d\tau = \varepsilon p, \quad d\eta/d\tau = \varepsilon q, \quad dp/d\tau = -qH, \quad dq/d\tau = pH. \quad (13)$$

The conservation of speed, $p^2 + q^2 = 1$, suggests the change of variables

$$p = \cos \vartheta, \quad q = \sin \vartheta,$$

so that (13) becomes

$$d\xi/d\tau = \varepsilon \cos \vartheta, \quad d\eta/d\tau = \varepsilon \sin \vartheta, \quad d\vartheta/d\tau = H(x, y). \quad (14)$$

Equations (14) are in a form suitable for the application of the generalized method of averaging (Krylov and Bogolyubov, 1934; Kruskal, 1962; Stern, 1970), as well as the generalized method of averaging using Lie transforms (Kamel 1970). In (14) the derivatives of the slow variables, ξ and η, depend on the fast variable ϑ. The essential idea of the generalized method of averaging is the introduction of a near-identity, time-varying, coordinate transformation to convert (14) into a system in which the slow variable may be integrated asymptotically without knowledge of the fast variables. Thus, we seek a transformation

$$\xi = z_1 + \varepsilon f_1(z) + \varepsilon^2 g_1(z) + O(\varepsilon^3),$$
$$\eta = z_2 + \varepsilon f_2(z) + \varepsilon^2 g_2(z) + O(\varepsilon^3), \quad (15)$$
$$\vartheta = z_3 + \varepsilon f_3(z) + \varepsilon^2 g_3(z) + O(\varepsilon^3),$$

such that to lowest order in ε, $dz_1/d\tau$ and $dz_2/d\tau$ are independent of z_3 and τ. In fact, it is not hard to show that the transformation (15) may be chosen so that

$$\frac{dz_1}{d\tau} = \frac{\varepsilon^2}{2H^2}\frac{\partial H}{\partial z_2} + O(\varepsilon^3),$$
$$\frac{dz_2}{d\tau} = \frac{\varepsilon^2}{2H^2}\frac{\partial H}{\partial z_1} + O(\varepsilon^3), \quad (16)$$
$$\frac{dz_3}{d\tau} = H + O(\varepsilon^2).$$

Thus, the first two equations in (16) form an autonomous subsystem of differential equations that gives the average drift motion of the electron.

We remark that (16) could just as well be derived from (12) by such other techniques as homogenization and the standard method of averaging. See Chin *et al.* (1984).

III. NUMERICAL METHODS FOR PROBLEMS WITH MULTIPLE SCALES

In Section II of this chapter we have seen that the presence of multiple scales in a physical system may make itself manifest in different ways, depending on the degree of interaction between the various scales. For many of these problems it is too costly, and it may even be impossible, to resolve all of the scales in a numerical computation. Therefore, we have to narrow our scope and focus on answering specific questions about the system. That is, we need to restrict our attention to certain objective functionals associated with physically meaningful aspects of the problem or to some particular and, therefore, restricted response of the problem. In either case we are interested in slow scales in the physical system, since they may be observed more accurately in experiments and are more amenable to computation. Thus, the numerical method must be designed to be attuned to the objective functionals desired.

These ideas are developed in the other chapters in this volume. For example, in the simulation of activated barrier crossing and methods for improving the convergence of Metropolis (MP) and Monte Carlo (MC) in Chapter

13, the objective functionals are the rate constants for barrier crossing. In Chapters 2 and 3, the emphasis is on the isolation of slowly evolving meteorologically important states and on the removal of the influences of the high-frequency motions. In Chapter 7 the objective is the simulation of slowly evolving plasma phenomena described by the magnetohydrodynamic (MHD) equations. The MHD equations, in turn, are low-frequency macroscopic equations obtained from the microscopic plasma equations. Chapters 8–11 restrict their numerical simulation to other low-frequency limits of the plasma equations. In the modeling of chemical kinetics (Chapter 5), the speed of propagation of the flame front is an easily measured quantity, which may be compared with computed estimates in order to evaluate a chemical model. In Chapter 12 the objective is the study of a gravitational N-body problem. Here, personal experience and an understanding of the problem are used to extract the aggregate motion from detailed numerical calculations.

In Section II we discussed a convection–diffusion equation involving boundary and internal layers. For this problem the rapidly varying behavior does not influence the slow components of the solution, so that it is a Class I problem. In Class I problems the location of the boundary and internal layers is of paramount importance. The presence of boundary and internal layers indicates where there are regions of transition, separating subdomains on which different scales of variation or different physical mechanisms are dominant. The reaction front in a reaction–diffusion problem is a good example of this phenomenon. The problem of electron motion discussed in Section II is of Class II. That is, it is an oscillatory problem in which the cumulative high-frequency motion significantly affects the slow motion. In Class II problems no such decomposition of the domain is possible. The effects of the interaction of the different scales accumulate over a long time or distance, so that seemingly insignificant local contributions gain significance with increasing time and distance. For this reason, the numerical methods designed for problems of Class I are not suitable for problems of Class II, and vice versa.

Let us begin our discussion of numerical methods for problems with multiple scales by an examination of methods for initial-value problems for systems of ordinary differential equations. The term *stiff system* is frequently associated with such problems. Originally, the concept of stiff system was applied only to Class I initial-value problems for ordinary differential equations (Dahlquist 1963). In this setting a stiff system is one in which there is a slowly varying solution such that some perturbations to it are rapidly damped. This original meaning of stiffness is still used in Shampine and Gear (1979). Other authors, however, have extended the notion of stiffness, and Miranker (1981) gives the following definition, which includes both Class I and II.

1. Solving Problems with Multiple Scales

DEFINITION. A collection of systems of differential equations is said to be *stiff* on an interval $[0, t]$, if there exists no positive constant M such that the variation of every component of every solution of every member of the collection is bounded by M.

The problem of selecting a numerical method for Class I ordinary differential equations with multiple scales is examined in this volume in Chapter 5. The difficulty with using classical explicit multistep methods is that they are stable only if the time step is small enough to resolve the fastest scale of motion. Outside a boundary layer near $t = 0$, however, the solution is smooth, and we would like to have the capability of computing on the relevant, slower scales. Thus, the antipodal requirements of accuracy and stability introduce a dilemma in the classical multistep methods for stiff initial-value problems for ordinary differential equations. The presence of this dilemma is explained by the following theorem (Dahlquist 1963).

THEOREM 1. *The numerical solution y_n of*

$$dy/dt = Ay$$

by a linear multistep method tends to zero as $n \to \infty$ for all values of the time step and for all constant matrices A such that all its eigenvalues have negative real parts only if the multistep method is an implicit scheme with order of accuracy at most 2.

Thus, if we use a multistep method without resolving the scales of rapid decay, accuracy must be sacrificed to obtain stability. Still, it is possible to write computer programs to solve systems of ordinary differential equations of Class I, using variable step size and methods of variable order. We may use a small step size in regions of rapid variation and an implicit multistep method in the smooth regions.

For problems of Class II for ordinary differential equations, the implicit multistep methods are of no help, for they are accurate only if the fastest scales are resolved. In this case we are faced with the choice of computing the fast motion in detail or making some use of such analytic techniques as the method of averaging, as discussed, for example, by Bogolyubov and Mitropolsky (1962). Krylov and Bogolyubov (1934), Sethna (1967), and Hale (1969). A survey of numerical methods induced by analytic techniques is given in Miranker (1981). Unfortunately, the algebraic manipulations required by a combined analytic–numerical treatment can become unwieldy, even for a system of equations of moderate size. We must, therefore, search for alternate means of solving these problems, as given in the rest of this section.

Kreiss (1978) has proposed a subspace projection, called the bounded derivative method, for computing the slowly varying component of the solution of a system of ordinary or partial differential equations of Class II. In this method the initial data are prepared to eliminate the fast components. Kreiss gives a detailed discussion of the method in this volume (Chapter 2). The following example given in Kreiss (1978) illustrates the underlying principle:

$$\varepsilon \, dy/dt = ay + e^{it}, \qquad 0 < t < T, \tag{17}$$

with $0 < \varepsilon \ll 1$, $|a| \approx 1$, and $y(0) = y_0$. The solution of (17) is the sum of rapidly and slowly varying functions:

$$y(t) = \frac{-1}{a - i\varepsilon} e^{it} + \left(y_0 + \frac{1}{a - i\varepsilon} \right) e^{iat/\varepsilon}.$$

The choice of initial data, $y_0 = -1/(a - i\varepsilon)$, enables us to eliminate the rapidly varying component of the solution, thus singling out the slowly varying component. The central issue in applying this method is the question of whether there exists a manifold of slow solutions that is the range of a projection operator. Some results in this direction were obtained by Dahlquist (1974), Karasalo (1975), and Kreiss (1978). In addition, Kreiss (Chapter 2) has shown that in linear systems that are critical (Hale, 1969), the fast solutions may be activated during the course of the computation. In particular, for a second-order equation with a turning point, after passing through a turning point, the solution may or may not return to the slow manifold, depending on the order of the turning point. Furthermore, for nonlinear problems internal and external resonances may activate the rapidly varying solutions (Sethna 1963). See also the paper by McNamara (1978) concerning the question of resonance. For such problems the main difficulty is the detection of recurring rapidly varying layers.

Let us mention some other methods for stiff, oscillatory, initial-value problems. Ziv and Amdursky (1977) proposed another subspace projection method for linear systems. An alternative to subspace projection is to use linear multistep methods on a coarse grid, introducing rapidly varying errors which are subsequently filtered out to leave a smooth solution. Methods of this kind have been proposed by Gragg (1965), Lindberg (1971), and Majda (1984). Some authors in this volume, including Brackbill and Forslund (Chapter 9), Jardin (Chapter 7), Kee et al. (Chapter 5), and Mason (Chapter 8), resort to the use of implicit linear multistep methods in which accuracy is sacrificed in order to improve stability.

Let us turn now to the question of numerical methods for two-point boundary-value problems with multiple scales. Strategies that are useful for initial-value problems for ordinary differential equations are equally applicable here. The need to resolve rapidly varying boundary and internal layers is as pressing. For this reason, adaptive mesh methods play a prominent role in effective numerical methods for stiff two-point boundary-value problems. The procedures for the selection of an adaptive mesh fall basically into two classes: (1) techniques based on the equidistribution of some positive weight function of the solution and (2) techniques based on coordinate transformations.

Equidistribution procedures have the following objective (Kautsky and Nichols, 1980): Given a positive weight function w defined on an interval, find a partitioning of the interval such that the integral of w over each subinterval takes the same value. The weight function w may be the absolute value of an estimate of the local truncation error, as in the works by Lentini and Pereyra (1977). Pereyra and Sewell (1975), and Russell and Christiansen (1978). For time-dependent problems with moving internal layers, it is difficult to find a stable method that selects a grid and finds the solution at the same time. Current methods rely on ad hoc stabilizing procedures, with parameters to be chosen by the user on the basis of experience (Miller and Miller, 1981; Miller, 1981). See also, Chapter 6.

In the coordinate transformation method the objective is to find a coordinate transformation so that in the transformed domain the solution is uniformly smooth. This objective is, however, rarely achieved. The method requires the addition of a supplementary differential equation to the system in order to define the coordinate transformation. Examples of such coordinate transformation methods may be found in the works by Dwyer (1983), Dwyer et al. (1982), and Brackbill and Saltzman (1982). For a detailed discussion of these and other adaptive-grid techniques, see Chapter 5 in this volume.

Finally, there are some fixed-mesh methods for two-point boundary-value problems with multiple scales. Methods based on finite differences may be found in the papers by Abrahamsson et al. (1974), Osher (1981), Kellog and Tsan (1978), and Kreiss and Nichols (1976). Other fixed-grid methods, based on finite elements or operator-compact implicit schemes or exponential fitting, are given by Berger et al. (1980), Hemker (1977), deGroen and Hemker (1979), and Flaherty and Mathon (1980).

Partial differential equations commonly exhibit multiple-scale phenomena, with local regions where the solution possesses large gradients. This necessitates the introduction of local fine grids to resolve the behavior of the solution, and it may be desirable to use different numerical methods on different local fine grids. Thus, in Section II we saw that a convection–diffusion equation may display hyperbolic, parabolic, or elliptic behavior in different subdomains, and a well-designed numerical method will adapt accordingly. Similarly, a change of type in the approximating equations may also take place in the radiative transport equation (Chapters 4 and 6). For evolution problems with moving fronts, it seems desirable to use local grids moving with the fronts (Davis and Flaherty, 1982; Harten and Hyman, 1983; White 1982). Most of the effort to date, however, has been devoted to the development of fixed-grid schemes for such problems.

We begin our discussion of numerical methods for partial differential equations with an examination of the difficulties to be faced in the use of a fixed grid for a hyperbolic equation. Here, even for linear equations, a solution may have discontinuities that propagate along characteristic surfaces (Courant and Hilbert, 1962). For such problems, when a finite-difference or finite-element method is used on a fixed grid, as the discontinuity moves across the grid, there is always a local region in which the numerical solution is changing rapidly. Because these rapid changes cannot be resolved, numerically induced dispersion results. In order to explain this numerical dispersion, consider the following initial-boundary-value problem for the wave equation:

$$u_t = \sigma_x, \qquad \sigma_t = u_x, \tag{18}$$

for $x > 0$ and $t > 0$, with $u(x, 0) = 0$, $\sigma(x, 0) = 0$, $u(0, t) = f(t)$, and $|u(x, t)|$ bounded as $x \to \infty$. Suppose we solve (18) using the method of lines:

$$du_n/dt = h^{-1}(\sigma_{n+1/2} - \sigma_{n-1/2}), \tag{19}$$

$$d\sigma_{n-1/2}/dt = h^{-1}(u_n - u_{n-1}), \qquad n = 1, 2, \ldots. \tag{20}$$

Upon eliminating $\sigma_{n-1/2}$ from (19) and (20), we obtain

$$d^2u_n/dt^2 = h^{-2}(u_{n+1} - 2u_n + u_{n-1}), \qquad n = 1, 2, \ldots. \tag{21}$$

This is the system of masses and springs (1) with $\mu/m = 1/h^2$, with the difference that in (21) the number of point masses is infinite. Thus, in this case the sum (8) reduces to a single term,

$$u_n(t) = g(t - n, n),$$

which is a dispersive wave. In fact, an investigation of the group velocities (Trefethen, 1982) shows that if in (21) we take $u_0(t) = H(t)$, the Heaviside step function, then the oscillations extend over the entire sector $0 < x < t$. To some extent, these oscillations may be reduced by the addition of an artificial viscosity (Chin and Hedstrom, 1978).

The argument in terms of group velocities is not adequate to describe the behavior of u_n near the wave front. To describe the motion there, we may make the following analysis. We assume that the grid is fine enough to resolve waves of interest. That is, we take h to be small compared with the wavelength. Therefore, with the notation $v(nh, t) = u_n(t)$, the function $v(x, t)$ is smooth and we may make Taylor series expansions

$$v((n \pm 1)h, t) = v \pm hv_x + (h^2/2)v_{xx} \pm \cdots.$$

1. Solving Problems with Multiple Scales

Thus, Eq. (21) becomes

$$\frac{\partial^2 v}{\partial t^2} = \frac{\partial^2 v}{\partial x^2} + \frac{h^2}{12}\frac{\partial^4 v}{\partial x^4} + \cdots. \tag{22}$$

The dispersive character of the waves in the chain appears here in the terms of order $O(h^2)$ and higher. We remark that (22) is known as the modified equation corresponding to (21) (Warming and Hyett, 1974). In order to analyze (22), we introduce a coordinate system moving with the wave, incorporating the expectation that in this coordinate system the temporal variation is slow compared with the spatial variation:

$$\sigma = h^2 t, \quad \eta = x - t.$$

After this transformation of (22), if we keep only the terms of order $O(h^2)$, we obtain the equation

$$(\partial^2 v/\partial\eta\,\partial\sigma) = (-1/24)(\partial^4 v/\partial\eta^4).$$

We integrate this equation with respect to η, setting the constant of integration $C(\sigma)$ equal to zero, as required by the physical situation. Thus, we have in wave coordinates

$$\partial v/\partial \sigma = (-1/24)(\partial^3 v/\partial \eta^3). \tag{23}$$

Upon solving (23) via a Fourier transform in η, we find that

$$v(\eta,\sigma) = \frac{1}{2\pi}\int_{-\infty}^{\infty} D(\varkappa)\exp\left(\frac{i\sigma\varkappa^3}{24} + i\eta\varkappa\right) d\varkappa.$$

Thus, when the source is a pulse, $D(\varkappa) = 1$, and v is an Airy function (Abramowitz and Stegun, 1965)

$$v(\eta,\sigma) = (2/\sigma^{1/3})\,\text{Ai}(-2\eta/\sigma^{1/3}).$$

In terms of the original variables, we see that when the source is a pulse, the smooth components of the wave travel with a profile given by

$$u_n(t) \approx (2/h^{2/3}t^{1/3})\,\text{Ai}(t - nh/h^{2/3}t^{1/3}).$$

We have seen that it is possible to analyze the wave motion of the lattice by using the group velocity away from the wave front and an approximation in terms of an Airy function near the front. It is also possible to carry out a uniform asymptotic analysis by the method of Chester, Friedman, and Ursell (Chester *et al.*, 1957: Olver, 1974). For the details we refer the interested reader to the paper by Chin (1975).

It seems reasonable to try to use more accurate methods, but even if we solve a linear hyperbolic system with a dissipative method with order of accuracy v, a discontinuity in the solution of the differential equation becomes a transition layer of width $O(N^{1/(v+1)})$ in the difference scheme, where $N = t/\Delta t$ (Chin and Hedstrom, 1978; Brenner and Thomée, 1971; Lax, 1978).

For nonlinear hyperbolic systems the situation is better in a way. This is because discontinuities may be resolved by the use of numerical methods that incorporate solutions to local Riemann problems. The first such method is that of Godunov (1959), but it has the disadvantage of being only first order accurate on smooth portions of the solution, because the numerical solution is represented locally as step functions. There has recently been an extensive development of improvements to Godunov's method, beginning with the series of papers by van Leer (1974, 1979). See the papers by Harten (1984), Harten et al. (1983), Roe (1981), and Woodward and Colella (1984). In Woodward and Colella (1984) the local representation of the numerical solution is in terms of parabolas, giving the opportunity of greater accuracy in smooth parts of the solution. Also, we may note that Harten (1984) suggests the creation of artificial nonlinearity in numerical methods for linear hyperbolic systems, in order to improve the resolution of discontinuities. In particular, such artificial nonlinearities improve the resolution of contact discontinuities in the computation of the solution of gas dynamics equations.

We now turn to a discussion of numerical methods for convection–diffusion equations. As we noted in Section II, the solutions of such problems may have regions of rapid variation, such as boundary layers and internal layers. Thus, we should not be surprised to find that, if the numerical computation is performed on a grid that is too coarse to resolve such layers, the solution may not be accurate. This question of resolution is usually expressed in terms of a parameter called the *cell Reynolds number*. In order to illustrate the cell Reynolds number and its effect on the numerical solution, consider a discrete approximation to (11) using central differences:

$$\text{Re}_h(v_{i+1,j} - v_{i-1,j}) = v_{i+1,j} + v_{i-1,j} + v_{i,j+1} + v_{i,j-1} - 4v_{i,j}, \quad (24)$$

for $i = 1, 2, \ldots, N-1$, $j = 1, 2, \ldots, N-1$, and $Nh = 1$. Here, $u_{i,j}$ is our numerical approximation to $u(ih, jh)$ and Re_h is the cell Reynolds number:

$$\text{Re}_h = h/2v.$$

As boundary data for (24) we use a discretization of the boundary data for (11); that is,

$$v_{i,0} = 0, \quad v_{i,N} = 0, \quad i = 1, 2, \ldots, N-1,$$

$$v_{N,j} = 0, \quad v_{0,j} = f(jh), \quad j = 1, 2, \ldots, N-1.$$

It is readily seen that (24) admits a one-parameter family of solutions, depending on the cell Reynolds number Re_h. From our discussion of (11) we have seen that the solution of (11) has parabolic boundary layers along $y = 0$

1. Solving Problems with Multiple Scales

and $y = 1$, and an exponential or ordinary boundary layer along $x = 1$. Depending on the behavior of the boundary data $f(y)$, there may be regions of birth of boundary layers at the corners $(0, 0)$ or $(0, 1)$.

The discrete approximation (24) may be analyzed by using the discrete Fourier transform (Hedstrom and Osterheld, 1980). It turns out that the effect of discretization is far more severe in the ordinary or exponential boundary layer than in the parabolic boundary layers. In fact, in the parabolic boundary layer region near the x-axis, the solution of (24) is asymptotic to the solution of the modified equation

$$u_x = vu_{yy} + v[1 + (\text{Re}_h^2/3)]u_{xx} \tag{25}$$

as $x/v \to \infty$, $y/x \to 0$, and $\text{Re}_h y/x \to 0$. Thus, in the parabolic boundary layer the effect of the discretization is to increase the viscosity in the direction of the characteristics. The effect of the ordinary boundary layer on the solution of (24) depends on how well (as measured by the cell Reynolds number Re_h) the layer is resolved. If the resolution is poor in the sense that $\text{Re}_h > 1$, then the solution $v_{i,j}$ of (24) contains a component,

$$\alpha^{N-1}(\alpha^{-i} - 1), \quad i = 1, 2, \ldots, N,$$

with

$$\alpha = (1 - \text{Re}_h)/(1 + \text{Re}_h).$$

Thus, the numerical solution contains an oscillation of period $2h$ which extends even into the smooth region where the reduced equation is approximately valid. These oscillations may be damped by using a method based on upwind differencing instead of (24). It may be noted that this need to use a low-order method when the boundary layer is not resolved is related to the ideas expressed in Theorem 1. For work on upwind schemes see, for example, Heinrich et al. (1977), Hughes et al. (1979), and Osher (1981). Another way to avoid these oscillations is to symmetrize Eq. (11) before discretizing it. This is done in Axelsson (1981), for example.

With this brief summary of standard numerical methods for multiple-scale problems, we see that there is a lot of room for improvement. Because of the great simplifications afforded by the analytic methods discussed in previous sections, one might expect to find numerical algorithms based on asymptotic methods, and such methods do exist. For examples of this approach, see Hoppensteadt and Miranker (1983), Miranker (1981), and Flaherty and O'Malley (1977). These algorithms have the advantage that they become more accurate as the different scales in the problem become more disparate. Thus, methods based on analytic techniques become more effective in just those situations when the standard numerical methods become more difficult to apply.

It is appropriate at this point to expand on some of the considerations required in the application of numerical methods based on perturbation analysis. To begin with, the equations must be cast in appropriate form, such as (17) for two-point boundary-value problems, or a generalization of (11) for singularly perturbed convection–diffusion equations,

$$\varepsilon \nabla^2 u = A(x,u) \cdot \nabla u + h(x,u),$$

or for modulated oscillatory problems using the standard method of averaging,

$$dx/dt = \varepsilon f(t,x,\varepsilon), \tag{26}$$

where f is periodic or almost periodic in t. This requires the identification of the pertinent small parameter or parameters, and for a complex physical system this may or may not be easy to do. Once we have identified the small parameters from physical considerations or by other means, we may proceed to reduce and decompose the original system of equations into the corresponding subspaces. Moreover, identification of the small parameters yields insight into the competing physical mechanisms. For a linear system of ordinary differential equations a systematic algorithm for performing this decomposition has been proposed in O'Malley and Anderson (1979). Still, it is clear that in problems with multiple scales the most difficult task is the identification of the small and large parameters.

We should also comment on the amount of work involved in the implementation of algorithms derived from analytic methods, as discussed in Section II. A straightforward implementation of such perturbation methods may involve costly evaluations of special functions and even more costly numerical quadratures. Furthermore, the algebra can quickly become overwhelming, even when it, too, is done by computer, such as with MACSYMA. Such considerations lead us to consider numerical algorithms based on the use of analytic methods to obtain only the gross features of the solution. We close this section with examples of numerical methods based on this approach.

Hoppensteadt and Miranker (1983) propose an extrapolation method for the numerical solution of singularly perturbed problems for ordinary differential equations of both boundary-layer and oscillatory type. The extrapolation procedures are based on knowledge of the form of the perturbation series as an expansion in the appropriate small parameter. No detailed information about the individual terms of the expansions is needed. It should be noted that this sort of extrapolation is quite different from the Richardson extrapolation, which is common in numerical analysis and is based on an expansion of the numerical error in powers of the mesh size. In a number of sample calculations, the authors obtain a significant increase in efficiency over a standard ordinary differential equation solver when a desired accuracy is prescribed.

1. Solving Problems with Multiple Scales

For modulated oscillatory problems Petzold (1981) used the observation that when a quasi-periodic function is sampled at multiples of the period of oscillation, the resulting sequence changes slowly, and she developed linear multistep formulas analogous to the Adams–Moulton methods to integrate this slowly varying amplitude modulation. This technique requires a knowledge of the period of the motion and is, therefore, conceptually akin to the method of averaging, even though no explicit averaging is done. To apply this method, an algorithm for finding the period is required, and the solution must be computed accurately over certain periods. It is only necessary to perform these operations at the points required by the linear multistep formula. In spite of the additional computations required to follow the slow changes of the period of oscillation, significant gains of efficiency are achieved over the standard ordinary differential equation solvers, because accurate computation over a period is done only locally.

These ideas were extended in Kirchgraber (1983) to produce an ordinary differential equation solver for oscillatory problems, incorporating the method of averaging. Equation (26) is averaged to form

$$dX/dt = \varepsilon F(X, t, \varepsilon),$$

where

$$F(X, t, \varepsilon) = \sum_{k=1}^{m} \varepsilon^k F_k(X) + \varepsilon^{m+1} R_m(x, t, \varepsilon). \tag{27}$$

Here, F_1 is the average:

$$F_1(X) = \lim_{T \to \infty} \frac{1}{T} \int_0^T f(t, X, \varepsilon)\, dt,$$

and the remaining F_k are obtained from a recursion relation (Perko 1968), (Kamel 1970). The key to the development is the observation that the function $F(X, t, \varepsilon)$ appearing in (27) may be calculated systematically by a formula of Runge–Kutta type,

$$Q^0 = X, \quad Q^1 = P(X),$$

$$Q^k = P\left(\sum_{j=0}^{k-1} \beta_j^k Q^j\right), \quad k = 2, 3, \ldots, m,$$

$$F(X, t, \varepsilon) = \sum_{k=0}^{m} \beta_k Q^k + O(\varepsilon^{m+1}), \tag{28}$$

where β_j^k and β_k are real numbers and $P(\varphi) = \Phi(t + T, \varphi(t))$ is the T-periodic Poincaré map. Moreover, if the step size used for the integration of the averaged system is chosen to be a multiple of the period, then the amplitude modulation is obtained automatically. Thus, this algorithm is a synthesis of three ideas: (1) Petzold's sampling at the period of the oscillation, (2) Lie transform techniques used to derive (27), and (3) a Runge–Kutta method for the evaluation of a polynomial expression (28).

For two-point boundary-value problems without a turning point, Chin and Krasny (1983) have developed a hybrid finite-element method for

$$\varepsilon^2 y'' = h(x, y), \quad a < x < b,$$
$$y(a, \varepsilon) = A, \quad y(b, \varepsilon) = B.$$

More precisely, it is required that there exist a positive number δ such that $\partial h/\partial y \geq \delta$. The method has three steps: (1) reduction to a sequence of linear problems through an interpolation of $h(x, y)$ by a piecewise linear function in y for each fixed x, (2) approximate solution of the resulting linear problems by asymptotic methods, and (3) patching together of the local solutions to obtain global continuity of the approximate solution and its first derivative. This method is a natural extension to nonlinear problems of an algorithm introduced for linear problems in Preuss (1973). The use of asymptotic methods to construct the local solutions enables us to achieve efficiency and accuracy by capturing the essential behavior of the solution. Moreover, the approximate solution becomes more accurate as the singular perturbation parameter ε gets smaller (while remaining positive). To further increase the efficiency, an adaptive-mesh strategy is introduced to distribute the error equally in interpolating h. It is easier to control the mesh selection for h than it is to try to find the solution y while at the same time trying to find a mesh optimal for resolving y.

Another example of a numerical method relying on asymptotic analysis to solve a problem with disparate scales is the method of Chin and Braun (1980) for flow in a porous medium, in which the solid matrix undergoes an endothermic chemical reaction. This problem is distinguished by the rapid variation of the density S of the reacting solid, as governed by a reaction law of Arrhenius type,

$$\partial S/\partial t = -AS \exp(-E/T(t)), \quad (29)$$

while the temperature $T(t)$ is slowly varying. Here, A and E are positive constants such that $E/T(t) \gg 1$. If $T(t)$ is approximated by a piecewise linear function, then (29) may be integrated exactly, yielding an approximation of S in terms of piecewise exponential integrals. The expense of evaluating the exponential integrals is recovered many times over by the use of step sizes commensurate with the smooth temperature variation. In fact, even the reaction front is resolved on the coarse grid, so that this method is more efficient than a stiffly stable implicit difference formula.

IV. SUMMARY AND PERSPECTIVES

In the preceding sections we have presented a quick tour of the problems with multiple scales, with their different properties, requiring different analytic and numerical solution methods. In this section we summarize the earlier discussion and offer some suggestions on the further development of numerical techniques for such problems.

The essence of multiple-scale problems lies in the existence of distinguishable competing mechanisms, acting at disparate rates and scales. In mathematical terms, we may say that for such problems the solution space may be decomposed into nearly orthogonal subspaces corresponding to distinguishable limits of the governing equations. The existence of these subspaces leads naturally to the concept of subspace projection, which plays a prominent role in all of our examples. We have also seen that it is occasionally necessary to transform a particular subspace, in order to bring out an effective approximation of the solution. Thus, in our analysis of waves in a system of masses and springs, we used the Poisson summation formula to transform between representations of the solution in terms of rays and modes. Another way to look at multiple-scale problems is in terms of distinguished limits of the governing equations, leading to model equations or reduced equations. The solutions of the model equations span a subspace that approximates the space of solutions of the original governing equations.

Because of the different characteristics of the different subspaces arising in multiple-scale problems, it is to be expected that effective solution methods will require a combination of many techniques. This is true whether we take an analytic or numerical approach, or a combination of the two. Note that the high-quality ordinary differential equation solvers, with their capability of dynamically selecting the step size and the difference scheme, are an example of this principle. Let us reiterate the point that effective numerical methods for multiple-scale problems must be capable of capturing the essential behavior of the governing equations, without generating extraneous information or destroying the desired behavior. That is, in a numerical method the emphasis should be on high resolution. In the context of nonlinear hyperbolic systems this point was made by Lax (1978), and for these problems the methods based on solutions of local Riemann problems are examples of such high-resolution methods. Another example in the context of stiff two-point boundary-value problems is the method of Preuss (1973) and Chin and Krasny (1983), in which a differential equation is approximated by another differential equation that is easier to solve analytically.

For the future development of numerical methods for multiple-scale problems, let us offer the following observations. Essentially these same points were also made in Hoppensteadt and Miranker (1983). In order to obtain high-resolution methods, the essential analytic properties of the solutions must be captured. Thus, the development of a numerical method must begin with some asymptotic analysis, since knowledge of the structure of the solution may be used as a guide in the construction of a discrete analog of the governing equations. This requires an accommodation to the competing physical mechanisms and, hence, the identification of the small parameters. This point of view is in direct conflict with the ideas behind general-purpose software, but it must be recognized that multiple-scale problems are just the ones for which the general-purpose software is apt to be inefficient or even to fail. As an example of how a minimal amount of analytic information may be used in the development of a numerical method, consider the singularly perturbed convection–diffusion equation, discussed in Section II. Because the diffusive term has a significant contribution in only a small region of the domain, it is reasonable to decompose the domain into subregions, using on each subdomain a numerical method that takes into account the asymptotic behavior of the solution on that subdomain. The local solutions must then be matched across the subdomain boundaries. This domain decomposition method is strongly reminiscent of the method of matched asymptotic expansions. There is an important difference, however, in that the analytic apparatus of matched asymptotic expansions with its tedious algebraic manipulations is replaced by an iterative numerical algorithm.

This leads to our second observation that a straightforward transcription of analytic methods to the computer is unnecessarily cumbersome and is to be avoided. Instead, we want to use only the structure and the essential properties of the asymptotic solutions in our construction of an algorithm. The result is a compromise between the need to capture the essential behavior of the solution using analytic techniques and the need to have an efficient numerical algorithm. Thus, the development of effective algorithms for multiple-scale problems requires a combination of asymptotic analysis, approximation theory from numerical analysis, and analysis of algorithms.

ACKNOWLEDGMENT

The work of the first two authors was supported by the Applied Mathematical Sciences Research Program, Office of Scientific Computing of the Office of Energy Research, U.S. Department of Energy and Lawrence Livermore National Laboratory, under contract number W-7405-Eng-48. The work of the third author was partially supported by the National Aeronautics and Space Administration under grant number NAG-2-268 and the National Science Foundation under grant number DMS-8319783.

REFERENCES

Abrahamsson, L. R., Keller, H. B., and Kreiss, H.-O. (1974). *Numer. Math.* **22**, 367.

Abramowitz, M., and Stegun, I. A. (1965). *Handbook of Mathematical Functions*, National Bureau of Standards Applied Math. Ser. No. 55. U.S. Gov. Printing Office, Washington, D.C.
Ahluwalia, D. S., and Keller, J. B. (1977). *Lect. Notes Phys.* **70**, 18.
Axelsson, O. (1981). *IMA J. Numer. Anal.* **1**, 329.
Bender, C. M., and Orszag, S. A. (1978). "Advanced Mathematical Methods for Scientists and Engineers." McGraw-Hill, New York.
Berger, A. E., Solomon, J. M., Ciment, M., Leventhal, S. H., and Weinberg, B. C. (1980). *Math. Comput.* **35**, 659.
Bogolyubov, N. N., and Mitropolsky, Y. A. (1962). "Asymptotic Methods in the Theory of Nonlinear Oscillations." Gordon & Breach, New York.
Brackbill, J. U., and Saltzman, J. S. (1982). *J. Comput. Phys.* **46**, 342.
Brenner, P., and Thomée, V. (1971). *Math. Scand.* **28**, 329.
Chang, K. W., and Howes, F. A. (1984). "Nonlinear Singular Perturbation Phenomena: Theory and Applications." Springer-Verlag, New York.
Chester, C., Friedman, B., and Ursell, F. (1957). *Proc. Cambridge Philos. Soc.* **53**, 599.
Chin, R. C. Y. (1975). *J. Comput. Phys.* **18**, 233.
Chin, R. C. Y., and Braun, R. L. (1980). *J. Comput. Phys.* **34**, 74.
Chin, R. C. Y., and Hedstrom, G. W. (1978). *Math. Comput.* **32**, 1163.
Chin, R. C. Y., and Krasny, R. (1983). *SIAM J. Sci. Stat. Comput.* **4**, 229.
Chin, R. C. Y., Hedstrom, G. W., and Howes, F. A. (1984). "A Survey of Analytical and Numerical Methods for Multiple-Scale Problems," Rep. UCRL-90971. Lawrence Livermore Nat. Lab., Livermore, California.
Courant, R., and Hilbert, D. (1953). "Methods of Mathematical Physics," Vol. I. Wiley (Interscience), New York.
Courant, R., and Hilbert, D. (1962). "Methods of Mathematical Physics," Vol. II. Wiley (Interscience), New York.
Dahlquist, G. (1963). *BIT* **3**, 27.
Dahlquist, G. (1974). *In* "Stiff Differential Systems" (R. Willoughby, ed.) p. 67. Plenum, New York.
Davis, S. F., and Flaherty, J. E. (1982). *SIAM J. Sci. Stat. Comput.* **3**, 6.
deGroen, P. P. N., and Hemker, P. W. (1979). *In* "Numerical Analysis of Singular Perturbation Problems" (P. W. Hemker and J. J. H. Miller eds.), p. 217. Academic Press, London.
Dwyer, H. A. (1983). "Grid Adaption for Problems with Separation, Cell Reynolds Number, Shock-Boundary Layer Interaction, and Accuracy," Rep. AIAA Pap. 83-0449, 21st Aerosp. Sci. Meet., January, Reno. Am. Inst. Aeron. Astron., New York.
Dwyer, H. A., Smooke, M. D., and Kee, R. J., (1982). *In* "Numerical Grid Generation" (J. F. Thompson, ed.), p. 339. North-Holland Publ., Amsterdam.
Eckhaus, W. (1979). "Asymptotic Analysis of Singular Perturbations." North-Holland Publ., Amsterdam.
Felsen, L. B., and Kamel, A. (1981). *Bull Seismol. Soc. Am*, **71**, 1763.
Flaherty, J. E., and Mathon, W. (1980). *SIAM J. Sci. Stat. Comput.* **1**, 260.
Flaherty, J. E., and O'Malley, R. E., Jr. (1977). *Math. Comput.* **31**, 66.
Godunov, S. K. (1959). *Mat. Sbornik* **47**, 271 (in Russian).
Gragg, W. B. (1965). *SIAM J. Numer. Anal.* **2**, 384.
Hale, J. K. (1969). "Ordinary Differential Equations." Wiley (Interscience), New York.
Harten, A. (1984). *SIAM J. Numer. Anal.* **21**, 1.
Harten, A., and Hyman, J. M., (1983). *J. Comput. Phys.* **50**, 235.
Harten, A., Lax, P. D., and van Leer, B. (1983). *SIAM Rev.* **25**, 35.
Hedstrom, G. W., and Osterheld, A. (1980). *J. Comput. Phys.* **37**, 399.
Heinrich, J. C., Huyakorn, P. S., Zienkiewicz, O. C., and Mitchell, A. R. (1977). *Int. J. Numer. Methods Eng.* **11**, 1831.

Hemker, P. W. (1977). Ph.D. Dissertation, Mathematisch Centrum, Amsterdam.
Hoppensteadt, F. C., and Miranker, W. L. (1983). *SIAM J. Sci. Stat. Comput.* **4**, 612.
Hughes, T. J. R., Liu, W. K., and Brooks, A. (1979). *J. Comput. Phys.* **30**, 1.
Kamel, A. A. (1970). *Celestial Mech.* **3**, 90.
Kamin, S. (1955). *Izv. Akad. Nauk SSSR, Ser. Mat.* **19**, 345 (in Russian).
Karasalo, I. (1975). "On Smooth Solutions to Stiff Nonlinear Analytic Differential Systems," Rep. TRITA-NA-7501, Royal Institute of Technology, Stockholm.
Kautsky, K., and Nichols, N. K. (1980). *SIAM J. Sci. Stat. Comput.* **1**, 499.
Kellog, R. B., and Tsan, A. (1978). *Math. Comput.* **32**, 1025.
Kevorkian, J., and Cole, J. D. (1981). "Perturbation Methods in Applied Mathematics." Springer-Verlag, New York.
Kirchgraber, U. (1983). "Dynamical System Methods in Numerical Analysis. Part I: An ODE Solver Based on the Method of Averaging," Rep. Res. Rep. 83-02. Seminar für Angewandte Mathematik, Eidgenössische Technische Hochschule, CH-8092 Zürich.
Kreiss, H.-O. (1978). In "Recent Advances in Numerical Analysis" (C. de Boor and G. H. Golub, eds.), p. 95. Academic Press, New York.
Kreiss, H.-O., and Nichols, N. (1976). *Lect. Notes Phys.* **58**, 544.
Kruskal, M. (1962). *J. Math. Phys.* **3**, 806.
Krylov, N. M., and Bogolyubov, N. N. (1934). "The Application of the Methods of Nonlinear Mechanics to the Theory of Stationary Oscillations," Publ. 8. Ukr. Akad. Sci., Kiev.
Lax, P. D. (1978). In "Recent Advances in Numerical Analysis" (C. de Boor and G. H. Golub, eds.), p. 107. Academic Press, New York.
Lentini, M., and Pereyra, V. (1977). *SIAM J. Numer. Anal.* **14**, 91.
Levinson, N. (1950). *Ann. Math.* **51**, 428.
Lighthill, J. (1978). "Waves in Fluids." Cambridge Univ. Press, London and New York.
Lin, C. C., and Segel, L. A. (1974). "Mathematics Applied to Deterministic Problems in the Natural Sciences," Macmillan, New York.
Lindberg, B. (1971). *BIT* **11**, 29.
McNamara, B. (1978). *J. Math. Phys.* **19**, 2154.
Majda, G. (1984), *SIAM J. Numer. Anal.* **21**, 535.
Meyer, R. E., and Parter, S. V., eds. (1980). "Singular Perturbations and Asymptotics." Academic Press, New York.
Miller, K. (1981). *SIAM J. Numer. Anal.* **18**, 1033.
Miller, K., and Miller, R. N., (1981). *SIAM J. Numer. Anal.* **18**, 1019.
Miranker, W. L. (1981). "Numerical Methods for Stiff Equations and Singular Perturbation Problems." Reidel, Dordrecht, Netherlands.
Nayfeh, A. H. (1973). "Perturbation Methods," Wiley (Interscience), New York.
Olver, F. W. J. (1974). "Asymptotics and Special Functions." Academic Press, New York.
O'Malley, R. E., Jr., and Anderson, L. R. (1979). In "Numerical Analysis of Singular Perturbation Problems" (P. W. Hemker and J. J. H. Miller, eds.), p. 317. Academic Press, New York.
Osher, S. (1981). In "Analytical and Numerical Approaches to Asymptotic Problems in Analysis" (O. Axelsson, L. S. Frank, and A. van der Sluis, eds.), p. 179. North-Holland Publ., Amsterdam.
Pereyra, V., and Sewell, E. G. (1975). *Numer. Math.* **23**, 261.
Perko, L. M. (1968). *SIAM J. Appl. Math.* **17**, 698.
Petzold, L. R. (1981). *SIAM J. Numer. Anal.* **18**, 455.
Preuss, S. A. (1973). *Math. Comput.* **27**, 551.
Roe, P. L. (1981). *Lect. Notes Phys.* **141**, 354.
Russell, R. D., and Christiansen, J. (1978). *SIAM J. Numer. Anal.* **15**, 59.
Sethna, P. R. (1963). In "Nonlinear Differential Equations and Nonlinear Mechanics" (J. P. LaSalle and S. Lefschetz, eds.), p. 58. Academic Press, New York.

Sethna, P. R. (1967). *Q. Appl. Math.* **25**, 205.
Shampine, L. F., and Gear, C. W. (1979). *SIAM Rev.* **21**, 1.
Stern, D. P. (1970). *J. Math. Phys.* **11**, 2771.
Trefethen, L. N. (1982). *SIAM Rev.* **24**, 113.
Van Dyke, M. (1975). "Perturbation Methods in Fluid Mechanics." Parabolic Press, Stanford, California.
van Kampen, N. G. (1983). "The Elimination of Fast Variables."
van Leer, B. (1974). *J. Comput. Phys.* **14**, 361.
van Leer, B. (1979). *J. Comput. Phys.* **32**, 101.
Warming, R., and Hyett, B. (1974). *J. Comput. Phys.* **14**, 159.
White, A. B., Jr. (1982). *SIAM J. Numer. Anal.* **19**, 683.
Whitham, G. B. (1974). "Linear and Nonlinear Waves." Wiley (Interscience), New York.
Woodward, P., and Colella, P. (1984). *J. Comput. Phys.* **54**, 115.

2

Problems with Different Time Scales

HEINZ-OTTO KREISS
Department of Applied Mathematics
California Institute of Technology
Pasadena, California

I.	Introduction	29
II.	Systems of Ordinary Differential Equations	32
III.	Numerical Methods for Ordinary Differential Equations	36
IV.	Partial Differential Equations	40
V.	Shallow Water Equations	44
VI.	Atmospheric Motions	49
VII.	Plasma Physics	55
	References	56

I. INTRODUCTION

Perhaps the simplest problem with different time scales is given by the initial-value problem for the ordinary differential equation

$$\varepsilon \, dy/dt = ay + e^{it}, \quad t \geq 0, \quad y(0) = y_0. \tag{1}$$

Here ε, a are constants with $0 < \varepsilon \ll 1$, $|a| = O(1)$, and Real $a \leq 0$. The solution of (1) is given by

$$y(t) = y_S(t) + y_R(t), \tag{2}$$

where

$$y_S(t) = e^{it}(-a + i\varepsilon)^{-1}, \quad y_R(t) = e^{(a/\varepsilon)t}(y_0 - y_S(0)).$$

Thus, it consists of the slowly varying part $y_S(t)$ and the rapidly changing part $y_R(t)$. There are two fundamentally different situations.

1. $a = -1$

In this case $y_R(t)$ decays rapidly, and outside a boundary layer the solution of (1) varies slowly. Many people have developed numerical methods to solve problems of this kind (see, e.g., Kreiss, 1978) and we do not consider this case.

2. $a = i$ Is Purely Imaginary

Now $y_R(t)$ does not decay and $y(t)$ is highly oscillatory everywhere. In many applications one is not interested in the fast time scale. Therefore, it is of interest to develop methods of preparing the initial data such that the fast time scale is suppressed. There are two ways to do this.

INITIALIZATION. One prepares the initial data in such a way that the fast time scale is not activated. In the preceding example we need only to choose

$$y_0 = y_S(0) = (-a + i\varepsilon)^{-1} = -a^{-1}(1 + i\varepsilon/a - (\varepsilon/a)^2 + \cdots). \tag{3}$$

Then $y_R(t) \equiv 0$, and the solution of our problem consists only of the slowly varying part $y_S(t)$. For more complicated problems one can determine $y_S(0)$ only approximately. The rapidly changing part will always be present, but we can reduce its amplitude to the size $O(\varepsilon^p)$, p a natural number. An effective way to do this is to use the "bounded derivative principle," which is based on the following observation:

If $y(t)$ varies on the slow time scale, then $d^v y/dt^v \sim O(1)$ for $v = 1, 2, \ldots, p$ where $p > 1$ is some suitable number. Therefore our principle is

Choose the initial value $y(0) = y_0$ such that for $t = 0$

$$d^v y/dt^v|_{t=0} \sim O(1), \quad v = 1, 2, \ldots, p. \tag{4}$$

Using the differential equation, we can express the derivatives at $t = 0$ by $y(0)$. Therefore, we can determine $y(0)$ such that (4) is satisfied without solving the differential equations.

Let us apply this principle to our example: $dy/dt|_{t=0} = O(1)$ if and only if

$$ay(0) = -1 + O(\varepsilon);$$

i.e.,

$$y(0) = -1/a + \varepsilon y_1, \quad dy/dt|_{t=0} = ay_1, \quad y_1 = O(1). \tag{5}$$

If we choose $y(0)$ according to (5), then

$$y_R(0) = y(0) - y_S(0) = -1/a + \varepsilon y_1 - 1/(-a + i\varepsilon) = O(\varepsilon);$$

2. Probems with Different Time Scales

i.e., the amplitude of $y_R(t)$ is $O(\varepsilon)$ for all times. We consider now the second derivative. The differential equation gives us

$$\varepsilon \, d^2y/dt^2 = a \, dy/dt + ie^{it}.$$

Thus $d^2y/dt^2|_{t=0} = O(1)$ if and only if

$$a \, dy/dt|_{t=0} = a^2 y_1 = -i + O(\varepsilon);$$

i.e.,

$$y_1 = -i/a^2 + \varepsilon y_2, \quad d^2y/dt^2|_{t=0} = a^2 y_2,$$

and by (5)

$$y(0) = -1/a(1 + i\varepsilon/a) + \varepsilon^2 y_2. \tag{6}$$

In this case we obtain for the amplitude

$$y_R(0) = y(0) - y_S(0) = O(\varepsilon^2).$$

The above procedure can be continued. If we choose the initial data such that the first p time derivatives are $O(1)$, then the amplitude of the fast part of the solution is $O(\varepsilon^p)$. We shall prove that results of this kind are valid for very general systems of linear and nonlinear ordinary and partial differential equations.

FILTERING. Instead of preparing the initial data, one can use a numerical method which automatically filters out the fast waves. Approximate, for example, the differential equation (1) by the implicit Euler method

$$\varepsilon[\tilde{y}(t + \Delta t) - \tilde{y}(t)]/\Delta t = a\tilde{y}(t + \Delta t) + e^{i(t+\Delta t)}, \quad \tilde{y}(0) = y_0, \tag{7}$$

where we assume that $a \, \Delta t \gg \varepsilon$. (If $a \, \Delta t \ll \varepsilon$, then the fast oscillation will not be damped out). We now derive an asymptotic expansion for $\tilde{y}(t)$. Define $\tilde{y}_1(t)$ by

$$\tilde{y}(t) = f_1(t) + \tilde{y}_1(t), \quad f_1(t) = -(1/a)e^{it}.$$

Then $\tilde{y}_1(t)$ satisfies the difference equation

$$\varepsilon[\tilde{y}(t + \Delta t) - \tilde{y}(t)]/\Delta t = a\tilde{y}_1(t + \Delta t) + \varepsilon(b/a)e^{i(t+\Delta t)},$$

$$b = (1 - e^{-i\Delta t})/\Delta t.$$

Therefore,

$$\tilde{y}_2(t) = \tilde{y}_1(t) - f_2(t), \quad f_2(t) = -\varepsilon(b/a)e^{it}$$

is the solution of

$$\varepsilon[\tilde{y}(t + \Delta t) - \tilde{y}(t)]/\Delta t = a\tilde{y}_2(t + \Delta t) + \varepsilon^2(b^2/a^2)e^{i(t+\Delta t)},$$

$$\tilde{y}_2(0) = \tilde{y}(0) - f_1(0) - f_2(0). \tag{8}$$

We can write $\tilde{y}_2(t)$ as $\tilde{y}_2(t) = v_T + v_F$ where v_T is the solution of the homogeneous equation (8), which we can write in the form

$$[1 - (a/\varepsilon)\,\Delta t]v_T(t + \Delta t) = v_T(t), \qquad (9)$$
$$v_T(0) = \tilde{y}_2(0),$$

and v_F satisfies the inhomogeneous equation with homogeneous initial data $\bar{v}_F(0) = 0$. By assumption, $|1 - (a/\varepsilon)\,\Delta t| \gg 1$, and therefore, the solution

$$v_T(t) = [1 - (a/\varepsilon)\,\Delta t]^{-t/\Delta t} v_T(0)$$

of (9) converges rapidly to zero with increasing t. By standard methods we get for $v_F(t)$ the estimate

$$|v_F(t)| = O(\varepsilon^2).$$

Therefore,

$$y_S(t) - \tilde{y}(t) = y_S(t) - f_1(t) - f_2(t) + O(\varepsilon^2) + O[|1 - (a/\varepsilon)\,\Delta t|^{-t/\Delta t}]$$
$$= O(\varepsilon^2) + O(\varepsilon\,\Delta t) + O[|1 - (a/\varepsilon)\,\Delta t|^{-t/\Delta t}].$$

This shows that, except for the first few time steps, the solution of the difference approximation is close to the slowly varying part of the solution of the differential equation. Thus, we do not need to treat the initial data in a special way.

The above result holds for linear problems. For nonlinear problems it is only valid if the fast part can be thought of as a small perturbation of the slowly varying solution. Otherwise, the error in the first few steps can shift the solution into another regime. We shall discuss this procedure for rather general ordinary and partial differential equations.

II. SYSTEMS OF ORDINARY DIFFERENTIAL EQUATIONS

In this section we consider systems of ordinary differential equations

$$dy/dt = \varepsilon^{-1} A(t)y + f(y, t), \qquad y(0) = y_0, \quad t \geq 0. \qquad (10)$$

Here $\varepsilon > 0$ is a small constant and $y = (y^{(1)}, y^{(2)}, \ldots, y^{(n)})^T$ is a vector function with n components, $A(t) = -A^*(t)$ is a skew Hermitian matrix and $f(y, t)$ is a vector function that depends nonlinearly on y and t.[†] We assume also that

---

[†] Notation: If g is a column vector, then $\bar{g}$ denotes the complex conjugate vector, g^T the transposed vector, and $g^* = \bar{g}^T$; $g^*f = \sum \bar{g}^{(i)} f^{(i)}$ denotes the scalar product, and $|g| = (g^*g)^{1/2}$ the Euclidean norm. Corresponding notation holds for matrices. In particular, $|A| = \sup_{|g|=1} g^*Ag$ denotes the operator norm.

$y_0 = O(1)$, that $A(t)$ and $f(y,t)$ have p derivatives with respect to all variables, and that these derivatives can be estimated by expressions $K_1|y|^m + K_2$, $K_j = O(1)$, $m > 0$ natural number. Thus, there are two time scales present, a slow $O(1)$ and a fast $O(1/\varepsilon)$.

We assume also that the eigenvalues $\varkappa$ of $A(t)$ are either identically zero or different from zero for all times; i.e., the rank of $A(t)$ is independent of t. This assumption is essential. Consider, for example, the differential equation

$$\varepsilon \, dy/dt = i(t - \tfrac{1}{2})y + \varepsilon^{1/2}, \qquad y(0) = y_0,$$

for $t \geq 0$. We can choose the initial data—using the bounded derivative principle—such that the solution varies on the slow scale for $t \leq \tfrac{1}{2} - O(\varepsilon^{1/2})$. However, this solution will be highly oscillatory for $t \geq \tfrac{1}{2} + O(\varepsilon^{1/2})$; i.e., the fast time scale is activated when passing through the turning point $t = \tfrac{1}{2}$.

In Kreiss (1979) we considered the more general systems

$$\varepsilon \, d\tilde{y}/dt = F(\tilde{y}, t).$$

The form (10) is often sufficiently general, because in many applications the fast oscillations can be considered as perturbations of a slow solution y_S. Thus, an ansatz

$$\tilde{y} = y_S + \varepsilon y, \qquad \varepsilon \, dy_S/dt = F(y_S, t)$$

leads to the system (10) with $A(t) = \partial F(y_S, t)/\partial y$.

The matrix $A(t)$ is often not skew Hermitian. One has to introduce new dependent variables to transform the system to the form (10). A typical example is given by

$$\frac{d}{dt}\begin{pmatrix} u \\ v \end{pmatrix} = \begin{pmatrix} 0 & \varepsilon^{-1} \\ -1 & 0 \end{pmatrix}\begin{pmatrix} u \\ v \end{pmatrix} + \begin{pmatrix} \varepsilon^{-1} f \\ g \end{pmatrix}.$$

Let $v = -f + v_1$, $\varepsilon^{1/2} u = u_1$, $g_1 = df/dt + g$. Then u_1, v_1 satisfy

$$\varepsilon^{1/2} \frac{d}{dt}\begin{pmatrix} u_1 \\ v_1 \end{pmatrix} = \begin{pmatrix} 0 & 1 \\ -1 & 0 \end{pmatrix}\begin{pmatrix} u_1 \\ v_1 \end{pmatrix} + \varepsilon^{1/2}\begin{pmatrix} 0 \\ g_1 \end{pmatrix},$$

which has the form (10).

We now show that the bounded derivative principle is valid for systems of type (10). Let us transform (10) into a suitable normal form. By assumption, there is a smooth unitary transformation $U(t)$ such that

$$U^*(t)A(t)U(t) = \begin{pmatrix} A_1(t) & 0 \\ 0 & 0 \end{pmatrix}, \qquad A_1^* = -A_1.$$

The eigenvalues of $A_1(t)$ are all different from zero, and therefore $A_1^{-1}(t)$

exists and varies slowly. Introduce into (10) a new variable $y = Uw$. Then (10) becomes

$$\frac{dw}{dt} = \frac{1}{\varepsilon}\begin{pmatrix} A_1(t) & 0 \\ 0 & 0 \end{pmatrix} w - U^* \frac{dU}{dt} w + U^*f,$$

which can also be written in the form

$$du/dt = \varepsilon^{-1} A_1(t)u + g(w, t),$$
$$dv/dt = h(w, t), \quad w = (u, v)^T, \quad w(0) = U^*(0)y(0). \tag{11}$$

We can now prove

THEOREM 1. *There is an interval $0 \le t \le T$ and a constant K_p such that*

$$\sup_{0 \le t \le T} \sum_{v=0}^{p} \left| \frac{d^v w(t)}{dt^v} \right| \le K_p \sum_{v=0}^{p} \left| \frac{d^v w}{dt^v} \right|_{t=0}.$$

Here $K_p = O(1)$ and T does not depend on ε.

PROOF. By assumption

$$|u^* g(w, t)| + |v^* h(w, t)| \le K_1 |w|^{m+1} + K_2 |w|, \quad \text{Real } u^* A_1 u = 0.$$

Therefore

$$\frac{d|w|^2}{dt} = \frac{d|u|^2}{dt} + \frac{d|v|^2}{dt} = 2 \operatorname{Real}\left(u^* \frac{du}{dt} + v^* \frac{dv}{dt} \right)$$
$$= 2 \operatorname{Real}(u^* g(w, t) + v^* h(w, t)) \le 2K_1 |w|^{m+1} + K_2 |w|.$$

Integrating the above inequality shows that $w(t)$ is $O(1)$ in an interval $0 \le t \le T$, where T does not depend on ε.

Let $u_1 = du/dt$, $v_1 = dv/dt$, and differentiate (11). Then we obtain a differential equation for the first derivative:

$$du_1/dt = \varepsilon^{-1} A_1(t) u_1 + \varepsilon^{-1}(dA_1/dt)u + dg/dt,$$
$$dv_1/dt = dh(w, t)/dt. \tag{12}$$

We can eliminate the large term $\varepsilon^{-1} dA_1/dt \, u$ by a change of variable:

$$u_2 = u_1 + A_1^{-1} dA_1/dt \, u, \quad v_2 = v_1, \quad w_2 = (u_2, v_2)^T. \tag{13}$$

Then we obtain a system of the same form as before:

$$du_2/dt = \varepsilon^{-1} A_1(t) u_2 + g_2(w_2, w, t),$$
$$dv_2/dt = h_2(w_2, w, t);$$

and we can estimate w_2 in terms of w and the initial data. This proves the theorem for $p = 1$. The general case is proved by repeated differentiation.

REMARK. It is clear that the theorem is also valid for the system (10), because "boundedness" is invariant under smooth transformations.

Theorem 1 tells us that the solution is slowly varying for $0 \le t \le T$ provided the derivatives are bounded at $t = 0$. In general, we cannot expect T to be large. However, in Kreiss (1979) we have shown that $T = O(\varepsilon^{-(p-2)/2})$ provided reasonable extra assumptions for g, h are made. We now discuss how to choose the initial data. Without restriction, we can assume that the system has the form (11); $dw/dt\,|_{t=0} = O(1)$ if and only if

$$A_1 u(0) = O(\varepsilon).$$

Thus

$$u(0) = \varepsilon q_1, \qquad q_1 = O(1), \tag{14}$$

and

$$du/dt\,|_{t=0} = A_1(0)q_1 + g(u = 0, v(0), 0) + O(\varepsilon). \tag{15}$$

Equation (14) guarantees that $d^2v/dt^2 = O(1)$. For d^2u/dt^2 we have

$$d^2u/dt^2\,|_{t=0} = \varepsilon^{-1}[A_1(0)\,du/dt + dA_1/dt\,u]|_{t=0} + O(1)$$
$$= \varepsilon^{-1}A_1(0)[A_1(0)q_1 + g(u = 0, v(0), 0)] + O(1).$$

Therefore, $d^2u/dt^2\,|_{t=0} = O(1)$ if and only if we choose

$$q_1(0) = -A_1^{-1}(0)g[u = 0, v(0), 0] + \varepsilon q_2, \qquad q_2 = O(1),$$

or

$$u(0) = -\varepsilon A_1^{-1}(0)g[u = 0, v(0), 0] + O(\varepsilon^2). \tag{16}$$

Equation (16) tells us that $u(0)$ is determined by $v(0)$ up to terms of order $O(\varepsilon^2)$. This process can be continued. If we demand that p time derivatives are bounded independently of ε, then $u(0)$ is determined by $v(0)$ up to terms of order $O(\varepsilon^p)$.

We can now use our results to derive reduced systems. We know that if we choose the initial data such that p time derivatives are bounded independently of ε, then u is determined by v up to terms of order $O(\varepsilon^p)$. This relation can be derived for every fixed t with $0 \le t \le T$. Thus, we can replace the differential equation for u by a relation between u and v. We obtain reduced systems that become more and more refined, depending on the number of time derivatives that stay bounded. The crudest system is

$$u \equiv 0, \qquad dv/dt = h(u, v, t), \tag{17}$$

and an improved system is given by

$$A_1(t)u(t) + \varepsilon g[u \equiv 0, v(t), t] = 0, \qquad dv/dt = h(u, v, t). \tag{18}$$

We have shown how to construct a smooth solution (u_S, v_S) of the system (11). We can choose $v_S(0)$ arbitrarily. Then $u_S(0)$ is uniquely determined up to terms of order $O(\varepsilon^p)$. Now consider the system (11) with perturbed initial data:

$$u(0) = u_S(0) + q_0, \qquad v(0) = v_S(0).$$

Here q_0 is sufficiently small such that linearization around u_S, v_S can be justified. We want to study the effect of the perturbation. To first approximation, $q(t) = u(t) - u_S(t)$, $r(t) = v(t) - v_S(t)$ are the solution of the linearized equations

$$dq/dt = \varepsilon^{-1} A_1 q + A_{11} q + A_{12} r,$$
$$dr/dt = A_{21} q + A_{22} r, \qquad q(0) = q_0, \qquad r(0) = 0, \tag{19}$$

where

$$A_{11} = \partial g(u_S, v_S, t)/\partial u, \qquad A_{12} = \partial g(u_S, v_S, t)/\partial v,$$
$$A_{21} = \partial h(u_S, v_S, t)/\partial u, \qquad A_{22} = \partial h(u_S, v_S, t)/\partial v.$$

To first approximation, we can replace (19) by the decoupled system

$$\varepsilon \, d\tilde{q}/dt = (A_1 + \varepsilon A_{11})\tilde{q}, \qquad d\tilde{r}/dt = A_{21}\tilde{q} + A_{22}\tilde{r}.$$

$\tilde{q}$ is highly oscillatory, and, therefore, $\tilde{r} = O(\varepsilon|q_0|)$. This gives us for the smooth part of the solution of (19)

$$q_S = O(\varepsilon^2|q_0|), \qquad r_S = O(\varepsilon|q_0|).$$

Thus, the effect of a highly oscillatory perturbation on the smooth solution is much smaller than the size of the perturbation. This means that in computations, one can be quite sloppy with the initialization and can apply a numerical filter after the calculation. A rigorous derivation of the above result for partial differential equations is given in Barker (1982).

III. NUMERICAL METHODS FOR ORDINARY DIFFERENTIAL EQUATIONS

In this section we consider systems of type (10) and approximate them by difference equations. All the approximations that we discuss will treat the term $\varepsilon^{-1} Ay$ implicitly. If we were to use an explicit method, we would have to restrict the time step to $\Delta t < \varepsilon$; i.e., we would resolve the fast time scale, which we are not interested in doing.

The first method we discuss is the backward Euler method. Let $\Delta t > 0$ denote the time step and use the notation $t_n = n \Delta t$, $y_n = y(t_n)$. We approxi-

2. Problems with Different Time Scales

mate (10) by

$$\frac{\tilde{y}_{n+1} - \tilde{y}_n}{\Delta t} = \frac{1}{\varepsilon} A_{n+1}\tilde{y}_{n+1} + f(\tilde{y}_{n+1}, t_{n+1}), \qquad \tilde{y}(0) = y(0) = y_0. \qquad (20)$$

To begin with, let us assume that we have initialized the data properly, i.e., $y_0 = y_S(0)$, and that the solution $y(x, t)$ of the differential equation has two derivatives of order unity. Then we can derive error estimates by standard methods. To obtain bounds for the truncation error, we introduce the solution of the differential equation into the difference approximation and get

$$\frac{y_{n+1} - y_n}{\Delta t} - \left(\frac{1}{\varepsilon} A_{n+1}y_{n+1} + f(y_{n+1}, t_{n+1})\right)$$

$$= \frac{y_{n+1} - y_n}{\Delta t} - \frac{dy_{n+1}}{dt} = \frac{1}{2}\Delta t R_n, \qquad |R_n| \leq \max_{t_n \leq \xi \leq t_{n+1}} \left|\frac{d^2 y}{dt^2}\right|. \qquad (21)$$

Subtracting (20) from (21) and neglecting quadratic terms, we obtain an approximate equation for the error $e_n = y_n - \tilde{y}_n$:

$$(e_{n+1} - e_n)/\Delta t = \varepsilon^{-1} A_{n+1} e_{n+1} + B_{n+1} e_{n+1} + \tfrac{1}{2}\Delta t R_n, \qquad e_0 = 0, \qquad (22)$$

where

$$B(t) = \partial f(y, t)/\partial y$$

is the Jacobian of $f(y, t)$. The growth properties of e_n are determined by the stability properties of the homogeneous equation

$$(w_{n+1} - w_n)/\Delta t = (\varepsilon^{-1} A_{n+1} + B_{n+1}) w_{n+1}, \qquad (23)$$

which we write as

$$[I - \Delta t(\varepsilon^{-1} A_{n+1} + B_{n+1})] w_{n+1} = w_n. \qquad (24)$$

We are interested in oscillatory problems. Therefore, it is reasonable to assume that

$$B + B^* \leq 0.$$

Then

$$|w_{n+1}|^2 \leq |[I - \Delta t(\varepsilon^{-1} A_{n+1} + B_{n+1})] w_{n+1}|^2 = |w_n|^2. \qquad (25)$$

Thus, (22) is completely stable, and a simple calculation gives us the error estimate

$$|e_n| \leq \tfrac{1}{2}\Delta t \cdot t_n \max_{0 \leq v \leq n} |R_v|. \qquad (26)$$

Assume now that we do not initialize the initial data properly; i.e., in terms of the transformed variables, we use

$$u_0 = u_S(0) + q_0, \quad v_0 = v_S(0).$$

Denote by $\tilde{y}_S$ the solution with initial data u_S, v_S. Then, to first approximation, $d_n = \tilde{y} - \tilde{y}_S$ satisfies the linearized equation (23). We use now the transformation U of the last section to separate the fast and the slow scale: $\tilde{d}_n = (\tilde{q}_n, \tilde{r}_n)^T = U^* d_n$ is the solution of

$$\frac{\tilde{d}_{n+1} - \tilde{d}_n}{\Delta t} = \frac{1}{\varepsilon}\begin{pmatrix} A_1(t_{n+1}) & 0 \\ 0 & 0 \end{pmatrix}\tilde{d}_{n+1} + \tilde{B}_{n+1}\tilde{d}_{n+1} + \tilde{C}_n \tilde{d}_n, \quad (27)$$

$$\tilde{q}_0 = q_0, \quad \tilde{r}_0 = 0,$$

where

$$\tilde{B}_{n+1} = U^*_{n+1} B_{n+1} U_{n+1}, \quad \Delta t \tilde{C}_n = U^*_{n+1}(U_{n+1} - U_n).$$

One can derive the following estimate for the solution of (27):

$$\begin{aligned} |\tilde{q}_n| &= O[(\varepsilon/\Delta t)^n |q_0|] + O(\varepsilon |q_0|), \\ |\tilde{r}_n| &= O(\Delta t |q_0|). \end{aligned} \quad (28)$$

If $A(t)$ does not depend on t, then $C_n = 0$, and one gains a factor $\varepsilon/\Delta t$; i.e.,

$$\begin{aligned} |\tilde{q}_n| &= O[(\varepsilon/\Delta t)^n |q_0|] + O(\varepsilon^2/\Delta t |q_0|), \\ |\tilde{r}_n| &= O(\varepsilon |q_0|). \end{aligned} \quad (29)$$

Equations (28) say that the error—away from a possible boundary layer—is of the same order of magnitude as the error e, which we commit if we have used properly initialized data.

The backward Euler method is only first order accurate. A better method is the backward differentiation method:

$$\tfrac{3}{2}\tilde{y}_{n+1} - 2\tilde{y}_n + \tfrac{1}{2}\tilde{y}_{n-1} = \Delta t[\varepsilon^{-1} A(t_{n+1})\tilde{y}_{n+1} + f(\tilde{y}_{n+1}, t_{n+1})].$$

It is second order accurate and has the same stability and damping properties as the backward Euler method (see, e.g., Kreth, 1977). However, if we do not initialize properly, we obtain error estimates similar to those of (28) and (29). Thus, the accuracy can be destroyed.

Another method is the trapezoidal rule

$$\frac{\tilde{y}_{n+1} - \tilde{y}_n}{\Delta t} = \frac{1}{2}\left[\frac{1}{\varepsilon}A_{n+1}\tilde{y}_{n+1} + f(\tilde{y}_{n+1}, t_{n+1}) + \frac{1}{\varepsilon}A_n\tilde{y}_n + f(\tilde{y}_n, t_n)\right].$$

However, it has unsatisfactory stability properties. Also, one has to initialize the data properly because the oscillatory part is not damped.

2. Problems with Different Time Scales

One could also use a semi-implicit method, for example, a combination of "backward Euler and leapfrog":

$$(\tilde{y}_{n+1} - \tilde{y}_{n-1})/2\Delta t = \varepsilon^{-1} A_{n+1} \tilde{y}_{n+1} + f(\tilde{y}_n, t_n).$$

In Guerra and Gustafsson (1982) it is shown that the method is second order accurate provided one initializes the data properly and A does not depend on t. Then it also has good stability properties. It is doubtful that it is second order accurate if A depends on t.

Instead of initializing the data and then using a difference approximation for the full set of equations, we can replace the full system by the reduced system, i.e., replace (10) by (18). The error we commit is $O(\varepsilon^2)$. This often leads to more accurate solutions and less work because we can approximate $dv/dt = h(u, v, t)$ by an explicit method.

Finally, we consider the case that $A(t)$ is not skew Hermitian. As an example, we consider the system

$$\frac{dy}{dt} = \begin{pmatrix} 0 & \varepsilon^{-1} \\ -1 & 0 \end{pmatrix} y + \begin{pmatrix} \varepsilon^{-1} f \\ g \end{pmatrix}, \quad y = \begin{pmatrix} u \\ v \end{pmatrix},$$

$$\frac{dw}{dt} = u + h,$$
(30)

and approximate it by the backward Euler method

$$\frac{\tilde{y}_{n+1} - \tilde{y}_n}{\Delta t} = \begin{pmatrix} 0 & \varepsilon^{-1} \\ -1 & 0 \end{pmatrix} \tilde{y}_{n+1} + \begin{pmatrix} \varepsilon^{-1} f_{n+1} \\ g_{n+1} \end{pmatrix}$$

$$\frac{\tilde{w}_{n+1} - \tilde{w}_n}{\Delta t} = \tilde{u}_{n+1} + h_{n+1}.$$
(31)

To begin with, we assume that $f = g = h = 0$. Then the smooth solution of (30) is determined by the initial data

$$u(0) = v(0) = 0, \quad w(0) = w_0;$$

i.e.,

$$u(t) = v(t) \equiv 0, \quad w(t) = w_0.$$

With these initial values the difference approximation gives the same answer. However, if we do not initialize properly, then we will, in general, make an error. We can write the homogeneous difference equations for $\tilde{y}_{n+1}$ in the form

$$\tilde{y}_{n+1} = \frac{\varepsilon}{\varepsilon + (\Delta t)^2} \begin{pmatrix} 1 & \Delta t/\varepsilon \\ -\Delta t & 1 \end{pmatrix} \tilde{y}_n.$$

Assuming that $\varepsilon = (\Delta t)^2$, we obtain

$$\tilde{y}_{n+1} = \begin{pmatrix} \frac{1}{2} & \frac{1}{2}(\Delta t)^{-1} \\ -\frac{1}{2}\Delta t & \frac{1}{2} \end{pmatrix} \tilde{y}_n. \tag{32}$$

If $\tilde{v}_0 \neq 0$, then $\tilde{u}_1 = \frac{1}{2}(\Delta t)^{-1} \tilde{v}_0 + \frac{1}{2}\tilde{u}_0 = O(1/\Delta t)$. Thus

$$\tilde{w}_1 = w_0 + \Delta t\, \tilde{u}_1 = w_0 + \frac{1}{2}\tilde{v}_0 + \frac{1}{2}\Delta t\, \tilde{u}_0$$

which shows that $\tilde{w}_1 - w(\Delta t) = O(\tilde{v}_0)$.

The same result holds if $\varepsilon < (\Delta t)^2$. Also, the stability properties of (32) are not satisfactory. Therefore, if $\varepsilon \ll \Delta t$, it might be better to use the reduced system. Using the bounded derivative principle up to the fourth derivative, we obtain

$$v(t) = -f(t) + \varepsilon[d^2 f(t)/dt^2 + dg(t)/dt],$$
$$u(t) = df/dt + g - \varepsilon[d^3 f(t)/dt^3 + d^2 g(t)/dt^2], \quad dw/dt = u + h.$$

The error we commit is $O(\varepsilon^2)$.

IV. PARTIAL DIFFERENTIAL EQUATIONS

In this section we want to generalize the results of Section II to partial differential equations. Let $x = (x_1, x_2, \ldots, x_s)$ be a point in the s-dimensional Euclidean space R_s, e_j the unit vector in the x_j direction, and $u = [u^{(1)}(x,t), \ldots, u^{(n)}(x,t)]^T$ a vector function with n components. We start with systems of the form

$$u_t = \varepsilon^{-1} P_0(\partial/\partial x)u + P_1(u, \varepsilon, \partial/\partial x)u + F(x, t), \tag{33}$$

with periodic boundary conditions

$$u(x + 2\pi e_j) = u(x), \quad j = 1, 2, \ldots, n.$$

Here $F(x, t)$ is a smooth function of x, t and the coefficients of

$$P_0 = \sum_{j=1}^{n} A_j\, \partial/\partial x_j, \quad A_j = A_j^* \quad \text{constant matrices,}$$
$$P_1(u, \varepsilon, \partial/\partial x) = \sum_{j=1}^{s} B_j(u, \varepsilon)\, \partial/\partial x_j, \quad B_j = B_j^* \quad \text{smooth functions of } u, \varepsilon, \tag{34}$$

are real symmetric matrices. For the Fourier transform of P,

$$\hat{P}_0(i\omega) = i \sum_{j=1}^{s} A_j \omega_j, \quad \omega_j = 0, \pm 1, \pm 2, \ldots, \tag{35}$$

2. Probems with Different Time Scales

we make

ASSUMPTION 1. For every fixed $\omega = (\omega_1, \ldots, \omega_s)$, ω_j integer, the eigenvalues $\varkappa_v$ of P_0 satisfy either

$$\varkappa_v = 0 \quad \text{or} \quad |\varkappa_v| \geq 1.$$

REMARK. Observe that the number of eigenvalues $\varkappa_v$ with $\varkappa_v = 0$ can depend on ω. Also, $|\varkappa_v| \geq 1$ can be replaced by $|\varkappa_v| \geq \delta > 0$ by redefining ε.

In Browning and Kreiss (1982b) we have shown that the bounded derivative principle is valid, i.e., we have shown

THEOREM 2. *Assume that all derivatives*

$$\partial^{|v|+\mu} u(x,t)/\partial x_1^{v_1} \cdots \partial x_s^{v_s} \partial t^\mu \big|_{t=0}, \qquad |v| + \mu := \sum v_i + \mu \leq p$$

are bounded independently of ε. Then we can estimate these derivatives independently of ε in a time interval $0 \leq t \leq T$, T independent of ε, provided $p \geq [\frac{1}{2}s] + 2$. Here $[\frac{1}{2}s]$ is the largest integer with $[\frac{1}{2}s] \leq s/2$.

We do not give the proof here, but instead make the result plausible. In many applications it is possible to describe the solution of the problem with the help of a finite number of waves; i.e.,

$$u(x,t) = \sum_{|\omega_j| \leq N} \hat{u}(\omega, t) e^{i\langle \omega, x\rangle}, \qquad \langle \omega, x\rangle = \sum \omega_j x_j. \tag{36}$$

Also, the coefficients of P_1 are polynomials in u. Introducing (36) into (33) and neglecting all combined frequencies with $|\omega_j| > N$ for some j, we obtain a system of ordinary differential equations

$$\frac{d\hat{u}(v,t)}{dt} = \frac{1}{\varepsilon} P_0(iv)\hat{u}(v,t) + G_v[\hat{u}(\omega, t), t], \qquad v = 0, \pm 1, \ldots, \pm N. \tag{37}$$

Here G_v is a smooth function depending on all amplitudes $\hat{u}(\omega, t)$. This system is of the same form as that in Section II, and we know that the bounded derivative principle is valid.

We now describe an initialization procedure based on Fourier expansion. For meteorological applications, it is developed in Baer (1977), Kasahara (1982), Leith (1980) and Machenhauer (1977). For every fixed ω there exist a unitary matrix $\hat{U} = \hat{U}(i\omega)$ such that

$$\hat{U}^*(i\omega)\hat{P}_0(i\omega)\hat{U}(i\omega) = \begin{pmatrix} \hat{R}(i\omega) & 0 \\ 0 & 0 \end{pmatrix}.$$

By assumption 4.1 $\hat{R}^{-1}$ exists and

$$|\hat{R}^{-1}| \leq 1. \tag{38}$$

Let L_2 denote the space of all functions

$$f = \sum_\omega \hat{f}(\omega)e^{i\langle\omega,x\rangle}, \quad \sum |\hat{f}(\omega)|^2 < \infty,$$

which can be expanded into a Fourier series, and let Q denote the projection

$$f^{\mathrm{I}} = Qf = \sum_\omega \hat{U}(i\omega)\begin{pmatrix} I_\omega & 0 \\ 0 & 0 \end{pmatrix} \hat{U}^*(i\omega) e^{i\omega x} \hat{f}(\omega).$$

Here I_ω is the unit matrix of the same dimension as $\hat{R}(i\omega)$, Q splits L_2 into two subspaces $L_2^{\mathrm{I}}, L_2^{\mathrm{II}}$ defined by

$$f^{\mathrm{I}} = Qf, \quad f^{\mathrm{II}} = (I - Q)f, \quad f = f^{\mathrm{I}} + f^{\mathrm{II}}.$$

Note that Q commutes with P_0; i.e., $QP_0 = P_0 Q$ because

$$QP_0 u = \sum_\omega \hat{U}(i\omega)\begin{pmatrix} I_\omega & 0 \\ 0 & 0 \end{pmatrix} \hat{U}^*(i\omega)\hat{U}(i\omega)\begin{pmatrix} \hat{R}(i\omega) & 0 \\ 0 & 0 \end{pmatrix} U^*(i\omega)\hat{u}(\omega)e^{i\langle\omega,x\rangle}$$

$$= \sum_\omega \hat{U}(i\omega)\begin{pmatrix} \hat{R}(i\omega) & 0 \\ 0 & 0 \end{pmatrix} \hat{U}^*(i\omega)\hat{u}(\omega)e^{i\langle\omega,x\rangle}$$

$$= \sum_\omega \hat{U}(i\omega)\begin{pmatrix} \hat{R}(i\omega) & 0 \\ 0 & 0 \end{pmatrix} \hat{U}^*(i\omega)\hat{U}(i\omega)\begin{pmatrix} I_\omega & 0 \\ 0 & 0 \end{pmatrix} \hat{U}^*(i\omega)\hat{u}(\omega)e^{i\langle\omega,x\rangle}$$

$$= P_0 Q u.$$

and

$$P_0 u^{\mathrm{II}} = P_0 (I - Q)u$$

$$= \sum_\omega \hat{U}(i\omega)\begin{pmatrix} \hat{R}(i\omega) & 0 \\ 0 & 0 \end{pmatrix}\left[I - \begin{pmatrix} I_\omega & 0 \\ 0 & 0 \end{pmatrix}\right] \hat{U}^*(i\omega)\hat{u}(\omega)e^{i\langle\omega,x\rangle} = 0.$$

We can define the inverse of P_0 on L_2^{I} uniquely by

$$P_0^{-1} u^{\mathrm{I}} = \sum_\omega \hat{U}(i\omega)\begin{pmatrix} \hat{R}^{-1}(i\omega) & 0 \\ 0 & 0 \end{pmatrix} \hat{U}^*(i\omega)\hat{u}(\omega)e^{i\langle\omega,x\rangle},$$

and (38) gives us

$$\|P_0^{-1} u^{\mathrm{I}}\| \leq \|u^{\mathrm{I}}\|. \tag{39}$$

Using the projection Q, we can now write the system (33) in the form

$$u_t^{\mathrm{I}} = \varepsilon^{-1} P_0 u^{\mathrm{I}} + [P_1(u^{\mathrm{I}}, u^{\mathrm{II}}, \varepsilon, \partial/\partial x)u]^{\mathrm{I}} + F^{\mathrm{I}} \tag{40a}$$

$$u = u^{\mathrm{I}} + u^{\mathrm{II}}, \tag{40b}$$

$$u_t^{\mathrm{II}} = [P_1(u^{\mathrm{I}}, u^{\mathrm{II}}, \varepsilon, \partial/\partial x)u]^{\mathrm{II}} + F^{\mathrm{II}}. \tag{40c}$$

Thus, the projection accomplishes the same thing as the transformation of (10) into (11). It splits the system into a fast and a slow part.

2. Problems with Different Time Scales

We can now initialize (40) in the same way as in Section II. We assume that $u(x,0)$ is a smooth function of x. The $du(x,0)/dt$ is bounded independently of ε if and only if

$$P_0 u^I = O(\varepsilon);$$

hence

$$u^I = \varepsilon u_1^I, \qquad u_1^I = O(1), \tag{41}$$

i.e. $u(x,0)$ has to first approximation no component in L_2^I. Differentiating (40) with respect to t and assuming that (41) holds gives us

$$u_{tt}^I = \varepsilon^{-1} P_0 u_t^I + O(1)$$
$$= \varepsilon^{-1} P_0 \{P_0 u_1^I + [P_1(0, u^{II}, 0, \partial/\partial x) u^{II}]^I + F^I\} + O(1),$$
$$u_{tt}^{II} = O(1).$$

Therefore, the second time derivative is bounded independently of ε if and only if

$$P_0 u_1^I + [P_1(0, u^{II}, 0, \partial/\partial x) u^{II}]^I + F^I = \varepsilon P_0 u_2^I = O(\varepsilon). \tag{42}$$

Thus u^I is determined by u^{II} up to terms of order $O(\varepsilon^2)$. This process can be continued.

If one wants to solve the preceding problem numerically, then one can either initialize the data or use the reduced system consisting of (40b), (40c), (41) and (42). If one uses the reduced system, then one should use spectral methods because it is necessary to use the Fourier representation to calculate the projection anyway. Even if one uses the full system, there are advantages in using spectral method. In this case one solves a system of ordinary differential equations, and the results of Section III apply.

There are no new difficulties in treating systems of the form

$$D(u, x, t) \partial u/\partial t = \varepsilon^{-1} P_0(\partial/\partial x) u + P_1(u, \varepsilon, \partial/\partial x) u + F. \tag{43}$$

Also P_0 may include lower order terms with constant coefficients. However, if $P_0 = P_0(x, t, \partial/\partial x)$ depends on x, t, then the theory becomes much more complicated. In this case one has to assume that the number M of nonzero eigenvalues of the symbol

$$P_0(x, t, i\omega) = i \sum_j A_j(x, t) \omega_j, \qquad \omega \neq 0,$$

does not depend on x, t, ω. In Kreiss (1980) the case $M = 2$ is treated; it is generalized in Tadmor (1982) to arbitrary M. Again it is shown that the bounded derivative principle applies and that one can either use the full system with initialized data or solve the reduced problem.

The assumption that the number M of eigenvalues $\varkappa \neq 0$ does not depend on x, t, ω leads to a particular normal form of the differential equations, which implies that the relations the slow solutions must satisfy can be expressed as elliptic partial differential equations. Therefore, the reduced system consists of a combined hyperbolic–elliptic system. This makes it possible to treat the initial-boundary-value problem also (Gustafsson and Kreiss, 1983; Gustafsson, 1980a,b; Guerra and Gustafsson, 1982; Kreiss, 1980). This theory with variable coefficients enables one to treat the fully nonlinear case as well (Browning and Kreiss, 1982b).

V. SHALLOW WATER EQUATIONS

Shallow water equations play a central role in geophysics. They govern two classes of motions with different time scales, consisting of low-frequency Rossby waves and high-frequency inertia gravity waves. Often, one is not interested in the inertia gravity waves, which have to be filtered out. Initialization procedures have been considered for a long time and we refer to Browning *et al.* (1980) for a more detailed account of their development. In Cartesian coordinates x and y, directed eastward and northward, respectively, the shallow water equations including the effect of gravity (see, e.g., Kasahara, 1974) are expressed by

$$u_t + uu_x + vu_y + gh_x - fv = 0,$$
$$v_t + uv_x + vv_y + gh_y + fu = 0, \qquad (44)$$
$$h_t + (uh)_x + (vh)_y + h_0(u_x + v_y) - (uH)_x - (vH)_y = 0.$$

Here t is time, u and v the velocity components in the x and y directions, h_0 the mean height of the homogeneous fluid, h the deviation from the mean height, and H the elevation of orography; $g \sim 10$ msec^{-2} is the constant gravity acceleration, and f denotes the Coriolis force. We use the beta plane approximation; i.e.,

$$f = 2\Omega(\sin \theta_0 + y/r \cos \theta_0), \qquad 2\Omega = 10^{-4} \text{ sec}^{-1}, \quad r = 10^7 \text{ m}, \qquad (45)$$

where r is the radius, Ω the angular speed for the earth, and θ_0 the latitude of the coordinate origin.

Before one can apply the bounded derivative principle, one has to scale the dependent and independent variables in such a way that the desired solution and a number of its derivatives are of order $O(1)$. We assume that the motion in the x and y direction have the same characteristics, i.e., they have the same length scale L and the velocity components have the same size U. Therefore, we introduce new variables by

$$\begin{aligned} x' &= x/L, & y' &= y/L, & t' &= t/T, & f' &= f/2\Omega, \\ u' &= u/U, & v' &= v/U, & \varphi' &= h/D, & \Phi &= H/H_0, \end{aligned} \qquad (46)$$

2. Problems with Different Time Scales

and obtain

$$u'_{t'} + \frac{UT}{L}(u'u'_x + v'u'_y) + \frac{gDT}{LU}\varphi'_{x'} - 2\Omega T f' v' = 0,$$

$$v'_{t'} + \frac{UT}{L}(u'v'_x + v'v'_y) + \frac{gDT}{LU}\varphi'_{y'} + 2\Omega T f' u' = 0, \quad (47)$$

$$\varphi'_{t'} + \frac{UT}{L}[(u'\varphi')_x + (v'\varphi')_y] + \frac{h_0 UT}{DL}(u'_{x'} + v'_{y'})$$

$$- \frac{UH_0 T}{DL}[(u'\Phi')_x + (v'\Phi')_y] = 0$$

We are interested in motions where the time derivatives are of the same size as the convective terms; i.e.,

$$UT/L = 1. \quad (48)$$

We are interested in the case in which the Coriolis force has a very strong influence on the motion, and we assume that

$$2\Omega T \gg 1. \quad (49)$$

Necessarily, gDT/LU must be of the same order of magnitude, because otherwise there would be no balance between the terms in the momentum equations. Thus we assume that

$$2\Omega T = gDT/LU, \quad \text{i.e.,} \quad D = 2\Omega UL/g. \quad (50)$$

Dropping the prime notation, Eq. (47) become

$$u_t + uu_x + vv_y + R_0(\varphi_x - fv) = 0$$
$$v_t + uv_x + vv_y + R_0(\varphi_y + fu) = 0$$
$$\varphi_t + (u\varphi)_x + (v\varphi)_y + R_1(u_x + v_y) - R_2[(u\Phi)_x + (v\Phi)_y] = 0.$$

Here

$$R_0 = 2\Omega T, \quad R_1 = h_0/D, \quad R_2 = H_0/D. \quad (51)$$

We now choose the parameters according to meteorological scales:

$$L = 10^6 \text{ m}, \quad U = 10 \text{ msec}^{-2}, \quad h_0 = 10^4 \text{ m}, \quad H_0 = 10^3 \text{ m}.$$

By (48) and (50)

$$T = 10^5 \text{ sec}, \quad D = 10^2 \text{ m},$$

and, therefore,

$$R_0 = \varepsilon^{-1}, \quad R_1 = \varepsilon^{-2}, \quad R_2 = \varepsilon^{-1}, \quad \varepsilon = 10^{-1}.$$

With this notation our system becomes

$$\frac{du}{dt} + \varepsilon^{-1}(\varphi_x - fv) = 0, \tag{52a}$$

$$\frac{dv}{dt} + \varepsilon^{-1}(\varphi_y + fu) = 0, \tag{52b}$$

$$\frac{d\varphi}{dt} + \varepsilon^{-2}(1 + \varepsilon^2\varphi)(u_x + v_y) - \varepsilon^{-1}((u\Phi)_x + (v\Phi)_y) = 0. \tag{52c}$$

Here

$$d/dt = \partial/\partial t + u\,(\partial/\partial x) + v\,(\partial/\partial y)$$

denotes the convective derivative and

$$f = f_0 + \varepsilon\beta y, \qquad f_0 = \sin\theta_0, \qquad \beta = \cos\theta_0. \tag{53}$$

We consider the "midlatitude" case; i.e., $f_0 \sim 1$.

The system (52) is not exactly of the form discussed earlier. However, the proofs in Browning and Kreiss (1982b) and Kreiss (1980) can be modified to cover this case because we can symmetrize the equations by introducing new variables:

$$u(1 + \varepsilon\Phi) = \varepsilon^{1/2}\tilde{u}, \qquad v(1 + \varepsilon\Phi) = \varepsilon^{1/2}\tilde{v}.$$

Let us first consider the case in which $\Phi = 0$ and replace f by f_0. Also, linearize the system around the state of rest:

$$u = v = \varphi = 0.$$

Then we obtain a system with constant coefficients:

$$\begin{aligned}
\bar{u}_t + \varepsilon^{-1}(\bar{\varphi}_x - f_0\bar{v}) &= 0, \\
\bar{v}_t + \varepsilon^{-1}(\bar{\varphi}_y + f_0\bar{u}) &= 0, \\
\bar{\varphi}_t + \varepsilon^{-2}(\bar{u}_x + \bar{v}_y) &= 0.
\end{aligned} \tag{54}$$

This system can be solved by Fourier analysis. Let

$$w = e^{i\langle\omega,x\rangle}\hat{w}(\omega, t), \qquad \hat{w} = (\hat{u}, \hat{v}, \hat{\varphi})^T.$$

Then $\hat{w}(\omega, t)$ is the solution of

$$d\hat{w}/dt + \varepsilon^{-1}\hat{P}_0(i\omega)\hat{w} = 0,$$

$$\hat{P}_0(i\omega) = \begin{pmatrix} 0 & -f_0 & i\omega_1 \\ f_0 & 0 & i\omega_2 \\ i\omega_1/\varepsilon & i\omega_2/\varepsilon & 0 \end{pmatrix}. \tag{55}$$

2. Probems with Different Time Scales

The general solution of (55) is of the form

$$\hat{w} = \sum_{j=1}^{3} \sigma_j \exp(-\varkappa_j t)\hat{w}_j,$$

where $\varkappa_j$, $\hat{w}_j$ are the eigenvalues and corresponding eigenvectors of

$$\varkappa_j \hat{w}_j = \varepsilon^{-1} \hat{P}_0(i\omega)\hat{w}_j.$$

A simple calculation shows that the eigenvalues are given by

$$\varkappa_1 = 0, \qquad \varkappa_{2,3} = \pm \varepsilon^{-3/2}\sqrt{\omega_1^2 + \omega_2^2 + \varepsilon f_0^2}.$$

Thus, there are always exactly two large eigenvalues, and the initial data for the slow solution must satisfy two constraints. One can also determine the slow variable. Let

$$S = (\hat{w}_1, \hat{w}_2, \hat{w}_3)$$

and transform (55) to diagonal form:

$$\frac{d\hat{\hat{w}}}{dt} = \frac{1}{\varepsilon} S^{-1}\hat{P}_0(i\omega)S\hat{\hat{w}} = \begin{pmatrix} 0 & 0 & 0 \\ 0 & \varkappa_2 & 0 \\ 0 & 0 & \varkappa_3 \end{pmatrix}\hat{\hat{w}}, \qquad \hat{\hat{w}} = \begin{pmatrix} \hat{\hat{u}} \\ \hat{\hat{v}} \\ \hat{\hat{\varphi}} \end{pmatrix} = S\hat{w}.$$

Thus, $\hat{\hat{u}}$ is the slow variable. A simple calculation shows that

$$\hat{\hat{u}} = \text{const}(i\omega_2 \hat{u} - i\omega_1 \hat{v} + \varepsilon f \hat{\varphi}).$$

In physical space this expression corresponds to the potential vorticity

$$\bar{\xi}_P = \bar{\xi} + \varepsilon f_0 \bar{\varphi}, \qquad \bar{\xi} = \bar{u}_y - \bar{v}_x.$$

That $\bar{\xi}_P$ changes slowly can be deducted immediately from (54). We have

$$\partial \bar{\xi}_P/\partial t = \bar{u}_{yt} - \bar{v}_{xt} + \varepsilon f_0 \bar{\varphi}_t$$
$$= \varepsilon^{-1} f_0(\bar{u}_x + \bar{v}_y) - \varepsilon^{-1} f_0(\bar{u}_x + \bar{v}_y) = 0$$

$\xi + \varepsilon f \varphi = u_y - v_x + \varepsilon f \varphi$ is also the slow variable for the general problem (52). In this case the vorticity equation is given by

$$d\xi/dt + (u_x + v_y)\xi - \varepsilon^{-1} f(u_x + v_y) - \beta v = 0, \tag{56}$$

and, therefore,

$$d(\xi + \varepsilon f \varphi)/dt + (u_x + v_y)(\xi + \varepsilon f \varphi) - f[(u\Phi)_x + (v\Phi)_y] - \varepsilon^2 \beta \varphi v = 0. \tag{57}$$

Differentiate (52) with respect to x and (52) with respect to y, and add the equations. Then we obtain an equation for the divergence $\delta = u_x + v_y$:

$$d\delta/dt + J(u,v) + \varepsilon^{-1}(\Delta \varphi + f\xi) + \beta u = 0, \tag{58}$$

where
$$J(u,v) = (u_x)^2 + 2u_y v_x + (v_y)^2, \qquad \Delta\varphi = \varphi_{xx} + \varphi_{yy}.$$

Instead of (52) we can use Eqs. (57), (58), and (52) to describe our motion. The advantage of this system is that we have separated the fast and slow components. We shall initialize this system: $d\varphi/dt$ and $d\delta/dt$ are bounded independently of ε if and only if

$$\delta - \varepsilon[(u\Phi)_x + (v\Phi)_y] = \varepsilon^2 c, \qquad c = O(1), \tag{59}$$

$$\Delta\varphi + f\xi = \varepsilon e, \qquad e = O(1). \tag{60}$$

Then (52) and (58) become

$$d\varphi/dt + \varphi\delta + c = 0,$$

$$d\delta/dt + J(u,v) + e + \beta u = 0.$$

Thus, $d^2\delta/dt^2, d^2\varphi/dt^2$ are bounded independently of ε if $de/dt, dc/dt$ have this property; i.e.,

$$d\delta/dt - \varepsilon d[(u\Phi)_x + (v\Phi)_y]/dt = O(\varepsilon^2), \tag{61}$$

$$d\,\Delta\varphi/dt + d(f\xi)/dt = O(\varepsilon). \tag{62}$$

Using (58) and (52), we can write (61) in the form

$$\Delta\varphi + f\xi + \varepsilon(J(u,v) + \beta u) + \varepsilon^2 g(u,v,\varphi) = O(\varepsilon^3), \tag{63}$$

which is a more accurate form of (60). Correspondingly (62) is a more accurate form of (59); i.e.,

$$\delta - \varepsilon[(u\Phi)_x + (v\Phi)_y] + \varepsilon^2 h(u,v,\varphi) = O(\varepsilon^3). \tag{64}$$

In practical applications one can, in most cases, neglect $O(\varepsilon^2)$ terms in the initialization procedure. Therefore, one chooses the data in such a way that

$$\Delta\varphi + f\xi + \varepsilon[J(u,v) + \beta u] = 0, \tag{65}$$

$$\delta - \varepsilon[(u\Phi)_x + (v\Phi)_y] = 0. \tag{66}$$

Practically, we can proceed in the following way. We give ξ. Then we calculate u, v from (66) and $u_y - v_x = \xi$. We obtain φ from (65).

Equations (65) and (66) must hold approximately for all times, and therefore, we can replace the full system by the reduced system consisting of (57), (65), and (66). The error committed is $O(\varepsilon^2)$.

For problems with periodic boundary conditions, there are no problems in solving the above system numerically. For other initial boundary value problems it can be easier to use the primitive variables u, v, p, because the boundary conditions are formulated in these variables. We shall show that (63) and (64) also initialize the data when using the system (52). Note that

2. Problems with Different Time Scales

du/dt, dv/dt, and $d\varphi/dt$ are bounded independently of ε if

$$\varphi_x - fv = \varepsilon a, \qquad \varphi_y + fu = \varepsilon b,$$
$$\delta - \varepsilon[(u\Phi)_x + (v\Phi)_y] = \varepsilon^2 c; \qquad (67)$$

i.e., to first approximation u, v, φ are in geostrophic balance and the divergence is small. We can also write the first two relations in the following form:

$$a_x + b_y = \varepsilon^{-1}(\Delta\varphi + f\xi) + \beta u, \qquad a_y - b_x = \varepsilon^{-1}f\delta - \beta v. \qquad (68)$$

Therefore, if (59) and (60) hold, then a, b are $O(1)$ and the first time derivatives are bounded independently of ε. Correspondingly, the second time derivatives are bounded if (63) and (64) hold.

We make an error of order $O(\varepsilon^2)$ if we use (65) and (66) to replace (68) by

$$a_x + b_y = -J(u,v), \qquad a_y - b_x = (u\Phi)_x + (v\Phi)_y. \qquad (69)$$

In Browning and Kreiss (1982a) we have made a detailed investigation of the reduced system consisting of (69) and the momentum equations

$$du/dt + a = 0, \qquad dv/dt + b = 0. \qquad (70)$$

Other applications of the shallow water equations are considered in Gustafsson (1980a, b). Stability discussions for boundary conditions of the reduced problem are presented in Gustafsson and Kreiss (1983).

VI. ATMOSPHERIC MOTIONS

In this section we consider three-dimensional atmospheric motions and discuss results presented in Browning and Kreiss (1983). In Cartesian coordinates x, y, and z directed eastward, northward, and upward, respectively, the Eulerian equations have the form

$$\begin{aligned}
&ds/dt = 0, \\
&d/dt = \partial/\partial t + u\, \partial/\partial x + v\, \partial/\partial y + w\, \partial/\partial z, \\
&dp/dt + \gamma p(u_x + v_y + w_z) = 0, \qquad \gamma = 1.4, \\
&\rho\, du/dt + p_x - f\rho v = 0, \\
&\rho\, dv/dt + p_y + f\rho u = 0, \qquad \rho = sp^{1/\gamma}. \\
&\rho\, dw/dt + p_z + \rho g = 0,
\end{aligned} \qquad (71)$$

Here s is the entropy, p the pressure, ρ the density, u, v, w the velocity components in the x, y, z directions, respectively, and $g \sim 10\ \text{msec}^{-2}$ the gravity acceleration. We again make the β-plan approximation; i.e., the Coriolis force f is given by (5.2).

Again we have to introduce scaled variables before we can apply the bounded derivative principle. We introduce new variables by

$$x = L_1 x', \quad y = L_2 y', \quad z = D z', \quad t = T t', \qquad (72)$$
$$u = U u', \quad v = V v', \quad w = W w'.$$

Density and pressure can be written in the form

$$p = P_0[p_0(z) + S_1 p'], \quad \rho = R_0[\rho_0(z) + S_1 \rho'], \quad 0 < S_1 \ll 1, \quad (73)$$

where

$$P_0 \, \partial p_0/\partial z + g R_0 \rho_0 = 0,$$
$$P_0 = 10^5 \text{ kg m}^{-1} \text{ sec}^{-2}, \quad R_0 = 1 \text{ kg m}^{-3}.$$

Equations (73) express the fact that a number of digits of the pressure and density are independent of x, y and that p and ρ are to first approximation in hydrostatic balance. Equations (6.3) also imply that

$$s = R_0 P_0^{-1/\gamma} \rho_0(z) [p_0(z)]^{-1/\gamma} (1 + S_1 \rho'/\rho_0)(1 + S_1 p'/p_0)^{-1/\gamma}$$
$$= R_0 P_0^{-1/\gamma} s_0(z)(1 + S_1 s'), \qquad (74)$$
$$s_0(z) = \rho_0(z)[p_0(z)]^{-1/\gamma}, \quad s' = \rho'/\rho_0 - (1/\gamma) p'/p_0 + O(S_1).$$

We assume that the scales in the x, y directions are the same, that $\partial u/\partial t$, $\partial v/\partial t$ balance the horizontal convection terms, and that the Coriolis force has a strong influence. This leads to the following relations:

$$U = V, \quad L_1 = L_2 = L,$$
$$UT/L = 1, \quad 2\Omega T = S_1 P_0/(R_0 U^2). \qquad (75)$$

These relations are not valid for special types of motions like jet streams, ultralong waves, and small-scale problems. For the treatment of these cases we refer to Browning and Kreiss (1983). Introducing the scaled variables into (71) gives us

$$\frac{ds'}{dt'} + S_1^{-1} S_2 \tilde{s}(z)(1 + S_1 s') w' = 0,$$

$$\frac{dp'}{dt'} + S_1^{-1} p_0 \left[\gamma \left(1 + \frac{S_1 p'}{p_0}\right)(u_x + v_y + S_2 w_z) + S_2 \tilde{p}(z) w' \right] = 0,$$

$$\frac{du'}{dt'} + S_3 \left[\rho_0^{-1} \left(1 + \frac{S_1 \rho'}{\rho_0}\right)^{-1} p'_{x'} - f' v' \right] = 0, \qquad (76)$$

$$\frac{dv'}{dt'} + S_3 \left[\rho_0^{-1} \left(1 + \frac{S_1 \rho'}{\rho_0}\right)^{-1} p'_{y'} + f' u' \right] = 0,$$

$$\frac{dw'}{dt'} + S_1 S_4 \rho_0^{-1} \left(1 + \frac{S_1 \rho'}{\rho_0}\right)^{-1} [p'_{z'} - \gamma^{-1} \tilde{p}(z) \rho' + S_5 \rho_0 s' + O(S_1)] = 0,$$

2. Probems with Different Time Scales

where
$$\tilde{p}(z) = (\ln p_0)_z, \qquad \tilde{s}(z) = (\ln s_0)_z;$$

typically,
$$\tilde{p} \sim -1.3, \qquad -3 \leq \tilde{s} \leq -1;$$
$$d/dt = \partial/\partial t' + u' \, \partial/\partial x' + v' \, \partial/\partial y' + S_2 w' \, \partial/\partial z';$$

and the parameters S_i are given by

$$\begin{aligned} S_2 &= D^{-1}TW, & S_3 &= 2\Omega T, \\ S_4 &= TP_0(DR_0W)^{-1}, & S_5 &= gDP_0^{-1}R_0. \end{aligned} \tag{77}$$

We now choose the parameters according to the so-called large-scale dynamics:

$$L = 10^6 \text{ m}, \quad D = 10^4 \text{ m}, \quad U = V = 10 \text{ m/sec}, \quad S_1 = 10^{-2}. \tag{78}$$

Also, we set $W = \varepsilon^n \text{ msec}^{-1}$, $\varepsilon = 10^{-1}$, where n will be determined by consistency conditions. Introducing these values into (77) and (78), we obtain, dropping the prime notation,

$$\frac{ds}{dt} + \varepsilon^{n-2}\tilde{s}(z)w = 0, \qquad \varepsilon = 10^{-1},$$

$$\frac{\varepsilon^2}{\gamma p_0} \frac{dp}{dt} + u_x + v_y + \varepsilon^{n-1}Lw = 0,$$

$$\varepsilon \frac{du}{dt} + \rho_0^{-1} p_x - fv = 0, \tag{79}$$

$$\varepsilon \frac{dv}{dt} + \rho_0^{-1} py + fu = 0,$$

$$\rho_0 \varepsilon^{n+4} \frac{dw}{dt} - L^*p + \rho_0 s = 0,$$

where
$$d/dt = \partial/\partial t + u \, \partial/\partial x + v \, \partial/\partial y + \varepsilon^{n-1} w \, \partial/\partial z,$$
$$Lw = w_z + \gamma^{-1}\tilde{p}(z)w, \qquad L^*p = -p_z + \gamma^{-1}\tilde{p}(z)p.$$

For simplicity only, we have neglected terms of order $O(S_1)$. Also, by (53),
$$f = f_0 + \varepsilon\beta y.$$

Here we only consider the midlatitude case $f_0 \sim 1$.

The system (79) is not exactly of the form described in Section IV. Rather likely the proof in Browning and Kreiss (1982b) that the bounded derivative principle is valid can be adapted to the half-plane problem $z \geq 0$ with periodic boundary conditions in x, y; i.e., the results apply for the flow on the whole globe. For limited areas in the horizontal x, y direction with the usual in- and outflow conditions, the proof cannot be adapted. In fact, there are serious difficulties. Fast waves will always be generated at the boundary (Oliger and Sundstrom, 1978; Browning and Kreiss, 1983). One could use a reduced system that does not have this difficulty, and we derive such a system.

The first time derivatives of u, v are bounded independently of ε if and only if

$$\rho_0^{-1} p_x - fv = \varepsilon a(x, y, z, t), \qquad \rho_0^{-1} p_y + fu = \varepsilon b(x, y, z, t), \qquad (80)$$

where a, b are smooth functions. Equation (80) says that u, v, p are approximately in geostrophic balance. With this notation we can write the horizontal momentum equations in the form

$$du/dt + a = 0, \qquad dv/dt + b = 0. \qquad (81)$$

Also, as in the last section, we can replace (80) by

$$a_x + b_y = \varepsilon^{-1}(\rho_0^{-1} \Delta p + f\xi) + \beta u, \qquad a_y - b_x = \frac{f}{\varepsilon}(u_x + v_y) + \beta v. \qquad (82)$$

Thus,

$$\delta = u_x + v_y = O(\varepsilon), \qquad \rho_0^{-1} \Delta p + f\xi = O(\varepsilon). \qquad (83)$$

Note that dp/dt is bounded independently of ε if and only if

$$u_x + v_y + \varepsilon^{n-1} Lw = \varepsilon^2 c.$$

Therefore, by (83)

$$\varepsilon^{n-1} Lw = O(\varepsilon);$$

i.e., $n \geq 2$. We want to show that $n = 2$.

Combining the horizontal momentum equations (81) gives us an equation for $\delta = u_x + v_y$,

$$d\delta/dt + J_0 + \varepsilon J_1 + a_x + b_y = 0, \qquad (84)$$

where

$$J_0 = (u_x)^2 + u_y v_x + (v_y)^2, \qquad J_1 = w_x u_z + w_y v_z.$$

Therefore, by (83),

$$a_x + b_y = -J_0(u_1 v) + O(\varepsilon), \qquad (85a)$$

or

$$\rho_0^{-1} \Delta p + f\xi + \varepsilon \beta u - \varepsilon J_0(u, v) = O(\varepsilon^2). \qquad (85b)$$

2. Problems with Different Time Scales

Equations (82) and (84) imply

$$a_y - b_x = \varepsilon^{n-2} fLw + \beta v + O(\varepsilon^{n-1}). \tag{86}$$

Assume now that $n > 2$. Then, to first approximation, the horizontal momentum equations (85) and (86) become

$$d_H u/dt + a = 0, \qquad d_H/dt = \partial/\partial t + u\partial/\partial x + v\partial/\partial y,$$

$$d_H v/dt + b = 0, \tag{87}$$

$$a_x + b_y = -J_0(u,v), \qquad a_y - b_x = \beta v;$$

i.e., to first approximation, there is no coupling between the layers $z = \text{const}$. Therefore, the z derivatives, due to shearing, would become large, and the scaling assumptions would be violated. This proves that $n = 2$. To first approximation, we obtain the following set of equations:

$$d_H u/dt + a = 0, \qquad d_H v/dt + b = 0,$$

$$a_x + b_y = -J_0(u,v), \qquad a_y - b_x = fLw + \beta v. \tag{88}$$

We now derive an equation for w. Note that dw/dt is bounded independently of ε if and only if

$$-L^* p + \rho_0 s = \varepsilon^6 h. \tag{89}$$

Thus the hydrostatic assumption

$$-L^* p + g\rho_0 s = 0 \tag{90}$$

is very well satisfied.

The second time derivative of w is bounded independently of ε if the same is true for $d_H h/dt$; i.e.,

$$-L^* d_H p/dt + \rho_0 d_H s/dt = O(\varepsilon^6).$$

By (79),

$$\frac{d_H p}{dt} = \frac{dp}{dt} - \varepsilon w p_z = -\gamma p_0 \varepsilon^{-2}(u_x + v_y + \varepsilon Lw) - \varepsilon w p_z,$$

$$\frac{d_H s}{dt} = \frac{ds}{dt} - \varepsilon w s_z = -\tilde{s}(z) w - \varepsilon w s_z;$$

therefore,

$$\gamma L^*[p_0(u_x + v_y)] + \varepsilon \gamma L^*(\rho_0 Lw) - \varepsilon^2 \rho_0 \tilde{s}(z) w = O(\varepsilon^3). \tag{91}$$

Neglecting the $O(\varepsilon^3)$ term, we obtain

$$\gamma L^*[p_0(u_x + v_y)] + \varepsilon \gamma L^*(\rho_0 Lw) - \varepsilon^2 \rho_0 \tilde{s}(z) w = 0, \tag{91'}$$

which is Richardson's equation.

REMARK. The primitive equations that are often used in weather prediction models consist of the entropy equation, the two horizontal momentum equations, the hydrostatic equation (90), and Richardson's equation (91'). One solves these equations numerically in the following way. Assume we know all variables at time t and earlier. We use the first three equations to compute s, u, v at time $t + \Delta t$. These values give us p and w with the help of (90) and (91), respectively. There are a number of difficulties associated with this procedure (see Browning and Kreiss, 1983). In particular, to determine w from (91') requires calculating δ/ε. By (84) and (82),

$$d\delta/dt + J_0 + \varepsilon J_1 + \varepsilon^{-1}(\rho_0^{-1}\Delta p + f\xi) + \beta u = 0. \tag{92}$$

Therefore, if we want to calculate w with two accurate figures, we have to calculate the balance expression

$$\rho_0^{-1}\Delta p + f\xi$$

to four figures. Remembering that the first two figures of the mean pressure are constant, this means that we have to know the total pressure accurately to six digits.

The second time derivative of p is bounded independently of ε if and only if

$$d(u_x + v_y)/dt + \varepsilon\, d(Lw)/dt = \varepsilon\, dc/dt = O(\varepsilon^2);$$

therefore, by (92),

$$\Delta p + f\rho_0\xi + \varepsilon J_2 + \varepsilon^2 J_3 = O(\varepsilon^3),$$
$$J_2 = \rho_0 J_0 + \beta\rho_0 u, \qquad J_3 = \rho_0 J_1 - \rho_0\, d(Lw)/dt. \tag{93}$$

If relations (80), (89), (91), and (93) hold, then the first and second time derivatives of all variables are bounded independently of ε. To obtain an equation for w, we need to bound the third time derivative of p.

Using the hydrostatic relation (90), we can write (93) as an equation for s:

$$\rho_0\Delta s - fL^*(\rho_0\xi) - \varepsilon L^*(J_2 + \varepsilon J_3) = O(\varepsilon^3). \tag{94}$$

If we know u, v, then we can calculate p and s from (90), and

$$\rho_0\Delta s - fL^*(\rho_0\xi) - \varepsilon L^*J_2 = 0. \tag{95}$$

The error we commit is $O(\varepsilon^2)$. The third time derivative of p is bounded independently of ε if

$$\frac{d}{dt}(\rho_0\Delta s) - f\frac{d}{dt}L^*(\rho_0\xi) - \varepsilon\frac{d}{dt}(L^*J_2 + \varepsilon L^*J_3) = O(\varepsilon^3).$$

Now

$$\frac{d}{dt}\rho_0 \Delta s = \rho_0 \frac{d}{dt}\Delta s + O(\varepsilon) = \rho_0 \Delta \frac{ds}{dt} + \rho_0 G(u,v,s) + O(\varepsilon)$$

$$= -\rho_0 \tilde{s}(z)\Delta w + \rho_0 G(u,v,s) + O(\varepsilon)$$

and

$$f\frac{d}{dt}L^*(\rho_0 \xi) = fL^*\rho_0 \frac{d\xi}{dt} + O(\varepsilon),$$

$$\frac{d\xi}{dt} = a_y - b_x + O(\varepsilon) = f\varepsilon^{-1}(u_x + v_y) + \beta v + O(\varepsilon)$$

$$= -fLw + \beta v + O(\varepsilon).$$

Therefore, we get, to first approximation, the following elliptic equation for w:

$$-\rho_0 \tilde{s}(z)\Delta w + f^2 L^* \rho_0 L w = -\rho_0 G(u,v,s) + f\beta L^*(\rho_0 v). \tag{96}$$

Our reduced system consists of the Eq. (88), (90), (95), and (96). It is a coupled hyperbolic elliptic system, and there are no difficulties in constructing boundary conditions such that the initial boundary value problem is well posed. We solve it numerically in the following way. Assume that all variables are known up to time t. Then we determine u, v with the help of the horizontal momentum equations at time $t + \Delta t$. This enables us to determine p and s from (89) and (95). Then w is given by (96), and a, b can be calculated from the last two equations of (88).

VII. PLASMA PHYSICS

The application of the bounded derivative principle to problems in plasma physics are not yet very well developed. Here we only discuss a simple example in order to show that potentially there is a wealth of applications. The equations of compressible magnetofluid dynamics in three space dimensions form an 8 × 8 system of equations:

$$ds/dt = 0,$$

$$dp/dt + \varphi \operatorname{div} \mathbf{v} = 0,$$

$$\rho \, d\mathbf{v}/dt + \varepsilon^{-2} \operatorname{grad} p = \eta^{-2}\mathbf{H} \times \operatorname{curl} \mathbf{H}, \tag{97}$$

$$d\mathbf{H}/dt + \mathbf{H} \operatorname{div} \mathbf{v} - \mathbf{H} \cdot \operatorname{grad} \mathbf{v} = 0,$$

$$d/dt = \partial/\partial t + u\, \partial/\partial x + v\, \partial/\partial y + w\, \partial/\partial z,$$

for the entropy s, pressure p, velocity vector $\mathbf{v} = (u, v, w)^T$, and magnetic field $\mathbf{H} = (H_1, H_2, H_3)^T$. Also, $\rho = \rho(p, s)$ and $\varphi = \varphi(p, s)$ are smooth positive functions of p and s, and ε, η represent the Mach number and Alvén number, respectively. We consider the system (97) in the cube $0 \leq x, y, z \leq 2\pi$ with periodic boundary conditions.

Combining the above equations results in

$$\frac{\partial}{\partial t} \operatorname{div} \mathbf{H} = 0,$$

i.e.,

$$\operatorname{div} \mathbf{H} = \psi(x, y, z),$$

and we assume for simplicity $\psi = 0$. Also

$$\int_0^{2\pi} \int_0^{2\pi} \int_0^{2\pi} \mathbf{H} \, dx \, dy \, dz = \mathbf{H}_0, \qquad \frac{\partial \mathbf{H}_0}{\partial t} = 0.$$

We assume now that

$$p = p_0 + \varepsilon p', \qquad \mathbf{H}_0 + \eta \mathbf{H}',$$

with

$p_0, \mathbf{H}_0$ const, $\quad H'_{1x} + H'_{2y} + H'_{3z} = 0, \quad \int_0^{2\pi} \int_0^{2\pi} \int_0^{2\pi} H' \, dx \, dy \, dz = 0.$

Then, omitting the prime sign, we obtain

$ds/dt = 0,$

$\varphi^{-1} \, dp/dt + \varepsilon^{-1} \operatorname{div} \mathbf{v} = 0,$

$\rho \, d\mathbf{v}/dt + \varepsilon^{-1} \operatorname{grad} p = \eta^{-1} \mathbf{H}_0 \times \operatorname{curl} \mathbf{H} + \mathbf{H} \times \operatorname{curl} \mathbf{H},$ (98)

$d\mathbf{H}/dt + \eta^{-1}(\mathbf{H}_0 \operatorname{div} \mathbf{v} - \mathbf{H}_0 \cdot \operatorname{grad} \mathbf{v}) + \mathbf{H} \operatorname{div} \mathbf{v} - \mathbf{H} \cdot \operatorname{grad} \mathbf{v} = 0.$

Equations (98) are of the form (43). Thus, we can apply the results of Section IV. In particular, we can study different combinations of ε, η like $\varepsilon = 1, 0 < y \ll 1, 0 < \varepsilon \ll 1, y = 1$ (see Browning and Kreiss, 1982b) and derive the different reduced systems.

REFERENCES

Baer, F. (1977). Adjustment of initial conditions required to suppress gravity oscillations in nonlinear flows. *Beitr. Phys. Atmos.* **50**, 350–366.

Barker, J. (1982). Interactions of fast and slow waves in problems with two time scales, Ph.D. Thesis, California Institute of Technology, Pasadena (will partially appear in *Math. Methods Appl. Sci.* and *SIAM J. Anal.*).

Browning, G. and Kreiss, H.-O. (1982a). Initialization of the shallow water equations with open boundaries by the bounded derivative method. *Tellus* **34**, 334–351.

Browning, G., and Kreiss, H.-O. (1982b). Problems with different time scales for nonlinear partial differential equations. *SIAM J. Appl. Math.* **42**, 704–718.

Browning, G., and Kreiss, H.-O. (1983). Scaling and computation of atmospheric motions. (to be published).

Browning, G., Kasahara, A., and Kreiss, H.-O. (1980). Initialization of the primitive equations by the bounded derivative method. *J. Atmos. Sci.* **37**, 1424–1436.

Guerra, J., and Gustafsson, B. (1982). "A Semi-implicit Method for Hyperbolic Problems with Different Time-scales," Rep. No. 90. Dept. of Computer Sciences, Uppsala University.

Gustafsson, B. (1980a). Asymptotic expansions for hyperbolic systems with different time scales. *SIAM J. Numer. Anal.* **17**, No. 5, 623–634.

Gustafsson, B. (1980b). Numerical solution of hyperbolic systems with different time scales using asymptotic expansions. *J. Comput. Phys.* **36**, 209–235.

Gustafsson, B., and Kreiss, H.-O. (1983). Difference approximations of hyperbolic problems with different time scales. I. The reduced problem. *SIAM J. Number, Anal.* (to be published).

Kasahara, A., (1974). Various vertical coordinate systems used for numerical weather prediction. *Mon. Weather Rev.* **102**, 509–522.

Kasahara, A. (1982). Nonlinear normal mode initialization and the bounded derivative method. *Rev. Geophys. Space Phys.* **20**, No. 3, 385–397.

Kreiss, H.-O. (1978). Difference methods for stiff ordinary differential equations. *SIAM J. Numer. Anal.* **15**, 21–58.

Kreiss, H.-O. (1979). Problems with different time scales for ODE. *SIAM J. Numer. Anal.* **16**, 980–998.

Kreiss, H.-O. (1980). Problems with different time scales for PDE. *Commun. Pure Appl. Math.* **33**, 399–439.

Kreth, H. (1977). Time-discretisations for nonlinear evolution equations. *Lect. Notes Math.* **679**.

Leith, C. E. (1980). Nonlinear normal mode initialization and quasi-geostrophic theory. *J. Atmos. Sci.* **37**, 954–964.

Machenhauer, B. (1977). On the dynamics of gravity oscillations in a shallow water model, with applications to normal mode initialization. *Beitr. Phys. Atmos.* **50**, 253–271.

Oliger, J., and Sundstrom, A. (1978). Theoretical and practical aspects of some initial-boundary value problems in fluid dynamics. *SIAM J. Appl. Math.* **35**, 419–446.

Tadmor, E. (1982). Hyperbolic systems with different time scales. *Commun. Pure Appl. Math.* **35**, 839–866.

3

Nonlinear Normal-Mode Initialization of Numerical Weather Prediction Models

C. E. LEITH

Lawrence Livermore National Laboratory
University of California
Livermore, California

I. Introduction 59
   A. The Noise Problem in Numerical Weather
      Prediction Models 59
   B. Linear Initialization 60
   C. Nonlinear Initialization 60
   D. Data Assimilation 61
   E. Mathematical Problems 61
II. Normal-Mode Analysis 62
III. Nonlinear Analysis 65
IV. Analysis of Observations 68
V. Remaining Problems 70
   References 71

I. INTRODUCTION

A. The Noise Problem in Numerical Weather Prediction Models

Global atmospheric models in current use for numerical weather prediction are based on the so-called primitive equations of atmospheric flow. In these, the only approximation made is that the pressure is related to the vertical temperature profile through a hydrostatic relation, and, in particular, vertical acceleration terms are ignored. The primitive equations are sufficiently general to describe the horizontal propagation of relatively high-frequency gravity waves, although these are found to be, at most, a minor component of observed atmospheric flow. The initial state for prediction by a numerical model is based on a large and rather heterogeneous set of observations of the atmosphere made with sondes, satellites, and aircraft. The inevitable instrumental and interpolation errors lead to an initial state with typically a much larger gravity-wave component than is natural for the atmosphere. A prediction made from such a state with a primitive equation model will have the slowly evolving meteorologically important state obscured by a completely erroneous high-frequency gravity-wave component, which was recognized many years ago by Hinklemann (1951) and called meteorological noise.

The earliest numerical models for weather prediction, devised during the 1950s, evaded the problem of meteorological noise by further approximations that reduced the primitive equations to some sort of balance equations, which would not support and describe gravity-wave motions. However, lack of confidence in the additional balance approximations that were required, as well as technical numerical problems, caused such models to be abandoned in favor of those using the primitive equations.

Although dissipative terms in the dynamics equations tend to damp the noise over a day or so of prediction time, the most satisfactory solution to the noise problem is to constrain the initial state so that the ensuing prediction is noise free. Such a constraining process is called initialization. The present chapter describes the recent work on initialization, which has gone a long way toward the understanding and solution of this old problem. An earlier review (Leith, 1978) describes the more general problems of objective weather prediction, and Daley (1981) has reviewed many aspects of the development of normal-mode initialization methods.

B. Linear Initialization

The natural first step is to linearize the dynamics of the model, for convenience about a state at rest, and to distinguish, as in classical tidal theory, the linear gravitational and rotational modes of the system. It is then a straightforward bit of linear algebra to decompose any tentative initial state into its linear gravitational and rotational modes and to keep only the latter as a linear initialization. Such a linear analysis, described in Section II, has been found to solve a large part, but not all, of the noise problem.

C. Nonlinear Initialization

Since the dynamics are essentially nonlinear, it is found that nonlinear coupling of linear rotational to linear gravitational modes induces gravitational oscillations in spite of linear initialization. Nonlinear initialization by

3. Numerical Weather Prediction Models

methods given in Section III has been introduced during the early 1980s as a way of balancing the induced nonlinear tendency toward gravitational modes by suitably chosen gravitational components. For sufficiently simple dynamic models, nonlinear initialization effectively solves the noise problem. In the phase space of the model dynamic system, a nonlinear manifold of suitable initial states is thereby defined. If an initial state is in this so-called slow manifold, then it evolves in it free of gravitational oscillations; if it is not, then rapid oscillations about the slow manifold occur during the course of the prediction.

D. Data Assimilation

The problem of data assimilation and initialization can be stated quite simply in the geometry of phase space. A set of observations of the atmosphere at a given time defines a point in phase space. Knowledge of the error variance associated with the observation and analysis process defines a probability distribution localized in phase space about the observed state. This distribution gives the likelihood of determining any phase point to be the true state on the basis of observations alone. But more is known. An earlier prediction of the state that is currently valid and knowledge of the error variance in the prediction process determine a probability distribution of the same sort, but now centered about the predicted state and confined to the slow manifold. Using a version of Bayes's theorem, the composite knowledge of the true state of the atmosphere is seen to be represented by the normalized product of these two probability distributions, which is again confined to the slow manifold. A repetition of this process for, say, a 12-h observing cycle leads to the assimilation of observations into the best definition of the evolving model state. Such, in principle, is the nature of the data assimilation process. We shall see that, in practice, this geometrically simple process can only be approximated owing to the extremely large dimension of a typical model phase space, on the order of 1 million, and the consequent computational complexity of finding manifolds tangent to the slow manifold. Details are provided in Section IV.

E. Mathematical Problems

Although the problem of nonlinear initialization has been satisfactorily solved for simple dynamic models, there remain a number of problems for more realistic models of the atmosphere, as discussed in Section V. Most atmospheric models define the vertical structure of the atmosphere in terms of a finite number of layers. The associated tidal theory reveals internal gravity waves with vertical structure. The higher vertical wave-number modes have horizontal propagation velocities and frequencies comparable to the rotational waves of meteorological importance. For multilayer models, therefore, the separation between rejected fast modes and accepted slow modes should not be based on the distinction between gravitational and rotational, but rather on a division of the frequency spectrum. Unfortunately, there may not be a natural gap in the frequency spectrum where the dividing frequency should be located. As a consequence, the choice of this frequency is somewhat arbitrary, there may not be much separation between the highest frequency kept and the lowest rejected, and the nonlinear iteration procedures commonly used have convergence difficulties. In this case there have been raised a number of basic mathematical problems concerning the definition and existence of the slow manifold and of the convergence and efficiency of algorithms for reaching it.

In addition to mathematical problems, there are a number of physical problems yet to be resolved. There is more to a weather prediction model than dynamics. Also modeled are important physical processes, such as latent heat release from water vapor condensation and precipitation, and radiative heating and cooling. Some of these processes are highly nonlinear and contribute to the definition of the slow manifold. But their complexity adds considerably to the difficulty of finding convergent initialization algorithms.

II. NORMAL-MODE ANALYSIS

In his studies of the tides, Laplace derived the linearized equations of motion for small disturbances of a thin layer of incompressible fluid, with depth H, at rest relative to a rotating smooth earth, to which it is held by gravity. He noted that these equations are also appropriate for isothermal oscillations of an isothermal atmosphere with suitable choice of an equivalent depth H. Lamb (1932) showed more generally that such an equivalence also exists for adiabatic oscillations of an isentropic atmosphere.

The linearized shallow water equations in one space dimension without curvature or rotational effects are

$$\partial u/\partial t + g\,\partial h/\partial x = 0, \tag{1}$$

$$\partial h/\partial t + H\,\partial u/\partial x = 0, \tag{2}$$

where $u(x,t)$ is the fluid velocity assumed to be independent of depth, and $h(x,t)$ the perturbation of the surface height from the mean height H. Equations (1) and (2) may be combined into a wave equation

$$\partial^2 h/\partial t^2 + gH\,\partial^2 h/\partial x^2 = 0, \tag{3}$$

with wave solutions

$$h(x,t) = f_1(x - ct) + f_2(x + ct) \tag{4}$$

3. Numerical Weather Prediction Models

for arbitrary f_1, f_2 determined by initial conditions. The wave velocities c are the two roots of

$$c^2 = gH. \tag{5}$$

These are called gravitational waves, and they propagate in either the $+x$ or $-x$ direction.

If we now add a second Cartesian coordinate dimension, the shallow water equations become

$$\partial u/\partial t + g\, \partial h/\partial x = 0, \tag{6}$$

$$\partial v/\partial t + g\, \partial h/\partial y = 0, \tag{7}$$

$$\partial h/\partial t + H[\partial u/\partial x + \partial v/\partial y] = 0. \tag{8}$$

These equations have another class of possible wave motions, in addition to the gravitational waves. These so-called rotational waves, such as $v = f_3(x)$ for arbitrary f_3, are nondivergent and in this framework not very interesting since they remain stationary.

In the case of a rotating spherical earth, Eqs. (6)–(8) become Laplace's tidal equations,

$$\frac{\partial u}{\partial t} - 2\Omega \sin \varphi\, v + \frac{g}{a \cos \varphi} \frac{\partial h}{\partial \lambda} = 0, \tag{9}$$

$$\frac{\partial v}{\partial t} + 2\Omega \sin \varphi\, u + \frac{g}{a} \frac{\partial h}{\partial \varphi} = 0, \tag{10}$$

$$\frac{\partial h}{\partial t} + \frac{H}{a \cos \varphi} \left[\frac{\partial u}{\partial \lambda} + \frac{\partial}{\partial \varphi} (\cos \varphi\, v) \right] = 0, \tag{11}$$

where λ and φ are the latitude and longitude, and Ω and a the angular velocity and radius of the earth.

The effect of rotation and sphericity is to modify propagation velocities. There remain, however, identifiable fast eastward and westward moving gravitational wave modes with characteristic velocities on the order of $(gH)^{1/2}$. The rotational waves now move slowly westward and are of dominant importance for weather prediction.

The relative impact of spherical rotational effects compared with inertial effects is given by the dimensionless Lamb's parameter

$$\varepsilon = 4\Omega^2 a^2/gH. \tag{12}$$

Longuet-Higgins (1968) tabulated and plotted in considerable detail the eigenfrequencies and modal structures of Laplace's tidal equations for various values of ε and longitudinal wave number. A new algorithm for calculation of the eigenvalues and eigenfunctions of the tidal operator, in terms of vector spherical harmonics, has recently been developed by Swarztrauber and Kasahara (1985), who provide many references to earlier efforts.

The tidal eigenfunctions, called Hough functions because Hough (1898) made some of the first calculations of these eigenfunctions, provide the dependences of u, v, and h on latitude φ. The longitudinal and time dependence is a simple wave of the form $\exp i(s\lambda + \sigma t)$, with longitudinal wave number s and eigenfrequency σ.

In the more realistic case of an atmosphere with a more general but convectively stable mean vertical temperature profile, Taylor (1936) showed that, in addition to the external mode, higher order internal vertical modes exist and satisfy a vertical-structure equation. From this analysis there emerges a separation constant that defines a decreasing sequence H_0, H_1, H_2, ... of equivalent depths for the associated horizontal Hough modes. The corresponding gravity wave speeds are also decreasing.

The vertical structure in this generalized tidal theory remains separable from the horizontal structure, and thus algebraic manipulations with the tidal operator are computationally feasible.

In any numerical model of the atmosphere, the continuous fields of tidal theory are replaced by a finite representation on grid points or spectral components. A linearized tidal theory can be worked out for any model as a finite-dimensional algebraic problem. There will be differences in detail from the results of continuous tidal theory, but it is the model tidal theory that is appropriate for any analysis of model behavior. In particular, the sequence of equivalent depths for a model is truncated to a finite number determined by the vertical resolution.

The results of a tidal theory analysis of a numerical model may be summarized by saying that it reduces the model dynamics equations to the form

$$x_t = i\mathscr{L}x + n(x). \tag{13}$$

Here x is a state vector, x_t its time derivative in the dynamic phase space of the model, $\mathscr{L}$ the self-adjoint linear tidal operator, and $n(x)$ a vector-valued function giving the effects of additional linear and nonlinear dynamic and physical processes. For the state vector x, defined in terms of the model Hough functions as a basis for representation of the model phase space, the operator $\mathscr{L}$ is diagonal. In any case, its eigenvalues are the real eigenfrequencies of the linear tidal problem in which $n(x)$ is set to zero. The dimension of the model phase space is large but finite.

To demonstrate the multiple time-scale problem in its simplest form, we examine in detail initialization of a one-layer model having a single equivalent depth $H = H_0$. In this case the eigenvalues of $\mathscr{L}$ fall into three well-separated classes of equal size, namely, rapidly propagating eastward gravity waves, rapidly propagating westward gravity waves, and slowly propagating westward rotational waves. We are primarily interested in this third class. The associated eigenvectors span a linear manifold in phase space, which we call the linear rotational manifold $\mathbb{R}$. All the rest of the eigenvectors, associated with gravitational waves, span the linear gravitational manifold $\mathbb{G}$, which is the orthogonal complement of $\mathbb{R}$. The phase space is thus decomposed into the orthogonal direct sum $\mathbb{R} \oplus \mathbb{G}$ through spectral decomposition of the self-adjoint tidal operator $\mathscr{L}$.

We may also characterize the linear manifolds $\mathbb{R}$ and $\mathbb{G}$ by their associated projection operators $\mathscr{R}$ and $\mathscr{G}$. These are orthogonal, commute with each other and with $\mathscr{L}$, and are such that $\mathscr{R} + \mathscr{G} = \mathscr{I}$, the identity operator. Any state vector x may be uniquely decomposed into two orthogonal components as

$$x = \mathscr{R}x + \mathscr{G}x. \tag{14}$$

If x is a tentative initial state for a model prediction, then linear initialization consists most simply of replacing x by $\mathscr{R}x$. We shall see later that a more suitable linear initialization is to find the state x in the $\mathbb{R}$ manifold, i.e., with $\mathscr{R}x = x$, that is most compatible with observations.

By construction $\mathbb{R}$ is an invariant manifold for linearized dynamics; i.e., if $n(x) = 0$ and $\mathscr{R}x = x$, then

$$\mathscr{R}x_t = i\mathscr{R}\mathscr{L}x = i\mathscr{L}\mathscr{R}x = i\mathscr{L}x = x_t \tag{15}$$

is also in $\mathbb{R}$, and the motion remains in $\mathbb{R}$. Since the eigenfrequencies in $\mathbb{R}$ are all small, the motion in $\mathbb{R}$ is slow and remains slow if $n(x) = 0$.

III. NONLINEAR ANALYSIS

Atmospheric dynamics are essentially nonlinear, the term $n(x)$ couples the manifolds $\mathbb{R}$ and $\mathbb{G}$, and linear initialization does not remove completely the unwanted rapid oscillations. For $x = \mathscr{R}x$ or $\mathscr{G}x = 0$ the dynamics equation (13) when projected by $\mathscr{G}$ gives

$$\mathscr{G}x_t = i\mathscr{G}\mathscr{L}\mathscr{R}x + \mathscr{G}n(x) = \mathscr{G}n(\mathscr{R}x) \neq 0, \tag{16}$$

and rapid motion components in the $\mathbb{G}$ manifold are seen to develop from purely rotational initial states. A first-order solution to this problem was proposed independently by Baer (1977) and by Machenhauer (1977). In the present notation, this consists of replacing the linear constraint $\mathscr{G}x = 0$ by the nonlinear constraint obtained from the $\mathscr{G}$ projection of Eq. (13):

$$\mathscr{G}x_t = i\mathscr{G}\mathscr{L}x + \mathscr{G}n(x) = 0. \tag{17}$$

Let $y = \mathscr{R}x$ and $z = \mathscr{G}x$ be, respectively, the rotational and gravitational components of an arbitrary state vector x; then the nonlinear constraint of Eq. (17) is satisfied if z is related to y in such a way that

$$i\mathscr{L}z + \mathscr{G}n(y + z) = 0 \tag{18}$$

or

$$z = i(\mathscr{L}\mathscr{G})^{-1}\mathscr{G}n(y + z), \tag{19}$$

where the inverse $(\mathscr{L}\mathscr{G})^{-1}$ is restricted to the $\mathbb{G}$ manifold. Since the eigenvalues of $\mathscr{L}$ are large on the $\mathbb{G}$ manifold, those of $(\mathscr{L}\mathscr{G})^{-1}$ are small and Eq. (19) lends itself to an iterative solution:

$$z^{(m)} = i(\mathscr{L}\mathscr{G})^{-1}\mathscr{G}n(y + z^{(m-1)}) \tag{20}$$

starting with $z^{(0)} = 0$. For any y for which the iteration converges, there is determined thereby a z such that $x = y + z$ satisfies the first-order nonlinear constraint of Eq. (17).

Note that in satisfying Eq. (17) the component z in $\mathbb{G}$ has been chosen in such a way that its linear dynamics in $\mathbb{G}$ exactly balance the nonlinear tendency in $\mathbb{G}$ induced by $n(x)$. We may thus call x a first-order nonlinearly balanced state.

First-order nonlinear initialization consists of finding a first-order balanced state to serve as the initial state for a model prediction. It has been observed in practice that, if this can be done, then unwanted gravity-wave oscillations are effectively removed, and higher order balancing, to be discussed later, provides little added benefit.

Temperton and Williamson (1981) and Williamson and Temperton (1981) describe the results of, respectively, linear and first-order nonlinear initialization applied to a grid-point prediction model, used for operational forecasts by the European Center for Medium-Range Weather Forecasts (ECMWF). Figure 1 is a composite of their results and shows an example of the relative benefits of linear and nonlinear initialization.

The first-order balanced states belong to a manifold $\mathbb{M}$ in phase space, which is called the first-order slow manifold. By its definition, it is clearly nonlinear and can be thought of as the result of a nonlinear warping of the linear rotational manifold $\mathbb{R}$.

There is a close relationship between the nonlinear analysis described here and the classical methods of atmospheric dynamics (Leith, 1980). The simplest dynamic relation used in the analysis of geophysical fluid flows is the geostrophic relation, by which the velocity field is related to the pressure field in such a way that pressure gradient terms are balanced by Coriolis terms in the acceleration equations. For sufficiently simple plane geometry, the geostrophic relation corresponds exactly to the rotational relation imposed when a state vector x is confined to the linear rotational manifold $\mathbb{R}$. The geostrophic relation, however, is a local differential relation that breaks down

3. Numerical Weather Prediction Models

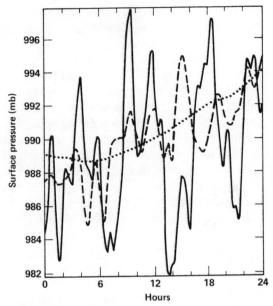

Fig. 1. Surface pressure traces at a particular grid point of the ECMWF model for three 24-hr forecasts using different initial states. The solid curve is without initialization; the dashed curve is with linear and the dotted curve with nonlinear initialization. [Data from Temperton and Williamson (1981) and Williamson and Temperton (1981).]

at the equator, whereas the rotational relation is global and is thus a suitable generalization.

It has long been realized that the geostrophic relation by itself gives uninteresting dynamics but that the next higher order analysis leads to important ageostrophic motions, such as vertical velocity, nonlinearly forced by geostrophic flows. The resulting so-called quasi-geostrophic theory has been moderately successful in describing the dynamics of the atmosphere.

Quasi-geostrophic theory corresponds in simple plane geometry to the first iteration of Eq. (20), thus giving the ageostrophic or gravitational component

$$z = i(\mathscr{L}\mathscr{G})^{-1}\mathscr{G}n(y) \tag{21}$$

forced by the geostrophic or rotational component y. The natural global generalization of quasi-geostrophic theory is called quasi-rotational theory. The relative success of quasi-geostrophic theory is consistent with the observation that the first iteration of Eq. (20) provides a fair approximation to the final result.

A more general approach to the problems of different time scales for systems of ordinary or partial differential equations has used the bounded-derivative method of Kreiss (1979, 1980), which is described in Chapter 2 of this volume. In this method constraints on initial states are deduced from requirements that time derivatives of various orders remain bounded in magnitude. Since the method does not depend explicitly on a decomposition into normal modes, it is applicable to such problems as limited-area atmospheric models with open lateral boundaries (Browning and Kreiss, 1982), for which normal-mode analysis is difficult. In general, it provides an alternative method for initialization of atmospheric models (Browning et al., 1980). Kasahara (1982) has shown that the bounded-derivative method also reduces in its lowest interesting order of approximation to classical quasi-geostrophic theory. The bounded-derivative method differs, therefore, from that of nonlinear normal-mode analysis only in higher orders, and choices between them should probably be based on practical computational considerations rather than on theoretical merit.

In normal-mode analysis the linear condition $\mathscr{G}x = 0$ and the nonlinear initial condition [Eq. (17)] $\mathscr{G}x_t = 0$ are only the first of a sequence of possible conditions involving higher time derivatives

$$\mathscr{G}x_{tt} = 0, \quad \mathscr{G}x_{ttt} = 0, \ldots . \tag{22}$$

Tribbia (1984) has shown that the imposition of such higher order conditions does not require calculation of the general Frechet derivatives of the nonlinear term $g = \mathscr{G}n(x)$ but rather only of its time derivatives along the evolving orbit in phase space. For the condition of order m, the required initial-value and $m - 1$ time derivatives of g can be estimated by polynomial fit to m successive time-step evaluations of the nonlinear term. The amount of arithmetic required is thus equivalent to that of m time steps in the model integration, for each iteration. Such higher order nonlinear initialization procedures are therefore feasible for any computationally feasible model.

Each of these higher order conditions defines its own slow manifold. It is not yet known whether these approach some sort of limit. It seems more likely that this sequence of conditions should be considered useful only in an asymptotic sense. Evidence for some improvement in initialization at second order is given by Tribbia (1984). Higher order initialization probably provides too little benefit to justify the increased amount of arithmetic and risk of divergence.

IV. ANALYSIS OF OBSERVATIONS

In numerical weather prediction it is, of course, necessary to determine as accurately as possible, on the basis of all available information, the initial state of the atmosphere. Observations of the atmosphere are derived from a heterogeneous collection of sources, including sounding balloons, satellite measurements, and aircraft reports. These measurements of temperatures, winds, and moisture are scattered erratically in space and time. They are reduced to fields that are globally defined at a particular time by a process known as analysis. Analysis can be thought of as a method of interpolation between observation points to a fixed array of grid points, but more generally it is a method of statistically optimal estimation of meteorological fields by multivariate analysis using observations as predictors.

Such statistically optimal methods require a priori statistical information about the estimated fields. This information was originally based on climatological information (Gandin, 1963), with the thermodynamic and hydrodynamic fields assumed to be in geostrophic balance. In the geometry of phase space this information corresponds to an a priori climate probability distribution defined in the ℝ manifold. The new observations with their inevitable errors can be represented in phase space by a probability density distributed about the observed point. According to Bayes's theorem in its most general form, the optimal combination of these two sources of information, as represented by these two probability distributions, is, in turn, represented by their renormalized product. In this case, therefore, the a posteriori probability distribution will be confined to the ℝ manifold and so also will be the most likely state.

The ℝ state generated by classical analysis is then modified to an 𝕄 state by nonlinear initialization. It is obviously more appropriate to combine the analysis and initialization processes into one in which a statistically optimal 𝕄 state is found. Williamson and Daley (1983) describe the benefits and difficulties in doing this.

In current practice it is recognized that a previous prediction valid at the present time is a better source of a priori information than the climate. This previous prediction, together with knowledge of prediction error statistics, is represented by an a priori probability density distributed about a predicted 𝕄 state. It is most satisfactory if the distribution is confined, as it should be, to the 𝕄 manifold, for then Bayes's theorem will guarantee the generation of an a posteriori most likely 𝕄 state suitable for starting a new prediction.

Unfortunately, there are practical computational problems in carrying out this program, which is so easy to describe geometrically. Multivariate analysis is essentially linear and, owing to the possible separation of variables described in Section II, can be easily done on the ℝ manifold. In carrying out an analysis constrained to the 𝕄 manifold, it would at least be useful to determine a linear manifold tangent to 𝕄. But this tangent is defined in terms of Frechet derivatives of the nonlinear term $g = \mathcal{G}n(x)$, discussed in Section III. It is not computationally feasible to impose a general linear tangent constraint in the analysis procedure owing to the massive dimensionality of the phase space and the complexity of the 𝕄 manifold. An iterative approximation that uses a linear manifold for analysis and that is parallel to but displaced from the $\mathbb{R}$ manifold by the $\mathscr{G}$ vector arising from initialization is feasible and has been shown to be effective by Williamson and Daley (1983).

V. REMAINING PROBLEMS

Although the introduction of nonlinear normal-mode initialization in recent years has significantly improved data analysis and weather prediction, there remain a number of outstanding problems to be resolved.

Initialization theory and algorithms are still largely based only on the nonlinear dynamics of the atmosphere. But there are many physical processes included in models, some of which, such as precipitation and latent heat release, are highly nonlinear. Initialization algorithms have tended to reject circulations driven primarily by latent heat and have, therefore, been particularly poor in the equatorial regions. Progress is being made in understanding this difficulty (e.g., Puri, 1983), but there are not yet completely systematic ways to balance nonlinear physics, as well as nonlinear dynamics.

There are also purely mathematical problems connected with the concept of an invariant slow manifold. The question of whether or in what sense the sequence of slow manifolds associated with higher order balancing conditions converges to a limiting manifold has already been discussed in Section III. There is evidence from a simple example given by Tribbia (1981) that a slow manifold might exist but, in some regions, be unstable for gravitational wave radiation. Warn (1983) has suggested that the slow manifold is, in fact, a thin fuzzy set. Lorenz (1980) has shown that, as expected, a chaotic attractor is embedded in a slow manifold for a low-order dynamic system, but topological details are still unclear.

Even if a slow manifold exists, there are problems in finding it. Tribbia (1981) and Phillips (1981) have studied the convergence of the nonlinear iteration algorithms in common use. Convergence, stability, and existence problems appear to be only loosely related.

These basic theoretical questions are being studied with very simple models, but consideration of more realistic multilevel models raises new questions. In Section I.E, it is pointed out that, for models with relatively slow internal gravitational modes, the dividing frequency between slow and fast is somewhat arbitrary. Thus the dimensionality of the slow manifold is also arbitrary. In fact, any sharp division of frequencies seems artificial and inconsistent with optimal filter design based on the observed frequency spectrum of atmospheric motions. The definition and practical application of such a filter, however, encounter the same problems of mixing linear regression methods and nonlinear constraints that are discussed in Section IV.

Although nonlinear initialization has been clearly beneficial and the concept of a slow manifold has great heuristic value, there remain many fundamental questions to answer about the details of the nonlinear dynamics of the atmosphere relevant to the problem of initialization.

REFERENCES

Baer, F. (1977). *Beitr. Phys. Atmos.* **50**, 350–366.
Browning, G., Kasahara, A., and Kreiss, H.-O. (1980). *J. Atmos. Sci.* **37**, 1424–1436.
Browning, G., and Kreiss, H.-O. (1982). *Tellus*, **34**, 334–351.
Daley, R. (1981). *Rev. Geophys. Space Phys.* **19**, 450–468.
Gandin, L. S. (1963). *In* "Objective Analysis of Meteorological Fields," Transl. TT65-50007. Dept. of Commerce, National Technical Information Service, Springfield, Virginia.
Hinkelmann, K. V. (1951). *Tellus* **3**, 285–296.
Hough, S. S. (1898). *Philos. Trans. R. Soc. London, Ser. A* **191**, 139–185.
Kasahara, A. (1982). *Rev. Geophys. Space Phys.* **20**, 385–397.
Kreiss, H.-O. (1979). *SIAM J. Numer. Anal.* **16**, 980–998.
Kreiss, H.-O. (1980). *Commun. Pure Appl. Math.* **33**, 399–439.
Lamb, H. (1932). In "Hydrodynamics," 6th ed. Dover, New York.
Leith, C. E. (1978). *Annu. Rev. Fluid Mech.* **10**, 107–128.
Leith, C. E. (1980). *J. Atmos. Sci.* **37**, 958–968.
Longuet-Higgins, M. S. (1968). *Philos. Trans. R. Soc. London, Ser. A* **262**, 511–607.
Lorenz, E. N. (1980). *J. Atmos. Sci.* **37**, 1685–1699.
Machenhauer, B. (1977). *Beitr. Phys. Atmos.* **50**, 253–271.
Phillips, N. A. (1981). *Mon. Weather Rev.* **109**, 2415–2426.
Puri, K. (1983). *Mon. Weather Rev.* **111**, 23–33.
Swarztrauber, P. N., and Kasahara, A. (1985). *SIAM J. Sci. Stat. Comput.*, April, 1985.
Taylor, G. I. (1936). *Proc. R. Soc. London, Ser. A* **156**, 318–326.
Temperton, C., and Williamson, D. L. (1981). *Mon. Weather Rev.* **109**, 729–743.
Tribbia, J. J. (1981). *Mon. Weather Rev.* **109**, 1751–1761.
Tribbia, J. J. (1984). *Mon Weather Rev.* **112**, 278–284.
Warn, T. (1983). unpublished note, Dept. of Meteorology, McGill University.
Williamson, D. L., and Daley, R. (1983). *Mon. Weather Rev.* **111**, 1517–1536.
Williamson, D. L., and Temperton, C. (1981). *Mon. Weather Rev.* **109**, 744–757.

4

The Diffusion-Synthetic Acceleration of Transport Iterations, with Application to a Radiation Hydrodynamics Problem

RAYMOND E. ALCOUFFE
BRADLEY A. CLARK
EDWARD W. LARSEN

Los Alamos National Laboratory
University of California
Los Alamos, New Mexico

I. Introduction	74
II. Transport Iteration Methods	75
A. Source-Iteration Method for the Transport Equation . .	75
B. Diffusion-Synthetic Acceleration Method for the Transport Equation	77
C. Synthetic Acceleration of the Multifrequency Diffusion Equation	80
D. Some Numerical Considerations	83
III. A Problem in Radiation Hydrodynamics	84
A. Problem Description	84
B. Discretization in Time	86
C. Discretization in Frequency	89
D. Discretization in Angle	89
E. A Simple Spatial Discretization	90
F. Iteration Method for the Discretized Transport Equation	92
IV. Time-Dependent Example Calculations	104
References	111

I. INTRODUCTION

Radiation hydrodynamics treats the interactions of thermal radiation with matter. Thermal radiation here is taken to mean electromagnetic radiation of atomic origin, obtained from the processes of scattering, absorption, and thermal emission. The matter is generally modeled as a fluid of electrons and ions whose motion is governed by the equations of hydrodynamics. The radiation field is described by an equation of transfer, which is a Boltzmann transport equation for photons. In radiation hydrodynamics one is primarily interested in the regime for which the radiation field and matter are intimately coupled to one another. That is, the radiation field has a dominant influence on the state of the matter, and thermal processes in the matter are the primary source of the radiation field. Generally, the modeling of these processes and interactions leads to a system of very complicated coupled equations, with intricate nonlinearities (Byrne, 1973; Pomraning, 1973, 1982; Wilson, 1973).

From a computational viewpoint, time-dependent radiative transfer problems in emitting media are notoriously difficult to solve. In general, their solution involves an equation with seven independent variables for the radiation field itself, along with additional equations describing the mass, momentum, and energy balance of the underlying medium. In this chapter, we describe new numerical methods for solving such problems.

We treat, in detail, three aspects of the solution method: the time dependence, the intimate coupling between the radiation field and the medium, and the iteration acceleration of the equation of transfer. The time dependence is treated implicitly; that is, the equation of transfer is written in terms of the dependent variables evaluated at the advanced time (sometimes called backward Euler time differencing). The solution within each time step is obtained by iteration; the use of an acceleration method, which generates this solution with a convergence rate independent of the time step, is the main focus of this chapter. The coupling of the radiation field to the medium is related to the implicit formulations of the transfer and energy balance equations. This coupling is cast so that the equation of transfer contains terms from the energy balance equation; thus, the iteration normally occurring between these equations is compatible with the iterations involved in solving the equation of transfer itself at each time step. These ideas lead to a numerical algorithm that is stable and efficiently solvable for all time steps.

The remainder of this chapter consists of three major sections. In the first (Section II) we describe iterative procedures for solving the equation of transfer, to which terms from the energy balance equation have been adjoined. We do this in a general form, without reference to a specific problem, to clearly present our method of iteration acceleration. Section III deals with a specific problem in radiation hydrodynamics; here we use material from Section II to develop our numerical solution method. Then, in Section IV we discuss how the choice of time steps and iteration convergence affects solution accuracy.

To conclude this introduction, we remark that many of our ideas on iteration acceleration are new in the area of radiation transport but are based on methods that have been developed and tested in recent years for other neutron transport and radiative transfer problems (Alcouffe, 1977; Alcouffe and O'Dell, 1983; Carlson and Lathrop, 1968; Cullen and Pomraning, 1980; Larsen, 1982, 1983; Morel et al., 1984). This prior body of work contains numerous discussions on the details of numerical discretizations. Owing to a lack of space, we cannot fully cover these discussions here, but we occasionally refer to them.

II. TRANSPORT ITERATION METHODS

A. Source-Iteration Method for the Transport Equation

In this section we discuss iteration methods for solving the equation of transfer, modified by terms from the energy balance equation. To begin, we describe the standard source-iteration (SI) method and apply a Fourier analysis to show that this method can converge arbitrarily slowly for arbitrarily large time steps. This motivates the subsequent development of the diffusion-synthetic acceleration (DSA) method, which converges at a rapid rate *independent* of the time step. The implicit time-differencing of the transport equation and this essential feature of the DSA method permit the calculation of time-dependent transport solutions on arbitrarily large or small time scales.

The equation of transfer, modified by the energy balance equation and time-differenced implicitly, is shown in Section III to have the general form

$$\mathbf{\Omega} \cdot \nabla I(\mathbf{r}, v, \mathbf{\Omega}) + [\sigma(v) + \tau]I(\mathbf{r}, v, \mathbf{\Omega})$$

$$= \frac{\chi(v)}{4\pi(1 + \alpha\tau)} \int_0^\infty \sigma(v')I_0(\mathbf{r}, v')\,dv' + Q(\mathbf{r}, v, \mathbf{\Omega}), \tag{1a}$$

$$I_0(\mathbf{r}, v) = \int I(r, v, \mathbf{\Omega})\,d^2\Omega. \tag{1b}$$

Here, the dependent variables are $\mathbf{r}$, v, and $\mathbf{\Omega}$, denoting position, frequency, and direction ($|\mathbf{\Omega}| = 1$); I is the specific intensity at the advanced time step; σ, χ, α, and Q are positive quantities described in Section III, with Q containing only information from the previous time step;

$$\tau = \frac{1}{c\,\Delta t}, \tag{2}$$

where c is the speed of light; and

$$\int_0^\infty \chi(v)\,dv = 1. \tag{3}$$

The standard SI method for iteratively solving Eqs. (1a, b) is defined by

$$\mathbf{\Omega} \cdot \nabla I^{(l+1/2)} + (\sigma + \tau)I^{(l+1/2)} = \frac{\chi}{4\pi(1+\alpha\tau)}\int_0^\infty \sigma I_0^{(l)}\,dv + Q, \tag{4a}$$

$$I_0^{(l+1/2)} = \int I^{(l+1/2)}\,d^2\Omega, \tag{4b}$$

$$I_0^{(l+1)} = I_0^{(l+1/2)}. \tag{4c}$$

To study the convergence properties of this method, we define

$$D^{(l+1/2)} = I^{(l+1/2)} - I^{(l-1/2)}, \tag{5a}$$

$$D_0^{(l)} = I_0^{(l)} - I_0^{(l-1)}. \tag{5b}$$

The functions $D^{(l+1/2)}$ and $D_0^{(l)}$, which denote the difference between the values of I and I_0 on successive iterations, satisfy Eqs. (4) with I replaced by D, I_0 replaced by D_0, and Q set equal to zero. The SI method converges if and only if $D^{(l+1/2)}$ and $D_0^{(l)}$ tend to zero as $l \to \infty$, and if the SI method does converge, its rate of convergence is equal to the rate at which $D^{(l+1/2)}$ and D_0 tend to zero.

To determine this rate, we introduce the Fourier decomposition

$$D^{(l+1/2)} = \omega^l D(\lambda, v, \mathbf{\Omega})e^{i\lambda \cdot \mathbf{r}}, \tag{6a}$$

$$D_0^{(l)} = \omega^l D_0(\lambda, v)e^{i\lambda \cdot \mathbf{r}}, \tag{6b}$$

$$\int_0^\infty \sigma D_0\,dv = 1 \quad \text{(normalization)}, \tag{6c}$$

into the equations satisfied by $D^{(l+1/2)}$ and $D_0^{(l)}$ to obtain Eq. (6c) and

$$(i\mathbf{\Omega} \cdot \lambda + \sigma + \tau)D = \chi/4\pi(1+\alpha\tau), \tag{7a}$$

$$\omega D_0 = \int D\,d^2\Omega. \tag{7b}$$

In Eqs. (6a, b), ω, which is termed the iteration eigenvalue, determines the growth or decay of the Fourier mode with increasing l. The largest value of $|\omega|$, denoted by ρ, is termed the spectral radius. If $\rho < 1$, then the SI method converges with an error multiplication factor per iteration bounded from above by ρ; if $\rho \geq 1$, then it is possible that the error does not tend to zero, and the SI method can diverge.

4. Diffusion-Synthetic Acceleration

Solving Eqs. (7a, b) for ω, we easily obtain

$$\omega = \frac{1}{1 + \alpha\tau} \int_0^\infty \frac{\sigma\chi}{\lambda} \arctan \frac{\lambda}{\sigma + \tau} \, dv, \tag{8}$$

where $\lambda = |\boldsymbol{\lambda}|$. This shows that ω is a monotone-decreasing function of λ, with its maximum value occurring at $\lambda = 0$:

$$\rho(\tau) = \omega|_{\lambda=0} = \frac{1}{1 + \alpha\tau} \int_0^\infty \frac{\sigma\chi}{\sigma + \tau} \, dv. \tag{9}$$

Thus, ρ is a monotone-decreasing function of τ, with its maximum value occurring at $\tau = 0$. Equations (3) and (9) imply

$$\rho < 1, \quad 0 < \tau < \infty, \tag{10a}$$

$$\lim_{\tau \to 0} \rho = 1. \tag{10b}$$

Therefore, for all $\tau > 0$ [i.e., by Eq. (2), all $0 \leq \Delta t < \infty$], the SI method is stable and convergent. However, as $\tau \to 0$ ($\Delta t \to \infty$), the error reduction per iteration can become arbitrarily close to 1. In other words, the SI method converges for all time steps, but for arbitrarily large time steps, this convergence can become arbitrarily slow.

This result holds for the special case of the spatially analytic transport equation in an infinite medium with constant coefficients (i.e., σ, α, and χ, independent of **r**). For the spatially discretized equations in an infinite medium, with constant coefficients and a uniform spatial mesh, one can perform a discrete Fourier analysis, virtually identical to that outlined above, to show that this result also holds for all values of the spatial mesh. Experience has shown that the same result is again true for finite-medium problems, discretized in any standard way, with variable coefficients and nonuniform spatial meshes.

B. Diffusion-Synthetic Acceleration Method for the Transport Equation

To summarize the above, in the standard SI method, one has a numerical iteration method that is always stable, but that can converge unacceptably slowly for large time steps. This difficulty can be overcome by observing, from Eq. (7a), that the eigenmodes that are most slowly converging (i.e., those for $\lambda \approx 0$) satisfy

$$D(\boldsymbol{\lambda}, v, \boldsymbol{\Omega}) = \frac{1}{4\pi(1 + \alpha\tau)} \frac{\chi}{\sigma + \tau} \left[1 - i\frac{\boldsymbol{\Omega} \cdot \boldsymbol{\lambda}}{\sigma + \tau} + O(\lambda^2) \right] \tag{11}$$

and, hence, are nearly linear functions of $\mathbf{\Omega}$. The DSA method uses this fact to modify the SI iteration method in the following way: *one retains Eqs. (4a,b) in the iteration procedure, but replaces Eq. (4c) by one that computes the exact solution in one iteration if it is a linear function of angle.* The goal is, simply, to force the slowly converging eigenfunctions to converge more quickly. To accomplish this, we compute the zeroth and first angular moments of Eq. (4a), obtaining

$$\nabla \cdot \mathbf{I}_1^{(l+1/2)} + (\sigma + \tau) I_0^{(l+1/2)} = \frac{\chi}{1 + \alpha\tau} \int_0^\infty \sigma I_0^{(l)} \, dv + Q_0, \tag{12}$$

$$\frac{1}{3} \nabla I_0^{(l+1/2)} + \frac{2}{3} \nabla \cdot \mathbf{I}_2^{(l+1/2)} + (\sigma + \tau) \mathbf{I}_1^{(l+1/2)} = \mathbf{Q}_1, \tag{13}$$

where

$$\mathbf{I}_1^{(l+1/2)} = \int \mathbf{\Omega} I^{(l+1/2)} \, d^2\Omega, \tag{14a}$$

$$\mathbf{I}_2^{(l+1/2)} = \int \frac{1}{2}(3\mathbf{\Omega}\mathbf{\Omega} - \mathbf{E}) I^{(l+1/2)} \, d^2\Omega, \tag{14b}$$

and $\mathbf{E}$ is the identity tensor. Eliminating $\mathbf{I}_1^{(l+1/2)}$ between Eqs. (12) and (13), we obtain

$$-\nabla \cdot \frac{1}{3(\sigma + \tau)} \nabla I_0^{(l+1/2)} + (\sigma + \tau) I_0^{(l+1/2)}$$

$$= \frac{\chi}{1 + \alpha\tau} \int_0^\infty \sigma I_0^{(l)} \, dv + Q_0 - \nabla \cdot \frac{1}{\sigma + \tau} \mathbf{Q}_1$$

$$+ \nabla \cdot \frac{2}{3(\sigma + \tau)} \nabla \cdot \mathbf{I}_2^{(l+1/2)}. \tag{15}$$

This equation is exactly satisfied by the solution of Eqs. (4a, b). Noting that $\mathbf{I}_2$ in this equation vanishes if I is a linear function of $\mathbf{\Omega}$, we define the DSA acceleration for $I_0^{(l+1)}$ as

$$-\nabla \cdot \frac{1}{3(\sigma + \tau)} \nabla I_0^{(l+1)} + (\sigma + \tau) I_0^{(l+1)}$$

$$= \frac{\chi}{1 + \alpha\tau} \int_0^\infty \sigma I_0^{(l+1)} \, dv + Q_0 - \nabla \cdot \frac{1}{\sigma + \tau} \mathbf{Q}_1 + \nabla \cdot \frac{2}{3(\sigma + \tau)} \nabla \cdot \mathbf{I}_2^{(l+1/2)}. \tag{16}$$

This equation agrees with the previous equation upon convergence, and it fulfills the condition stated above: if I is a linear function of angle, then $\mathbf{I}_2 = 0$, and Eq. (16) determines $I_0^{(l+1)}$ exactly. For calculational purposes, it is sometimes convenient to define

$$F_0^{(l+1)}(\mathbf{r}, v) = I_0^{(l+1)}(\mathbf{r}, v) - I_0^{(l+1/2)}(\mathbf{r}, v) \tag{17}$$

4. Diffusion-Synthetic Acceleration

and subtract Eq. (15) from Eq. (16) to obtain the following equation for $F_0^{(l+1)}$:

$$-\nabla \cdot \frac{1}{3(\sigma + \tau)} \nabla F_0^{(l+1)} + (\sigma + \tau) F_0^{(l+1)}$$

$$= \frac{\chi}{1 + \alpha\tau} \int_0^\infty \sigma F_0^{(l+1)} \, dv + \frac{\chi}{1 + \alpha\tau} \int_0^\infty \sigma [I_0^{(l+1/2)} - I_0^{(l)}] \, dv. \quad (18)$$

The "source" term for F_0 is now much simpler than that for I_0.

The DSA iteration procedure consists of Eqs. (4a, b), (18), and (17). [Equivalently, one could use Eqs. (4a), (14b), and (16).] Operationally, we now use a frequency-dependent diffusion equation, rather than Eq. (4c), to obtain $I_0^{(l+1)}$. A Fourier analysis for this iteration method, identical in form to the one described above for the SI method, leads to the following expression for the iteration eigenvalue:

$$\omega = \frac{\lambda^2 \int_0^\infty \left\{ \chi \left[\frac{\sigma(\sigma + \tau)}{\lambda^2 + 3(\sigma + \tau)^2} \right] \left[\int_0^1 \frac{1 - 3\mu^2}{(\sigma + \tau)^2 + \lambda^2 \mu^2} \, d\mu \right] \right\} dv}{1 + \alpha\tau - \int_0^\infty \chi \left[\frac{3\sigma(\sigma + \tau)}{\lambda^2 + 3(\sigma + \tau)^2} \right] dv}. \quad (19)$$

This expression tends to zero as $\lambda \to 0$ for all values of $\tau \geq 0$. Therefore, the $\lambda \approx 0$ eigenvalues, which converge slowly for the SI method, converge rapidly for the DSA method. This, of course, was the aim in deriving the DSA method.

Numerical evaluations of Eq. (19) for the analytic expressions

$$\chi(v) = a_0 \exp(-a_0 v), \quad (20a)$$

$$\sigma(v) = a_1 [1 - \exp(-a_0 v)]/v^3, \quad (20b)$$

with $a_n > 0$, $\lambda \geq 0$, and $\tau \geq 0$, leads to the result $|\omega| \leq 0.164$; hence, for all such χ and σ and all Δt, the DSA method is stable with a spectral radius (error multiplication factor per iteration)

$$\rho \leq 0.164. \quad (21)$$

Two questions remain to be addressed. First, regardless of whether one chooses to use Eq. (16) or (18) in the DSA method, it is necessary to solve one frequency-dependent diffusion equation of the form

$$-\nabla \cdot \frac{1}{3(\sigma + \tau)} \nabla \Phi(\mathbf{r}, v) + (\sigma + \tau) \Phi(\mathbf{r}, v)$$

$$= \frac{\chi(v)}{1 + \alpha\tau} \int_0^\infty \sigma \Phi(\mathbf{r}, v') \, dv' + S(\mathbf{r}, v), \quad (22)$$

for each transport iteration. How can one do this efficiently? Second, how does one discretize the entire process to preserve the property that the spectral radius is bounded less than 1 for all values of the discretization parameters?

C. Synthetic Acceleration of the Multifrequency Diffusion Equation

Let us first consider the question of efficiently solving the frequency-dependent diffusion equation (22). In general, this must be done iteratively, and, therefore, *within* the "outer" iteration process described above (one step of which consists of a multifrequency transport calculation followed by a multifrequency diffusion calculation) is an "inner" iteration process for solving the frequency-dependent diffusion equation. This inner iteration process is the focus of the following analysis; it uses ideas very similar to those used in developing the outer iteration process.

The standard SI method for solving Eq. (22) is described by

$$-\nabla \cdot \frac{1}{3(\sigma + \tau)} \nabla \Phi^{(l+1/2)}(\mathbf{r}, v) + (\sigma + \tau)\Phi^{(l+1/2)}(\mathbf{r}, v)$$

$$= \frac{\chi(v)}{1 + \alpha\tau} \Phi_0^{(l)}(\mathbf{r}) + S(\mathbf{r}, v), \tag{23a}$$

$$\Phi_0^{(l+1/2)}(\mathbf{r}) = \int_0^\infty \sigma(v)\Phi^{(l+1/2)}(\mathbf{r}, v) \, dv, \tag{23b}$$

$$\Phi_0^{(l+1)}(\mathbf{r}) = \Phi_0^{(l+1/2)}(\mathbf{r}), \tag{23c}$$

and the corresponding iteration eigenvalue is

$$\omega = \frac{1}{1 + \alpha\tau} \int_0^\infty \frac{3\sigma(\sigma + \tau)\chi}{\lambda^2 + 3(\sigma + \tau)^2} \, dv. \tag{24}$$

The largest value of $|\omega|$ (the spectral radius) occurs at $\lambda = 0$,

$$\rho = \omega|_{\lambda=0} = \frac{1}{1 + \alpha\tau} \int_0^\infty \frac{\sigma\chi}{\sigma + \tau} \, dv, \tag{25}$$

and ρ is a monotone decreasing function of τ, taking its maximum value at $\tau = 0$:

$$\rho < 1, \quad 0 < \tau < \infty, \tag{26a}$$

$$\lim_{\tau \to \infty} \rho = 1. \tag{26b}$$

4. Diffusion-Synthetic Acceleration

Therefore, just as with the transport equation, the SI method converges for all values of Δt; but for arbitrarily large Δt, the method can converge arbitrarily slowly.

To construct an acceleration method for Eq. (22) that is closely analogous to the DSA method for Eqs. (1a, b), we integrate Eq. (23a) over frequency and rearrange to obtain

$$-\nabla \cdot \left[\int_0^\infty \frac{1}{3(\sigma+\tau)} \nabla \theta^{(l+1/2)} \, dv \right] \Phi_0^{(l+1/2)} + \left[1 + \tau \int_0^\infty \theta^{(l+1/2)} \, dv \right] \Phi_0^{(l+1/2)}$$

$$= \frac{1}{1+\alpha\tau} \Phi_0^{(l)} + \int_0^\infty S \, dv$$

$$+ \int_0^\infty \left[\nabla \cdot \frac{1}{3(\sigma+\tau)} \nabla - \tau \right] \left[\Phi^{(l+1/2)} - \theta^{(l+1/2)} \Phi_0^{(l+1/2)} \right] dv. \qquad (27)$$

Here $\theta^{(l+1/2)}(\mathbf{r}, v)$ is a "shape" function, normalized by

$$\int_0^\infty \sigma \theta^{(l+1/2)} \, dv = 1, \qquad (28)$$

but otherwise arbitrary and to be determined. Equation (27) is exactly satisfied by the solution of Eqs. (23a, b) for any choice of $\theta^{(l+1/2)}$. We define the acceleration equation for $\Phi_0^{(l+1)}$ as

$$-\nabla \cdot \left[\int_0^\infty \frac{1}{3(\sigma+\tau)} \nabla \theta^{(l+1/2)} \, dv \right] \Phi_0^{(l+1)} + \tau \left[\frac{\alpha}{1+\alpha\tau} + \int_0^\infty \theta^{(l+1/2)} \, dv \right] \Phi_0^{(l+1)}$$

$$= \int_0^\infty S \, dv + \int_0^\infty \left[\nabla \cdot \frac{1}{3(\sigma+\tau)} \nabla - \tau \right] \left[\Phi^{(l+1/2)} - \theta^{(l+1/2)} \Phi_0^{(l+1/2)} \right] dv. \qquad (29)$$

This equation agrees with Eq. (27) upon convergence, and a proper choice of $\theta^{(l+1/2)}$ will ensure that it actually accelerates the iteration procedure. For calculational purposes, it is sometimes convenient to define

$$\Psi_0^{(l+1)}(\mathbf{r}) = \Phi_0^{(l+1)}(\mathbf{r}) - \Phi_0^{(l+1/2)}(\mathbf{r}), \qquad (30)$$

and subtract Eq. (27) from Eq. (29) to obtain the following equation for $\Psi_0^{(l+1)}$:

$$-\nabla \cdot \left[\int_0^\infty \frac{1}{3(\sigma+\tau)} \nabla \theta^{(l+1/2)} \, dv \right] \Psi_0^{(l+1)} + \tau \left[\frac{\alpha}{1+\alpha\tau} + \int_0^\infty \theta^{(l+1/2)} \, dv \right] \Psi_0^{(l+1)}$$

$$= \frac{1}{1+\alpha\tau} [\Phi_0^{(l+1/2)} - \Phi_0^{(l)}]. \qquad (31)$$

The "source" term for $\Psi_0^{(l+1)}$ is now much simpler than that for $\Phi_0^{(l+1)}$.

The acceleration procedure for solving Eq. (22) is now given by Eqs. (23a, b), (31), and (30). [Equivalently, one could use Eqs. (23a, b) and (29).] Operationally, we use a frequency-independent (i.e., "grey") equation, rather than Eq. (23c), to accelerate the convergence of the multifrequency diffusion equation.

To completely specify the acceleration method, we must prescribe the shape function $\theta^{(l+1/2)}$, whose only constraint is Eq. (28). We have considered two possibilities, leading to linear and nonlinear *multifrequency–grey* (MFG) acceleration methods.

To obtain the linear MFG method, we note that the iteration eigenvalue for the SI method corresponding to λ is given by Eq. (24); the corresponding eigenfunction is

$$\Phi = \frac{1}{1 + \alpha\tau} \frac{3(\sigma + \tau)\chi}{\lambda^2 + 3(\sigma + \tau)^2} e^{i\lambda \cdot \mathbf{r}}, \tag{32}$$

and for $\lambda = 0$ (the most slowly converging mode) its frequency variation is described by the shape function

$$\theta(v) = \frac{\chi(v)}{\sigma(v) + \tau} \left[\int_0^\infty \frac{\sigma(v')\chi(v')}{\sigma(v') + \tau} dv' \right]^{-1}, \tag{33}$$

which is normalized to satisfy Eq. (28). When we use this definition in Eq. (29) or (31), θ is independent of the iteration superscript, and the driving term from the $l + \frac{1}{2}$ calculation is on the right side of Eq. (29) and is proportional to $\Phi^{(l+1/2)} - \theta\Phi_0^{(l+1/2)}$. This acceleration method has the property that it computes the exact solution in one iteration if the solution's frequency variation at each point is described by the most slowly converging eigenfunction $\theta(v)$, as defined by Eq. (33). In other words, the driving term from the $l + \frac{1}{2}$ calculation in Eq. (29) has the most slowly converging eigenfunction projected out of it, since this (normalized) eigenfunction is described by θ and

$$\int_0^\infty \sigma[\Phi^{(l+1/2)} - \theta\Phi_0^{(l+1/2)}] dv = 0. \tag{34}$$

To obtain the nonlinear MFG method, we take

$$\theta^{(l+1/2)}(\mathbf{r}, v) = \Phi^{(l+1/2)}(\mathbf{r}, v)/\Phi_0^{(l+1/2)}(\mathbf{r}). \tag{35}$$

Then, the driving term from the $l + \frac{1}{2}$ calculation on the right side of Eq. (29) vanishes, but $l + \frac{1}{2}$ quantities appear within the leakage and removal terms. This method has the property that if $\Phi^{(l+1/2)}$ has the correct frequency variation at each point (but the incorrect amplitude), then $\theta^{(l+1/2)}$ is exact; and the exact Φ_0, computed in one iteration, is given by $\Phi_0^{(l+1)}$. Thus, this nonlinear MFG method is based on the hypothesis that the frequency variation of the iterative solution converges more quickly than the amplitude variation.

4. Diffusion-Synthetic Acceleration

The linear MFG method is amenable to a Fourier stability analysis, from which one obtains the following expression for the iteration eigenvalue:

$$\omega = \frac{\dfrac{1}{1+\alpha\tau}\int_0^\infty \dfrac{\chi}{\sigma+\tau}\left[\dfrac{\lambda^2}{3(\sigma+\tau)}+\tau\right]\left[\dfrac{\int_0^\infty \dfrac{3\sigma(\sigma+\tau)\chi}{\lambda^2+3(\sigma+\tau)}dv'}{\int_0^\infty \dfrac{\sigma\chi}{\sigma+\tau}dv'} - \dfrac{3(\sigma+\tau)^2}{\lambda^2+3(\sigma+\tau)^2}\right]dv}{\dfrac{\lambda^2}{3}\dfrac{\int_0^\infty \dfrac{\chi}{(\sigma+\tau)^2}dv}{\int_0^\infty \dfrac{\sigma\chi}{\sigma+\tau}dv} + \tau\left[\dfrac{\alpha}{1+\alpha\tau} + \dfrac{\int_0^\infty \dfrac{\chi}{\sigma+\tau}dv}{\int_0^\infty \dfrac{\sigma\chi}{\sigma+\tau}dv}\right]}$$

(36)

This formula tends to zero as $\lambda \to 0$ for all values of $\tau \geq 0$. Therefore, the $\lambda \approx 0$ modes, which converge most slowly for the SI method, converge rapidly for the linear MFG method.

Numerical evaluation of Eq. (36) using the analytic expressions (20) for σ and χ lead to the result $|\omega| \leq 0.867$ for all values of $a_n > 0$, $\alpha \geq 0$, $\lambda \geq 0$, and $\tau \geq 0$. Hence, with these definitions for σ and χ, the linear MFG method is stable with the spectral radius

$$\rho \leq 0.867, \tag{37}$$

which is bounded <1 for all Δt. The upper bound $\rho = 0.867$ occurs for $\tau = 0$ ($\Delta t = \infty$), and unfortunately is not smaller in value. Nevertheless, the linear MFG method will produce a solution of Eq. (22), with a convergence rate independent of the time step.

A stability analysis of the MFG method based on Eq. (35) is not possible, owing to the method's inherent nonlinearities. Nevertheless, numerical studies (described in the following section) indicate that the performance of the linear and nonlinear MFG methods in actual physical problems are very similar.

This completes our discussion of the outer acceleration process, which iterates between the multifrequency transport and multifrequency diffusion equations, and the inner acceleration process, which iterates between the multifrequency diffusion and grey equations. It remains to discuss the effect of discretization on the stability and performance of the overall method.

D. Some Numerical Considerations

It turns out that the effects of discretization can be easily summarized: If the basic principles leading to the analytic forms of the outer and inner acceleration methods are carefully followed in the development of the discretized acceleration methods, then the discretized methods will be stable and perform at least as well as indicated above; otherwise, for sufficiently large spatial meshes, they will generally be unstable. Therefore, the discretized multifrequency diffusion equation must be one that produces the exact numerical transport solution in one iteration if that transport solution is a linear function of the (discretized) angular variable. Likewise, for the linear MFG method, the discretized grey equation must be one that produces the exact discretized multifrequency diffusion solution in one iteration if this solution at each point has a (discretized) frequency variation equal to that of the local value of θ, defined by Eq. (33). (Here θ can be a function of position, as well as frequency.) An analogous condition holds for the nonlinear MFG method.

These requirements all amount to stating that given the discretization of the multifrequency transport equation, one must choose *consistent* discretizations of the multifrequency and grey diffusion equations. These consistent discretizations can be derived by duplicating the derivations described above with the analytic equations, but by using the discretized multifrequency transport equation as the starting point. Boundary conditions for the multifrequency and grey diffusion equations can be derived from multifrequency transport boundary conditions by following the basic principles enunciated above. In general, the discretization problem is essential, and we discuss aspects of it in detail in the next section. For further information on discretization procedures and their effects, we refer the reader to Alcouffe (1977), Larsen (1982, 1983), and Morel *et al.* (1984).

III. A PROBLEM IN RADIATION HYDRODYNAMICS

As discussed in Section I, the complete modeling of radiation hydrodynamic processes leads to a system of very complicated equations with intricate nonlinear couplings. However, the essential numerical considerations involved in the solution of these equations can be demonstrated by using a greatly simplified problem. Through this problem, described below, we demonstrate how the ideas presented in Section II can be implemented for an efficient numerical solution.

A. Problem Description

Our simplifying assumptions are as follows:

(a) The background matter is stationary.
(b) The thermal radiative emission is governed by local thermodynamic equilibrium (LTE) in the matter.

4. Diffusion-Synthetic Acceleration

(c) Scattering processes are ignored.
(d) Thermal conduction in the matter is ignored.

With these assumptions, the mathematical description of the radiation transport problem can be written as follows (Pomraning, 1973):

$$\frac{1}{c}\frac{\partial}{\partial t} I(\mathbf{r}, v, \mathbf{\Omega}, t) + \mathbf{\Omega} \cdot \nabla I(\mathbf{r}, v, \mathbf{\Omega}, t) + \sigma(\mathbf{r}, v, T) I(\mathbf{r}, v, \mathbf{\Omega}, t)$$
$$= \sigma(\mathbf{r}, v, T) B(v, T), \tag{38}$$

$$\frac{\partial}{\partial t} u_e(\mathbf{r}, t) = \beta(\mathbf{r}, T) \left[\int_0^\infty \sigma(\mathbf{r}, v', T) I_0(\mathbf{r}, v', t) \, dv' - c\sigma_p(\mathbf{r}, T) u_e(\mathbf{r}, t) + W(\mathbf{r}, t) \right], \tag{39}$$

$$B(v, T) = (2hv^3/c^2)(e^{hv/kT} - 1)^{-1}, \tag{40}$$

$$u_e(\mathbf{r}, t) = aT^4(\mathbf{r}, t) \tag{41}$$

$$I_0(\mathbf{r}, v, t) = \int I(\mathbf{r}, v, \mathbf{\Omega}, t) \, d^2\Omega, \tag{42}$$

$$\sigma_p(\mathbf{r}, T) = \int_0^\infty B(v, T)\sigma(\mathbf{r}, v, T) \, dv \Big/ \int_0^\infty B(v, T) \, dv. \tag{43}$$

Here $I(\mathbf{r}, v, \mathbf{\Omega}, t)$ is the specific intensity of radiation, defined such that $\mathbf{n} \cdot \mathbf{\Omega} I \, dv \, d^2\Omega \, dS \, dt$ equals the amount of radiative energy transported across a surface element dS with normal $\mathbf{n}$ at space point $\mathbf{r}$, in the frequency range dv about v, in the solid angle $d^2\Omega$ about $\mathbf{\Omega}$, and in the time interval dt about t. Also, $T(\mathbf{r}, t)$ is the material temperature at $\mathbf{r}$ and t, and $\sigma(\mathbf{r}, v, T)$ is the macroscopic absorption coefficient or cross section, defined such that if the material temperature is T, then the probability that a photon at $(\mathbf{r}, v)$ will be absorbed in a distance ds is $\sigma \, ds$. In addition, B is the Planck function, satisfying

$$\int_0^\infty B \, dv = (ac/4\pi)T^4 = (c/4\pi)u_e,$$

where a is the radiation constant, u_e the equilibrium radiative energy density, σ_p the Planck mean absorption coefficient (or cross section), and the right side of Eq. (38) is the thermal emission source. Finally, we take $W(\mathbf{r}, t)$, an external source of energy to the matter, and $\beta^{-1}(\mathbf{r}, t) = \partial u_m/\partial u_e$, where u_m is the material energy density, to be known functions.

Equation (38) is the equation of radiative transfer under the assumption of LTE, and Eq. (39) expresses the energy balance between the matter and the radiation field. We note that the Planck function emission source and cross sections are functions of temperature, which is obtained from Eqs. (39) and (41). These couplings can be conveniently divided into two categories, *strong* and *weak*. Strong coupling refers to the temperature dependence of the emission source in Eq. (38) and the intensity dependence of the temperature in Eqs. (39) and (41); weak coupling refers to the temperature dependence of the cross sections σ and σ_p. It turns out that an efficient solution strategy is sensitive to the strong coupling but is relatively insensitive to the weak. Thus, we can formulate an iteration strategy that focuses on the strong coupling and treats the weak coupling as secondary.

In the formulation of a solution method for Eqs. (38) and (39), we have the following three objectives:

(a) We wish to combine the transfer and energy balance equations to obtain a single transfer-like equation for I, which contains information from the energy balance equation.

(b) We want the condition of radiative equilibrium $[I(\mathbf{r}, v, \mathbf{\Omega}, t) = B(v, T)]$ to be satisfied at all steps if the conditions so merit.

(c) We want the effectiveness of the solution techniques to be independent of the time-step size; this means that the equations must be solved implicitly in time. Here the concepts of strong and weak coupling, discussed above, play an essential role.

B. Discretization in Time

To accommodate these objectives, we use a backward Euler time differencing of Eqs. (38) and (39) to obtain

$$\frac{1}{c\,\Delta t^n}\left[I^{n+1}(\mathbf{r},v,\mathbf{\Omega}) - I^n(\mathbf{r},v,\mathbf{\Omega})\right] + \mathbf{\Omega}\cdot\nabla I^{n+1}(\mathbf{r},v,\mathbf{\Omega}) + \sigma(\mathbf{r},v,T^{n*})I^{n+1}(\mathbf{r},v,\mathbf{\Omega})$$

$$= \sigma(\mathbf{r},v,T^{n*})B(v,T^{n+1}), \qquad (44)$$

$$\frac{1}{\Delta t^n}\left[u_e^{n+1}(\mathbf{r}) - u_e^n(\mathbf{r})\right]$$

$$= \beta(\mathbf{r},T^{n*})\left[\int_0^\infty \sigma(\mathbf{r},v',T^{n*})I_0^{n+1}(\mathbf{r},v')\,dv' - c\sigma_p(\mathbf{r},T^{n*})u_e^{n+1}(\mathbf{r}) + W^n(\mathbf{r})\right],$$

$$(45)$$

and

$$u_e^{n+1}(\mathbf{r}) = a[T^{n+1}(\mathbf{r})]^4, \qquad (46)$$

where $\Delta t^n = t^{n+1} - t^n$ is the time step, and $T^{n*}(\mathbf{r})$ is an approximation to $T^{n+1}(\mathbf{r})$ described below. Equations (44)–(46) are three coupled equations in the three $n+1$ unknowns, I^{n+1}, T^{n+1}, and u_e^{n+1}. To rewrite these equations

4. Diffusion-Synthetic Acceleration

in a more suitable form, we define

$$B(v, T) = (c/4\pi)u_e(\mathbf{r}, t)b(v, T), \tag{47a}$$

where $b(v, T)$ is the Planckian, normalized by

$$\int_0^\infty b(v, T)\, dv = 1. \tag{47b}$$

Now, using the fact that the T variation of b is weaker than that of B, we introduce the approximation

$$B(v, T^{n+1}) \approx (c/4\pi)u_e^{n+1}(\mathbf{r})b(v, T^{n*})$$

in Eq. (44), solve Eq. (45) for u_e^{n+1}, and then eliminate u_e^{n+1} in the first equation. This results in the system

$$\boldsymbol{\Omega} \cdot \nabla I^{n+1}(\mathbf{r}, v, \boldsymbol{\Omega}) + [\sigma(\mathbf{r}, v, T^{n*}) + \tau^n]I^{n+1}(\mathbf{r}, v, \boldsymbol{\Omega})$$
$$= \frac{\chi(\mathbf{r}, v, T^{n*})}{4\pi[1 + \tau^n \alpha(\mathbf{r}, T^{n*})]}$$
$$\times \left[\int_0^\infty \sigma(\mathbf{r}, v', T^{n*})I_0^{n+1}(\mathbf{r}, v')\, dv' + W^n(\mathbf{r}) + \frac{u_e^n(\mathbf{r})}{\Delta t^n \beta(\mathbf{r}, T^{n*})}\right] + \tau^n I^n(\mathbf{r}, v, \boldsymbol{\Omega}), \tag{48}$$

$$u_e^{n+1}(\mathbf{r}) = \frac{1}{c\sigma_p(\mathbf{r}, T^{n*})[1 + \tau^n \alpha(\mathbf{r}, T^{n*})]}$$
$$\times \left[\int_0^\infty \sigma(\mathbf{r}, v', T^{n*})I_0^{n+1}(\mathbf{r}, v')\, dv' + W^n(\mathbf{r}) + \frac{u_e^n(\mathbf{r})}{\Delta t^n \beta(\mathbf{r}, T^{n*})}\right], \tag{49}$$

$$T^{n+1}(\mathbf{r}) = [a^{-1}u_e^{n+1}(\mathbf{r})]^{1/4}, \tag{50}$$

where

$$\tau^n = 1/c\,\Delta t^n, \qquad \alpha(\mathbf{r}, T^{n*}) = 1/\sigma_p(\mathbf{r}, T^{n*})\beta(\mathbf{r}, T^{n*}),$$

and

$$\chi(\mathbf{r}, v, T^{n*}) = \sigma(\mathbf{r}, v, T^{n*})b(v, T^{n*})/\sigma_p(\mathbf{r}, T^{n*})$$
$$= \sigma(\mathbf{r}, v, T^{n*})B(v, T^{n*}) \bigg/ \int_0^\infty \sigma(\mathbf{r}, v', T^{n*})B(v', T^{n*})\, dv'.$$

We note that for all $\mathbf{r}$ and T^{n*},

$$\int_0^\infty \chi\, dv = 1.$$

Equations (48)–(50) enable us to solve the coupled problem in a manner that meets all of the objectives stated above. That is, Eq. (48) contains the energy balance equation explicitly, the equilibrium solution exists if the conditions so merit, the backward Euler time differencing guarantees stability of the solution independent of the time step, and finally, Eqs. (48)–(50) are in a computationally advantageous form for treating the strong and weak couplings discussed above.

This final item occurs for three reasons. First, Eqs. (48)–(50) are in a triangular form for the $n + 1$ unknowns; i.e., if Eq. (48) is solved for I^{n+1}, then Eq. (49) explicitly gives u_e^{n+1}, and then Eq. (50) explicitly gives T^{n+1}. Second, Eq. (48) is amenable to the acceleration methods described in Section II. (Therefore, the strong couplings can be treated efficiently.) Third, the weak temperature couplings of various parameters in Eqs. (48)–(50) are depicted by use of the symbol T^{n*}, and so the following temperature iteration can be devised. We initially choose $T^{n*} = T^n$, evaluate the parameters, solve Eqs. (48)–(50) to obtain (the first estimate for) T^{n+1}, set T^{n*} equal to this estimate, and then repeat the process until some convergence criterion is met.[†] When the process is fully converged, the final values of T^{n*} and T^{n+1} are equal. We refer to this iteration procedure as the parameter iteration, and we shall return to it when we describe the solution procedure for the fully discretized equations. We now discuss the method of solving Eq. (48), assuming that the parameters evaluated at T^{n*} are held fixed.

To simplify the discussion, we write Eq. (48) as

$$\mathbf{\Omega} \cdot \nabla I(\mathbf{r}, v, \mathbf{\Omega}) + [\sigma(\mathbf{r}, v) + \tau] I(\mathbf{r}, v, \mathbf{\Omega})$$

$$= \frac{\chi(\mathbf{r}, v)}{4\pi[1 + \alpha(\mathbf{r})\tau]} \int_0^\infty \sigma(\mathbf{r}, v') I_0(\mathbf{r}, v') \, dv' + Q(\mathbf{r}, v, \mathbf{\Omega}), \qquad (51)$$

where

$$Q^{n*}(\mathbf{r}, v, \mathbf{\Omega}) = \frac{\chi(\mathbf{r}, v, T^{n*})}{4\pi[1 + \alpha(\mathbf{r}, T^{n*})\tau^n]} \left[W^n(\mathbf{r}) + \frac{u_e^n(\mathbf{r})}{\Delta t^n \beta(\mathbf{r}, T^{n*})} \right] + \tau^n I^n(\mathbf{r}, v, \mathbf{\Omega}),$$

and where, in Eq. (51), we dropped the indices n^* and $n + 1$, and have defined the quantity $Q(\mathbf{r}, v, \mathbf{\Omega})$, which is known during the time step. We are now in a position to use the developments presented in Section II to solve Eq. (51). However, in computing such a solution, we work with the equation discretized on a mesh of the independent variables, and then we apply the iteration procedures of Section II. Thus, in the following, we briefly present the considerations we have used to discretize Eq. (51), and then we describe the procedure used to solve the resulting discretized equations.

[†] Alternatively, one can extrapolate (linearly, logarithmically) a value of T^{n*} from previous time steps and obtain a more accurate initial solution.

4. Diffusion-Synthetic Acceleration

C. Discretization in Frequency

The frequency variation of Eq. (51) is discretized by the so-called multifrequency (multigroup) method, which involves application of the operator

$$\int_{v_{g+1/2}}^{v_{g-1/2}} (\cdot) \, dv$$

to the equation, where $v_{g+1/2}$, $g = 0, 1, \ldots, G$ are the group boundaries, G is the number of groups, and we use the convention that $v_{1/2}$ is the highest and $v_{G+1/2}$ the lowest frequency. Performing this operation and making certain approximations, we obtain the following set of equations:

$$\Omega \cdot \nabla I_g(\mathbf{r}, \mathbf{\Omega}) + [\sigma_g(\mathbf{r}) + \tau] I_g(\mathbf{r}, \mathbf{\Omega})$$

$$= \frac{\chi_g(\mathbf{r})}{4\pi[1 + \alpha(\mathbf{r})\tau]} \sum_{g'=1}^{G} \sigma_{g'}(\mathbf{r}) I_{0g'}(\mathbf{r}) + Q_g(\mathbf{r}, \mathbf{\Omega}), \tag{52}$$

where

$$I_g(\mathbf{r}, \mathbf{\Omega}) = \int_{v_{g+1/2}}^{v_{g-1/2}} I(\mathbf{r}, v, \mathbf{\Omega}) \, dv,$$

$$\sigma_g(\mathbf{r}) = \int_{v_{g+1/2}}^{v_{g-1/2}} \sigma_g(\mathbf{r}, v) \omega(v) \, dv \bigg/ \int_{v_{g+1/2}}^{v_{g-1/2}} \omega(v) \, dv,$$

$$\chi_g(\mathbf{r}) = \int_{v_{g+1/2}}^{v_{g-1/2}} \chi(\mathbf{r}, v) \, dv,$$

$$Q_g(\mathbf{r}, \mathbf{\Omega}) = \int_{v_{g+1/2}}^{v_{g-1/2}} Q(\mathbf{r}, v, \mathbf{\Omega}) \, dv,$$

and $\omega(v)$ is a spectral weighting function, usually chosen to have a similar frequency variation to that presumed for I.

A wealth of experience has shown that the accuracy of the multifrequency discretization depends on the choice of the weighting functions that define the multigroup parameters (Pomraning, 1973; Cullen and Pomraning, 1980). It is beyond the scope of this chapter to delve into the details of the choice of this weighting function. Of course, the number of groups also affects the accuracy of the approximation; increasing the number of groups tends to reduce the requirements upon the weighting function for cross sections that may vary smoothly in v. For cross sections with lines (resonances or edges), special considerations are needed to use the weighting functions effectively and to choose group boundaries (Cullen and Pomraning, 1980).

D. Discretization in Angle

We discretize Eq. (52) over the angular variable $\mathbf{\Omega}$ by subdividing the unit sphere into cones C_m, $1 \leq m \leq M$, choosing a representative direction (unit vector) $\mathbf{\Omega}_m$ in each C_m, and defining the weights w_m as the area of the part of

C_m on the unit sphere. The number and choice of the discrete directions should be such that using only these directions, the angular distribution of the flux is adequately represented. If we define

$$I_{g,m}(\mathbf{r}) = \frac{1}{w_m} \int_{C_m} I_g(\mathbf{r}, \mathbf{\Omega}) \, d^2\Omega,$$

then we can write the so-called discrete-ordinates form of Eq. (52) as

$$[\mathbf{\Omega} \cdot \nabla I_g(\mathbf{r}, \mathbf{\Omega})]_m + [\sigma_g(\mathbf{r}) + \tau] I_{g,m}(\mathbf{r})$$

$$= \frac{\chi_g(\mathbf{r})}{4\pi[1 + \alpha(\mathbf{r})\tau]} \sum_{g'=1}^{G} \sigma_{g'}(\mathbf{r}) I_{0g'}(\mathbf{r}) + Q_{g,m}(\mathbf{r}), \quad (53)$$

$$1 \leq m \leq M, \quad 1 \leq g \leq G;$$

$$I_{0g}(\mathbf{r}) = \sum_{m=1}^{M} I_{g,m}(\mathbf{r}) w_m;$$

$$4\pi = \sum_{m=1}^{M} w_m.$$

We have left the discrete-ordinates representation of the streaming term in Eq. (53) unspecified because its form depends on the spatial geometry and is usually treated in detail with the spatial discretization method. Much has been written on the appropriate choice of the discrete directions and the associated weights (Carlson and Lathrop, 1968; Alcouffe and O'Dell, 1983), and we do not go into the detail required for complete understanding of these choices in this chapter. However, the general criteria used to select a discrete-ordinates set can be summarized as follows:

(a) The physical symmetries of the problem should be preserved as much as possible.

(b) A maximal number of spherical harmonics should be integrated exactly.

(c) Angular derivatives in curvilinear geometries should be simply approximated.

Other criteria may be added to account for special circumstances. The accuracy of the resulting representation is a function of both the choice of discrete ordinates and their number, whereas the cost of the transport solution increases in direct proportion to the number of angles M.

E. A Simple Spatial Discretization

To conclude the discretization process, we must specify a geometry and a method of spatial discretization. To explain most simply the principles

4. Diffusion-Synthetic Acceleration

involved in solving Eq. (53) by our numerical algorithm, we choose one-dimensional slab (planar) geometry. In this case the discrete-ordinates representation of the streaming term is especially simple; i.e.,

$$[\mathbf{\Omega} \cdot \nabla I_g(\mathbf{r}, \mathbf{\Omega})]_m = \mu_m (\partial/\partial x) I_{g,m}(x), \tag{54}$$

where μ_m is the cosine of the angle between $\mathbf{\Omega}_m$ and the positive x axis, and x the distance variable. We define a spatial mesh $\{x_{i+1/2}, 0 \leq i \leq IT\}$ such that the ith mesh cell is defined by $x_{i-1/2} \leq x \leq x_{i+1/2}$, and we integrate Eq. (53) together with Eq. (54) over this cell to obtain

$$\frac{\mu_m}{\Delta x_i}(I_{g,m,i+1/2} - I_{g,m,i-1/2}) + (\sigma_{gi} + \tau)I_{g,m,i}$$

$$= \frac{\chi_{g,i}}{2(1+\alpha_i \tau)} \sum_{g'=1}^{G} \sigma_{g',i} I_{0g',i} + Q_{g,m,i}, \tag{55}$$

$$1 \leq i \leq IT, \quad 1 \leq m \leq M, \quad 1 \leq g \leq G,$$

where

$$\Delta x_i = x_{i+1/2} - x_{i-1/2},$$

$$I_{g,m,i} = \frac{1}{\Delta x_i} \int_{x_{i-1/2}}^{x_{i+1/2}} I_{g,m}(x)\, dx,$$

$$\sigma_{g,i} = \frac{1}{\Delta x_i} \int_{x_{i-1/2}}^{x_{i+1/2}} \sigma_g(x, T_i)\, dx,$$

$$\chi_{g,i} = \frac{1}{\Delta x_i} \int_{x_{i-1/2}}^{x_{i+1/2}} \chi_g(x, T_i)\, dx,$$

$$\alpha_i = \frac{1}{\Delta x_i} \int_{x_{i-1/2}}^{x_{i+1/2}} \alpha(x, T_i)\, dx,$$

$$T_i = \left(a^{-1} \int_{x_{i-1/2}}^{x_{i+1/2}} u_e(x)\, dx\right)^{1/4},$$

$$Q_{g,m,i} = \frac{1}{\Delta x_i} \int_{x_{i-1/2}}^{x_{i+1/2}} Q_{g,m}(x)\, dx,$$

and

$$I_{0g,i} = \sum_{m=1}^{M} I_{g,m,i} w_m,$$

where the weights w_m have now been normalized in the conventional manner for one-dimensional quadrature sets; i.e.,

$$2 = \sum_{m=1}^{M} w_m.$$

Equation (55), which expresses a balance condition over each mesh cell, is the fundamental equation of transport in the discretized phase space. Approximations have been made, for example, in defining the parameters to be evaluated at the cell-averaged temperature T_i, which itself is obtained from a discretized energy balance equation. The boundary condition for Eq. (55) can be either a specified incident intensity

$$I_{g,m,i_b} = \Lambda_{g,m} \quad \text{for} \quad \mu_m \quad \text{directed into the system} \tag{56}$$

or specular reflection

$$\mu_m = -\mu_n \rightarrow I_{g,m,i_b} = I_{g,n,i_b},$$

where i_b is the problem boundary. (This latter condition illustrates one example of the symmetry required for discrete-ordinates sets; in this case, the directions must be antisymmetric about $\mu = 0$.)

We still do not have sufficient information to solve Eq. (55), since it is one equation with two unknowns per cell: the cell-averaged intensity and the outgoing cell-edge intensity. (The incoming cell-edge intensity is known either from the boundary conditions or from the preceding cell calculation.) A particularly simple way to resolve this is to add an equation expressing an assumed (approximate) relationship between the cell-edge and cell-average intensities. In particular, we assume a linear (also called "diamond") relationship such that

$$I_{g,m,i} = \tfrac{1}{2}(I_{g,m,i-1/2} + I_{g,m,i+1/2}). \tag{57}$$

Equations (55) and (57), together with the boundary conditions, completely specify our discretized problem.

F. Iteration Method for the Discretized Transport Equation

1. Development

We are now at the point where we can apply the methodology developed in Section II. We can write an SI method for Eqs. (55) and (57) as

$$\frac{\mu_m}{\Delta x_i}(I^{l+1/2}_{g,m,i+1/2} - I^{l+1/2}_{g,m,i-1/2}) + (\sigma_{gi} + \tau)I^{l+1/2}_{g,m,i}$$

$$= \frac{\chi_{g,i}}{2(1+\alpha_i\tau)} \sum_{g'=1}^{G} \sigma_{g',i}I^{l}_{0g',i} + Q_{g,m,i}, \tag{58a}$$

4. Diffusion-Synthetic Acceleration

$$I_{g,m,i}^{l+1/2} = \frac{1}{2}(I_{g,m,i+1/2}^{l+1/2} + I_{g,m,i-1/2}^{l+1/2}), \tag{58b}$$

$$I_{0g,i}^{l+1} = I_{0g,i}^{l+1/2} \equiv \sum_{m=1}^{M} I_{g,m,i}^{l+1/2} w_m, \tag{59}$$

where l is an iteration index. This iteration procedure has the same convergence properties as those noted for Eqs. (4a–c), and hence there is a need for convergence acceleration as $\tau = 1/c \, \Delta t$ becomes small. We shall use the same ideas that led to the development of Eq. (16) to obtain our discretized acceleration equation. Thus, we take the first two angular moments of Eq. (58a) to obtain

$$\frac{1}{\Delta x_i}(I_{1g,i+1/2}^{l+1/2} - I_{1g,i-1/2}^{l+1/2}) + (\sigma_{gi} + \tau)I_{0g,i}^{l+1/2}$$

$$= \frac{\chi_{g,i}}{1 + \alpha_i \tau} \sum_{g'=1}^{G} \sigma_{g',i} I_{0g',i}^{l} + Q_{0g,i}, \tag{60a}$$

$$\frac{1}{3\Delta x_i}(I_{0g,i+1/2}^{l+1/2} - I_{0g,i-1/2}^{l+1/2}) + \frac{2}{3\Delta x_i}(I_{2g,i+1/2}^{l+1/2} - I_{2g,i-1/2}^{l+1/2})$$

$$+ (\sigma_{g,i} + \tau)I_{1g,i}^{l+1/2} = Q_{1g,i}. \tag{60b}$$

Here we have defined

$$I_{ng}^{l+1/2} = \sum_{m=1}^{M} P_n(\mu_m) I_{mg}^{l+1/2} w_m, \quad n = 0, 1, 2, \tag{60c}$$

where P_n is the nth Legendre polynomial. For our acceleration equations [which replace Eq. (59)], we define

$$\frac{1}{\Delta x_i}(I_{1g,i+1/2}^{l+1} - I_{1g,i-1/2}^{l+1}) + (\sigma_{gi} + \tau)I_{0g,i}^{l+1}$$

$$= \frac{\chi_{g,i}}{1 + \alpha_i \tau} \sum_{g'=1}^{G} \sigma_{g',i} I_{0g',i}^{l+1} + Q_{0g,i}, \tag{61a}$$

$$\frac{1}{3\Delta x_i}(I_{0g,i+1/2}^{l+1} - I_{0g,i-1/2}^{l+1}) + \frac{2}{3\Delta x_i}(I_{2g,i+1/2}^{l+1/2} - I_{2g,i-1/2}^{l+1/2})$$

$$+ (\sigma_{g,i} + \tau)I_{1g,i}^{l+1} = Q_{1g,i}, \tag{61b}$$

$$I_{ng,i}^{l+1} = \frac{1}{2}(a_{ng,i}^{l+1/2} I_{ng,i+1/2}^{l+1} + b_{ng,i}^{l+1/2} I_{ng,i-1/2}^{l+1}), \quad n = 0, 1, \tag{61c}$$

where the final two equations (61c) were derived from Eq. (58b), with coefficients $a_{ng,i}^{l+1/2}$ and $b_{ng,i}^{l+1/2}$ introduced to account for the possibility that Eq.

(58b) is not satisfied for all $g, m,$ and i. If Eq. (58b) is strictly satisfied, then $a_{ng,i}^{l+1/2} = b_{ng,i}^{l+1/2} = 1$. The full definition of these coefficients, together with reasons why one might not always wish to invoke Eq. (58b), are given later.

We can now combine Eqs. (61a–c) and (60b) and eliminate unknowns to obtain the following discretized diffusion equation for $I_{0g,i+1/2}^{l+1}$:

$$-\frac{1}{3}\left[\frac{1}{\delta_{1g,i+1}}(I_{0g,i+3/2}^{l+1} - I_{0g,i+1/2}^{l+1}) - \frac{1}{\delta_{1g,i}}(I_{0g,i+1/2}^{l+1} - I_{0g,i-1/2}^{l+1})\right]$$

$$+\frac{1}{2}[\beta_{1g,i+1}\hat{\sigma}_{g,i+1}(a_{0g,i+1}I_{0g,i+3/2}^{l+1} + b_{0g,i+1}I_{0g,i+1/2}^{l+1})$$

$$+ (1-\beta_{1g,i})\hat{\sigma}_{gi}(a_{0g,i}I_{0g,i+1/2}^{l+1} + b_{0g,i}I_{0g,i-1/2}^{l+1})]$$

$$-\frac{1}{2}\left[\beta_{1g,i+1}\Delta x_{i+1}\frac{\chi_{g,i+1}}{1+\alpha_{i+1}\tau}\sum_{g'=1}^{G}\sigma_{g',i+1}\right.$$

$$\times (a_{0g',i+1}I_{0g',i+3/2}^{l+1} + b_{0g',i+1}I_{0g',i+1/2}^{l+1})$$

$$+ (1-\beta_{1g,i})\Delta x_i \frac{\chi_{g,i}}{1+\alpha_i\tau}\sum_{g'=1}^{G}\sigma_{g',i}(a_{0g',i}I_{0g',i+1/2}^{l+1} + b_{0g',i}I_{0g',i-1/2}^{l+1})\right]$$

$$= \beta_{1g,i+1}\Delta x_{i+1}Q_{0g,i+1} + (1-\beta_{1g,i})\Delta x_i Q_{0g,i}$$

$$-[\beta_{1g,i+1}I_{1g,i+3/2}^{l+1/2} + (1-\beta_{1g,i+1}-\beta_{1g,i})I_{1g,i+1/2}^{l+1/2} - (1-\beta_{1g,i})I_{1g,i-1/2}^{l+1/2}]$$

$$-\frac{1}{3}\left[\frac{1}{\delta_{1g,i+1}}(I_{0g,i+3/2}^{l+1/2} - I_{0g,i+1/2}^{l+1/2}) - \frac{1}{\delta_{1g,i}}(I_{0g,i+1/2}^{l+1/2} - I_{0g,i-1/2}^{l+1/2})\right],$$

(62a)

where

$$\beta_{1g,i} = a_{1g,i}/(a_{1g,i} + b_{1g,i}), \quad (62b)$$

$$\delta_{1g,i} = \tfrac{1}{2}\Delta x_i(\sigma_{g,i} + \tau)(a_{1g,i} + b_{1g,i}), \quad (62c)$$

$$\hat{\sigma}_{g,i} = (\sigma_{g,i} + \tau)\Delta x_i. \quad (62d)$$

Equations (62a–d) are the discrete analog to Eq. (16), and it can be shown that the iteration comprising Eqs. (58), (60c), and (62) (with $a_{ng,i}^{l+1/2} = b_{ng,i}^{l+1/2} = 1$; the case in which this is not true is discussed below) has an eigenvalue whose expression is the discrete analog of that given in Eq. (19). Also, it is easily shown that for any choice of $a_{1g,i} > 0$ and $b_{1g,i} > 0$, Eq. (62a) has the solution $I_{0g,i+1/2}^{l+1} = I_{0g,i+1/2}^{l+1/2}$, provided that the balance equation is satisfied in every cell. Therefore, if we take $a_{1g,i} = b_{1g,i} = 1$ for all cells [which we do below, even though it is incorrect for cells for which Eq. (58b) is not satisfied], the acceleration Eqs. (62a–d) still admit the correct transport solution.

2. Numerical Results Assuming Diamond Differencing

To reiterate, for the situation with continuous independent variables, the analysis performed in Section II predicts a spectral radius of 0.164 for an infinite homogeneous medium with opacities described by Eqs. (20a, b). In order to depict the spectral radius for the discretized equations and a finite, inhomogeneous medium, a test problem is solved using Eqs. (58), (60c), and (62) with $a_{ng,i} = b_{ng,i} = 1$. The test problem is a 20-cm slab of material with the Eq. (20) opacities, $a_0 = 1/kT$, and $a_1 = 27$. The initial material temperature is 1 eV. For $t \geq 0$, a 1-keV blackbody (i.e., Planckian) radiation source emits in the positive μ directions at $x = 0$ and a 1-eV blackbody source emits in the negative μ directions at $x = 20$ cm.

The solution is obtained for an infinite time step ($\tau = 0$), where we expect the slowest convergence for linear transport. To calculate temperatures for an opacity evaluation, an initial 1-group diffusion calculation is performed. The final temperature distribution is shown in Fig. 1; clearly, this problem is an inhomogeneous one. The empirical spectral radii derived from the observed convergence rates for various multigroup and S_N discretizations are shown in Table I.

We see from the results in Table I that the spectral radius of convergence is not adversely affected by the discretization of the problem, or by the temperature and opacity gradients illustrated in Fig. 1. In fact, all the empirical spectral radii are smaller than the value obtained by the Fourier analysis. We note that the spectral radius is insensitive to the number of frequency groups and increases with increasing S_N order.

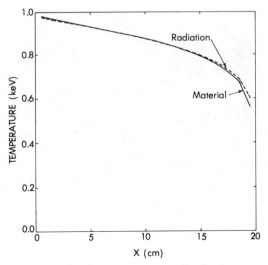

Fig. 1. Steady-state temperature distributions.

Table I
Spectral Radii for an Accelerated Linear Transport Problem[a]

Number of frequency groups	S_N Order		
	S_4	S_8	S_{16}
20	0.111	0.139	0.146
40	0.111	0.137	0.143
80	0.110	0.135	0.142

[a] The theoretical radius from Fourier analysis is 0.164. All of the spectral radii for unaccelerated iterations are greater than 0.93.

3. Numerical Results Using Diamond with Fixup

We now discuss a practical added complication to this formalism. The simplicity of the diamond assumption of Eq. (58b) is marred by the fact that the resulting extrapolated boundary intensity is not guaranteed to be positive. For one-dimensional slab geometry, it is readily demonstrated that if the space and angular mesh are such that $(\sigma_{g,i} + \tau) \cdot \Delta x_i / |\mu_m| > 2$, then a negative angular flux is possible. In radiation hydrodynamic situations, the source is generally strong enough to prevent negative intensities, the common exceptions being at thermal wave fronts, where steep temperature gradients exist, and at small times when computing the evolution of an initial condition into cold material. However, it is essential that a positive solution be obtained in every situation; thus, we have devised methods to retain much of the simplicity of diamond difference and yet guarantee a positive transport solution.

To illustrate the method we use to correct negative intensities, we return to the method used to develop a solution to Eqs. (58a, b). If $\mu_m > 0$, then $I_{g,m,i-1/2}^{l+1/2}$ is known from the solution of the $i - 1$ cell. We use Eq. (58b) to eliminate the unknown boundary intensity $I_{g,m,i+1/2}^{l+1/2}$ in Eq. (58a), an equation for the cell-averaged intensity $I_{g,m,i}^{l+1/2}$, which is nonnegative assuming that $I_{g,m,i-1/2}^{l+1/2} \geq 0$ and the right-hand side is nonnegative. We then obtain the exiting boundary intensity $I_{g,m,i+1/2}^{l+1/2}$ from Eq. (58b). If this value is negative, then we resort to what we call a negative intensity fixup. Our experience has shown that for the time-dependent radiation transport equation, the fixup scheme should be based on a quantity $\xi_i \equiv c \Delta t / \Delta x_i$ to ensure accuracy and stability in our iteration acceleration methods. Thus, if the intensity extrapolates negative from Eq. (58b), we replace Eq. (58b) with the following:

$$I_{g,m,i+1/2}^{l+1/2} = \begin{cases} 0 & \text{if } \xi_i < 1, \\ I_{g,m,i}^{l+1/2} & \text{if } \xi_i \geq 1. \end{cases} \tag{63}$$

4. Diffusion-Synthetic Acceleration

The solution is completed by substituting Eq. (63) into Eq. (58a) and solving again for the cell-averaged intensity $I_{g,m,i}^{l+1/2}$, thus ensuring particle conservation. A further complication is that for $1 \leq \xi \leq 30$, the use of Eqs. (58b) or (63), depending on the circumstances, can set up oscillations in the value of the intensities from one iteration to the next. In order to avoid such oscillatory behavior, we have observed that, if after a given number of iterations l^* we always use Eq. (63) for $l > l^*$ for each (g, m, i) when it was used at iterate l^* or at any time afterward, we then obtain convergence. We illustrate this in our next example.

In order to incorporate this fixup method into our transport iteration acceleration method, we return to Eq. (61c) and define our values for the coefficients a and b appearing therein. From a multitude of possible choices, we have found that the following definitions are, at present, the most effective:

$$a_{0g,i}^{l+1/2} = \frac{I_{0g,i}^{l+1/2}}{I_{0g,i+1/2}^{l+1/2}}, \quad b_{0g,i}^{l+1/2} = \frac{I_{0g,i}^{l+1/2}}{I_{0g,i-1/2}^{l+1/2}}, \quad a_{1g,i}^{l+1/2} = b_{1g,i}^{l+1/2} = 1, \quad (64a)$$

if a negative fixup is used in cell i for group g, or

$$a_{ng,i}^{l+1/2} = b_{ng,i}^{l+1/2} = 1, \quad n = 0, 1, \quad (64b)$$

if no negative fixup is used in cell i and group g.

Definition (64), used in conjunction with the diffusion acceleration equations (62), accounts for the use of negative intensity fixups. We note that Eqs. (63) and (64) are nonlinear and that these nonlinearities appear in Eq. (62a) through $\delta_{1g,i}$, $\beta_{1g,i}$, and $a_{0g,i}$, $b_{0g,i}$ (which, for simplicity, we have not indexed by $l + \frac{1}{2}$). Therefore, the stability of the resulting iteration method cannot be analytically analyzed, and so we present some convergence results, obtained numerically (in Figs. 3 and 4) for problems where the fixup is used.

Thus, we use a second example problem to illustrate the convergence behavior of our transport iterations when a negative fixup is employed. We use the same 20-cm slab of material as before.

The problem is discretized with 20 equally spaced spatial zones, 30 multifrequency groups, and S_8 discrete ordinates. The initial conditions assume an exponential temperature spatial shape ranging from 1 keV to 1 eV, left to right, in the slab; at the left boundary there is a 1-keV blackbody radiation source and at the right boundary there is a 1-eV radiation source. Our example transport solution is obtained after one time step where the opacities have been evaluated at some intermediate temperature T_o^*, determined from a 1-group diffusion solution for that time step (see Fig. 2). We chose this as a test problem because it resembles the situation in the evolution of time-dependent problems with a thermal wave. We can now choose a variety of time-step sizes (as we shall indicate in our time-dependent examples) that trigger various interesting situations in the employment of the negative fixup

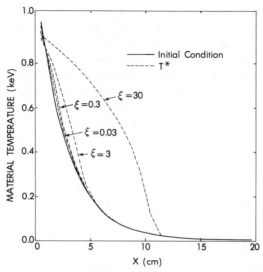

Fig. 2. Initial condition and 1-group diffusion results for the time-dependent problem.

method. Thus, in this example, we vary the size of $\xi_i = c\,\Delta t/\Delta x_i$ in the range $0.03 \leq \xi_i \leq 30$ to illustrate the behavior of the fixup.

Because the fixup process is nonlinear, it is not useful to characterize the iteration convergence behavior by a single number, the spectral radius. That is, the convergence will not always be described by a constant factor from iteration to iteration; hence, we show the convergence behavior as a plot of the error as a function of iteration. In our case, we define the error at iterate l as

$$E^l = \max_{g,m,i} |(I_{g,m,i}^{l+1} - I_{g,m,i}^l)/I_{g,m,i}^l|. \tag{65}$$

In Fig. 3 we present a summary of our results for this example with values of $\xi = 0.03, 0.3, 3, 30$. In all cases fixup has been used. We show both the accelerated (solid lines) and the unaccelerated (dashed lines) convergence behavior as a function of iteration, and at this time we take note of the great improvement offered by acceleration. In fact, the accelerated convergence rates for this example seem to be bounded by a spectral radius of 0.16 for all values of ξ except $\xi = 3$. As we elaborate below, this is a special case because it falls into the range $1 \lesssim \xi \lesssim 30$. For the case of small time step ($\xi \lesssim 1$) and large time step ($\xi \gtrsim 30$), the fixup apparently does not interfere with the model problem convergence behavior.

As mentioned in the discussion following Eq. (63), we use a special procedure to stabilize the fixup when $1 \lesssim \xi \lesssim 30$. To illustrate the possible difficulties, we present further results on the $\xi = 3$ case in Fig. 4, where we give

4. Diffusion-Synthetic Acceleration

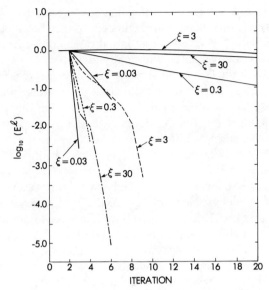

Fig. 3. Error reduction for accelerated and unaccelerated iterations.

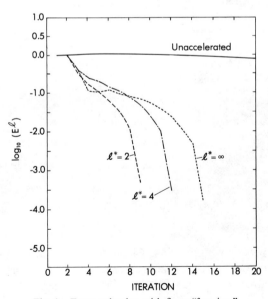

Fig. 4. Error reduction with fixup "freezing."

the convergence history for a variety of values of l^* from 2 to ∞ to demonstrate the effect of negative fixup and the impact of the l^* strategy. For this case, the problem with $l^* = \infty$ shows an initial rapid convergence, followed by a region of slower convergence, and finally rapid convergence. A similar pattern is repeated for the $l^* = 2$ and 4 cases, but convergence is faster. Thus, for the $l^* = 2$ case, we save almost a factor of 2 in iterating to our convergence criterion of 10^{-3} relative error. In our experience, this is typical of convergence behavior when $1 \leq \xi \lesssim 30$, i.e., we always get improved convergence behavior when $l^* \cong 2$. In fact, in some cases the iterations do not converge when $l^* = \infty$ because of fixup oscillations, but they will converge much as in Fig. 4 when $l^* = 2$.

We conclude that the convergence behavior, when fixup is used, is markedly different from our model problem predictions (Section II) when $1 \leq \xi \lesssim 30$, and this is caused by oscillations in the use of fixup from iteration to iteration. This latter is a highly nonlinear process and, thus, may yield unpredictable results; however, our strategy embodied in Eq. (63) yields acceptable results and leads to a satisfactory acceleration method. This is made clear from the unaccelerated convergence behavior shown in Figs. 3 and 4.

4. Acceleration of the Discretized Diffusion Equation

We now apply the formalism developed in Section II to the acceleration of the iterations involved in solving Eq. (62a) above. For our purposes here, we rewrite the SI convergence procedure for this diffusion equation as

$$-\frac{1}{3}\left[\frac{1}{\delta_{1g,i+1}}(I_{0g,i+3/2}^{k+1/2} - I_{0g,i+1/2}^{k+1/2}) - \frac{1}{\delta_{1g,i}}(I_{0g,i+1/2}^{k+1/2} - I_{0g,i-1/2}^{k+1/2})\right]$$

$$+ \frac{1}{4}\left[\hat{\sigma}_{g,i+1}(a_{0g,i+1}I_{0g,i+3/2}^{k+1/2} + b_{0g,i+1}I_{0g,i+1/2}^{k+1/2})\right.$$

$$\left. + \hat{\sigma}_{g,i}(a_{0g,i}I_{0g,i+1/2}^{k+1/2} + b_{0g,i}I_{0g,i-1/2}^{k+1/2})\right]$$

$$= \frac{1}{4}\left[\frac{\Delta x_{i+1}\chi_{g,i+1}}{1+\alpha_{i+1}\tau}\sum_{g'=1}^{G}\sigma_{g',i+1}(a_{0g',i+1}I_{0g',i+3/2}^{k} + b_{0g',i+1}I_{0g',i+1/2}^{k})\right.$$

$$\left. + \frac{\Delta x_i \chi_{g,i}}{1+\alpha_i\tau}\sum_{g'=1}^{G}\sigma_{g',i}(a_{0g',i}I_{0g',i+1/2}^{k} + b_{0g',i}I_{0g',i-1/2}^{k})\right] + S_{g,i+1/2}.$$

(66)

In writing Eq. (66), we have used Eqs. (64a, b) to set $\beta_{1g,i}$ of Eq. (62) to $\frac{1}{2}$ and have gathered the noniterated terms coming from the transport corrections and the other source terms into $S_{g,i+1/2}$. The unaccelerated SI procedure takes

$$I_{0g,i+1/2}^{k+1} = I_{0g,i+1/2}^{k+1/2}, \quad i = 0,\ldots,IT, \quad g = 1,\ldots,G. \quad (67)$$

4. Diffusion-Synthetic Acceleration

As developed in Section II.C, an acceleration procedure is obtained for Eq. (66) by summing over the groups with a particular weighting function. To achieve results that are discrete analogs of Eqs. (23a, b) and (29), we make the following replacements:

$$f_{g,i}^{k*} \equiv (a_{0g,i}I_{0g,i+1/2}^{k*} + b_{0g,i}I_{0g,i-1/2}^{k*}) \to \theta_{g,i}^{k+1/2}(a_{0,i}I_{0,i+1/2}^{k+1} + b_{0,i}I_{0,i-1/2}^{k+1})$$
$$+ f_{g,i}^{k+1/2} - \theta_{g,i}^{k+1/2}(a_{0,i}I_{0,i+1/2}^{k+1/2} + b_{0,i}I_{0,i-1/2}^{k+1/2}), \tag{68a}$$

$$g_{g,i}^{k+1/2} \equiv (I_{0g,i+1/2}^{k+1/2} - I_{0g,i-1/2}^{k+1/2}) \to \varphi_{g,i}^{k+1/2}(I_{0,i+1/2}^{k+1} - I_{0,i-1/2}^{k+1})$$
$$+ g_{g,i}^{k+1/2} - \varphi_{g,i}^{k+1/2}(I_{0,i+1/2}^{k+1/2} - I_{0,i-1/2}^{k+1/2}), \tag{68b}$$

where $k^* = k + \frac{1}{2}$ or k as appropriate,

$$a_{0,i} = \sum_g a_{0g,i}I_{0g,i+1/2}^{k+1/2}/I_{0,i+1/2}^{k+1/2}, \tag{68c}$$

$$b_{0,i} = \sum_g b_{0g,i}I_{0g,i-1/2}^{k+1/2}/I_{0,i-1/2}^{k+1/2}, \tag{68d}$$

$$I_{0,i+1/2}^{k+1/2} = \sum_g I_{0g,i+1/2}^{k+1/2},$$

and $\theta_{g,i}^{k+1/2}$, $\varphi_{g,i}^{k+1/2}$ are specified spectral functions that can depend upon the iteration if desired.

Replacement in Eqs. (68a–d) refers to the process of putting these expressions into Eq. (66), summed over the groups. (Note that for algebraic convenience, our unknowns, $I_{0,i+1/2}^{k+1}$, are not σ-weighted sums, as they were in Section II.C). To obtain the acceleration equation, we make these substitutions and obtain:

$$-\frac{1}{3}\left[\sum_{g=1}^{G}\left(\frac{\varphi_{g,i+1}^{k+1/2}}{\delta_{1g,i+1}}\right)(I_{0,i+3/2}^{k+1} - I_{0,i+1/2}^{k+1}) - \sum_{g=1}^{G}\left(\frac{\varphi_{g,i}^{k+1/2}}{\delta_{1g,i}}\right)(I_{0,i+1/2}^{k+1} - I_{0,i-1/2}^{k+1})\right]$$
$$+ \frac{1}{4}\left[\zeta_{i+1}(a_{0,i+1}I_{0,i+3/2}^{k+1} + b_{0,i+1}I_{0,i+1/2}^{k+1}) + \zeta_i(a_{0,i}I_{0,i+1/2}^{k+1} + b_{0,i}I_{0,i-1/2}^{k+1})\right]$$
$$= \sum_{g=1}^{G} S_{g,i+1/2} + \frac{1}{3}\left[\sum_g\left(\frac{I_{0g,i+3/2}^{k+1/2} - I_{0g,i+1/2}^{k+1/2}}{\delta_{1g,i+1}}\right)\right.$$
$$- \sum_g\left(\frac{I_{0g,i+1/2}^{k+1/2} - I_{0g,i-1/2}^{k+1/2}}{\delta_{1g,i}}\right) - \sum_g\left(\frac{\varphi_{g,i+1}^{k+1/2}}{\delta_{1g,i+1}}\right)(I_{0,i+3/2}^{k+1/2} - I_{0,i+1/2}^{k+1/2})$$
$$\left. + \sum_g\left(\frac{\varphi_{g,i}^{k+1/2}}{\delta_{1g,i}}\right)(I_{0,i+1/2}^{k+1/2} - I_{0,i-1/2}^{k+1/2})\right]$$
$$- \frac{1}{4}\left[\sum_g \zeta_{g,i+1}f_{g,i+1}^{k+1/2} + \sum_g \zeta_{g,i}f_{g,i}^{k+1/2}\right.$$
$$- \zeta_{i+1}(a_{0,i+1}I_{0,i+3/2}^{k+1/2} + b_{0,i+1}I_{0,i+1/2}^{k+1/2})$$
$$\left. - \zeta_i(a_{0,i}I_{0,i+1/2}^{k+1/2} + b_{0,i}I_{0,i-1/2}^{k+1/2})\right], \tag{69a}$$

where

$$\zeta_{g,i} = \left(\frac{\alpha_i \sigma_{g,i}}{1 + \alpha_i \tau} + 1\right)\tau \Delta x_i, \tag{69b}$$

$$\xi_i = \sum_g \zeta_{g,i} \theta_{g,i}^{k+1/2}. \tag{69c}$$

These equations determine the quantities $I_{0,i+1/2}^{k+1}$. Using Eq. (68a), we obtain

$$\sum_{g=1}^{G} \sigma_{g,i} f_{g,i}^{k+1} = \sum_{g=1}^{G} \sigma_{g,i}[\theta_{g,i}^{k+1/2}(a_{0,i} I_{0,i+1/2}^{k+1} + b_{0,i} I_{0,i-1/2}^{k+1})$$

$$+ f_{g,i}^{k+1/2} - \theta_{g,i}^{k+1/2}(a_{0,i} I_{0,i+1/2}^{k+1/2} + b_{0,i} I_{0,i-1/2}^{k+1/2})], \tag{69d}$$

which specifies the right side of Eq. (66) for the next iteration.

Now, if we choose the spectral functions

$$\theta_{g,i}^{k+1/2} = \varphi_{g,i}^{k+1/2} = \frac{\chi_{g,i}/(\sigma_{g,i} + \tau)}{\sum_g \chi_{g,i}/(\sigma_{g,i} + \tau)}, \tag{70}$$

then we obtain a linear acceleration method that is a discrete analog of Eq. (27). For our model problem,[†] using Eqs. (69a–d) with Eq. (70) to accelerate Eq. (66), we compute an accelerated spectral radius of 0.65, which compares favorably with the theoretical value of 0.886 (of Section II). The unaccelerated convergence rate is 0.995.

For nonmodel problems (temperature discontinuities), we find difficulties with the linear method. That is, for large temperature discontinuities, it appears that the linear method is unstable. [However, in a different treatment of the linear method, based on accelerating a central, rather than edge-differenced diffusion equation, these difficulties are not observed; see Morel et al. (1984)]. To alleviate this problem, we use a nonlinear method, as suggested in Section II. That is, we set

$$\theta_{g,i}^{k+1/2} = f_{g,i}^{k+1/2}/(a_{0,i} I_{0,i+1/2}^{k+1/2} + b_{0,i} I_{0,i-1/2}^{k+1/2}), \tag{71a}$$

$$\varphi_{g,i}^{k+1/2} = g_i^{k+1/2}/(I_{0,i+1/2}^{k+1/2} - I_{0,i-1/2}^{k+1/2}). \tag{71b}$$

When this is done, Eq. (69a) becomes quite simple in form and is, in fact, a discretized form of the nonlinear multifrequency–gray method. That is, Eq. (69a) becomes

$$-\frac{1}{3}\left[\sum_{g=1}^{G}\left(\frac{\varphi_{g,i+1}^{k+1/2}}{\delta_{1g,i+1}}\right)(I_{0,i+3/2}^{k+1} - I_{0,i+1/2}^{k+1}) - \sum_{g=1}^{G}\left(\frac{\varphi_{g,i}^{k+1/2}}{\delta_{1g,i}}\right)(I_{o,i+1/2}^{k+1} - I_{0,i-1/2}^{k+1})\right]$$

$$+ \frac{1}{4}[\zeta_{i+1}(a_{0,i+1} I_{0,i+3/2}^{k+1} + b_{0,i+1} I_{0,i+1/2}^{k+1})$$

$$+ \zeta_i(a_{0,i} I_{0,i+1/2}^{k+1} + b_{0,i} I_{0,i-1/2}^{k+1})]$$

$$= \sum_{g=1}^{G} S_{g,i+1/2}. \tag{72}$$

[†] Fifty groups based on analytic opacities, ten spatial mesh points with $\Delta x = 1.0$.

Table II
Convergence Behavior of the Linear and Nonlinear Diffusion Equation Accelerations for a Nonmodel Problem

T_D	Acceleration		Nonaccelerated
	Nonlinear	Linear	
1.0	0.668	0.594	0.997
0.9	0.678	0.610	0.997
0.8	0.683	0.592	0.998
0.7	0.688	0.541	0.999
0.6	0.690	0.762	0.999+
0.5	0.701	*	0.999+
0.4	0.699	*	0.999+
0.3	0.672	*	0.999+
0.2	0.769	*	0.999+
0.1	0.992	*	0.999+
0.09	0.954	*	0.999+
0.08	0.965	*	0.999+
0.07	0.976	*	0.999+
0.06	*	*	0.999+

Notes:
$0 \leq x \leq 10.0$.
T_D is the temperature in $3.0 < x \leq 10.0$.
$T_0 = 1.0$ is the temperature in $0 \leq x < 3.0$.
* $\Rightarrow$ divergence.

In Table II we present the convergence behavior of the acceleration methods (with $a_{0,i} = b_{0,i} = 1$) based on Eq. (69a) in the linear and nonlinear forms for a problem with a step change in temperature. As the magnitude of the step is increased, the performance of the acceleration methods changes until at some point they diverge. We note that the nonlinear method performs better than the linear method for this problem, and that the unaccelerated convergence rate is exceedingly slow. The divergence of the acceleration methods seems to be due to the nature of the differencing of the diffusion equation and, to some extent, to the artificial nature of the test problem. In practice, we have not experienced this divergence in time-dependent problems, although it does remain a concern. We have concluded that this 1-group acceleration procedure is useful for real problems, and we do employ it as the accelerator for the multigroup diffusion equation (which itself is the accelerator for the transport iterations). In the time-dependent problems to be presented next, we find that our technique reduces the computational effort required to solve the diffusion equation by a factor of 4.

IV. TIME-DEPENDENT EXAMPLE CALCULATIONS

To demonstrate the efficacy of the numerical procedure outlined above, in this section we present results of radiation hydrodynamics calculations based on this procedure. Our major point has been that an implicit treatment in time, coupled with the use of efficient iteration convergence techniques, leads to a more effective and flexible method than is obtained by using strict time-step controls to ensure convergence and accuracy. In the following, we demonstrate this and other aspects of our method.

To begin, we establish criteria for obtaining a base solution, i.e., a solution that is accurate for a given problem having a prescribed spatial mesh[†] and group structure. First, we set a convergence criterion of a 1% maximum relative error in the angular intensities in each frequency group and space point. Thus, we allow all the iterations to proceed until this criterion is met. (We note that this procedure includes convergence of the nonlinear parameter iteration outlined in Section III.B.) Second, to control the accuracy of the solution as a function of time, we employ a time-step control that restricts Δt so that the temperature in every spatial cell changes by no more than 5% from one time step to the next. We have verified that a tighter time-step control does not change the solution within the convergence criterion of 1%.

Our criteria for obtaining base solutions are rather inefficient because many iterations are wasted in converging relatively unimportant regions of the solution. However, we are confident that these criteria do provide base, or benchmark, solutions of the spatially and frequency-discretized equations, which can be used in judging other iteration strategies.

In the following we present two slab-geometry radiation transport problems. These problems are "diffusionlike" in the sense that diffusion theory gives an acceptable answer. For this reason alone these are exceedingly difficult problems in transport theory, and so they serve as an excellent test for

---

[†] We select a group structure and spatial grid that are fine enough so that refinements in the group structure by a factor of 2 or in the spatial grid by a factor of 10 lead to less than a 1% change in the solution. Although this gives confidence that our overall numerical solution is accurate as compared with the analytic solution, a few words of caution are appropriate here. In the lower frequency groups, the opacities are so large ($\sigma \approx 10^6$) that no practical discretization is possible for which the optical thickness of a cell, $\sigma \Delta x$, is less than 1, as is normally required for accurate diamond-differenced transport solutions. Consequently, there is some question that our base solution is accurate, even with the criteria stated above. This difficulty exists with all numerical procedures for solving the equations of radiation hydrodynamics; it has never been satisfactorily addressed in the literature, and its consideration is beyond the scope of this paper. To summarize, we believe that our solutions are accurate, but we cannot assert this rigorously. Our approach is to select fixed spatial and frequency discretizations satisfying the stated criteria, and then focus on the errors generated by relaxing the iteration criteria within each time step.

4. Diffusion-Synthetic Acceleration

our methods. A second difficult aspect of these problems is that they contain a thermal wave front moving through a medium whose mean-free-path dimensions are measured in the millions. Such problems have led to our rather elaborate treatment of the negative intensity fixup outlined in Section III.F; thus, these problems will provide a good test for the fixup.

Our first time-dependent problem is the 20-cm slab of material with the analytic opacities of Eq. (20), with $a_0 = 1/kT$ and $a_1 = 27$, and with an initial temperature of 1 eV. We impose blackbody boundary conditions at both sides of the slab; on the right side ($x = 20$ cm) the boundary temperature is 1 eV, and on the left side ($x = 0$ cm) the boundary temperature is a linear function of time, 1 eV at $t = 0$ and 1 keV at $t = 0.1$ ns. After $t = 0.1$ ns, the left boundary temperature remains constant at 1 keV. The opacity within the slab is described by the analytic model in Eqs. (20a, b). The specific heat capacity of the material is 8.118×10^5 J keV^{-1} cm^{-3}. All problems discussed below utilize the S_8 quadrature set, 20 equal spatial zones, and 30 frequency groups; the convergence criterion used within each time step is identical to the base case; i.e., we allow at most a 1% relative change in the intensities.

The base solution, obtained with full parameter iteration, is shown in Fig. 5 as the graphs of temperature in thousand electron volts versus position, at the indicated times. As discussed above, this is our accuracy standard, and we compare approximate solution strategies against it.

For our first approximate solution, we do not perform the parameter iteration within each time step. Thus, we solve the problem using parameters

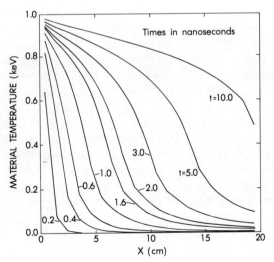

Fig. 5. Base case temperature distributions.

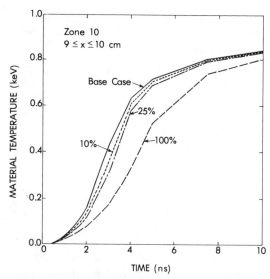

Fig. 6. Temperature in zone 10 for various temperature-change criteria.

evaluated at $T^{n^*} = T^n$, while we vary the time-step limitation criterion within the range of 10 to 100% temperature change. The results are summarized in Fig. 6, in which we plot the average material temperature in zone 10 ($10 \leq x \leq 11$) versus time for various time-step criteria. These results indicate that a 25% temperature change per time step provides acceptable accuracy without the parameter iteration. Larger time steps, such as those in the 100% temperature-change case, clearly require parameter iteration. In these examples we observed that the number of iterations required for convergence within each time step is relatively insensitive to the time-step size. That is, the average number of transport source iterations varied from 3.3 for the 10% temperature-change case to 4.1 for the 100% temperature-change case.

Since the number of iterations is insensitive to the time-step size, we next investigate the effect of limiting the number of transport iterations per time step while choosing a time-step control tight enough so that parameter iteration is unnecessary. Thus, the above problem was recomputed with an artificial upper limit of two transport iterations and five multigroup diffusion iterations (using nonlinear MFG acceleration) per transport iteration; the calculational results are shown in Fig. 7. In Fig. 8, we present the number of transport iterations (8a) and diffusion iterations (8b) for the unlimited iteration and artificially limited cases. In Fig. 8c, we present the time-step size as a function of time as a typical illustration of the range encountered in these problems. We conclude that when the time step is suitably limited so

4. Diffusion-Synthetic Acceleration

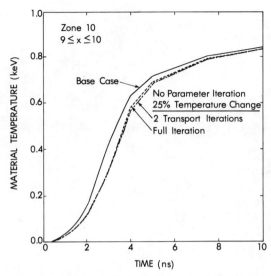

Fig. 7. Temperature in zone 10 for limited and full transport iteration.

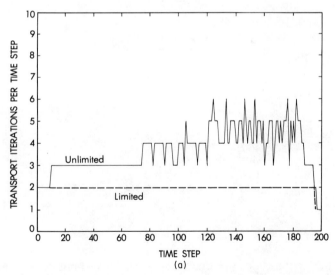

Fig. 8. (a) Transport iterations for full and limited iteration schemes. (b) DSA iterations for full and limited iteration schemes. (c) Time steps. Figure continues on p. 108.

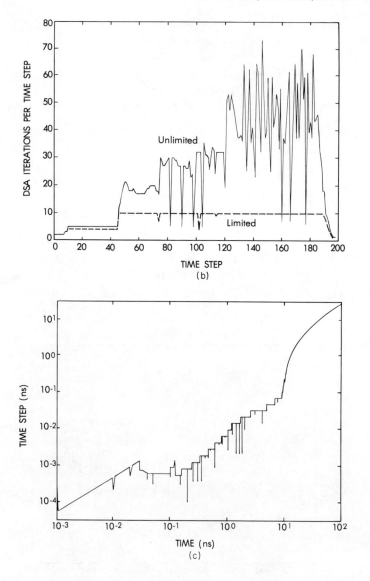

that parameter iterations are unnecessary, an upper limit of two transport iterations and five multigroup diffusion calculations per transport iteration, per time step, are sufficient to obtain acceptably accurate solutions.

Next, we come to the crucial issue of the trade-off between iterating and simply choosing smaller time steps to achieve an acceptable accuracy. We studied this by solving the above test problem with the iteration strategy

4. Diffusion-Synthetic Acceleration

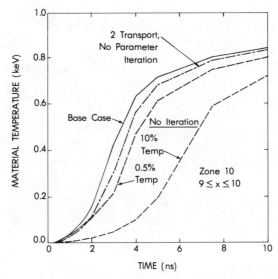

Fig. 9. Temperature in zone 10—comparison of iterated and uniterated calculations.

limited to one transport calculation, one MFG-accelerated multigroup diffusion calculation, and no parameter iterations per time step. The time-step criterion was then reduced to attempt to obtain acceptably accurate results as measured against our base solution. In Fig. 9, we depict the results of two of these runs. As can be seen, the uniterated run with a 0.5% temperature change per time step is less accurate than the iterated run with a 25% temperature-change criterion. Therefore, iteration allows one to take temperature changes larger by at least a factor of 50, while retaining comparable accuracy. In our test problem, the unaccelerated solution required 45 × more time steps. As another measure of the computational work, the uniterated run required a factor of 22 more transport calculations, 5 × more multigroup diffusion calculations, and 45 × more opacity calculations. Therefore, it is clearly advantageous to take large time steps and iterate rather than small time steps and not iterate.

Finally, we present a second example to show that our iteration strategy can accommodate inhomogeneous regions. This problem is the same as the previous time-dependent problem, except that in the region $10 \text{ cm} \leq x \leq 12 \text{ cm}$ the opacities are multiplied by a factor of 10^3. The base solution for this problem, obtained with full parameter iteration, is depicted in Fig. 10. A comparison of zone-averaged temperatures in the zone immediately beyond the "wall," $12 \leq x \leq 13$, is shown in Fig. 11. This figure compares the base solution with the solution obtained by the strategy that limits to two transport iterations, five multigroup diffusion iterations per transport iteration, and no

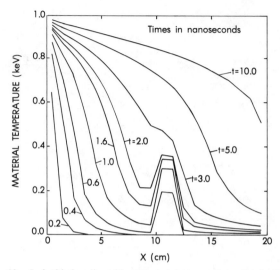

Fig. 10. Imbedded wall problem; material temperature distributions.

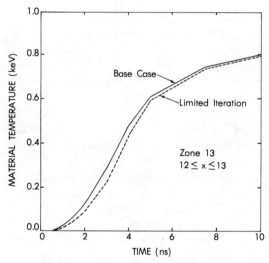

Fig. 11. Temperature in zone 13, imbedded wall problem—comparison of full and limited iteration strategies.

parameter iterations with a 25% temperature-change criterion. It illustrates the acceptable accuracy obtained using our iterative method on this more difficult problem.

With the results of these examples (and others not shown) we conclude that the iteration strategy presented in this article provides a reliable and flexible solution method for radiation hydrodynamics problems. Thus, accurate transport solutions that are reliable and accurate with respect to time-step size can be obtained.

REFERENCES

Alcouffe, R. E. (1977). Diffusion synthetic acceleration methods for the diamond-differenced discrete-ordinates equations. *Nucl. Sci. Eng.* **64**, 344.

Alcouffe, R. E., and O'Dell, R. D. (1983). Transport calculations for nuclear reactors. *Los Alamos Nat. Lab.* [*Rep.*] **LA-UR-83-1722**.

Byrne, R. N. (1973). Numerical techniques in two- and three-dimensional radiation hydrodynamics. *Prog. High Temp. Phys. Chem.* **5**, 157.

Carlson, B. G., and Lathrop, K. D. (1968). Transport theory—the method of discrete ordinates. *In* "Computing Methods in Reactor Physics" (H. Greenspan, C. N. Kelber, and D. Okrent, eds.), p. 171–266. Gordon & Breach, New York.

Cullen, D. E., and Pomraning, G. C. (1980). The multiband method in radiative transfer calculations. *J. Quant. Spectrosc. Radiat. Transfer* **24**, 97.

Larsen, E. W. (1982). Unconditionally stable diffusion-synthetic acceleration methods for the slab geometry discrete-ordinates equations. Part 1. Theory. *Nucl. Sci. Eng.* **82**, 47.

Larsen, E. W. (1983). Diffusion-synthetic acceleration methods for the discrete-ordinates equations. *Proc. Am. Nucl. Soc. Top. Meet., Adv. React. Comput., 1983* Vol. 2, p. 705.

Morel, J. E., Larsen, E. W. and Matzen, M. K. (1984). A synthetic acceleration scheme for radiative diffusion calculations. *J. Quant. Spectrosc. Radiat. Transfer* (to appear).

Pomraning, G. C. (1973). "The Equations of Radiation Hydrodynamics." Pergamon, Oxford.

Pomraning, G. C. (1982). Radiation hydrodynamics; notes from a short course given at Los Alamos. *Los Alamos Nat. Lab* [*Rep.*] **LA-UR-2625**.

Wilson, H. L. (1973). A new numerical technique for solving the time-dependent radiation transport equation. *Prog. High Temp. Phys. Chem.* **5**, 125.

5

Implicit Methods in Combustion and Chemical Kinetics Modeling

ROBERT J. KEE, LINDA R. PETZOLD,
MITCHELL D. SMOOKE, and JOSEPH F. GRCAR

Applied Mathematics Division
Sandia National Laboratories
Livermore, California

I. Introduction	113
II. Stiffness and Implicit Methods	115
III. The Method of Lines	122
A. The Method of Lines as Differential–Algebraic Equations	123
IV. Adaptive Meshing	127
A. Coordinate Transformation Methods	128
B. Variable Mesh Methods	130
V. Solution of the Nonlinear Equations	138
References	142

I. INTRODUCTION

Our purpose in modeling chemical kinetics is to predict and explain the evolution of chemical species in a reacting flow. Typical of the combustion processes that we consider are systems consisting of 30 to 40 chemical species and over 100 chemical reactions. Potentially, there is an enormous payoff to be gained by understanding the complex chemical kinetics processes that govern the production and control of pollutants in combustion. For example, recent application of computational models developed at our laboratory has made a significant contribution to the successful application and operation of a process to reduce nitric oxide pollutants from coal-burning power plants. Together, new computational algorithms and new computing hardware are contributing to rapid growth in this field by both industrial and academic researchers.

We first describe why the chemical kinetics equations are *stiff*, and why implicit numerical algorithms are needed for their solution. The need for implicit methods to model chemistry efficiently was discovered in the early 1950s by Curtiss and Hirschfelder (1952). However, it was not until the early 1970s that computational methods for stiff ordinary differential equations (ODEs) were put on firm theoretical ground and implemented in usable software by Gear (1971) and Hindmarsh (1972, 1983). Today, the solution of systems of chemical kinetics equations without spatial dependencies is accomplished routinely by use of well-tested and documented software. The challenge for the future is to model the coupled effects of chemical kinetics and fluid transport.

Chemical systems are usually characterized by the presence of widely disparate time scales. This is because some chemical species (such as active free radicals like hydrogen atoms) have reaction times that are extremely short compared with the reaction times of others (such as relatively inert species like molecular nitrogen). Equations governing these systems are called stiff, and stiff equations suffer from the fact that their solution by an explicit numerical method is especially inefficient. Since explicit methods are easier and less expensive to implement, the solution of differential equations on early computers was by explicit methods. Nevertheless, today the presence of stiffness in ODEs is well understood and the solution is obtained using appropriate methods. We devote the following section of this paper to a discussion of stiffness and its computational consequences.

When chemical kinetics systems are considered alone, the governing equations are stiff systems of ODEs. However, when fluid transport is also considered, the governing equations become systems of stiff partial differential equations (PDEs). They may include hyperbolic, parabolic, or elliptic equations, depending on the particular transport physics. The solution of PDEs that couple transport with chemical kinetics is more difficult than that of chemical kinetics equations alone. We discuss here what we believe are sound general approaches to the computational solution of these coupled equations.

A tremendous body of knowledge and experience has been gained from solving stiff ODEs and elliptic boundary-value problems, and it should be carried over into the development of methods for solving stiff PDEs. Therefore, we advocate a general method-of-lines solution. In this way we take

5. Combustion and Chemical Kinetics Modeling

advantage of the techniques that have been developed for measuring truncation error, choosing time steps, and selecting spatial discretizations. Nevertheless, significant problems still remain regarding adaptive spatial meshing and solution of large nonlinear algebraic systems.

The method of lines is a computational approach for solving PDEs wherein some of the derivatives are discretized and the resulting lower dimensional system is solved. Most often, the approach has found use in solving parabolic PDEs wherein the spatial derivatives are discretized (for example, by finite-difference approximations) and the resulting initial-value ODE system is solved using software for initial-value problems. However, we show that considering the discretized problem as a system of differential–algebraic equations (DAEs) rather than ODEs is a more general approach than that found in the traditional method of lines. In addition, we discuss an alternative view of the method of lines, in which the time derivatives are discretized, and the resulting boundary-value problem is solved. This viewpoint is helpful in developing adaptive meshing techniques.

The same disparate time-scale characteristics that cause the kinetics equations to be stiff contribute to the presence of disparate length scales in problems with spatial dependence. For example, it is well known that flames are thin relative to combustion chamber dimensions. Therefore, we must be concerned in our spatial discretization schemes to develop adaptive-mesh methods that adequately resolve the flame itself, yet still provide sufficient resolution in the remainder of the combustion chamber. We address this general topic of adaptive meshing in Section IV.

The final problem we consider is the numerical solution of the large systems of nonlinear algebraic equations that are present in any implicit method. For systems of ODEs and even for systems of one-dimensional PDEs, this is usually not a problem. However, for two- or three-dimensional systems the matrices that describe the linearized systems typically do not fit in the memory of even the largest modern scientific computers. Therefore, we need to develop iterative methods that are more storage efficient and faster than direct methods, but do not compromise the accuracy of the solution. We devote the last section of the chapter to this topic.

II. STIFFNESS AND IMPLICIT METHODS

In this section we describe the phenomenon of stiffness, giving an example from chemical kinetics. We explain why stiff problems cause difficulties for classical explicit numerical ODE methods, and why implicit methods overcome many of these difficulties. Finally, we outline some of the methods that are currently available for solving stiff systems of equations. Our discussion here is in terms of ODEs because the salient points are easier to see; the conclusions carry to the solution of PDEs.

What is stiffness? Perhaps the easiest way to understand the concept is via a sketch of the families of solutions to a stiff equation. Consider, for example, the linear equation

$$y' = f(t, y) = -\lambda[y - (t^2 + 1)] + 2t, \qquad \lambda \gg 0. \tag{1}$$

For the initial value $y(0) = 1$, this equation has the slowly varying solution $y = t^2 + 1$. However, for any other initial value, the solution tends very rapidly toward the smooth solution $y = t^2 + 1$. A sketch of the family of solution curves for Eq. (1) for different initial values is given in Fig. 1. We see that neighboring solutions vary much more rapidly than the smooth solution $y = t^2 + 1$. After a short time, which is known as the transient region, these solutions coincide with the smooth solution. The problem is not considered to be stiff in the transient region. In general, we can think of a problem as being stiff when nearby solution curves tend very rapidly toward a smooth, slowly varying solution curve.

In the chemistry problems that we consider, stiffness has two sources. First, chemical kinetics equations are almost always very stiff; an example demonstrates this point. Second, the spatial discretization is often a source of stiffness when PDEs are solved by the method of lines; we touch on this problem briefly at the end of this section.

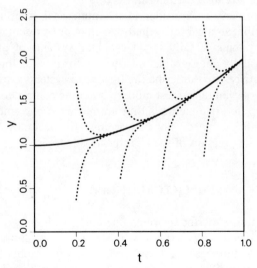

Fig. 1. A sample stiff problem showing solutions beginning from different initial conditions rapidly coverging toward a smooth solution.

5. Combustion and Chemical Kinetics Modeling

To illustrate stiffness in combustion kinetics, we choose the relatively simple reaction of hydrogen and air. Assuming that the mixture reacts adiabatically (i.e., with no heat loss) and at constant pressure, the governing equations are

$$\frac{dT}{dt} = \frac{1}{\rho c_p} \sum_{k=1}^{K} h_k \dot{\omega}_k W_k, \tag{2}$$

$$\frac{dY_k}{dt} = \frac{\dot{\omega}_k W_k}{\rho}, \quad k = 1, \ldots, K. \tag{3}$$

The independent variable is time t, and the dependent variables are the temperature T and the mass fractions of the K species Y_k. The other variables are the mass density ρ, specific heat c_p, molecular weight W_k, specific enthalpy of the species h_k, and the molar production rate of species by chemical reaction $\dot{\omega}_k$. The specific heat and enthalpy depend on temperature, whereas the density and chemical production depend on both temperature and mass fraction. The chemical production rates depend on the chemical reactions being considered; they are exponential in temperature and bilinear in species concentrations. For this example we take the initial conditions to be those of a stoichiometric (with sufficient oxidizer to consume all the fuel, and vice versa) mixture of hydrogen and air at a temperature of 1000 K. We describe the chemical species production $\dot{\omega}_k$ by a reaction sequence of 16 elementary reactions that involve 10 species (Kee et al., 1980).

The solutions shown in Fig. 2 are typical of combustion processes like ignition. Initially, the temperature remains essentially unchanged (actually it decreases slightly) while bonds in the fuel, H_2, and oxidizer, O_2, break and free radical species (such as H, O, OH) form. When sufficient concentrations of the radical species have been produced, chain branching reactions (those that produce two radicals for each one consumed) begin to dominate the chemistry and the temperature rises very rapidly; i.e., an explosion occurs. During this period most of the fuel is consumed and most of the principal product, H_2O, is formed. Also, the radical species "overshoot" their final or equilibrium values. Then, for a relatively long period of time the radicals recombine to produce a little more of the product species. It is often in this period that the fate of pollutant species is determined. For example, in this problem, some of the nitrogen that is present in the air is oxidized to nitric oxide (NO). The nitrogen chemistry proceeds on significantly slower time scales than the hydrogen–oxygen chemistry. We see the concentration of nitric oxide continuing to grow long after the "combustion" of the fuel is complete.

This problem is stiff everywhere except in the transient region, where the chain branching occurs. The transient region, seen in Fig. 2, is the region

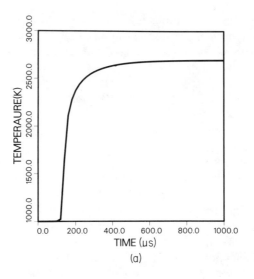

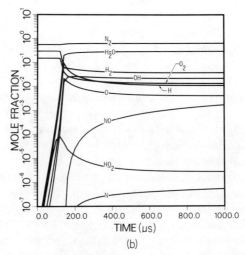

Fig. 2. Temperature and mole fraction histories for the reaction of a stoichiometric mixture of hydrogen and air beginning at 1000 K.

where the fuel and oxidizer are changing most rapidly. Of course, it is not possible to determine from the figure alone whether the system is stiff. To get a better understanding of why and where the system is stiff, chemical kinetics equations are often written in the form

$$dy_k/dt = y_k/\tau_k + c_k, \tag{4}$$

where c_k and τ_k depend on the species concentrations, and y_k is the kth species. The coefficient $1/\tau_k$ often plays a role very similar to that of $-\lambda$ in Eq. (1). That is, species for which $-1/\tau_k$ is very large generally correspond to stiff solution components. For our example problem, the parameter τ_k can be as small as 10^{-7}, whereas the time interval can be as large as 10^{-3} sec. Thus some of the neighboring solutions for this problem vary much more rapidly than the smooth solutions pictured in Fig. 2.

The first class of methods likely to be attempted in solving a system of ODEs is the class of explicit methods, for example, explicit multistep or Runge–Kutta methods. In comparison with implicit methods, these are easier to implement, easier to understand, and much less costly when they are applicable. Unfortunately, the cost advantage disappears for stiff equations. Consider the model equation

$$y' = -\lambda y + g(t) \tag{5}$$

and the explicit Euler method

$$y_{n+1} = y_n + h_n y'_n = y_n + h_n[-\lambda y_n + g(t_n)], \tag{6}$$

where $h_n = t_{n+1} - t_n$ is the step size. The analytic solution satisfies

$$y(t_{n+1}) = y(t_n) + h_n[-\lambda y(t_n) + g(t_n)] - [y(t_n) + h_n y'(t_n) - y(t_{n+1})]. \tag{7}$$

The second bracketed expression is the local truncation error d_n, which for the explicit Euler method is given more generally by $d_n = (h_n^2/2)y'' + O(h_n^3)$. Substracting Eq. (7) from Eq. (6) yields a recurrence equation for the global error $e_n = y_n - y(t_n)$,

$$e_{n+1} = (1 - h_n \lambda)e_n + d_n. \tag{8}$$

The error is amplified unless $|1 - h_n\lambda| \leq 1$. Thus the method is stable (it does not amplify errors) provided that $h_n \leq 2/\lambda$. But the method is accurate (the truncation error is less than some tolerance ε) provided that

$$\|h_n^2 y''/2\| \leq \varepsilon \tag{9}$$

or

$$h_n \leq (2\varepsilon/\|y''\|)^{1/2}. \tag{10}$$

For large λ, the requirement on h_n for stability is much more stringent than for accuracy. In other words, even though the solution that we seek is slowly varying, the step size of the explicit method must be chosen as small as the nearby rapidly varying solutions would require in order to maintain stability.

Another way to view this dilemma is that the explicit Euler method uses the tangent to the solution at the current point to find a new approximation at a future point. If we are even slightly off the smooth solution, then in a stiff problem the tangent is very steep, so that step sizes that would be "practical" for the smooth solution lead us even further away. Although we have explained the problem in detail only for the explicit Euler method, the problem is the same for any explicit method. Moreover, a similar situation occurs with systems of equaions, in which case the eigenvalues of the locally linear system (the eigenvalues of the Jacobian matrix) determine the stiffness.

The solution to these stability difficulties is to use an implicit method. The backward Euler method is the simplest of these. When applied to Eq. (5) it has the form

$$y_{n+1} = y_n + h_n y'_{n+1} = y_n - h_n \lambda y_{n+1} + g_{n+1}. \tag{11}$$

The local truncation error d_n is as before, but unlike Eq. (8) the error now satisfies

$$e_{n+1} = e_n - h_n \lambda e_{n+1} + d_n. \tag{12}$$

Here the error does not grow provided that $|1/(1 + h_n \lambda)| < 1$, so the method is stable for arbitrarily large positive λ. Just from looking at Fig. 1 one would expect intuitively that the implicit method should work better for stiff systems because it uses the tangent at the future point instead of the present point. Even if the present point lies slightly off the slowly varying smooth solution, the tangent at the future point is likely to be much closer to the smooth solution's mild tangent because all solutions decay so rapidly to the smooth solution. With a tangent of small slope, the implicit method can select a large time step and yet remain near the smooth solution.

Implicit methods cure the problem of stiffness, but they introduce other difficulties. Because the method is implicit, a nonlinear system of equations must be solved at each step. Simple functional iteration can be used for this purpose in solving nonstiff problems. For the backward Euler method, this is

$$y_{n+1}^{(m+1)} = y_n + hf(t_{n+1}, y_{n+1}^{(m)}). \tag{13}$$

The iteration converges if $\|h(\partial f/\partial y)\| < 1$. However, for stiff systems with practical step sizes, $\|h(\partial f/\partial y)\| \gg 1$. Thus stiff problems require the more robust modified Newton iteration,

$$P(y_{n+1}^{(m+1)} - y_{n+1}^{(m)}) = y_{n+1}^{(m)} - h\beta_0 f(t_n, y_{n+1}^{(m)}) - y_n, \tag{14}$$

in which P is an approximation to the Jacobian matrix of the algebraic system to be solved:

$$P \approx I - h_n J, \quad J \approx \partial f/\partial y. \tag{15}$$

The Jacobian matrix is usually approximated by finite differences.

All modern codes for solving ODEs, whether stiff or nonstiff, vary the step size to maintain some norm of the local truncation error below a user-defined

5. Combustion and Chemical Kinetics Modeling

error tolerance. The exact mechanism for accomplishing this task varies, depending mostly on the basic formulas being used (i.e., Runge–Kutta, multistep, extrapolation), but the idea is usually to estimate a norm of the first term of the local truncation error [for example, $(h^2/2)y''$ for explicit Euler] and then to try to select the step size so that this error is nearly equal to some user-defined error tolerance ε. For example, for the explicit Euler method, the derivative y''_{n+1} can be estimated by the second divided difference of y, and the next step size would then be chosen so that

$$\|h_{n+1}^2 y''_{n+1}/2\| \approx \varepsilon \tag{16}$$

or

$$h_{n+1} \approx \sqrt{2\varepsilon/\|y''_{n+1}\|}. \tag{17}$$

This error control mechanism prevents nonstiff codes from "blowing up" on stiff problems. Instead, they just use very small step sizes. This occurs because the error estimate is composed of the difference of the computed solution plus the difference of the global error. When instability becomes a problem, either because of stiffness or for some other reason, the part of the error estimate involving the global error increases, causing the step size to be reduced. This and other aspects of stiffness are discussed in more detail in Shampine and Gear (1979).

Despite their added complexity, implicit methods are much more efficient than explicit methods for stiff problems. We computed the solution to the chemical kinetics example discussed earlier by both the stiff and nonstiff options of the code LSODE (Hindmarsh, 1983). The stiff option uses implicit backward differentiation formulas combined with Newton iteration, whereas the nonstiff option employs Adams formulas with functional iteration. Both options vary the time step and the order of the method. For a very small time interval, the implicit method finished the problem in 25 sec on a VAX 11/780 computer, whereas the explicit method required 15 min. When the problem was carried out to 10^{-2} sec (at which time the nitric oxide concentration still grows), the implicit method required 36 sec and the explicit method nearly 3 hr!

Today, systems of chemical kinetics equations are solved routinely using well-tested and well-documented software based on implicit methods and Newton iterations. Some relatively new ideas for improving the efficiency of solution include switching to methods for nonstiff equations in the transient region (Shampine, 1981, 1982a; Petzold, 1983b) and employing diagonal approximations to the often diagonally dominant Jacobian matrices of chemical kinetics systems (Shampine, 1982b). These ideas are important, for example, in operator splitting techniques, where many systems of chemical kinetics equations must be solved.

Software for solving systems of chemical kinetics equations coupled with fluid transport equations is less well developed. The nonlinear systems that must be solved at each time step are larger; the formation of the Jacobian matrices and the solution of the linear systems are more time consuming. Additionally, the spatial discretization can be an additional source of stiffness independent of chemistry. For example, if we solve the diffusion equation

$$\partial y/\partial t = \partial^2 y(t,x)/\partial x^2 \qquad (18)$$

by replacing the second derivative with a centered difference, we find that the resulting system of ODEs has a Jacobian matrix [Eq. (15)] whose largest eigenvalue is proportional to $\Delta t/(\Delta x)^2$. Thus for small Δx this system is stiff.

Many initial-value problems resulting from the method of lines do not fit easily into the form $y' = f(t,y)$ required by most ODE software. We have found a particularly convenient formulation to be that of the differential–algebraic system $F(t,y,y') = 0$. These systems include the standard ODEs as a special case, but they also include many systems that exhibit mathematical properties different from ODEs. Their solution requires a rethinking of ODE methods and error estimation strategies (Petzold, 1982b; Gear and Petzold, 1983). We have written a computer program, DASSL (Petzold, 1982a), that is appropriate for solving most of these initial-value problems that arise from the method of lines. The advantages of this nonstandard formulation are discussed in detail in the next section.

III. THE METHOD OF LINES

Numerical methods for solving PDEs usually involve the replacement of all derivatives by discrete difference approximations. The method of lines for time-dependent problems does this also, but in a special way that takes advantage of existing software. The method of lines discretizes the derivatives with respect to some but not all of the independent variables. However, if a software package can solve the lower dimensional problem, then the human effort required to solve the original system of PDEs is reduced.

Most implementations of the method of lines are more restrictive than our definition. These solve parabolic PDEs by approximating the spatial derivatives by finite differences on fixed meshes, and then solving the resulting ODE initial-value problems using standard software such as LSODE (Hindmarsh, 1983). The success and popularity of this "longitudinal" method stem from the availability of quality software for initial-value problems. In view of the static spatial discretization, the method is a good one if the solution is smooth in the spatial variables or at least if the locations of steep fronts are known a priori. We solve many combustion problems for which this method works well. Its shortcoming is evident in problems whose solutions exhibit significantly different spatial profiles from one time to another. A traveling flame front is illustrative of such a situation. In these cases a fixed spatial meshing is not adequate.

Just as ODE software for initial-value problems dynamically chooses time steps, PDE software should, in addition, adapt the spatial mesh to the problem. We consider this task in the general framework of the *transverse* method of lines, which discretizes the time derivatives and then applies methods for elliptic problems to the resulting equations. If software for two-point boundary-value problems had been available earlier, then this transverse method might be in wider use today. In any case, the growing literature on the use of adaptive meshing in the solution of boundary-value problems is applicable to the solution of PDEs. We combine techniques for choosing spatial meshes with methods for choosing time steps. In the following section on adaptive meshing, we explore some of the ideas for achieving such a fully adaptive method.

In the remainder of this section we discuss an extension of the traditional longitudinal method of lines. We find significant advantages to posing the initial-value problem as a system of differential–algebraic equations (DAEs) rather than as a standard ODE initial-value problem. The recent development of software to solve DAEs has made this alternative formulation feasible and attractive (Petzold, 1982a).

A. The Method of Lines as Differential–Algebraic Equations

We use two examples to explain how the discretization of PDEs naturally leads to systems of equations in the DAE format and how the latter are solved by DAE software. The first example is the simple heat equation that we considered earlier [Eq. (18)]. Recall that the general form for DAEs is $F(t, y, y') = 0$. In the case of PDEs, it is clearer if we include the spatial derivatives in the DAE form, and write $F(t, x, y, y_x, y_{xx}, y') = 0$. As a specific example, consider the heat equation

$$F(t, x, y, y_x, y_{xx}, y') = -\partial^2 y/\partial x^2 + y' = 0. \tag{19}$$

After discretizing in space by finite differences on a uniform mesh, we have

$$F_i = -[(y_{i-1} - 2y_i + y_{i+1})/\Delta x^2] + y'_i = 0. \tag{20}$$

The boundary conditions are handled in a similar way. For example, at $x = 0$,

$$F_1 = y_1 - y(t, 0) = 0. \tag{21}$$

The software for solving DAEs is designed in such a way that the user writes a subroutine in which, given the y and y' vectors, the residual F vector is computed.

The DASSL code solves the DAEs with variable-order, variable-step, backward differentiation formula (BDF) methods. However, to obtain an idea of the equations the code solves at each time step, consider how the equations would look if a backward Euler method were used. Equation (20) would become

$$F_i^{n+1} = -\left(\frac{y_{i-1}^{n+1} - 2y_i^{n+1} + y_{i+1}^{n+1}}{\Delta x^2}\right) + \frac{y_i^{n+1} - y_i^n}{h_n} = 0. \tag{22}$$

This particular system of equations is linear, but, in general, the equations are nonlinear and are solved by some variant of Newton's method.

Of course, the spatially discretized heat equation can be posed as a standard system of ODEs by incorporating the spatial boundary conditions into the differential equations. This traditional approach removes the algebraic equation $F_1 = 0$ and redefines F_2 as

$$F_2 = dy_2/dt - [y(t, 0) - 2y_i + y_{i+1}]/\Delta x^2. \tag{23}$$

However, in many situations the algebraic equations cannot be discarded so easily; the DAE formalism and DASSL are essential to solving such problems. Among these are the boundary layer equations describing fluid flow in a pipe. Our interest in this problem stems from the modeling of fluid mixing and chemical reactions in flow tube reactors (Kee and Miller, 1981; Branch et al., 1982; Coltrin et al., 1984). Here we confine our discussion to the relatively simple steady flow of a compressible fluid. Nevertheless, the example illustrates the essential points relating to the DAE formulation.

The equations governing flow in a pipe are

$$\frac{\partial \rho u r}{\partial z} + \frac{\partial \rho v r}{\partial r} = 0, \tag{24}$$

$$\rho u \frac{\partial u}{\partial z} + \rho v \frac{\partial u}{\partial r} = -\frac{dp}{dz} + \frac{1}{r}\frac{\partial}{\partial r}\left(r\mu \frac{\partial u}{\partial r}\right), \tag{25}$$

$$\rho u c_p \frac{\partial T}{\partial z} + \rho v c_p \frac{\partial T}{\partial r} = \frac{1}{r}\frac{\partial}{\partial r}\left(r\lambda \frac{\partial T}{\partial r}\right), \tag{26}$$

$$p = \rho R T. \tag{27}$$

The first three equations are statements of mass, momentum, and energy conservation, and the fourth is a perfect gas equation of state. These so-called boundary layer equations are derived from the more general Navier–Stokes equations by making two approximations. The first replaces the radial momentum equation with the assumption that pressure is uniform across the radius of the tube and is a function only of distance down the tube. The second approximation neglects diffusive transport in the axial direction (the principal flow direction) because it is small compared with convective transport. This assumption removes all second derivatives with respect to z and thus changes the characteristics of the system from elliptic to parabolic. The dependent variables are the axial velocity u, the radial velocity v, the temperature T, and the pressure p. We do not consider the density ρ to be a dependent variable because we can use the equation of state to eliminate ρ from the other equations. The viscosity and conductivity are given by μ and λ. The independent variables are the radial coordinate r and the axial coordinate z. In these boundary layer equations the axial coordinate is "timelike." We therefore specify initial conditions at $z = 0$ and compute the solution marching down the length of the pipe.

We could solve the equations as stated above; however, some of the points that we want to make regarding DAEs are more easily seen if we first make the von Mises coordinate transformation. We define a stream function ψ that exactly satisfies the mass continuity equation:

$$\rho u r = \partial \psi / \partial r, \qquad \rho v r = -\partial \psi / \partial z. \tag{28}$$

In the r, z coordinate system a line of constant ψ is a streamline or a particle flow path; thus in the ψ, z coordinate system the flow is parallel to lines of constant ψ. The transformed equations are given by

$$\rho u \frac{\partial u}{\partial z} + \frac{dp}{dz} = \rho u \frac{\partial}{\partial \psi} \left(\rho u \mu r^2 \frac{\partial u}{\partial \psi} \right), \tag{29}$$

$$\rho u c_p \frac{\partial T}{\partial z} = \rho u \frac{\partial}{\partial \psi} \left(\rho u \lambda r^2 \frac{\partial T}{\partial \psi} \right), \tag{30}$$

$$0 = \frac{\partial r^2}{\partial \psi} - \frac{2}{\rho u}, \tag{31}$$

$$0 = \frac{\partial p}{\partial \psi}. \tag{32}$$

Notice that although v has been eliminated as a dependent variable and the mass continuity equation is gone, the radial coordinate r must be obtained from the relationship between ψ and r in Eq. (31). Therefore, we still have four equations and four dependent variables.

As initial conditions for the original equations, we specify profiles for only axial velocity, temperature, and pressure. In the physical coordinates it is difficult to see that all the dependent variables cannot be specified independently. This is clear from the transformed equations, where r can be determined from the initial p, u, and T by first obtaining ρ from Eq. (27) and then by solving Eq. (31). A similar situation occurs with the radial velocity, but the equation describing its initial value is more complex.

We now call attention to two facts that make the solution of the boundary layer equations by the method of lines and standard ODE software troublesome, and perhaps impossible. These points regard both the form of the equations and the specification of the boundary conditions. In contrast, the DAE formulation is straightforward.

For the initial conditions we specify profiles for axial velocity, temperature, and pressure. However, even though radial velocity is a dependent variable, it may not be specified independently. This is difficult to see in the physical coordinates, but the fact that all the dependent variables cannot be specified independently can be seen readily in the transformed equations. There, the fourth dependent variable r can be determined from Eq. (31) and the specified p, u, and T profiles. Unlike the situation in standard-form ODEs, where all initial conditions are independent, in this case there is only one r profile that is consistent with the other initial conditions. In the physical coordinates, a similar situation occurs with the radial velocity, but the equation describing its specification is a complicated expression that comes from substituting all the other equations into the continuity equation. Failure to specify consistent initial conditions will prevent a solution with DAE software.

The boundary conditions in physical terms are that no flow crosses the center line and there is no transfer of energy or momentum across the center line. In the transformed coordinates the no-flow condition is replaced by the condition $r = 0$ at the center line. Thus, at the center line $\psi = 0$ we have

$$r = 0, \quad \partial u/\partial \psi = 0, \quad \partial T/\partial \psi = 0. \tag{33}$$

At the pipe boundary, the boundary conditions are that the velocity is zero and the temperature is specified. In the transformed coordinates, the boundary conditions are

$$r = r_0, \quad T = T_0, \quad u = 0. \tag{34}$$

Application of standard ODE software would be impeded by the fact that there is no explicit boundary condition on pressure, even though it is one of the dependent variables. Instead, there are two conditions on r. The problem is well posed since the two first-order equations, Eqs. (31) and (32), have two boundary conditions, but both are conditions on r. The more general DAE formulation handles such problems easily because boundary conditions are just additional algebraic equations that the solution must satisfy.

Moreover, the traditional method of lines requires that each of the partially discretized equations have one timelike derivative of one dependent variable. In our example's transformed coordinates, the momentum equation has two timelike derivatives, $\partial u/\partial z$ and $\partial p/\partial z$, whereas Eqs. (31) and (32) have none. Solution via standard ODE software for initial-value problems is impossible, whereas this distribution of time derivatives is routine in a DAE setting.

As a final note, we mention some programming conveniences that we have found useful when using the method of lines. Note that we solve Eqs. (30) and (31) in their stated form instead of treating the pressure as a global variable and evaluating r by an integral. Our motivation is to simplify the structure of the Jacobian matrix in the linear system of equations that must be solved at each step of Newton's method. We could consider r to be a coefficient in the diffusion terms of the momentum and energy equations, and we could evaluate it from the integral equation that is equivalent to Eq. (31). However, doing so expands the bandwidth of the Jacobian and results in an inefficient solution procedure. Therefore, we introduce r as a dependent variable at each mesh point and solve for it with all the other variables. The system of equations is larger, but in a form more amenable to efficient solution. A similar argument is made for the evaluation of pressure p. The pressure appears in every other equation because it is needed to evaluate the density ρ from the equation of state. By solving Eq. (32), we introduce pressure as a variable at each mesh point and solve a trivial equation to ensure that all pressures are all equal. Once again this allows the Jacobian matrix to be banded, but at the expense of creating more dependent variables, i.e., the pressure at each mesh point. Treating simple variables like pressure in this way is a programming convenience. If treated as a global variable, the Jacobian would have a few dense rows and columns but would not be essentially more difficult to solve. However, as there is no standard software for these matrices, we prefer to avoid writing a special-purpose linear equation solver.

IV. ADAPTIVE MESHING

We have already discussed the stiffness that occurs in combustion problems owing to the widely varying chemical production rates. Solutions to these stiff problems are characterized by regions of very rapid change (the transient region) and regions of relatively smooth behavior (the approach to equilibrium). The spatial dependencies that occur in the solution of PDEs arising from combustion models often exhibit analogous behavior. Here chemical reactions have shorter time scales than the transport processes of convection and diffusion. Just as codes for solving initial-value problems must be efficient in choosing small time steps when the solution is changing rapidly and longer ones when it is not, PDE software must choose fine spatial meshes in regions of sharp change. In a transient PDE solution the spatial mesh could potentially change at each time step.

Several approaches for solving parabolic PDEs in one space dimension with adaptive spatial grids have recently appeared in the literature. Among these are the moving finite-element method of Miller and Miller (1981) and Gelinas et al. (1981), the adaptive finite-element method of Davis and Flaherty (1982), and the arc-length approach of White (1983) and Dwyer et al. (1980, 1982). In these and others a fixed number of grid points are repositioned at each time level in regions where the spatial activity of the solution is highest. Bolstad (1982), Berger and Oliger (1983), and Smooke and Koszykowski (1983) have all developed methods that vary the number of mesh points as the solution progresses.

In the nomenclature of the two-point boundary-value problem (Kautsky and Nichols, 1980), adaptive-mesh selection can be viewed as equi- or subequidistributing the definite integral of a positive weight function $w(x)$ which depends in some way upon the solution. That is, we seek mesh points so that in any mesh interval

$$\int_{x_j}^{x_{j+1}} w \, dx \leq W, \quad j = 1, \ldots, J - 1. \tag{35}$$

Several weight functions have been suggested. Each one emphasizes a different aspect of the solution. For example, an estimate of the local truncation error in the spatial discretization is a natural choice for w that parallels the time-step selection for initial-value problems. Another alternative is to highlight the spatial complexity of the solution by equidistributing the solution's arc length or total variation, in which case

$$w = \sqrt{1 + (dy/dx)^2} \quad \text{or} \quad w = |dy/dx|. \tag{36}$$

These choices tend to concentrate mesh points where the solution changes rapidly.

Two different approaches to mesh adaptation, depending on whether a fixed or variable number of mesh points are used, follow. In the case of a fixed number, the weight function is equally distributed among the mesh intervals. If a variable number of mesh points are used, then the weight function can be both equidistributed and brought below a given tolerance, $W \leq \varepsilon$. We have used both approaches, depending on the problem.

A. Coordinate Transformation Methods

When using a fixed number of points, we pose the problem as one of a coordinate transformation. For example, the following transformation is one that we have used successfully:

$$\frac{\partial \eta}{\partial x} = \frac{1 + b_1 |\partial y/\partial x| + b_2 |\partial^2 y/\partial x^2|}{\int_0^X (1 + b_1 |\partial y/\partial x| + b_2 |\partial^2 y/\partial x^2|) \, dx}, \tag{37}$$

5. Combustion and Chemical Kinetics Modeling

where b_1 and b_2 are scalar constants that determine the relative importance of adaptation criteria (Dwyer, 1983). The three terms correspond to the adaptation criteria based on (1) independent variable variation, i.e., geometry of the domain; (2) dependent variable variation, i.e., changes in the solution itself; and (3) the dependent variable's slope variation. Depending on the choices for b_1 and b_2, the available grid points can be apportioned to emphasize one of the adaptation criteria. The integral in the denominator serves to normalize the new coordinate η between 0 and 1.

Some limiting cases can help illuminate how this transformation method distributes mesh points. First, take the trivial case $b_1 = b_2 = 0$; clearly this choice results in a uniform grid. Next, take only $b_2 = 0$, and evaluate b_1 from the following equation:

$$b_1 \int_0^x \left|\frac{\partial y}{\partial x}\right| dx = \int_0^x dx. \tag{38}$$

This equality implies that the grid points are assigned equally between changes in the solution y and the independent variable x. More generally, we could prescribe the weight given to each factor in the equidistribution in terms of the following ratio:

$$R_1 = \frac{b_1 \int_0^X |\partial y/\partial x|\, dx}{\int_0^X (1 + b_1 |\partial y/\partial x|)\, dx}. \tag{39}$$

Once the ratio R_1 is chosen to set the relative importance of solution and geometry variation, then b_1 can be determined from the above equation. For example, if R_1 is large, then points will be concentrated in regions where the solution varies rapidly. In the limit of very large R_1, the transformation becomes one that equidistributes the dependent variable. A similar ratio can be found to define the relative importance of the solution's slope compared with its curvature. In this way we determine a priori how the mesh points will be distributed. A complete discussion of the approach is found in Dwyer (1983).

To further illustrate how the transformation method works, take the heat equation as an example. After transformation it becomes

$$\frac{\partial y}{\partial t} - \frac{\partial \eta}{\partial x}\frac{\partial y}{\partial \eta}\frac{\partial x}{\partial t} = \frac{\partial \eta}{\partial x}\frac{\partial}{\partial \eta}\left(\frac{\partial \eta}{\partial x}\frac{\partial y}{\partial \eta}\right). \tag{40}$$

The second term on the left accounts for the motion of the spatial mesh. In addition to solving the transformed equation, we have to solve the equation for the transformation itself. Since η is now the independent variable, Eq. (37) is written as

$$\frac{\partial x}{\partial \eta} = \frac{\int_0^X (1 + b_1|\partial y/\partial x| + b_2|\partial^2 y/\partial x^2|)\, dx}{1 + b_1|\partial y/\partial x| + b_2|\partial^2 y/\partial x^2|}. \tag{41}$$

Moreover, another equation must be solved to evaluate the normalization integral that now appears in the numerator of the above equation.

Unfortunately, there are some difficulties in the application of transformation methods. For one, even if we begin with a linear PDE, it is transformed into a nonlinear system. Therefore, we expect to have more difficulty solving problems in the transformed coordinates. Another problem that we and others (Flaherty et al., 1983) have observed is that the fully coupled simultaneous solution of equations similar to Eqs. (40) and (41) appears to be unstable for many problems.

The stability problem in the implicit application of the coordinate transformation is not well understood. However, those who use mesh transformation techniques circumvent the problem in various ways. Evidently, it is necessary to take some measure that modifies the desired coordinate transformation. For example, Winkler et al. (1984) imposes as "stiffness factor" that controls the ratio of adjacent mesh intervals. To a similar end, Brackbill and Saltzman (1982) control the "global smoothness of the mapping," which may enhance the stability properties. Dwyer et al. (1980) have implemented methods that compute the transformation explicitly and hold it fixed for the time step.

The results of a two-dimensional adaptive-mesh computation to predict the flow over a burning droplet are shown in Fig. 3 (taken from Dwyer, 1983). Initially a spherical flame begins to move away from the droplet. As the flame moves further from the droplet, it sweeps downstream with the flow. The flame lies in the narrow region where the temperature increases rapidly and the flow can be seen from its vorticity contours. Since the greatest temperature change and the largest vorticity occur at different locations, two meshes were used, one for the flow field and one for the temperature and chemistry. We held the radial mesh lines fixed and used coordinate transformations between time steps to select the mesh points along each radial line.

B. Variable-Mesh Methods

When the mesh adaptation strategy varies not only the locations but also the numbers of mesh points, then we find it convenient to view the discretization process in terms of the transverse method of lines mentioned at the outset of Section III. In this case we approximate the solution to the general system of PDEs,

$$F(t, x, y, y_x, y_{xx}, y') = 0, \quad t \geq 0, \quad x \in [a, b], \tag{42}$$

by a sequence of approximate solutions $y^{n+1}(x)$ to two-point boundary-value problems at advancing times t^{n+1}.

Given $y^n(x)$ at time t^n, we obtain $y^{n+1}(x)$ at time t^{n+1} by approximately solving the boundary-value problem that results from replacing the time

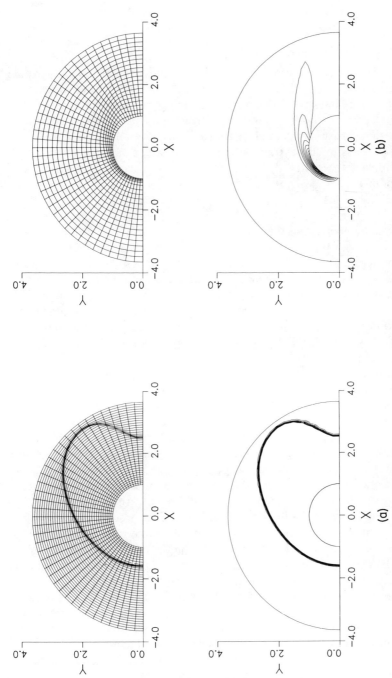

Fig. 3. Coordinate systems, isotherms, and isovorticity contours for the computation of flow around a burning droplet. The problem began as a spherical flame at the droplet surface, but by the time shown here the flame is beginning to be swept downstream.

derivative by a first-order backward difference:

$$F\left(t^{n+1}, x, y^{n+1}, y_x^{n+1}, y_{xx}^{n+1}, (y^{n+1} - y^n)/h_n\right) = 0. \tag{43}$$

A higher order time discretization could be employed to achieve larger time intervals. However, we hope to avoid time discretizations that depend on many previous steps because computer memory capacity can severely limit the problem size. We are currently investigating the use of higher order one-step methods, such as the implicit Runge–Kutta formulas. In any case, our purpose here is to concentrate on the spatial aspects of the solution.

We replace the spatial derivatives by finite differences on a sequence of mesh points:

$$a = x_1^{n+1} < x_2^{n+1} < \cdots < x_J^{n+1} = b, \tag{44}$$

where the number J and the location of the points vary with the time step. Alternatively, for two-point boundary-value problems we could use software that employs collocation or the finite-element method. We have chosen to use second-order finite differences so that we can discretize the transport equations to ensure mass and energy conservation. If M is the number of dependent variables, then there results a system of $M \times J$ nonlinear equations to be solved for each of the M components of $y^{n+1}(x)$ at each of the J mesh points. We extend y^{n+1} by linear interpolation to a function defined upon all of $[a, b]$.

In contrast to coordinate transformation approaches, or any approach that uses a fixed number of points, the variable-mesh method allows considerable freedom in selecting the mesh points at each time level. Both the number and location of the points can be varied as accuracy requirements dictate. If the elliptic problems of the last few time steps employ identical numbers of points, then we presume that the combustion front moves steadily through space, and we locate the points for the next time step by linear extrapolation from the last two steps. We can progress through several time steps in this fashion, provided that the solutions to the elliptic boundary-value problems satisfy our accuracy criteria. This process eventually fails either because the extrapolation does not follow the front precisely or because the combustion process changes. When the new points cannot be obtained by extrapolation, we then borrow some points from the last successful time step and insert new points as needed until the criteria are satisfied at the present time level. In this way the number of points may increase or decrease from one time step to the next.

Unfortunately, there is no convenient general relationship to guide the selection of the points x_j^{n+1} to achieve a desired level of error in the solution to nonlinear two-point boundary-value problems solved by finite differences. Pereyra and Sewell (1975) adaptively add points based on estimates of the

5. Combustion and Chemical Kinetics Modeling

local truncation errors. We prefer to distribute the points according to the geometry of the solution. In order to resolve fronts, we bound the variation in the solution components:

$$\int_{x_{j\pm 1}^{\eta}}^{x_j^{\eta+1}} \left|\frac{dy_m}{dx}\right| \leq \delta(\max y_m - \min y_m); \tag{45}$$

and to resolve the curvature of peaks, we bound the variation in the solution's derivatives:

$$\int_{x_{j\pm 1}^{\eta}}^{x_j^{\eta+1}} \left|\frac{d^2 y_m}{dx^2}\right| \leq \gamma\left(\max \frac{dy_m}{dx} - \min \frac{dy_m}{dx}\right). \tag{46}$$

The scalar constants δ and γ can be adjusted to vary the importance of slope or curvature in the adaptation process. The scaling by the maximum change in the varying quantities ensures that the number of points devoted to each spatial feature remains roughly the same even when new fronts or peaks develop.

Changing mesh points between time steps to suit the elliptic boundary-value problem impacts the selection of the time steps. As in Eq. (9), we choose t^{n+1} so that at each point x_j^{n+1} the local truncation error d_j^{n+1} falls below some predetermined number ε. However, the interpolation error introduces an additional term that is not present in Eq. (9). The local truncation error d_j^{n+1} is bounded by the expression

$$2d_k^{n+1} \leq h_n^2 \tau + (x_k^n - x_j^{n+1})(x_j^{n+1} - x_{k-1}^n)\lambda; \tag{47}$$

in which x_{k-1}^n and x_k^n are the points from the previous time level that surround x_j^{n+1}, and

$$\chi = \max_{[x_j^{\eta}, x_j^{\eta+1}]} |y_{xx}(t^n, x)|, \qquad \tau = \max_{[t^n, t^{n+1}]} |y_{tt}(t, x_j^{n+1})|.$$

The first term in Eq. (47) is the local truncation error of the backward Euler formula [Eq. (9)], and the second term is the approximation error of the linear interpolation for $y^n(x_j^{n+1})$. If the points happen to remain unchanged from the previous time level, then the bound on d_j^{n+1} reduces to that for the backward Euler method. The expression for the maximum time step that accounts for both sources of discretization error is

$$h_n \leq \frac{\varepsilon}{\tau} + \sqrt{\left(\frac{\varepsilon}{\tau}\right)^2 - (x_k^n - x_j^{n+1})(x_j^{n+1} - x_{k-1}^n)\frac{\chi}{\tau}}. \tag{48}$$

These time-step bounds, one for each x_k^n, are estimated a posteriori and the approximate solution $y^{n+1}(x)$ recomputed with a new time step if necessary. The discriminants, necessarily positive, restrict the spatial step size needed to obtain a given local truncation error in the time discretization. This is expressed most simply as

$$x_k^n - x_{k-1}^n \leq 2\varepsilon(|y_{tt}(t^n, x_j^{n+1})| |y_{xx}(t^n, x_j^{n+1})|)^{-1/2}. \tag{49}$$

The constraint can be relaxed by employing a more accurate spatial interpolation formula. Its present form poses no difficulty for the problems we have considered.

To illustrate the application of variable node adaptation methods, we compute the temperature and mass fractions of an idealized unsteady propagating flame. The governing reaction–diffusion equations use a one-step chemical reaction scheme (see Peters and Warnatz, 1983) and can be written in the following form:

$$\partial T/\partial t = \partial^2 T/\partial x^2 + r(T, Y), \tag{50}$$

$$\partial Y/\partial t = (1/\text{Le})(\partial^2 Y/\partial x^2) - r(T, Y), \tag{51}$$

where $T(t, x)$ is a nondimensional temperature, and $Y(t, x)$ a normalized mass fraction of the reactant. The normalized reaction rate r is given by

$$r(T, Y) = \frac{\beta^2 Y}{2 \text{ Le}} \exp\left[-\frac{\beta(1 - T)}{1 - \alpha(1 - T)}\right], \tag{52}$$

in which Le is the Lewis number, β a nondimensional activation energy, and α a nondimensional heat release term. The initial conditions are $T(0, x) = \exp x$ and $Y(0, x) = 1 - \exp(\text{Le } x)$ for $x \leq 0$, and $T(0, x) = 1$ and $Y(0, t) = 0$ for $x > 0$. The boundary conditions require $T = 0$ and $Y = 1$ at $-\infty$ and $T_x = 0$ and $Y_x = 0$ at $+\infty$. In our example ∞ is taken as 50; the other parameters are set as $\alpha = 0.8$, $\beta = 20$, and Le $= 2$.

This problem has a solution in which the flame propagates in an unsteady, oscillating fashion. The flame accelerates and decelerates in a repeatable cycle. Such behavior is often observed in flames that are near their flammability limit (Margolis and Matkowsky, 1983). Accurate predictions of this oscillating behavior require that mesh points be concentrated in the vicinity of the flame. They must be spaced in such a way that both the solution's slope and curvature are accurately resolved. Failure to provide adequate resolution results in solutions that miss the period and amplitude of the oscillations (Smooke and Koszykowski, 1983).

Solutions to the example problem are shown in Figs. 4 and 5. Figure 4 is a plot of the flame speed as a function of time; it shows that the flame oscillates with a period of about 10 sec. The flame acceleration coincides with a rise in temperature, which Fig. 5 depicts through one oscillation. As expected, the variable-mesh method uses the greatest numbers of points during the velocity change, and it uses the least when the flame moves with nearly constant velocity.

5. Combustion and Chemical Kinetics Modeling

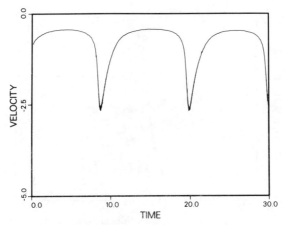

Fig. 4. Flame velocity as a function of time for the example problem.

We have modeled this flame by several methods. As a basis for comparison, we found that 4000 equispaced mesh points were required to predict the correct oscillating behavior. When the variable-mesh methods were used, the solutions required between 100 and 200 points. Since the memory requirement for implicit methods is proportional to the number of mesh points, the savings in storage with the adaptive methods can be enormous. In fact, for a problem with 30–40 chemical species, we require about a half million words of storage for a problem requiring 100 mesh points. Such a problem could not be solved reasonably on a mesh of 4000.

Even though the number of required points was reduced by factors of 20 to 40, the computer times were reduced only by factors of 2 to 4. Thus the computer time savings, relative to fixed-mesh methods, are less than might be expected. The principal reason for this is that the Jacobian has to be recomputed every time the mesh changes, and Jacobian evaluations are expensive. The fixed-mesh calculations can typically use the same Jacobian for about 10 time steps. Some of this inefficiency can be recovered by detecting periods of the solution in which the flame moves with reasonably constant velocity and by projecting the mesh points ahead from one time to another. In this way the Jacobian does not need to be recomputed at each step. In our implementations, for this example such projections cut computer time by about a factor of 2.

For the sake of comparison, we also computed the solution to this example using an arc-length transformation. We found that 3000 points were required to accurately resolve the oscillating behaviour. The surprisingly high number here is because the arc-length distribution does not account for the effect of the solution's curvature on its accuracy. Presumably, many

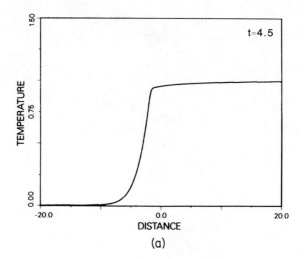

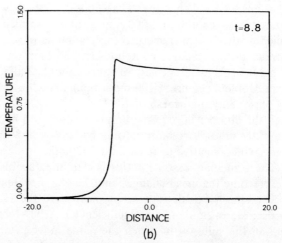

Fig. 5. Temperature profiles at various times during an oscillation of an unsteady propagating flame. (a) $t = 4.5$; (b) $t = 8.8$; (c) $t = 9.4$; (d) $t = 15.6$.

5. Combustion and Chemical Kinetics Modeling

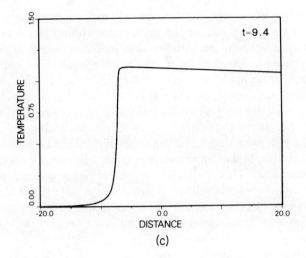

(c)

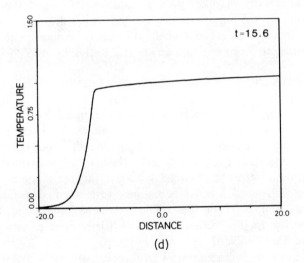

(d)

fewer points would be required if the transformation also took into account the effects of curvature, but we did not test that conjecture on this example.

We now consider how the variable-mesh methods extend to two or more spatial dimensions. The most straightforward extension is to retain a logically rectangular network and add mesh lines in the same way as mesh points were added in one dimension. On one hand, these "tensor product" meshes are inefficient in that they place an entire line of mesh points even though only a few points on the line are required to satisfy the equidistribution criteria. On the other hand, such meshes are efficient because the data structure is simple, that is, any mesh point's neighbors can be determined by a simple index shift. This is especially important on a vector computer like the Cray because complicated data structures usually prohibit vectorization. Nevertheless, even though the data structure is simple, one has to deal with many more points than would be required just to satisfy the equidistribution criteria.

We can reduce the number of mesh points by placing them only where they are required. However, this procedure requires the computation of pointers that identify any mesh point's nearest neighbors. Although this significantly reduces the number of mesh points, it is still unclear whether the inefficiency incurred as a result of the indirect referencing required by the data structure can be overcome. We are still testing both methods.

V. SOLUTION OF THE NONLINEAR EQUATIONS

We have explained how to select time steps and spatial coordinates at which the solutions to systems of PDEs are approximated by discrete values that satisfy systems of finite-difference equations. The difference equations relate the approximations at each time level to those of a few previous levels, so the approximations can be obtained at all levels by marching in time. However, because chemical kinetics equations are stiff and because transport equations are parabolic, the finite-difference equations define the discrete values implicity. In this section we discuss briefly these equations and the methods we use to solve them.

Implicit methods for nonlinear PDEs repeatedly solve systems of nonlinear, nondifferential equations $f(y) = 0$. In this section we simplify the notation by omitting reference to the time level. It is customary to refer to these equations as nonlinear algebraic equations in order to emphasize that finite differences replace exact differentials. Strictly speaking, the equations are not algebraic because they include transport coefficients and chemical kinetics relationships that can be complicated transcendental functions of the dependent variables. The solution of the nonlinear equations is a vector containing one value at each spatial coordinate for each dependent vari-

5. Combustion and Chemical Kinetics Modeling

able in the system of PDEs. Although the number of discrete variables equals the number of algebraic equations, the nonlinear equations may, in general, have multiple "nonphysical" solutions (Schreiber and Keller, 1983). We do not expect to find these because we seek solutions near those of previous time levels.

Since combustion models on interest today routinely have 30–50 chemical species, even modest spatial resolutions produce very large systems of nonlinear equations. Implicit methods have often employed " operator splittings" to break the large systems of nonlinear algebraic equations into many smaller systems (Yanenko, 1971). This process alternately solves the kinetics equations or the transport equations over consecutive time intervals. The kinetics equations at each spatial coordinate are an initial-value problem independent of all other locations; similarly, the transport of each chemical species is parabolic problem only weakly dependent on other species. By repeatedly solving many small problems in this way, it is possible to match the accuracy of the fully coupled finite-difference equations.

Unfortunately, when operator splitting methods are used with combustion models, the stiffness of the kinetics equations greatly increases the expense of solving the initial-value problems at each spatial coordinate. The transport steps (which neglect the chemistry) produce initial values for the subsequent kinetics steps that do not lie on asymptotic solutions of the kinetics equations. As a result, numerical solution of the initial-value problems then requires many small intermediate time steps in order to track the transient solutions into the asymptotic regime. The kinetics equations have solutions that are similar to the dashed lines in Fig. 1. The computational effort required to solve for these transients increases with the problem's stiffness. In our experience, all but a small fraction of the total computation time may be spent passing through these artificially induced transient phases. Successful implementations of operator splittings therefore reformulate the problems to treat the chemistry asymptotically (Oran and Boris, 1981). We have preferred not to make the assumptions on the chemistry models that are required in order to use the asymptotic methods.

Our preferred method of solving the fully coupled system of nonlinear equations $f(y) = 0$ is a defect correction iteration of the form

$$y^{(m+1)} = y^m - A_m^{-1} y^{(m)}, \qquad m = 0, 1, \ldots, \tag{53}$$

in which A_m is a suitably chosen matrix. Our starting iterate $y^{(0)}$ is derived from the previous time level and is usually near the solution of the nonlinear algebraic equations. The very fast convergence of Newton's method occurs by choosing A_m equal to the Jacobian J of f evaluated at $y^{(m)}$:

$$J_{i,j} = \frac{\partial f_i}{\partial y_j}. \tag{54}$$

The costs of repeatedly evaluating the Jacobian make this choice impractical. It is more economical to fix A_m not only through several steps of the defect correction iteration but also through several time levels. At each iteration the change in the norm of successive modified Newton steps is measured, and it is used to determine when a new A_m is needed. For fixed A_m equal to an approximate Jacobian, Smooke (1983) establishes a computable test necessary for the convergence of the sequence $\{y^{(m)}\}$ to the solution of $f(y) = 0$. Similar tests are implemented in most general-purpose software for stiff ODEs. As an alternative to the Jacobian matrix, the success of operator splittings recommends choosing A_m to be a product of the simpler Jacobians of the individual kinetics and transport equations.

Since the Jacobians are central to the solution of the nonlinear equations, it is appropriate to describe in detail the matrices that arise in combustion models. By grouping together the discrete variables and the difference equations associated with the same spatial location, the Jacobians matrices assume a blocked structure. The kinetics equations treat species at a common point, so the differencing of these contributes only to the diagonal blocks. Finite-difference approximations to transport equations mingle species at adjacent points and therefore produce off-diagonal blocks, one above and one below the diagonal for each pair of adjacent points. As noted earlier, global variables, such as pressure, can be handled to preserve the sparse, blocked structure of the Jacobian matrices. This structure is essential to the efficient storing and manipulating of such large matrices. In one spatial dimension the resulting matrices are block tridiagonal. We are currently solving one-dimensional problems whose Jacobians contain on the order of one-half million nonzero entries. In two spatial dimensions the matrices are block pentadiagonal for the orthogonal spatial meshes we employ. Chemical models in two or three dimensions, with sufficient complexity to be of interest to combustion chemists, yield Jacobian matrices with too many nonzero entries to fit into the central memories of existing scientific computers.

The Jacobian also reflects the qualitative physical differences between the kinetics and transport equations. On one hand, the chemical production equations are highly nonlinear and tightly coupled. They contribute dense diagonal blocks whose entries range over several orders of magnitude and dominate the matrix. On the other hand, diffusion terms such as

$$\partial/\partial x \, (\alpha_k \, \partial y_k/\partial x)$$

only loosely couple the species because the diffusion coefficients α_k depend strongly on the kth species and weakly on the others. Thus, each of the off-diagonal blocks of the Jacobian tend to be diagonally dominant. Convective transport terms of the form

$$u \, \partial y_k/\partial x$$

are nonlinear in that the velocity is itself a dependent variable. Such first-order derivatives contribute to the Jacobian's nonsymmetry. As might be expected from the disparate time scales of combustion models, the Jacobian matrices are quite ill conditioned. We have seen cases in which the problems become "too stiff". If some species react extremely rapidly compared with others, then the Jacobian can have a condition number that is large enough to prevent its accurate factorization with available machine precision. In these cases some rescaling of the problem is required.

Evaluation of the Jacobian is one of the greater expenses of solving the nonlinear equations. We do not compute the partial derivatives analytically because the difference equations are sufficiently complex to make differentiation arduous and prone to error. Moreover, coefficients within the equations are often based on tables of experimental data for which no derivatives exist. We approximate each entry of the Jacobian by a one-sided finite-difference formula,

$$J_{i,j} \approx f_i(y_j + \delta) - f_i(y_j)/\delta, \qquad (55)$$

in which the perturbation is of the usual form:

$$\delta = r \times y_j + a. \qquad (56)$$

We do not use the more accurate centered difference formulas because they require double the number of expensive function evaluations. Whatever errors result from the Jacobian evaluation usually impact the defect correction process Eq. (53) much less than the retention of A_m over many steps. Choosing the perturbation δ is the greatest source of difficulty in forming the Jacobian. It must be small enough for the difference quotient to represent the derivative accurately, yet it must be large enough for the subtraction not to be affected by computer precision. In Eq. (56) we take the relative perturbation r as the square root of the computer's unit round-off. The absolute perturbation a is problem dependent and should vary somewhat with the magnitude of y_j and f_i. In ODE software for initial-value problems, a is related to the accuracy tolerances set by the user. Curtis and Reid (1974) discuss the numerical evaluation of Jacobian matrices in more detail.

The task of constructing the Jacobian is eased by the possibility of producing all the entries in several columns "at once." This occurs because the software for evaluating the vector-valued function f is most efficient when all components are computed together to exploit common subexpressions and the speed of vector arithmetic on today's scientific computers. Since the Jacobian matrix is banded, each column contains mostly zeros. That is, many of the component functions of f are functions of mutually exclusive sets of variables. When two or more columns of the Jacobian share no non-zero rows, then all can be evaluated numerically by a single perturbation of the vector of variables. This greatly reduces the number of function evaluations required to approximate the Jacobian. These ideas were first exploited by Curtis, et al. (1974). Even with these savings, we have found the Jacobian matrix evaluation to account for as much as 90% of the total computation time in some one-dimensional flame problems involving large reaction mechanisms.

The solution of the linear equations in the defect correction iteration can also account for a major portion of the overall computation time. In models with one spatial dimension, the most effective solution technique is Gaussian elimination, that is, triangular factorization followed by substitution. This makes efficient use of time because the cost of one factorization can be amortized over the several defect correction steps that employ the same matrix. It also uses memory efficiently because the triangular factors exactly replace the nonzero entries of the block-tridiagonal matrix. The matrices of two-dimensional problems cannot be factored without creating additional nonzero entries. Iterative methods are attractive for these. For orthogonal arrays of spatial points, we use both the alternating-direction implicit method and the block-line successive overrelaxation method (Young, 1971). The major work of each is the solution of several one-dimensional problems along designated rows or columns of the mesh. These methods work best when spatial features of the solution, such as flame fronts, lie perpendicular to the designated rows or columns.

More generally, Elman's survey (1982) and the recent work of Faber and Manteuffel (1984) make it appear unlikely that general-purpose algorithms that can be applied universally to nonsymmetric systems of linear equations will be found. It will be necessary to develop specialized methods that take advantage of the chemical, physical, and spatial structure of the problem. One promising alternative has been proposed by Bank (1981), who uses multigrid methods to solve the linear equations that arise in the Newton iteration.

REFERENCES

Bank, R. E. (1981). A multi-level iterative method for nonlinear elliptic equations. *In* "Elliptic Problem Solvers" (M. Schultz, ed.), pp. 1–16. Academic Press, New York.
Berger, M., and Oliger, J. (1983). "Adaptive Mesh Refinement for Hyperbolic Partial Differential Equations," Rep. NA-83-02. Computer Science Department, Stanford University, Stanford, California.
Bolstad, J. H. (1982). An adaptive finite difference method for hyperbolic systems in one space dimension. Ph.D. Thesis, Stanford University, Stanford, California.
Brackbill, J. U., and Saltzman, J. S. (1982). Adaptive zoning for singular problems in two dimensions. *J. Comput. Phys.* **46**, 342–368.

Branch, M. C., Kee, R. J., and Miller, J. A. (1982). A theoretical investigation of mixing effects in the selective reduction of nitric oxide by ammonia. *Combust. Sci. Technol.* **29**, 147–165.

Coltrin, M. E., Kee, R. J., and Miller, J. A. (1984). A mathematical model of the coupled fluid mechanics and chemical kinetics in a chemical vapor deposition reactor. *J. Electrochem. Soc.* **131**, 425–434.

Curtis, A. R., and Reid, J. K. (1974). The choice of step length when using differences to approximate Jacobian matrices. *J. Inst. Math. Appl.* **13**, 121–126.

Curtis, A. R., Powell, M. J., and Reid, J. K. (1974). On the estimation of sparse Jacobian matrices. *J. Inst. Math. Appl.* **13**, 117–119.

Curtiss, C. F., and Hirschfelder, J. O. (1952). Integration of stiff equations. *Proc. Natl. Acad. Sci. U.S.A.* **38**, 235–243.

Davis, S. F., and Flaherty, J. E. (1982). An adaptive finite element method for initial-boundary value problems for partial differential equations. *SIAM J. Sci. Stat. Comput.* **3**, 6–27.

Dwyer, H. A. (1983). "Grid Adaption for Problems with Separation, Cell Reynolds Number, Shock-Boundary Layer Interaction, and Accuracy," AIAA Pap. 83-0449, 21st Aerosp. Sci. Meet., January, Reno. Am. Inst. Aeron. Astron., New York.

Dwyer, H. A., Kee, R. J., and Sanders, B. R. (1980). Adaptive grid methods for problems in fluid mechanics and heat transfer. *AIAA J.* **18**, 1205–1212.

Dwyer, H. A., Smooke, M. D., and Kee, R. J. (1982). Adaptive gridding for finite difference solutions to heat and mass transfer problems. *In* "Numerical Grid Generation" (J. F. Thompson, ed.), pp. 339–356. North-Holland, Publ., Amsterdam.

Elman, H. C. (1982). Iterative methods for large, sparse, nonsymmetric systems of linear equations. Ph.D. Thesis, Res. Report No. 229. Department of Computer Science, Yale University, New Haven, Connecticut.

Faber, V., and Manteuffel, T. A. (1984). Conjugate gradient methods. *SIAM J. Numer. Anal.* **21**, No. 2, 352–362.

Flaherty, J. E., Coyle, J. M., Ludwig, R., and Davis, S. F. (1983). Adaptive finite element methods for parabolic partial differential equations. *In* "Proceedings of the Army Research Office Workshop on Adaptive Methods for Partial Differential Equations," (I. Babuška, J. Chandra, J. E. Flaherty, eds.), pp. 144–164. SIAM, Philadelphia.

Gear, C. W. (1971). "Numerical Initial Value Problems in Ordinary Differential Equations." Prentice-Hall, Englewood Cliffs, New Jersey.

Gear, C. W., and Petzold L. R. (1983). ODE methods for the solution of differential/algebraic systems. *SIAM J. Numer. Anal.* **21**, No. 4, 716–728.

Gelinas, R. J., Doss, S. K., and Miller, K. (1981). The moving finite element method: Application to general partial differential equations with multiple large gradient. *J. Comput. Phys.* **40**, 202–249.

Hindmarsh, A. C. (1972). "Linear Multistep Methods for Ordinary Differential Equations," Rep. UCRL-51186. Lawrence Livermore Lab., Lawrence, California.

Hindmarsh, A. C. (1983). ODEPACK, a systematized collection of ODE solvers. *In* "Scientific Computing" (R. S. Stepleman *et al.*, eds.), Vol. 1, pp. 55–64. *IMACS Trans. on Scientific Computation*, North-Holland Publ., Amsterdam.

Kautsky, K., and Nichols, N. K. (1980). Equidistributing meshes with constraints. *SIAM J. Sci. Stat. Comput.* **1**, 499–511.

Kee, R. J., and Miller, J. A. (1981). "A Computational Model for Chemically Reacting Flow in Boundary Layers, Shear Layers, and Ducts," Sandia Natl. Lab. Rep. SAND 81-8241. Sandia Lab., Albuquerque, New Mexico.

Kee, R. J., and Miller, J. A. (1984). Computational modeling of flame structure. *Physica D*, **12D**, 198–211.

Kee, R. J., Miller, J. A., and Jefferson, T. H. (1980). "CHEMKIN: A General-Purpose, Problem-Independent, Transportable, Fortran Chemical Kinetics Code Package," Sandia Natl. Lab. Rep. SAND 80-8003. Sandia Lab., Albuquerque, New Mexico.

Margolis, S. B., and Matkowsky, B. J. (1983). Nonlinear stability and bifurcation in the transition from laminar to turbulent flame propagation *Combust. Sci. Technol.* **34**, 45–77.

Miller K., and Miller, R. (1981). Moving finite elements, I. *SIAM J. Numer. Anal.* **18**, 1019–1032.

Oran, E. S., and Boris, J. P. (1981). Detailed modeling of combustion systems. *Prog. Energy Combust. Sci.* **7**, 1–72.

Pereyra, R., Sewell, E. G. (1975). Mesh selection for discrete solution of boundary value problems in ordinary differential equations. *Numer. Math.* **23**, 261–268.

Peters, N., and Warnatz, J., eds. (1983). "Numerical Methods in Laminar Flame Propagation." Vieweg, Wiesbaden.

Petzold, L. R. (1982a). "A Description of DASSL: A Differential-Algebraic System Solver," Sandia Natl. Lab. Rep. SAND 82-8637. Sandia Lab., Albuquerque, New Mexico.

Petzold, L. R. (1982b). Differential/algebraic equations are not ODEs. *SIAM J. Sci. Stat. Comput.* **3**, 367–384.

Petzold, L. R. (1983a). "Multistep Methods: An Overview of Methods, Codes, and New Developments," Sandia Natl. Lab. Rep. SAND 83-8673. Sandia, Albuquerque, New Mexico.

Petzold, L. R. (1983b). Automatic selection of methods for solving stiff and nonstiff systems of ODES. *SIAM J. Sci. Stat. Comput.* **4**, 136–148.

Schreiber, R., and Keller, H. B. (1983). Spurious solutions in driven cavity calculations. *J. Comput. Phys.* **49**, 165–172.

Shampine, L. F. (1981). Type-insensitive ODE codes based on implicit A-stable formulas. *Math. Comput.* **36**, 499–510.

Shapine, L. F. (1982a). Type-insensitive codes based on implicit A(α)-stable formulas. *Math. Comput.* **39**, 109–123.

Shampine, L. F. (1982b). Solving ODEs quasi steady state. *In* "Numerical Integration of Differential Equations and Large Linear Systems" (J. Hinze, ed.), 234–245. Springer-Verlag.

Shampine, L. F., and Gear, C. W. (1979). A user's view of solving stiff ordinary differential equations. *SIAM Rev.* **21**, No. 1.

Smooke, M. D. (1982). Solution of burner-stabilized premixed laminar flames by boundary value methods. *J. Comput. Phys.* **48**, 72.

Smooke, M. D. (1983). An error estimate for the modified Newton method with application to the solution of nonlinear two-point boundary value problems. *J. Opt. Theory Appl.* **39**, 489–511.

Smooke, M. D., and Koszykowski, M. L. (1983). Fully adaptive one-dimensional mixed initial-boundary value problems with applications to unstable problems in combustion. *SIAM J. Sci. Stat. Comput.* (to be published).

Smooke, M. D., Miller, J. A., and Kee, R. J. (1983a). Numerical solution of burner stabilized pre-mixed laminar flames by an efficient boundary value method. *In* "Numerical methods in Laminar Flame Propagation" (N. Peters and J. Warnatz eds.), pp. 112–129. Vieweg, Wiesbaden.

Smooke, M. D., Miller, J. A., and Kee, R. J. (1983b). Determination of adiabatic flame speeds by boundary value methods. *Combust. Sci. Technol.* **34**, 79.

White, A. B. (1983). On the numerical solution of initial/boundary-value problems in one space dimension. *SIAM J. Numer. Anal.* **19**, 683–697.

Winkler, K. H. A. Norman, M. L., and Newman, M. J. (1984). Adaptive mesh techniques for fronts in star formation. *Physica D (Amsterdam)* **12D**, 408–425.

Yanenko, N. N. (1971). "The Method of Fractional Steps." Springer-Verlag, Berlin and New York.

Young, D. M. (1971). "Iterative Solution of Large Linear System." Academic Press, New York.

6

Implicit Adaptive-Grid Radiation Hydrodynamics

KARL-HEINZ A. WINKLER*, MICHAEL L. NORMAN*,
and DIMITRI MIHALAS[†]

Max-Planck-Institut für Physik und Astrophysik
Institut für Astrophysik
D-8046 Garching bei München
Federal Republic of Germany

I. Introduction	146
II. Physical Equations	148
III. Adaptive-Mesh Equations	153
IV. Numerical Equations	155
V. The Adaptive Mesh	157
VI. Numerical Techniques	160
A. Tensor Artificial Viscosity	160
B. Artificial Mass and Heat Diffusion	162
C. On the Use of $\Delta\xi$, Δm, and $(\delta m/\delta t)$	163
D. Advection	164
E. Solution Procedure and Control	166
VII. Ordinary Gas Dynamics: Shock Tubes	169
VIII. Radiation Hydrodynamics: A Supercritical Shock	173
IX. A "Hilbert Program" for Nonlinear Radiation Hydrodynamics	179
References	183

---

* Present address: Max-Planck-Institut Für Astrophysik, D-8046 Garchingbei München, Federal Republic of Germany.
† Present address: High Altitude Observatory, National Center for Atmospheric Research, Boulder, Colorado.

I. INTRODUCTION

The term *radiation hydrodynamics* refers to flows in which radiation contributes significantly to, or totally dominates, the energy and momentum balance in the flow. In such flows radiation plays at least three important roles: (1) It can drive the *dynamics* of the flow via direct energy and momentum input. These processes may be highly nonlocal because radiation produced at one region in the flow (e.g., a star) may be absorbed at a remote (i.e., many characteristic flow lengths away) location in a transparent medium (e.g., a stellar wind) in the vicinity of the source. (2) Radiation can drive the *kinetics* of the flow via radiation-induced changes in the internal state of the material (e.g., in dissociation–ionization fronts). In the most general case one may have to account for intricate chains of radiative and collisional processes that determine the occupation numbers in bound and free states of the material. (3) Finally, radiation provides the *diagnostics* from which an outside observer can infer the internal properties and the dynamic state of the fluid. Here, very elaborate calculations may be required to describe complex mechanisms (e.g., angle-frequency-coupled redistribution by scattering) of the radiation transport in the moving material. In this chapter we consider only the first of the three items just mentioned.

Radiation-coupled flows occur in a wide variety of physical arenas ranging from microscopic phenomena in the laboratory (e.g., laser–fusion experiments) to experiments on a macroscopic scale (e.g., nuclear fireballs), to astrophysical flows on all scales up to the size of the Universe itself. Consistent with our professional backgrounds, we focus mainly on problems occurring in astrophysical contexts. These are, in fact, ideal for the purpose because they span such an immense range of physical parameters, from the extremely rarefied conditions in the intergalactic medium, through the interstellar medium, to stellar envelopes, into the stellar interior, to compact objects at nuclear densities (e.g., neutron stars), and, finally, to collapsed objects where general relativistic effects come into play and the flow recedes beyond the event horizon.

In attempting to treat radiating flows under such vastly different conditions, one encounters extremely challenging physical and numerical problems because of the enormous range of length and time scales present. Thus, typical length scales in a single flow may range from a characteristic size L of the whole medium in regions where the density distribution is flat to a gravitational length R in regions having a power-law density distribution, to a scale height H in an exponentially stratified region, to a photon-mean-free path λ_p, and finally down to a particle-mean-free path λ. Likewise, if l is a typical flow length, we find time scales ranging from photon flight times λ_p/c (shakes) to radiation-flow times l/c (milliseconds to seconds) to photon diffusion times $l^2/\lambda_p c$ (days to years); and from fluid-flow time scales l/v_{sound}

(minutes) to pulsation times R/v_{sound} (hours to days), to gravitational collapse times $(R^3/GM)^{1/2}$ (thousands of years), to Kelvin–Helmholtz contraction times (GM^2/RL) (10^7 yr), up to nuclear burning evolution times (10^{10} yr). Furthermore, one may encounter the entire range of flow velocities from subsonic to supersonic to extreme relativistic.

From the macroscopic point of view, it is useful to consider the *radiating fluid* as a composite gas comprising material particles and photons. At this level several distinctive (compared with ordinary hydrodynamics) features enter the problem. First, there is the fact that typically $\lambda_p \gg \lambda$, whereas $c \gg v_{\text{sound}}$ implies that *radiant energy transport always dominates the total energy flow within the system* and, therefore, must be treated with special care. Second, the mere fact that $c \gg v_{\text{sound}}$ means that the time evolution of the radiation field will generally be very stiff compared with the flow-induced evolution of the material. Third, the radiation energy and momentum equations have a hyperbolic character in transparent material (where they combine to yield the wave equation) but become parabolic in opaque material (where they combine to yield the time-dependent diffusion equation). Fourth, unlike the material component, for which we can usually assume isotropy of the microscopic distribution function (or at least a Chapman–Enskog-like solution) in order to close the hierarchy of moments and evaluate transport coefficients, the photon distribution function can become extremely anisotropic in transparent regions of the fluid, where photon-mean-free paths may approach (or exceed) all characteristic structural lengths in the flow. Here we must *construct* a closure relation, just as one must for the free-molecule flow of material gases. An unpleasant consequence of this fact is that, rigorously speaking, the dimensionality of the problem increases because we must somehow account for the angular variation of the radiation field.

The situation becomes even more complex at the microscopic level, where we encounter a bewildering array of microphysics and a wide variety of relaxation processes. Even if we assume that the material remains in local thermodynamic equilibrium (LTE), we require detailed equation-of-state and opacity data, and perhaps a host of nuclear reaction rates. In many cases internal processes, such as excitation, dissociation, and ionization, produce drastic variations in material properties, which are extremely difficult to cope with numerically. Moreover, the physics of the material–radiation interaction can be extremely complex, and the dimensionality of the problem is enlarged even further if we attempt to account for the spectral characteristics of the radiation field (which may be far from equilibrium) and material cross sections. In some cases (e.g., general Compton scattering or relativistic flows) we may have an inextricable angle–frequency coupling. Finally, the problem rapidly becomes unmanageable if we must treat non-LTE processes in any detail.

In the face of such difficulties, it is obvious from the outset that one requires a robust algorithm for the computation of astrophysical (and other!)

radiation-hydrodynamics problems. In this chapter we describe some of the major elements of the numerical techniques implemented thus far in the one-dimensional radiation-hydrodynamics code WH80s, which is especially designed to handle the problems arising from the multiple time and length scales described above.

The code employs an implicit, adaptive-mesh, finite-difference technique. The *adaptive mesh* is constructed in such a way as to resolve features in the flow, so that the "average" change in flow structure (measured in a sense to be described later) from grid zone to grid zone is the same throughout the entire grid. Local grid refinement of up to a factor of a million is achieved. Further, the mesh is designed to track features by solving the grid equation implicitly, along with the equations for the flow variables. Thus, a very steep and narrow flow feature can travel many (up to 50) times its own width in physical space, while moving less than one zone relative to the adaptive mesh. The equations governing the grid motion contains several key elements:

(a) a function that provides an objective measure of *structure* in the flow;
(b) an *elliptic operator* that determines a globally consistent mesh in response to the structure in the flow;
(c) a *stiffness operator* that prevents tangling of the mesh;
(d) a parabolic term employing *asymmetric time-filtering*, which ensures a smooth time evolution of the grid.

Other essential numerical techniques used in the code include a tensor artificial viscosity using either a fixed length or a fixed relative length; a second-order accurate, implicit, monotonic, upwind, interface-centered, flux-differencing algorithm for advection; and a numerically rugged treatment of mass–flux terms.

In this chapter we concentrate on the present state of the art. But it is extremely important to keep current efforts in perspective, lest we lose sight of the ultimate goal by becoming too preoccupied with, and enmeshed in, a thicket of ephemeral detail. We therefore briefly sketch a preliminary attempt to formulate what one might call a "Hilbert program" for computational nonlinear radiation hydrodynamics. Only the first step of this program is achievable on present-day machines (CRAY-1 class machines). The next level is within reach on machines expected imminently. Subsequent steps must await a new generation of machines with massive memories and parallel-processing capabilities. We expect to refine and enlarge this preliminary program as our experience grows and our insight deepens.

II. PHYSICAL EQUATIONS

In this section we state the physical conservation relations governing the dynamics of one-dimensional, spherically symmetric, nonrelativistic flows of

6. Implicit Adaptive-Grid Radiation Hydrodynamics

a radiating fluid. Magnetohydrodynamic effects are ignored, and the material is assumed to be a single fluid in local thermodynamic equilibrium (LTE). Derivations of the equations and discussion of the underlying physics can be found in Castor (1984), Mihalas and Weaver (1982), Mihalas and Mihalas (1984), and Pomraning (1973).

The equations governing the dynamics of the radiation field follow ultimately from the *radiation transport equation*, which, in the Eulerian frame, can be written

$$\left(\frac{1}{c}\frac{\partial}{\partial t} + \mathbf{n}\cdot\boldsymbol{\nabla}\right)I(\mathbf{x},t;\mathbf{n},v) = \eta(\mathbf{x},t;\mathbf{n},v) - \chi(\mathbf{x},t;\mathbf{n},v)I(\mathbf{x},t;\mathbf{n},v). \tag{1}$$

This equation describes the space–time evolution of the photon distribution function, as measured in the (stationary) laboratory frame. Here $\mathbf{n}$ is the direction of photon propagation and v the frequency, both measured in the lab frame; χ is the total opacity and η the total emissivity of the material, both per unit volume. In Eq. (1) the lab frame opacity and emissivity are anisotropic, even though all material properties are isotropic in the comoving fluid frame because of Doppler shifts and aberration between frames. These seemingly small velocity-dependent terms have important physical consequences, particularly in the diffusion limit, and cannot be neglected (see, e.g., Mihalas and Mihalas, 1984, Sections 93 and 94). Their presence significantly complicates the radiation momentum and energy equations derived from Eq. (1).

A physically more sensible approach is to use the *Lagrangian radiation transport equation* (Castor, 1972; Buchler, 1979; Mihalas and Mihalas, 1984, Section 95):

$$\frac{\rho}{c}\frac{D}{Dt}\left[\frac{I_0(\mu_0,v_0)}{\rho}\right]$$
$$+ \frac{\partial}{\partial \mathrm{Vol}}\left[\mu_0 r^2 I_0(\mu_0,v_0)\right]$$
$$+ \frac{\partial}{\partial \mu_0}\left[(1-\mu_0^2)\left\{\frac{1}{r} + \frac{\mu_0}{c}\left(\frac{3u}{r} + \frac{D\ln\rho}{Dt}\right) - \frac{a}{c^2}\right\}I_0(\mu_0,v_0)\right]$$
$$- \frac{\partial}{\partial v_0}\left[v_0\left\{(1-3\mu_0^2)\frac{u}{cr} - \frac{\mu_0^2}{c}\frac{D\ln\rho}{Dt} + \frac{\mu_0 a}{c^2}\right\}I_0(\mu_0,v_0)\right]$$
$$= \eta_0(v_0) - \left[\chi_0(v_0) + \left\{(1-3\mu_0^2)\frac{u}{cr} - \frac{\mu_0^2}{c}\frac{D\ln\rho}{Dt} + \frac{2\mu_0 a}{c^2}\right\}\right]I_0(\mu_0,v_0). \tag{2}$$

In Eq. (2) all quantities, including angles and frequencies, are measured in the comoving fluid frame. As discussed in the references cited, the physical advantages of this approach are substantial. In particular, we now work in the proper frame of the material, in which it is easiest to handle microphysical processes governing emission, absorption, and scattering of radiation, and the coupling of the radiation field to the equations determining the occupation numbers of various states in the material. Furthermore, Eq. (2) leads to radiation energy and momentum equations that admit a clear physical interpretation and have a structure closely parallel to their material counterparts.

Thus, taking the zeroth moment of Eq. (2) over angle, and integrating over frequency, we obtain the *radiation energy equation*

$$\rho \frac{D}{Dt}\left(\frac{E_0}{\rho}\right) + \frac{1}{r^2}\frac{\partial}{\partial r}(r^2 F_0) - P_0 \frac{D \ln \rho}{Dt} + (E_0 - 3P_0)\frac{u}{r} + \frac{2aF_0}{c^2}$$
$$= \rho[4\pi\varkappa_P B_0(T) - c\varkappa_E E_0]. \tag{3}$$

Here E_0, F_0, and P_0 are the comoving-frame radiation energy density, flux, and pressure; u and a the fluid velocity and acceleration; $B(T) = \sigma T^4/\pi$ the integrated Planck function; and $\varkappa_P$ and $\varkappa_E$, respectively, the *Planck mean opacity* (per gram),

$$\varkappa_P \equiv \int_0^\infty \varkappa_0(v_0) B(v_0, T)\, dv_0 / B_0(T), \tag{4}$$

and *absorption-mean opacity* (per gram),

$$\varkappa_E \equiv \int_0^\infty \varkappa_0(v_0) E_0(v_0)\, dv_0 \bigg/ \int_0^\infty E_0(v_0)\, dv_0. \tag{5}$$

In writing Eq. (3) we invoked LTE, which implies that the thermal emissivity $\eta_0^t(v_0) = \varkappa_0(v_0)B(v_0, T)$; in addition, we assumed that any scattering terms are coherent and isotropic, in which case they cancel out exactly from Eq. (3).

Combining Eq. (3) with the material internal energy equation, we obtain a *radiating-fluid internal energy equation*:

$$\rho\left[\frac{D}{Dt}\left(e + \frac{E_0}{\rho}\right) + (P + P_0)\frac{D}{Dt}\left(\frac{1}{\rho}\right)\right] + \frac{1}{r^2}\frac{\partial}{\partial r}[r^2(F_0 + F_c)]$$
$$+ \frac{(E_0 - 3P_0)u}{r} + \frac{2aF_0}{c^2} = \rho\varepsilon_N. \tag{6}$$

Here e and P are the material internal energy and pressure, F_c the convective flux (henceforth ignored), and ε_N the rate of thermonuclear energy release per unit mass.

Taking the first moment of Eq. (2) over angle and integrating over frequency, we obtain the *radiation momentum equation*

$$\frac{\rho}{c^2} \frac{D}{Dt}\left(\frac{F_0}{\rho}\right) + \frac{\partial P_0}{\partial r} + \frac{(3P_0 - E_0)}{r} + \frac{F_0}{c^2}\frac{\partial u}{\partial r} + \frac{a}{c^2}(E_0 + P_0) = -\rho \frac{\varkappa_F}{c} F_0. \quad (7)$$

Here the *flux-mean opacity* (per gram),

$$\varkappa_F \equiv \int_0^\infty \varkappa_0(\nu_0) F_0(\nu_0)\, d\nu_0 \bigg/ \int_0^\infty F_0(\nu_0)\, d\nu_0, \quad (8)$$

must include the contribution (if any) from scattering. Equation (7) couples directly to the *material momentum equation*

$$\rho \frac{Du}{Dt} + \frac{\partial P}{\partial r} + \frac{4\pi G m \rho}{r^2} = \rho \frac{\varkappa_F}{c} F_0. \quad (9)$$

Here m is the total mass (per steradian) contained inside radius r:

$$m(r) = \int_0^r \rho(x) x^2\, dx. \quad (10)$$

Equation (10) is equivalent to a solution of Poisson's equation in spherical symmetry. The system is completed by the *material continuity equation*

$$D\rho/Dt + (\rho/r^2)(\partial/\partial r)(r^2 u) = 0. \quad (11)$$

Equations (3)–(11) completely determine the dynamics of the flow provided that we are given (1) the closure between P_0 and E_0, and (2) the mean opacities $\varkappa_p$, $\varkappa_E$, and $\varkappa_F$ as functions of material state variables (say ρ and T). To relate P_0 to E_0, we must know the angular distribution of the radiation field. To construct mean opacities, we must know the spectral profiles

$$e(\nu_0) \equiv E_0(\nu_0)/E_0 \quad (12)$$

and

$$f(\nu_0) \equiv F_0(\nu_0)/F_0 \quad (13)$$

describing the frequency distribution of the radiation field. In practice, neither of these distributions are known a priori. Two important ideas for overcoming this difficulty were suggested by the developers of the VERA code (Freeman et al., 1968).

First, to obtain closure between P_0 and E_0, introduce the *variable Eddington factor*

$$f \equiv P_0/E_0 = \int_{-1}^{1} I_0(\mu_0) \mu_0^2\, d\mu_0 \bigg/ \int_{-1}^{1} I_0(\mu_0)\, d\mu_0, \quad (14)$$

which, in general, is a function of frequency. For isotropic radiation, $f = \frac{1}{3}$. For radially streaming radiation, $f = 1$. The assumption is then made that during a hydrodynamic time step this ratio is fixed and will subsequently be updated in a subiteration on the angle-dependent transport equation by the use of current values of material properties (hence, source–sink terms).

Second, Freeman et al. (1968) devised the *multifrequency–gray* approach, in which it is assumed that during a hydrodynamic time step the spectral profiles in Eqs. (12) and (13) remain fixed so that one can solve the equivalent "gray" (more precisely, frequency-integrated) equations (3) and (7), and then subsequently update the profiles by solving frequency-dependent (multigroup) transport or moment equations, again using current values of the material properties.

These procedures have been used with considerable success in a variety of astrophysical problems. For nonrelativistic flows they are, in fact, astute splittings of the problem, because the Eddington factor depends mainly on geometry (e.g., distance from boundary surfaces and/or intense sources), but only relatively weakly on local source distributions.

For the computation of Eddington factors in many practical situations, it is often adequate to drop the velocity- and time-dependent terms in Eq. (2), and simply solve the static transport equation for a static snapshot of the radiation field. In this limit the frequency derivative vanishes and each frequency can be considered independently. The present code computes the angular distribution of the radiation field from a ray solution of Eq. (2), omitting the time- and velocity-dependent terms (Yorke, 1979, 1980). Alternative procedures are discussed in Sections 83 and 98 of Mihalas and Mihalas (1984). To avoid this step, a large number of ad hoc schemes have been suggested for estimating the Eddington factor. Some of these offer fair success in simple cases. Our view is that such schemes are at best problematical and will inevitably fail for hard problems, an excellent example of which is encountered in protostar calculations—see Fig. 6d of Winkler and Newman (1980).

Execution of the multifrequency step is often hindered by a lack of detailed frequency-dependent opacities, and, in practice, it is often necessary to simply skip the multigroup calculation and to replace $\varkappa_E$ by $\varkappa_P$ and $\varkappa_F$ by the *Rosseland mean* $\varkappa_R$. This procedure is followed in the present code. Other possibilities are outlined in Sections 82 and 97 of Mihalas and Mihalas (1984). It should be noted that if one wishes to derive spectral profiles by solving the monochromatic moment equations derivable from Eq. (2), then it is necessary to retain the frequency derivative term (which accounts for photon blue and red shifts induced by adiabatic compression or expansion of the radiating fluid). The equations are then frequency coupled, even in the absence of frequency redistribution effects in source terms (e.g., by Compton scattering—see Axelrod et al., 1984).

III. ADAPTIVE-MESH EQUATIONS

We now reexpress the radiation and fluid equations of Section II on the adaptive mesh, which moves both in space, and with respect to fluid elements (Winkler et al., 1984c). Let $(\partial/\partial t)$ denote the Eulerian time derivative taken with respect to fixed lab frame coordinates, (D/Dt) the Lagrangian time derivative taken following the motion of a definite material element, and (d/dt) the adaptive-mesh time derivative taken with respect to fixed values of the adaptive-mesh coordinate. The fluid velocity is then $u \equiv (Dr/Dt)$, the grid velocity is $u_{\text{grid}} \equiv (dr/dt)$, and the relative velocity of the fluid with respect to the grid is $u_{\text{rel}} \equiv u - u_{\text{grid}}$. Then, using the adaptive-mesh transport theorem (Winkler et al., 1984b), one can immediately rewrite the equations of Section II in integral conservation form on the adaptive mesh as the adaptive mesh radiation energy equation:

$$\frac{d}{dt}\left[\int_{V(t)} \frac{E_0}{\rho} \rho \, d\text{Vol}\right] + \int_{\partial V} \frac{E_0}{\rho} (\rho \mathbf{u}_{\text{rel}} \cdot d\mathbf{S}) + \int_V \frac{\partial}{\partial \text{Vol}} [r^2 F_0] \, d\text{Vol}$$

$$+ \int_V \left[P_0 \frac{\partial(r^2 u)}{\partial \text{Vol}} + (E_0 - 3P_0)\frac{u}{r} + \frac{2aF_0}{c^2}\right] d\text{Vol}$$

$$= \int_V [4\pi \kappa_p B_0(T) - c\kappa_E E_0]\rho \, d\text{Vol}; \tag{15}$$

the internal energy equation:

$$\frac{d}{dt}\left[\int_{V(t)} \left(e + \frac{E_0}{\rho}\right)\rho \, d\text{Vol}\right] + \int_{\partial V}\left(e + \frac{E_0}{\rho}\right)(\rho \mathbf{u}_{\text{rel}} \cdot d\mathbf{S})$$

$$+ \int_V \frac{\partial}{\partial \text{Vol}} [r^2(F_0 + F_c)] \, d\text{Vol}$$

$$+ \int_V \left[(P + P_0)\frac{\partial(r^2 u)}{\partial \text{Vol}} + (E_0 - 3P_0)\frac{u}{r} + \frac{2aF_0}{c^2}\right] d\text{Vol}$$

$$= \int_V \varepsilon_N \rho \, d\text{Vol}; \tag{16}$$

the adaptive mesh radiation momentum equation:

$$\frac{d}{dt}\left[\int_{V(t)} \frac{F_0}{\rho c^2} \rho \, d\text{Vol}\right] + \int_{\partial V} \frac{F_0}{\rho c^2} (\rho \mathbf{u}_{\text{rel}} \cdot d\mathbf{S})$$

$$+ \int_V \left[\frac{\partial P_0}{\partial r} + \left(\frac{3P_0 - E_0}{r}\right)\right] d\text{Vol} + \int_V \left[\frac{F_0}{c^2}\frac{\partial u}{\partial r} + \frac{a}{c^2}(E_0 + P_0)\right] d\text{Vol}$$

$$= -\int_V \frac{\kappa_F}{c} F_0 \rho \, d\text{Vol}; \tag{17}$$

the momentum equation:

$$\frac{d}{dt}\left[\int_{V(t)} u\rho\, d\text{Vol}\right] + \int_{\partial V} u(\rho \mathbf{u}_{\text{rel}} \cdot d\mathbf{S}) + \int_V \left[\frac{\partial P}{\partial r} + \frac{4\pi Gm}{r^2}\rho\right] d\text{Vol}$$
$$= \int_V \frac{\varkappa_F}{c} F_0 \rho\, d\text{Vol}; \tag{18}$$

the continuity equation:

$$\frac{d}{dt}\left[\int_{V(t)} \rho\, d\text{Vol}\right] + \int_{\partial V} \rho(\mathbf{u}_{\text{rel}} \cdot d\mathbf{S}) = 0; \tag{19}$$

the Poisson equation:

$$dm = \rho\, d\text{Vol}. \tag{20}$$

In Eqs. (15)–(20), $d\text{Vol} \equiv \frac{1}{3}d(r^3)$ denotes an adaptive-mesh volume element specified by fixed values of the adaptive-mesh variable, and $d\mathbf{S}$ is an element of the surface ∂V enveloping the adaptive volume $V(t)$.

To write an adaptive-mesh radiation transport equation, we introduce adaptive grids in angle and frequency and, hence, in the tangent space attached to each event on a photon world line. The result (Winkler *et al.*, 1984b) is shown in the adaptive mesh radiation transfer equation

$$\frac{d}{dt}\left[\int_{\tilde{V}(t)} \frac{I_0(\mu_0, \nu_0)}{c} d\widetilde{\text{Vol}}\right] + \int_{\partial \tilde{V}} \frac{I_0(\mu_0, \nu_0)}{c}(\tilde{\mathbf{u}}_{\text{rel}} \cdot d\tilde{\mathbf{S}})$$

$$+ \int_{\tilde{V}} \frac{\partial}{\partial \text{Vol}} \left[\mu_0 r^2 \qquad\qquad\qquad\qquad I_0(\mu_0, \nu_0)\right] d\widetilde{\text{Vol}}$$

$$+ \int_{\tilde{V}} \frac{\partial}{\partial \mu_0} \left[(1-\mu_0^2)\left\{\frac{1}{r} + \frac{\mu_0}{c}\left(\frac{3u}{r} - \frac{\partial(r^2 u)}{\partial \text{Vol}}\right) - \frac{a}{c^2}\right\} I_0(\mu_0, \nu_0)\right] d\widetilde{\text{Vol}}$$

$$- \int_{\tilde{V}} \frac{\partial}{\partial \nu_0} \left[\nu_0 \quad \left\{(1-3\mu_0^2)\frac{u}{cr} + \frac{\mu_0^2}{c}\frac{\partial(r^2 u)}{\partial \text{Vol}} + \frac{\mu_0 a}{c^2}\right\} I_0(\mu_0, \nu_0)\right] d\widetilde{\text{Vol}}$$

$$= \int_{\tilde{V}} \left(\eta_0(\nu_0) - \left[\chi_0(\nu_0) + \left\{(1-3\mu_0^2)\frac{u}{cr} + \frac{\mu_0^2}{c}\frac{\partial(r^2 u)}{\partial \text{Vol}} + \frac{2\mu_0 a}{c^2}\right\}\right] I_0(\mu_0, \nu_0)\right) d\widetilde{\text{Vol}} \tag{21}$$

In Eq. (21), $\tilde{\mathbf{u}}_{\text{rel}}$ is the generalization of u_{rel} to the three-dimensional space (r, μ_0, ν_0):

$$\tilde{\mathbf{u}}_{\text{rel}} \equiv \left(u - \frac{dr}{dt}, -\frac{d\mu_0}{dt}, -\frac{d\nu_0}{dt}\right); \tag{22}$$

likewise

$$d\widetilde{\text{Vol}} \equiv d\text{Vol}\, d\mu_0\, d\nu_0, \tag{23}$$

and $d\tilde{S}$ is the corresponding surface element. We remind the reader that we do not yet actually solve Eq. (21) in WH80S, but only the static time-independent transport equation.

IV. NUMERICAL EQUATIONS

Difference equation representations of the adaptive-mesh equations above are as follows:

$$\Delta m - \rho \, \Delta \text{Vol} = 0; \tag{24}$$

$$\frac{\delta[\Delta \xi]}{\delta t} + \Delta[r^2 u_{\text{rel}} \rho] = \Delta\left[r^2 \sigma_\rho \frac{\Delta \rho}{\Delta r}\right], \tag{25}$$

$$\frac{\delta[(e + E_0/\rho) \Delta \xi]}{\delta t} - \Delta\left[\frac{\delta m}{\delta t}\left(e + \frac{E_0}{\rho}\right) - r^2(F_0 + F_c)\right] + (P + P_0) \Delta(r^2 u)$$

$$+ \left[(E_0 - 3P_0)\frac{u}{r} + \frac{2aF_0}{c^2}\right] \Delta \text{Vol} = (\varepsilon_N + \varepsilon_Q) \Delta \xi + \Delta\left[r^2 \sigma_e \frac{\Delta e}{\Delta r}\right], \tag{26}$$

$$\frac{\delta[(E_0/\rho) \Delta \xi]}{\delta t} - \Delta\left[\frac{\delta m}{\delta t}\left(\frac{E_0}{\rho}\right) - r^2 F_0\right] + P_0 \Delta(r^2 u)$$

$$+ \left[(E_0 - 3P_0)\frac{u}{r} + \frac{2aF_0}{c^2}\right] \Delta \text{Vol} = [4\pi \varkappa_P B_0 - c \varkappa_E E_0] \Delta \xi, \tag{27}$$

$$\frac{\delta[u \, \Delta \xi]}{\delta t} - \Delta\left[\frac{\delta m}{\delta t} u\right] + r^2 \Delta P + \frac{4\pi G m}{r^2} \Delta \xi = \frac{\varkappa_F}{c} F_0 \Delta \xi + u_Q \Delta \text{Vol}, \tag{28}$$

$$\frac{\delta[(F_0/\rho c^2) \Delta \xi]}{\delta t} - \Delta\left[\frac{\delta m}{\delta t}\left(\frac{F_0}{\rho c^2}\right)\right] + r^2\left(\Delta P_0 + \frac{F_0}{c^2} \Delta u\right)$$

$$+ \left[\frac{3P_0 - E_0}{r} + \frac{a}{c^2}(E_0 + P_0)\right] \Delta \text{Vol} = -\frac{\varkappa_F}{c} F_0 \Delta \xi. \tag{29}$$

The independent variables in these equations are time index n and adaptive-mesh index j; δ denotes a time difference at fixed j, and Δ a spatial difference with respect to j at fixed t. The primary dependent variables are r, m, u, ρ, e, E_0, and F_0; secondary variables derivable from the primary set are u_{rel}, a, P, P_0, T, B_0, and the opacities. In Eqs. (24)–(29) $\Delta \xi$ is a shorthand notation for $\rho \, \Delta \text{Vol}$ where $\Delta \text{Vol} \equiv \frac{1}{3} \Delta(r^3)$. The spatial centering of the primary and secondary (in parentheses) variables is sketched in Fig. 1. Clearly tensors of odd rank are centered at interfaces; tensors of even rank are zone centered.

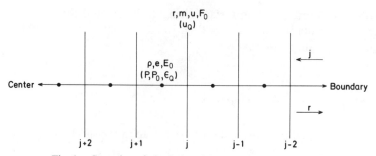

Fig. 1. Centering of physical variables on the adaptive grid.

In Eq. (25) the term on the right-hand side (normally zero) is used to produce artificial mass diffusion when required; similarly, in Eq. (26) the term containing σ_e is used to produce artificial diffusion of internal energy when required. In Eqs. (26) and (28) the terms ε_Q and u_Q represent, respectively, energy dissipation and momentum deposition by artificial viscosity; they are computed according to the algorithms described in Section VI. Equations (31a, b) are the grid equations that determine the adaptive mesh; these equations are discussed in detail in Section V.

With the exceptions of Eqs. (24) and (31a, b), all of which apply at the advanced time level, all physical variables are first centered in time as

$$\langle x \rangle \equiv zx^{n+1} + (1-z)x^n; \quad (30)$$

z lies within the range $0.5 < z \leq 1.0$, with a typical value of 0.55. Thus, the code uses the differential operators on time-averages rather than time averages of the differential operators. With the exception of advected quantities (see below), simple spatial averages are used where necessary for centering; nonlinear terms are represented as products of averages rather than as averages of products.

The advection scheme is described in detail in Section VI, including reasons for using $(-\delta m/\delta t)$ instead of $r^2 u_{rel}\rho$, which is formally identical, in Eqs. (26)–(29). Suffice it to say here that in Eqs. (26) and (27) $e + (E_0/\rho)$ and (E_0/ρ) are evaluated at interfaces using a monotonized interpolation scheme, as are u and $(F_0/c^2\rho)$ in Eqs. (28) and (29). These quantities are then advected according to the sign of $(-\delta m/\delta t)$ at the interface in Eqs. (26) and (27), and according to the sign of the zone-centered spatial average of $(-\delta m/\delta t)$ in Eqs. (28) and (29). In Eq. (29), the zone-centered density ρ_j is used at interface j (instead of an interface-centered spatial average) in order to avoid enlarging the stencil from 5 to 7 points; in nonrelativistic flows this term is too small for the miscentering to have any practical consequences. In Eqs. (26) and (27) we use the spatial averages of $\Delta\xi$, ΔVol, E_0, P_0, and $\varkappa_F$. All other quantities are automatically correctly centered on the staggered mesh.

V. THE ADAPTIVE MESH

The grid distribution $\{r_j\}$ as a function of grid index j is determined from the grid equations (31)–(47a, b), which are solved implicitly with the radiation-hydrodynamics equations. As shown by Eqs. (31a, b), the grid equation contains three key elements, each of which plays a distinctive role. First, the zone-centered *structure function* f^s [see Eqs. (36)–(47a, b)] provides an objective measure of the "structure" of the solution at a given time level. The term Δf^s thus acts as an interface-centered driving term to the rest of the equation, which is globally elliptic but may be locally parabolic (Winkler *et al.*, 1984a). Next, the *radial function* f^r [see Eqs. (34a, b) and (35a, b)] contains the terms that define mesh intervals, stabilize the mesh, and set both a ceiling and a floor on the mesh spacing. As discussed by Winkler *et al.* (1984c), these powerful operators allow mesh points to move instantly to those parts of the computational domain where new structures are developing and better grid resolution is needed, and have proved successful in a wide variety of applications.

In some flows, structure may temporarily disappear locally in some part of the flow, only to reappear a few time steps later, for example, when a shock reflects from a wall. In this event, an instantaneous response of the mesh in that part of the flow is neither necessary nor desirable. Therefore, to avoid an unnecessary (and perhaps catastrophic) double rearrangement of the mesh in such circumstances, we introduce a parabolic term $(\delta r_j/\delta t)$, which is allowed to be operative only in certain parts of the flow. The coefficient of this term is constructed in such a way as to contain "memory" about the history of the structure in the flow, and is used to provide an *asymmetric time-filtering* of the grid motion in the sense that it is set to zero in those parts of the flow where the structure function is increasing (hence, instantaneous grid response is needed) and to a positive number, with a fadeout over a specified number of time steps, in those parts of the flow where the structure function is decreasing. The effect of this operation is to allow grid points to diffuse out of newly structureless regions only slowly and, hence, to keep grid points available temporarily should they suddenly be needed a few time steps later. This method has proved very effective in controlling transient runaways of the grid distribution and has improved overall performance of the code.

Details of the asymmetric time-filtering algorithm are shown in the following:

$$-\text{CVMGP}(0., 1., \delta f)\left(\frac{\delta f}{f^n}\right)^2 \frac{T_{\text{scale}}}{R_{\text{scale}}} \frac{\delta r}{\delta t} + \Delta f^r + \Delta f^s = 0. \tag{31a}$$

$$-\text{CVMGP}(0., 1., \delta f)\left(\frac{\delta f}{f^n}\right)^2 \frac{T_{\text{scale}}}{r} \frac{\delta r}{\delta t} + \Delta f^{rl} + \Delta f^s = 0. \tag{31b}$$

where

$$\delta f = f^s - f^n, \tag{32}$$

$$f^{n+1} = f^s + (1 - \varepsilon)(f^n - f^s)\,\text{CVMGP}(0., 1., \delta f). \tag{33}$$

Here CVMGP is the vector-merge operation defined as $\text{CVMGP}(x, y, z) = x$ if $z > 0$; $= y$ if $z \le 0$.

Equation (31a) is used when the range of the spatial domain is small, and Eq. (31b) is used when the mesh has to cover many orders of magnitude in radius. Both f^s and f^r, as well as the operator Δ are dimensionless; hence, from dimensional arguments, one sees that the coefficient of $(\delta r/\delta t)$ must have units $[T/L]$. This ratio provides a measure of (the reciprocal of) the characteristic diffusion speed of the mesh relative to grid index. The parameter ε specifies the number of time steps over which old structure is to be "remembered." The quadratic factor $(\delta f/f)^2$ varies between 0 and 1, and provides a smooth (differentiable) off–on switch for the time-derivative term. The spatial operators defining, conditioning, and stabilizing the radial distribution of the mesh are shown in detail as follows:

$$\begin{aligned} f^r_j = &\; W_r\,\Delta r_j & \text{T1} \\ &+ W_{rz}\,(\Delta r_j)^2/(\Delta r_{j-1}\,\Delta r_{j+1}) & \text{T2} \\ &+ W_{r_{\max}}\,(\Delta r_j/\Delta r_{\max})^n & \text{T3} \\ &- W_{r_{\min}}\,(\Delta r_{\min}/\Delta r_j)^n, & \text{T4} \end{aligned} \tag{34a}$$

$$\begin{aligned} f^{rl}_j = &\; W_{rl}\,\Delta r\,l_j & \text{T1} \\ &+ W_{rlz}\,(\Delta r\,l_j)^2/(\Delta rl_{j-1}\,\Delta rl_{j+1}) & \text{T2} \\ &+ W_{rl_{\max}}\,(\Delta r\,l_j/\Delta rl_{\max})^n & \text{T3} \\ &- W_{rl_{\min}}\,(\Delta rl_{\min}/\Delta rl_j)^n, & \text{T4} \end{aligned} \tag{34b}$$

where

$$\Delta r_j = (r_j - r_{j+1})/R_{\text{scale}}, \tag{35a}$$

$$\Delta r\,l_j = (r_j - r_{j+1})/(r_j + r_{j+1}). \tag{35b}$$

In Eqs. (34a, b) the W's represent weights assigned to each term (see Winkler et al., 1984c, for examples and discussion), and R_{scale}, $\Delta r_{\max}$, and $\Delta r_{\min}$ are scale factors. The term T1 sets the spatial scale, the spatially symmetric operator in term T2 stabilizes the mesh and also guarantees positivity of our metric, and terms T3 and T4 set, respectively, upper and lower limits on the mesh spacing. The exponent n is a free parameter and, typically, is set to 4.

Thus far we have discussed only the "defensive" measures (existence, positivity, stability) that we build into the grid equation. In the absence of any driving terms the grid equation with appropriate boundary conditions would

6. Implicit Adaptive-Grid Radiation Hydrodynamics

result in a grid distribution equidistant in either r or $\ln r$. Therefore the success of the ability of the grid to detect and resolve arbitrary structures in the flow depends heavily on the quality of the structure function f^s, which drives the grid into the desired distribution. This function is designed to resolve all important flow variables equally well in all parts of the mesh. Terms included in the structure function to date are the following:

$$\begin{aligned}
f_j^s = &\; W_m \, \Delta m_j && \text{T1} \\
&+ W_{ml} \, \Delta m \, l_j && \text{T2} \\
&+ W_\mu \, \Delta \mu_j && \text{T3} \\
&+ W_{\rho l} \, \Delta \rho \, l_j && \text{T4} \\
&+ W_{el} \, \Delta e \, l_j && \text{T5} \\
&+ W_{Pl} \, \Delta P \, l_j && \text{T6} \\
&+ W_{\varkappa \rho l} \, \Delta \varkappa \, \rho l_j && \text{T7} \\
&+ W_{E_0 l} \, \Delta E_0 \, l_j && \text{T8} \\
&+ W_{\varepsilon_N l} \, \Delta \varepsilon_N l_j && \text{T9} \\
&+ W_{ql} \, \Delta q \, l_j && \text{T10} \\
&+ W_u \, \Delta u_j && \text{T11} \\
&+ W_{ul} \, \Delta u \, l_j && \text{T12} \quad (36)
\end{aligned}$$

where

$$\Delta m_j = (m_j - m_{j+1})/M_{\text{scale}}, \tag{37a}$$

$$\Delta m \, l_j = (m_j - m_{j+1})/(m_j + m_{j+1}), \tag{37b}$$

$$\Delta \mu_j = (\mu_{j-1} - \mu_{j+1})^2, \tag{38}$$

$$\Delta \rho \, l_j = [(\rho_{j-1} - \rho_{j+1})/\rho_j]^2, \tag{39}$$

$$\Delta e \, l_j = [(e_{j-1} - e_{j+1})/e_j]^2, \tag{40}$$

$$\Delta P \, l_j = [(P_{j-1} - P_{j+1})/P_j]^2, \tag{41}$$

$$\Delta \varkappa \rho \, l_j = [(\varkappa_{j-1}\rho_{j-1} - \varkappa_{j+1}\rho_{j+1})/\varkappa_j \rho_j]^2, \tag{42}$$

$$\Delta E_0 \, l_j = [(E_{0_{j-1}} - E_{0_{j+1}})/E_{0_j}]^2, \tag{43}$$

$$\Delta \varepsilon_N \, l_j = [(\varepsilon_{N_{j-1}} - \varepsilon_{N_{j+1}})/(\varepsilon_{N_j} + \varepsilon_N \text{ floor})]^2, \tag{44}$$

$$\Delta q \, l_j = [(q_{j-1} - q_{j+1})/q_j]^2, \tag{45}$$

$$q_j = (P_{Q_j}/P_j)^2 + q \text{ floor}, \tag{46}$$

$$\Delta u_j = [u_j - u_{j+1}]^2 \, \text{CVMGP}(1., 0., u_j - u_{j+1}), \tag{47a}$$

$$\Delta u \, l_j = [(u_j - u_{j+1})^2/(u_j^2 + u_{j+1}^2 + c_{ul})] \, \text{CVMGP}(1., 0., u_j - u_{j+1}). \tag{47b}$$

The term T1 (or T2) tries to resolve m (or $\ln m$); T3 the mean molecular weight μ; T4, T5, and T6 the quantities $\ln \rho$, $\ln e$, and $\ln P$; T7 the logarithm of the opacity; T8 the logarithm of the radiation energy density $\ln E_0$; and T9 the logarithm of the nuclear energy generation rate. T10 pulls grid points into shocks by resolving the ratio of viscous pressure to gas pressure. T11 focuses only on rarefaction fans in planar geometry, and T12 is used as a special-purpose tool to resolve the head of the rarefaction fan in our shock-tube example in Section VII.

The evolution of our thinking concerning the structure function f^s can be traced in Winkler (1976), Tscharnuter and Winkler (1979), Winkler and Newman (1980), and Winkler et al. (1984c). The last reference contains an extensive discussion of the spatial stabilization term. Further details and a thorough investigation of the basic concept of an adaptive mesh, including the analysis of a variety of possible operators, as well as space stabilization and time-filtering, can be found in Winkler et al. (1984a). It should also be noted that a diffusion equation for dynamic rezoning of a Lagrangian mesh has been used by Castor et al. (1977). Our algorithm differs significantly from theirs in two important respects: (1) Our elliptic operators have a precise and specifiable meaning and function, in contrast to their diffusion term, which was constructed heuristically. (2) We use our diffusive term in a highly discriminating manner, whereas it is always present in their method.

VI. NUMERICAL TECHNIQUES

A. Tensor Artificial Viscosity

Shock waves that arise in the flow are not treated as fluid discontinuities but rather are spread out over a desired length by artificial viscosity. Our formulation differs significantly from the familiar artificial viscosity of von Neumann and Richtmyer (1950) in order to satisfy additional requirements imposed by the use of an adaptive grid and the particular nature of the protostar problem, see Tscharnuter and Winkler (1979). Our starting point is to define a viscous stress tensor, which is symmetric, trace free, and independent of coordinate system and frame of reference, as follows:

$$\mathbf{Q} = \rho \mu_Q [(\nabla \mathbf{u}) - \tfrac{1}{3} \nabla \cdot \mathbf{u} \mathbf{I}], \tag{48}$$

where $(\nabla \mathbf{u})$ is the symmetrized velocity gradient tensor, $\mathbf{I}$ the unit tensor, and μ_Q the coefficient of artificial viscosity given by

$$\mu_Q = -c_1 l c_s + \min(c_2 l^2 \nabla \cdot \mathbf{u}, 0). \tag{49}$$

Here c_1 and c_2 are constants of order unity, c_s the adiabatic sound speed of the gas, and $4-5l$ the physical thickness over which the shock is spread. The $\mathbf{Q}$ is constrained to be trace free so that there will be no viscous heating in a homologous contraction—a desirable property when computing the Kelvin–Helmholtz contraction of a pre-Main Sequence star (Tscharnuter and Winkler, 1979).

The reader will recognize Eq. (48) as the standard expression for the shear stress in a fluid of kinematic viscosity μ_Q. Therefore, our artificial viscosity treatment is "artificial" only as regards our choice of μ_Q. The coefficient of artificial viscosity given in Eq. (49) is a sum of two terms generally referred to as linear and quadratic multipliers in the sense that substituting Eq. (49) for μ_Q in Eq. (48) yields a viscous stress tensor composed of terms that are linear and quadratic in the rate of strain (see, e.g., Noh, 1976). Notice that μ_Q is a scalar invariant (i.e., independent of coordinate system and frame of reference), as required by the adaptive mesh. The quadratic multiplier vanishes in regions of expansion in analogy with the von Neumann–Richtmyer approach. The linear multiplier is chosen to be small but nonzero in order to damp small amplitude oscillations that occur predominantly in stagnant (with respect to the mesh) parts of the flow.

The viscous momentum transfer $\mathbf{u}_Q$ and heating ε_Q are given by

$$\mathbf{u}_Q \equiv -\nabla \cdot \mathbf{Q} \qquad \text{per unit volume,} \tag{50}$$

$$\varepsilon_Q \equiv -\rho^{-1} \mathbf{Q} \cdot (\nabla \mathbf{u}) \qquad \text{per gram.} \tag{51}$$

Specializing to spherical symmetry, we have the following expressions for the quantities appearing in Eqs. (26) and (28) (for a derivation, see Tscharnuter and Winkler, 1979):

$$u_Q = -\frac{2}{3}\frac{1}{r}\frac{\Delta\left[r^3 \rho \mu_Q\left(\frac{\Delta u}{\Delta r} - \frac{u}{r}\right)\right]}{\Delta \text{Vol}} \tag{52}$$

and

$$\varepsilon_Q = -\frac{2}{3}\mu_Q\left(\frac{\Delta u}{\Delta r} - \frac{u}{r}\right)^2, \tag{53}$$

where

$$\mu_Q = -c_1 l c_s + \min\left[c_2 l^2 \frac{\Delta(r^2 u)}{\Delta \text{Vol}}, 0\right]. \tag{54}$$

Here we have discretized in space by substituting the finite-difference symbol Δ (cf. Section IV) for the partial derivative symbol ∂. The centering of u_Q and ε_Q on the grid, as well as the rules for space- and time-averaging of r, ρ, and u are given in Section IV.

The length l appearing in Eq. (54) governs the thickness of the shock front and is chosen to be either a *fixed length* (i.e, $l = $ constant) or a fixed *relative length* (i.e., l/r = constant), depending on whether the domain of interest extends over one or many orders of magnitude in radius, respectively. The constant is chosen to be as small as possible in order to produce sharp shock fronts, yet large enough to avoid convergence difficulties. When used in conjunction with the adaptive mesh described in Section V, our artificial viscosity method yields well-resolved (in a numerical sense) shock transitions with a typical thickness of 10^{-2} to 10^{-3} as the average zone size. This is possible because the adaptive mesh can achieve a local mesh refinement of up to a factor of a million.

The examples of flows with *isolated* strong shocks, shown in Section VII, demonstrate that our approach is both accurate and rugged. When strong nonlinear waves *interact*, however, additional techniques are required if the inviscid solution is sought. We now turn our attention to this issue.

B. Artificial Mass and Heat Diffusion

It is well known that the use of artificial viscosity in the solution of inviscid gas-flow problems leads to spurious results whenever strong shocks interact with walls, contact discontinuities, and other strong shocks (e.g., Cameron, 1966). This is a simple consequence of using viscous equations to solve an inviscid problem, and no amount of manipulation of the finite-difference approximations to the viscous equations can alter this fundamental fact. What can be done is to construct special-purpose fixes, which are invariably calibrated by knowledge of the desired solution.

The errors produced in regions containing shock interactions are errors in the final distribution of entropy after the waves have ceased to interact. Because entropy production in weak shocks scales as the third power of the pressure difference across the front (cf., Section 83 in Landau and Lifshitz, 1959), these entropy errors are noticeable only in interactions involving strong shocks; that is, $\eta \equiv p_2/p_1 \gg 1$. In such cases, the error typically appears as a spike of higher temperature gas in pressure equilibrium with the surrounding gas. The classical example of this effect is known as wall heating—the spurious overheating and consequent decompression when a strong shock reflects from a rigid wall (e.g., Noh, 1976). The spatial width of the feature is approximately equal to the thickness of the smeared-out shock wave.

Our use of a fixed-length (see Section VI.A) artificial viscosity in combination with the adaptive grid allows us to reduce the smeared-out shock thickness to a small fraction of an average zone size, typically $l = 10^{-2}$–$10^{-3}\ \overline{\Delta x}$. This means that the fractional amount of mass involved in the erroneous region is extremely small. Nevertheless, the error shows up in the solution as a conspicuous and undesirable spike. Our strategy for eliminating entropy spikes is to diffuse them away. Often, the numerical diffusion implicit in the advection procedure is enough to prevent the spike from appearing in the first place. In extreme cases, however, this is not sufficient (such is the case in the interacting blast-wave test problem of Woodward and Colella, 1983). We then resort to introducing mass-diffusion and heat-conduction terms explicitly into the continuity and energy equations, respectively. The form of these terms is shown on the right-hand sides of Eqs. (25) and (26).

In most applications the transport coefficients σ_ρ and σ_e are set to zero. Otherwise, we use the simple prescription $\sigma_\rho = \sigma_e = l^2/\tau$, where l is the shock thickness, and τ the characteristic time for the shock to propagate an average zone width $\overline{\Delta x}$. Since $l \ll \overline{\Delta x}$, this ensures that steady shock solutions are unaffected.

We have found that the use of the asymmetric time filter in the adaptive mesh (cf. Section V) virtually eliminates the effect of wall heating in our calculations. This is accomplished because during the moment of shock reflection, the adaptive mesh automatically refines the zoning adjacent to the wall to a size of order l. The effect of the time filter is to maintain this fine zoning for the duration of the reflection process (cf. Section VII). Thus the entropy spike has a width of a few l when it is produced. As the reflected shock recedes, however, the zoning next to the wall expands back up to an average zone size of 10^2–$10^3 l$. In the process the miniscule amount of mass with incorrect entropy is averaged with 10^2–$10^3 \times$ as much mass with the correct entropy, thus reducing the relative error accordingly and consigning it to the single zone adjacent to the reflecting boundary. Thus, no explicit diffusion or heat conduction is required to eliminate numerical wall heating. In principle, the same procedure should work for internal boundaries, for example, contact discontinuities represented numerically as discontinuous jumps.

C. On the Use of $\Delta\xi$, Δm, and $(\delta m/\delta t)$

At first glance, adopting both the gas density ρ and the integrated mass m as primary dependent variables and solving the implicit equations (24) and (25) for them may seem unnecessary and redundant, since ρ can easily be derived from m given that the radius r as a function of the zone index j is known. For doing so, however, there are practical reasons that are largely specific to the computation of gravitational collapse problems describing accretion onto compact objects.

Integrated mass is carried as a variable in addition to ρ; otherwise, $m(r)$ would have to be evaluated by the spatial integration $m(r) \equiv \int_0^r \rho x^2 \, dx$, which would destroy the band structure of the matrix to be inverted for the solution of the implicit equations. On the other hand, ρ is carried as a variable in addition to m; otherwise the differential mass of a zone $\Delta\xi \equiv \rho \, \Delta\text{Vol}$

would have to be approximated by Δm and would, therefore, be susceptible to round-off errors. For example, it is often the case that $\Delta \xi < 10^{-14} m$ over part of an accreting envelope in free fall. Differencing m to get the mass of a zone in such a region for use in the time derivatives of Eqs. (25)–(29) would lead to unacceptably large errors on a machine such as a CRAY-1, which has only 14 significant figures in a single-precision floating-point word. Thus the *spatial* properties of gravitational flow require both ρ and m to be carried as dependent variables.

Likewise, the *temporal* properties of gravitational flows necessitate using the quantity $(-\delta m/\delta t)$ in the advection terms of Eqs. (26)–(29) instead of $r^2 u_{\text{rel}} \rho$, which is algebraically, but not numerically, equivalent. We find this to be essential for numerical convergence whenever one is computing with a time step such that the flux of mass through a zone in one time step, δm, greatly exceeds the mass in the zone, $\Delta \xi$. For example, when integrating on the evolutionary time scale of an accreting compact object such as a protostellar core, it is not unusual to encounter $\delta m \approx 10^{12} \Delta \xi$ in certain parts of the free-falling envelope. In such a circumstance, the physical solution to which one is attempting to converge is quasi-steady, characterized by $\delta m_j =$ constant to at least 12 decimal places from zone to zone.

Now consider what would happen if, during the Newton–Raphson iteration procedure, the net mass flux in a single zone were computed to only 11 decimal places of accuracy. Then, in a single iteration the mass of that zone could change by a factor of 10 and prevent numerical convergence. Our experience is that representing $(-\delta m/\delta t)$ by the product $r^2 u_{\text{rel}} \rho$ in the difference equations leads to convergence difficulties of this nature. On the other hand, we find that using the corrections δm to the integrated mass directly avoids these problems.

D. Advection

The second terms in Eqs. (25)–(29) describe the advection of mass, total internal energy, radiation energy, gas momentum, and radiation momentum, respectively. In this section we describe in detail how this advection is performed numerically and give some rationale for our formulation. The formulation draws together a number of elements that we have found useful for constructing a stable, yet accurate, advection scheme. The scheme is implicit, second order accurate, and upwind.

We begin by noticing that Eqs. (25)–(29) have one of two basic forms:

$$\delta(q \, \Delta \text{Vol}) + \Delta(\delta t \, r^2 u_{\text{rel}} q_a) + \text{other terms} = 0, \qquad (55a)$$

or

$$\delta(q \, \Delta \xi) + \Delta(-\delta m \, q_a) + \text{other terms} = 0. \qquad (55b)$$

6. Implicit Adaptive-Grid Radiation Hydrodynamics

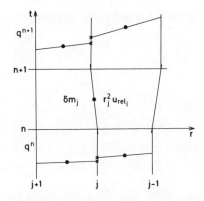

Fig. 2. Schematic of second-order, upwind, advection scheme.

The numerical reasons for the form used in Eq. (55b) have been given in Section VI.C. Here q is the zone average of any quantity such as ρ, u, or E_0/ρ, and q_a the corresponding advected value to be defined below. Equations (55) say that, apart from other terms, the total amount of q in a zone is updated in time by differencing the fluxes at the zone boundaries. Figure 2 illustrates how the flux at a single interface is computed.

The volume flux $\delta t\, r^2 u_{\text{rel}}$ and mass flux $-\delta m$ are time centered according to

$$(\delta t\, r^2 u_{\text{rel}})_j \equiv \delta t \langle r_j^2 \rangle_t [\langle u_j \rangle_t - (r_j^{n+1} - r_j^n)/\delta t] \tag{56a}$$

and

$$(\delta m_j) \equiv m_j^{n+1} - m_j^n, \tag{56b}$$

where $\langle x_j \rangle_t$ is defined in Eq. (30). To compute q_a, we first construct monotonic piecewise-linear distributions of q at both the old and new time levels using the van Leer (1979) prescription for the slopes:

$$dq_j = \begin{cases} \dfrac{c_3\, \Delta q_j\, \Delta q_{j+1}}{\Delta q_j + \Delta q_{j+1}}, & \text{if } \Delta q_j\, \Delta q_{j+1} > 0, \\ 0, & \text{otherwise} \end{cases} \tag{57}$$

Here $\Delta q_j \equiv q_{j-1} - q_j$, and the constant c_3, which controls the order of the interpolation, is 2 or 0 for second-order or first-order accuracy, respectively. Then q_a is computed by time-averaging old and new interpolated values evaluated on the *upwind* side of the zone interface in question; thus,

$$q_{aj} = (0.5 + s_j)[z(q_j^{n+1} + 0.5\, dq_j^{n+1}) + (1-z)(q_j^n + 0.5\, dq_j^n)] \\ + (0.5 - s_j)[z(q_{j-1}^{n+1} - 0.5\, dq_{j-1}^{n+1}) + (1-z)(q_{j-1}^n - 0.5\, dq_{j-1}^n)] \tag{58}$$

Here s_j is a logical switch that chooses the upwind-sided value of the q_a in question according to

$$s_j = \text{CVMGP}(+0.5, -0.5, u_{\text{rel }j}) \tag{59a}$$

or

$$s_j = \text{CVMGP}(+0.5, -0.5, -\delta M_j), \tag{59b}$$

depending on the case.

When updating the face-centered quantities u and F_0 (cf. Fig. 1), $\Delta \xi$ and Δm appearing in Eq. (55b) are appropriately centered with respect to the "momentum" cell by straight arithmetic averages.

The space–time centering and construction of the fluxes described above is the obvious (P. Woodward, private communication, 1981) second-order-accurate finite-difference approximation to flux terms of the form

$$\int_t^{t+\delta t} dt \oint_{\partial v} q \mathbf{u}_{\text{rel}} \cdot d\mathbf{S} \tag{60}$$

in Eqs. (15)–(19). The time-centering constant z is kept as close as possible to 0.5 for reasons of accuracy but is typically no smaller than 0.55 for reasons of stability. Each flux is upwind with respect to the interface to assure numerical stability.

E. Solution Procedure and Control

The implicit finite-difference equations for the radiation hydrodynamics [Eqs. (24)–(29)] and the adaptive mesh [Eqs. (31)–(47)], together with closures, constitute a complete set of coupled, nonlinear, algebraic equations for the vector of seven unknowns,

$$\mathbf{x}_j \equiv x_{i,j} \equiv (r_j, m_j, \rho_j, e_j, u_j, E_{0j}, F_{0j}), \tag{61}$$

as a function of zone index j and time level n. This nonlinear system is solved iteratively by the Newton–Raphson method. The functional dependence of the advection terms in Eqs. (25)–(29) on the neighboring zones [cf. Eq. (58)] and the spatial structure of the adaptive-mesh equation [cf. Eqs. (34a,b)] define a five-point spatial stencil at both the old and new time levels. Thus, upon linearizing, the matrix equation for the solution corrections $\delta \mathbf{x}_j \equiv \delta x_{i,j}$ will have a block pentadiagonal structure, where each block is a (7 × 7) full submatrix. Exploiting the band structure of the complete matrix, we make an *LU* decomposition and solve the linear system directly by block Gaussian elimination. We iterate over the entire system until numerical convergence, as defined below, is achieved.

6. Implicit Adaptive-Grid Radiation Hydrodynamics

The Newton–Raphson iterations are controlled in the following way. If l is the iteration count and $\mathbf{x}_j^l$ the previous iterate, then the lth correction $\delta\mathbf{x}_j^l$ is applied according to

$$\mathbf{x}_j^{l+1} = \mathbf{x}_j^l + f_1^l\, \delta\mathbf{x}_j^l, \tag{62}$$

where f_1^l is the fraction by which the corrections are reduced if they are too large and is given by the formula

$$f_1^l = \min\bigl[1, \min_{i,j}(X_i^{\max}/X_{i,j}^l)\bigr]. \tag{63}$$

Here i is the variable index, allowed to range only over the unsigned variables; $X_i^{\max}$ the desired maximum relative change for the variable x_i; and $X_{i,j}^l \equiv \delta x_{i,j}^l/x_{i,j}^l$ the computed relative change. We say that the solution has converged when

$$\max_{i,j}(X_{i,j}^l) < \varepsilon \quad \text{and} \quad (l_{\min} \leq l \leq l_{\max}). \tag{64}$$

Typical values are $\varepsilon = 10^{-5}$, $l_{\min} = 2$, and $l_{\max} = 20$. Under normal conditions convergence is reached in four–six iterations. We then apply a "boost" iteration, which boosts the final relative corrections into the range of 10^{-8} to 10^{-10}. Since the Newton–Raphson method has second-order convergence, the equations themselves have also been solved to this level of accuracy.

As in any implicit system, the time step is determined by accuracy, not stability, considerations. To compute the new time step, we multiply the old time step by some factor f_2, which is itself recomputed at each time step. The reason for this approach is that in applications involving gravitational collapse or explosions, the characteristic time scales typically decrease or increase as a function of time in some smooth way. It is when the time step *deviates* from this systematic trend that changes show up in the iteration procedure. Therefore, we increase or decrease the time-step *factor*, depending on whether the largest cumulative relative correction X is smaller or larger, respectively, than $X_{\max}$, the maximum relative correction allowed. Thus, defining

$$X = \max_{i,j}\Bigl(\sum_l |\delta x_{i,j}^l|/x_{i,j}^n\Bigr) \tag{65}$$

and

$$\varphi_1 = \begin{cases} (X_{\max}/X)^{1/2} & \text{if } (X/X_{\max}) < 1, \\ (X_{\max}/X)^2 & \text{if } (X/X_{\max}) > 1, \end{cases} \tag{66}$$

we set

$$\delta t^{n+1} = f_2^{n+1}\, \delta t^n, \tag{67}$$

where

$$f_2^{n+1} = \min\{[f_2^n \max(0.1, \varphi_2)]^{1/2}, 10\}, \tag{68}$$

and

$$\varphi_2 = \min[2, \min(\varphi_1, 1.25 f_2^n)]. \tag{69}$$

Typical applications are run with $0.05 \le X_{\max} \le 0.15$. The nonlinearity in Eq. (66) prevents f_2 from jumping between two extremes. The seemingly complicated procedure given above can be thought of as a nonlinear algorithm for computing time as a function of time index n.

If the solution fails to converge in the sense of Eq. (64), the entire iteration procedure is repeated with a smaller time step, determined according to the following simple prescription:

$$\delta t_K = \max(0.1, 0.8 - 0.2K)\,\delta t_{K-1}, \qquad K = 1, \ldots, 6 \tag{70}$$

where K counts the number of attempts at finding the converged solution. The effect of this procedure is to cut back the time step by a small factor for the first few attempts, but thereafter to cut back by factors of 0.1, for a maximum of six attempts. If the solution has failed to converge after six attempts, we say that it has diverged and we terminate the computation.

A characteristic of divergence is the unbounded growth of the corrections $\delta \mathbf{x}^l$ during the Newton–Raphson iteration procedure. This situation usually arises when, for either physical or numerical reasons, the solution is not sufficiently steady with respect to the adaptive mesh. Examples of difficulties related to physics are: (1) the equation-of-state and opacity tables are not smooth enough; (2) the space–time resolution of the physical phenomena is not fine enough; (3) the physical nonlinearities induce a sudden change in the character of the solution. Examples of difficulties related to numerics are (4) the adaptive mesh is not stabilized or is not tracking the flow features well enough; (5) round-off errors and/or unsophisticated mathematical library subroutines prevent the solution of linear systems with enormous condition numbers.

We have found that the use of monotonic splines reduces difficulties associated with (1). Divergence caused by difficulties (2)–(4) can usually be overcome by suitable adjustment of input constants and weight factors. The use of better linear algebra routines for matrix operations (e.g., the robust implementation of LINPACK by T. L. Jordan at Los Alamos National Laboratory) eases some of the difficulties associated with (5), although round-off errors seem to set a practical limit to what can be accomplished, especially regarding mesh refinement.

VII. ORDINARY GAS DYNAMICS: SHOCK TUBES

As an illustration of the use of implicit adaptive-mesh techniques on a problem of some familiarity, we compute the breakup of an initial pressure discontinuity in a shock tube into its three constituent nonlinear waves (i.e., a Riemann problem). Gravitation and radiation terms are assumed to be absent in this calculation, so that the flow is described by the equations of ideal gas dynamics. Planar geometry and a gamma-law equation of state $P = (\gamma - 1)\rho e$ is assumed, with $\gamma = 1.4$.

The initial conditions are those used by Sod (1978); namely, in the constant states on the left and right sides of the discontinuity, L and R, respectively, we assume

$$P_L = 1.0, \quad \rho_L = 1.0, \quad u_L = 0 \quad \text{for} \quad 0.0 \leq x \leq 0.5,$$
$$P_R = 0.1, \quad \rho_R = 0.125, \quad u_R = 0 \quad \text{for} \quad 0.5 \leq x \leq 1.0. \tag{71}$$

Reflection boundary conditions are assumed at $x = 0$ and $x = 1$. Our computational mesh consists of 100 grid points, whose motion is governed by Eqs. (31a), (32), (33), (34), (35a), and (36) with the following constants and weight factors: $W_r = W_{r_{max}} = W_{r_{min}} = 1.$, $W_{rz} = 0.5$, $\Delta r_{max} = 0.04$, $\Delta r_{min} = 10^{-5}$, $n = 4$, $R_{scale} = 1.$, $W_{\rho l} = 6$, $W_{el} = 3$, $W_{pl} = 1$, $W_{ql} = 3$, $W_u = 9$, $W_{ul} = 3$, $q_{floor} = 1$, and $C_{ul} = 10^{-3}$. All other weight factors are set to zero. The fixed length l used in the artificial viscosity—cf. Eq. (49)—was set to 2×10^{-2} of a normal (equidistant) zone width of $\overline{\Delta x} = 10^{-2}$, that is, to $l = 2 \times 10^{-4}$. The constants in Eq. (49) were set to $c_1 = 0.3$ and $c_2 = 3.0$, which yields a shock thickness of about 10^{-3}.

At $t = 0$ we imagine that a diaphragm that separates the two constant states is removed, allowing the fluid discontinuity to resolve itself into a right-facing shock, a left-facing rarefaction wave, and a contact discontinuity situated between them (Fig. 3a). For $t \leq 0.42$, this Riemann problem admits an analytic solution (e.g., Sod, 1978) against which one can compare one's numerical results. The shock jump in Fig. 3b is correct to 0.1%. We evolve the system to $t = 1$, which allows the waves to reflect from the boundaries (Fig. 3c) and to interact with each other in the interior (Fig. 3d). This exercise has proved useful in refining the treatment of the adaptive mesh, as illustrated in Fig. 4.

Figure 4a shows the space–time evolution of the adaptive mesh for the shock-tube test problem without the use of the locally parabolic asymmetric time filter [i.e., the first term in Eq. (31a)]. In the figure the spatial position of every second grid point is plotted as a function of time. The head of the rarefaction wave, the contact discontinuity, and the shock wave are rendered

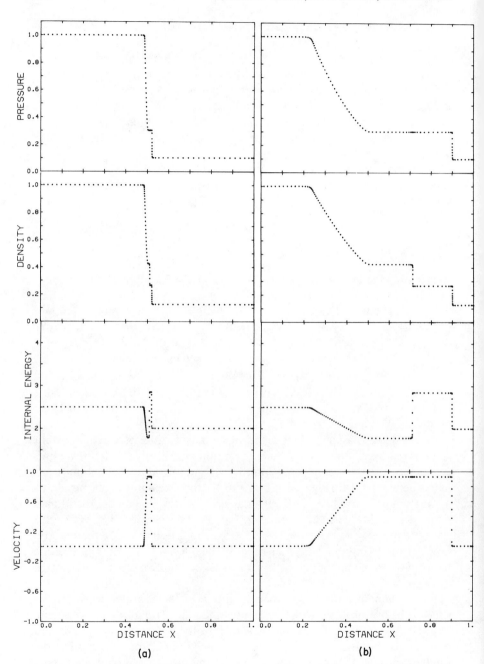

Fig. 3. Numerical solution of Sod's shock-tube test problem at times $t = 0.0108, 0.228, 0.364$, and 1.0. Grid points are shown individually.

6. Implicit Adaptive-Grid Radiation Hydrodynamics

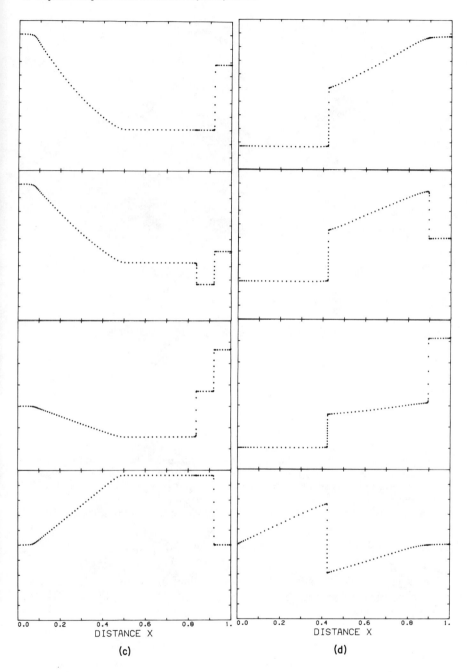

(c)   (d)

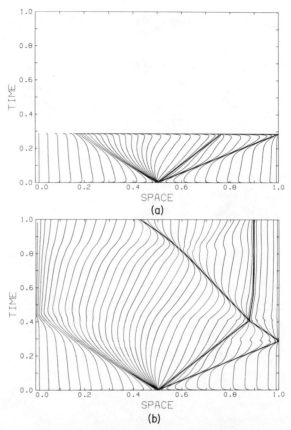

Fig. 4. Numerical space–time evolution of the adaptive mesh for Sod's shock-tube test problem. Solutions have 100 mesh points; the trajectory of every second mesh-point is plotted. Strong shocks and the contact discontinuity are made visible by the dense concentration of grid points. (a) No time filtering. As shown by the horizontal line, grid points run away from the wall at the instant of first shock reflection and the code fails to converge. (b) With time filtering. Grid point runaway is completely controlled, and solution progresses smoothly without incident.

visible by a concentration of grid points. One can see that the use of a purely elliptic operator yields good results prior to shock reflection at $t \approx 0.28$. However the singular behavior of the grid lines at the instant of reflection is undesirable and shows the rapid leftward migration of grid points to other parts of the flow as gradients at the wall temporarily disappear. Immediately after the instant of shock reflection, the grid points race back toward the right-hand boundary in order to resolve the newly created left-facing shock front.

In this case the sudden redistribution of grid points had a deleterious effect on convergence, and the calculation terminated. However it is possible, with a suitable modification of the right-hand boundary condition, to carry on the computation without the time filter. An example of such a computation can be found in Winkler et al. (1984c, Figure 3).

Clearly, this rapid redistribution of grid points with little or no net effect is undesirable and should be avoided by modifying the grid equation in such a way that fine-resolution zones are kept at the wall for at least the duration of the reflection process. The time filter described in Section V was developed with these requirements in mind and provides the necessary "memory" by introducing locally a parabolic term into the globally elliptic grid equation. Its effect on the evolution of the computational mesh is shown in Fig. 4b.

The concentration of grid lines in Fig. 4b shows the shock reflection at the right-hand boundary at $t \approx 0.28$, its interaction with the contact discontinuity at $t \approx 0.41$, and its subsequent propagation and acceleration leftward. The input constants appearing in Eqs. (31a) and (33) for the asymmetric time filter were set to $T_{\text{scale}} = 5 \times 10^{-3}$ and $\varepsilon = 0.03$. Inspection of Fig. 4b shows that the general effect of the time filter is to reduce the amount of grid redistribution in the vicinity of interacting waves, and to maintain a continuous space–time evolution of grid points in regions far from the wave-interaction zones. Its practical effects on the computation, besides the obvious one of permitting further integration, are twofold. The first, but not immediately recognizable, benefit of the time filtering is the smoothing of fine-scale grid jitter, which results from numerous and unaccountable numerical effects. One notes a small but significant improvement in the sturdiness and overall convergence properties of the system of equations. The second benefit is that of a substantial increase in the efficiency of the entire computation for problems involving interacting waves. Previously (Winkler et al., 1984c), wave interactions were computed at the expense of a considerable increase in the number of time steps with very small values of δt, which were needed to maintain convergence during the interaction phase as grid points were being redistributed rapidly. The time filter essentially eliminates this effect, and it has reduced the execution time for the shock-test problem by an order of magnitude. The evolution in Fig. 4b required approximately 300 time steps to compute, where the execution time per time step on the CRAY-1 computer is between 0.1 and 0.2 sec, depending on the number of iterations required to achieve convergence.

VIII. RADIATION HYDRODYNAMICS: A SUPERCRITICAL SHOCK

Adding the ingredients of self-gravity and radiation to the hydrodynamic shock example of the previous section considerably enlarges the wealth and complexity of the solution space of the equations. One particularly striking example is provided by the formation of a supercritical shock in an accreting protostellar envelope.

Consider a homogeneous isothermal interstellar cloud of $100 M_\odot$, embedded in a sphere of radius 10^{18} cm, in radiative equilibrium with an imposed external radiation field having a radiation temperature of 60 K. In this initially transparent configuration gravitational forces overpower pressure gradients. There ensues a gravitational collapse, which is finally stopped when the density in the central parts of the cloud becomes high enough to form an opaque core and trap the radiation. From that time onward, the energy gained from the fall into the gravitational potential well is no longer directly radiated away but goes instead into heating the gas. Pressure then builds up until a quasi-hydrostatic core forms, surrounded by an accretion shock.

The total luminosity of such an object has two contributions. Part of it is generated in a slow contraction of the quasi-hydrostatic core. The remainder is emitted directly from the accretion shock within which the kinetic energy of the infalling material is transformed primarily into outgoing radiation.

In order to get a clean test problem for our radiation-hydrodynamics method, we avoided complicated equation-of-state and opacity tables and used instead the ideal equation of state, with $\gamma = \frac{5}{3}$, and the following opacity law:

$$\varkappa = 10^{\{2\ \tanh[(T - 5 \times 10^4)/(2.5 \times 10^4)] - 2\}}. \tag{72}$$

Although this highly idealized problem does not provide a realistic model of the star-formation process, it nevertheless gives us the flavor of the correct physical phenomena with their multiple length and time scales. In the context of this volume, that is, describing a numerical method and demonstrating its ability to handle tough physical problems, one could view the physical setup as simply a natural way of generating families of supercritical accretion shocks. Therefore, in what follows we do not discuss the astrophysical implications at all but focus exclusively on the physics and numerics.

Figure 5 gives an overview of the large-scale structure in which the radiating shock is embedded. The flow in the protostellar cloud near the central embryo star is essentially in free fall until it slams into the core and is brought virtually to rest in the accretion shock. This can be seen in Fig. 5a, where the fluid velocity is shown as a function of radius. As in all other plots, each of the 300 grid points used in the computation is shown individually. Figure 5c shows the corresponding density distribution, and Fig. 5b shows the parameter

$$\alpha \equiv |\partial \ln \rho / \partial \ln r|, \tag{73}$$

which gives a measure of the local density gradient. Clearly α is largest where the accretion shock connects to the exponential stellar atmosphere (which is distinct from the free-falling protostellar envelope) of the central core. The Eddington factor f, the ratio of the radiation pressure to radiation energy

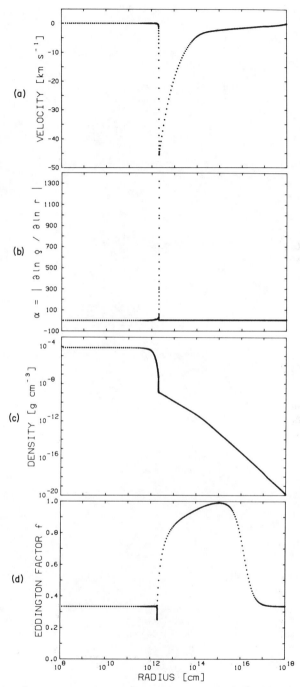

Fig. 5. Large-scale structure of protostellar accretion shock. (a) Fluid velocity. (b) Radial density gradient $|\partial \ln \rho / \partial \ln r|$. (c) Density distribution, showing power-law outer envelope, exponential atmosphere, and condensed core. (d) Variation of Eddington factor showing isotropic inner core, side-peaked radiation field within shock, radially streaming radiation in extended envelope, and return to isotropy within ambient interstellar medium.

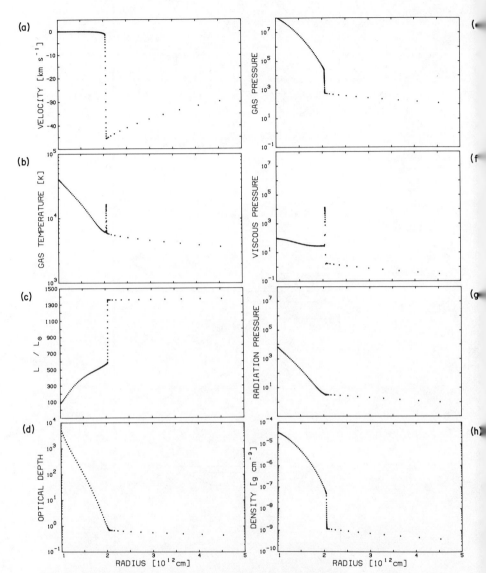

Fig. 6. Medium-scale flow structure in vicinity of supercritical shock. (a) Velocity. (b) Gas temperature showing optically thin postshock temperature spike. (c) Luminosity variation showing "discontinuity" from emission by radiating shock. (d) Optical depth. (e) Gas pressure showing discontinuity across shock. (f) Viscous pressure showing spike within shock front. (g) Radiation pressure. (h) Density showing jump of factor of 60 across shock.

density, shows four distinctive regions in Fig. 5d. First, the inner region where $f = \frac{1}{3}$ is characteristic of the isotropic radiation field of the stellar core. Next, the narrow zone with $0.25 \leq f \leq \frac{1}{3}$ results from the strongly sideways-peaked distribution of intensity in the supercritical accretion shock. Here luminosity is generated in an extremely narrow, optically thin (radially), spherical shell of material; see also Fig. 6d, which shows optical depth through the radiating shock. Third, the region with $\frac{1}{2} \leq f \leq 1$ is characteristic of the strongly outward-peaked intensity in the large transparent outer envelope, where essentially all the radiation originates from a central source (the protostellar core) of small angular size. Finally f drops back from 1 to $\frac{1}{3}$ in the extreme outer regions, where the protostellar cloud merges into the ambient external interstellar medium with its isotropic radiation field.

More detailed medium-scale structure of the solution at 1.236 initial free-fall times is shown in Fig. 6. The shock transition in velocity shown in Fig. 6a is at the same position where the gas temperature overshoots by a factor of 2.56, as shown in Fig. 6b. This optically thin postshock temperature spike is a dramatic example of the phenomenon first recognized by Zel'dovich (1957) and Raizer (1958), and subsequently analyzed in further detail by Heaslet and Baldwin (1963).

An immense luminosity is generated within the accretion shock. As shown in Fig. 6c, it rises from about $600L_\odot$, emitted by the contracting core, to about $1400L_\odot$ in the transparent envelope, where it remains constant. In these early phases of the evolution, the optical depth, shown in Fig. 6d, increases primarily as a result of increases in density, shown in Fig. 6h, and not as a result of changes in the opacity.

Figs. 6e,f,g show that except in the shock itself, gas pressure is dominant over radiation and viscous pressure. The radiation pressure never plays a significant role in these early phases of the evolution, and viscous pressure (from the artificial viscosity) is important only within the shock front where, of course, it sets the spatial scale of the shock's structure.

Figure 7 zooms in on the small-scale structure of the shock and fully resolves it. Here one sees the internal "machinery" of a supercritical shock in action. It is especially noteworthy that *all* physical variables have been resolved by the adaptive grid.

In the region where the material is brought nearly to rest (see Fig. 7a) the gas temperature develops its sharp peak, whereas the radiation temperature remains continuous, smooth, and nearly constant (see Fig. 7b). These results are in complete agreement with the predictions of Zel'dovich (1957) and provide, to our knowledge, one of the best examples of this effect available in the literature. As shown in Fig. 7c, the total width of the shock in optical depth is 0.025; thanks to our adaptive grid it is more than adequately resolved with 25 grid points.

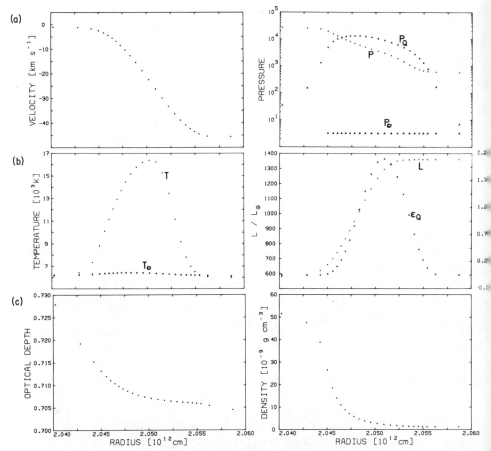

Fig. 7. Small-scale flow structure within shock; notice that all variables are resolved. (a) Fluid velocity. (b) Gas and radiation temperature; notice that the radiation temperature is constant, in accordance with the analysis of Zel'dovich (1957). (c) Optical depth. (d) Gas, viscous, and radiation pressure. (e) Viscous energy dissipation and radiant luminosity. (f) Density.

Figure 7d shows the run of the different forms of pressure within the shock front. The viscous energy dissipation ε_Q (shown in Fig. 7e, right-hand scale) temporarily stores a large part of the kinetic energy of the infalling gas; the remainder is temporarily stored in the internal energy of the gas (see the gas temperature in Fig. 7b). Then, in the downstream relaxation zone of the radiating shock, the gas temperature comes back into equilibrium with the radiation temperature and the temporarily stored energy is dumped into outgoing luminosity, shown in Fig. 7e, which escapes freely from the optically thin shock. Finally, Fig. 7f shows the rapid increase in density needed to maintain a steadily increasing total pressure in the face of the return of the gas temperature back to its upstream value in the postshock relaxation zone. This large density rise is exactly what one expects, because, on a scale on which the relaxation zone is simply lumped together with the dissipation zone, the shock is effectively *isothermal* because the temperature jump across it is nearly zero.

The solution shown in Figs. 5–7 was computed with the methods described in Sections II–VI. The following weight factors were used in the adaptive-mesh equation: $W_{rl} = 1$, $W_{rlz} = 60$, $W_{rl_{\max}} = 1$, $n = 4$, $\Delta rl_{\max} = 0.06$, $W_{rl_{\min}} = 1$, $\Delta rl_{\min} = 10^{-8}$, $W_m = 30$, $W_{ml} = 3$, $W_{Ql} = 15$, $W_{el} = 12$, $W_{\rho l} = 6$, $W_{Pl} = 6$, $W_{Eol} = 9$, $W_{\varkappa \rho l} = 30$, and $q_{\text{floor}} = 1$. All other parameters were set to zero. The parameters used for the artificial viscosity were a fixed relative length $l = 10^{-3} r$ and constants $c_1 = 0.3$, $c_2 = 3$.

IX. A "HILBERT PROGRAM" FOR NONLINEAR RADIATION HYDRODYNAMICS

As should be clear from the preceding discussion, radiation-hydrodynamics calculations are not easy. To perform them well is a demanding task, even for simple physics. It is, therefore, not surprising that many investigators, desiring to model physically complex situations (e.g., with realistic microphysics and/or multidimensional flow), have attempted to take shortcuts by introducing simplifying assumptions about the mathematical structure of the system of equations to be solved, usually in the form of ad hoc parameterizations or fixups. Unfortunately, the solutions thus obtained have an unknown level of reliability; in some cases they may be reasonable, and in others they may be fatally, if inconspicuously, flawed. If one takes seriously the dictum that the goal of computation is *insight*, not mere numbers, and that we wish to simulate real physics, not merely explore the pathology of a model of physics, then one must face the fact that, in the end, it will be necessary to be able to state the *quality* and *reliability* of one's results. Put another way, it is as much the responsibility of the computational physicist to state the error bars associated with his or her numerical experiments as it is of the experimentalist to state the error bars on his or her laboratory work.

It is our opinion (admittedly in the minority) that if rational and economical use is to be made of both human manpower and machine resources, then it is imperative to have a careful plan of action. We feel that no matter how compelling the reasons may be to attempt to model, however badly, some arbitrarily complicated system (e.g., because it is encountered observationally or in a real experiment), for the sake of improving our computational skill it is of yet higher priority to define and perform a logical progression of definitive benchmark calculations that provide genuine milestones against which progress can be measured. Taking this view, how should one proceed?

From our experience with a variety of demanding problems we have arrived at two important precepts that assist genuine progress: First, one should always seek to "overkill" the numerics (which is not to say that the calculation need be expensive—quite the contrary!). The strategy is to ensure that the calculation is done so well that, when one examines the results, one is presented with the *physical* behavior of the system, not numerical artifacts; only then can one learn new physics from the calculation with confidence. Second, one must always push the methodology to the most extreme cases physically allowable. It is rarely of much help, except as an initial learning experience, to do "easy" problems, however perfectly. Instead, the goal should be to develop methods that will always work at a statable level of accuracy, within the arena appropriate to the physical assumptions initially adopted. This goal can be achieved only by testing the method against a broad selection of astutely chosen "hard" problems that push the algorithm until it either breaks or (preferably) survives in the face of all physically sensible obstacles. In this vein, the method described in this chapter can be forced over the edge when propagating radiation fronts occur in the flow. The Eddington factors obtained from the time-independent static transfer equation then cease to be adequate and serious errors are made in the computation of the luminosity (radiant energy flux). At this point one requires our "step 2a" described below.

A consequence of the demand for a logical progression in the development of the methodology is that one is obliged to exhaust the possibilities at each level of physical complexity. This is not to say that several lines of exploration cannot proceed in parallel; however, it does imply that ultimately one must not settle for anything less than a complete solution of the problem posed at each level of complexity addressed.

In radiation-hydrodynamics calculations the computational effort is driven by the amount of information needed to specify the radiation field. So long as particle-mean-free paths are short compared with characteristic lengths in the flow, it suffices to describe the material component of the radiating fluid with a relatively small number of *moments* of the particle distribution function and some transport coefficients. For an n-dimensional (in space) flow we usually desire to specify the $(n + 2)$ variables $(\rho, e, \mathbf{u})$ on an $(n + 1)$-dimensional space–time. Thermal heat conduction is dwarfed by radiation transport, as is molecular viscosity by artificial viscosity, so those quantities can be ignored. In contrast, in a radiating flow with boundary surfaces, photon-mean-free paths will inevitably approach or exceed characteristic structural and flow lengths, and we have no choice but to specify the full photon *distribution function*. To do so introduces up to three new independent dimensions (angles and frequency) ascribable to the photon-momentum tangent space attached to each space–time event in the flow. Thus, even for one-dimensional flow, the radiation field is inherently four dimensional (including time). In general, the angle–frequency dimensions of the photon distribution function are strongly coupled, and we are required to follow its (stiff) time evolution fully implicitly. Even for one-dimensional flows the general problem is beyond the capacity of existing computers; for spatially multidimensional flows the situation is even worse.

For these reasons, we are of the opinion that for the logical development of radiation hydrodynamics it is pointless, at the present time, to attempt to consider multidimensional (in space) radiating flows (in contrast to the situation for ordinary—i.e., nonradiating—hydrodynamics, where such problems are the natural domain of contemporary research). Rather, we feel that it is more fruitful to examine what are probably the minimum steps in a "Hilbert program" aimed at a systematic, exhaustive development of one-dimensional radiation hydrodynamics in the next decade. As we shall see, the effort is nontrivial.

The lowest level of approximation (step 1) is to use the multifrequency–gray method with Eddington factors, as discussed earlier in this chapter. The computational effort to solve the equations of hydrodynamics on the one hand, and to update the Eddington factors and spectral profiles on the other, are roughly comparable. For Newtonian flows the procedure can be made internally consistent by accounting for velocity-dependent terms in both the angle-dependent and multigroup calculations (see, e.g., Mihalas and Mihalas, 1984). Without major effort one can also elaborate the microphysics to the extent of allowing a multifluid gas (e.g., electrons + ions + atoms), each constituent being considered to be in local thermodynamic equilibrium (Maxwellian velocity distribution, Boltzmann–Saha excitation) at a distinct temperature. A more demanding exercise, still achievable without enlarging the dimensionality of the system, is presented by gray relativistic flows, including relativistic effects in the material properties themselves. Although this problem can, and should, be solved on existing computers, one recognizes that the strong angular peaking of the radiation field in such flows makes the computation of meaningful Eddington factors more and more difficult with increasing flow speeds, unless one resorts to very dense and carefully tailored angle meshes. Inevitably, the whole approach becomes unmanageable and can be expected to fail.

The next level of approximation (step 2a) might therefore be to solve the gray angle-dependent transport problem directly, using a fully relativistic formulation on adaptive angle and radial meshes, as described in Winkler et al. (1984b) and Mihalas et al. (1984). The full angle coupling resulting from relativistic aberration and solid-angle effects is then taken into account on an angle mesh designed to adapt automatically to the angular distribution of the radiation field. Because the number of variables required to describe the adaptive angle mesh and the angular variation of the symmetric and antisymmetric parts of the radiation field greatly exceeds the number of fluid variables to be calculated, the hydrodynamics can be simply "piggybacked" on top of the radiation transport calculation with essentially no increase in effort. It should be noted that, because angles and radii are located in separate spaces, the problem is in a sense "$1\frac{1}{2}$-D" instead of fully 2-D (though the number of variables is just the same) because no mesh tangling is possible; it thus provides a good starting point for learning how to extend adaptive-mesh techniques to several dimensions without immediately having to confront problems caused by shear and vorticity.

In principle, one could again attempt to account for nongrayness by supplying the profile functions needed to evaluate $\varkappa_E$ and $\varkappa_F$ from a solution of the moment equations. However, in relativistic flows the Doppler shifts become so large that the coupling among frequency bins is quite strong; hence, we actually have a new problem (step 2b) of qualitatively the same kind as just described. That is we must solve frequency-dependent moment equations on adaptive radial and frequency meshes, the latter designed to give automatically an optimal sampling of the spectrum at each point in the flow. As in step 2a, the dimensionality of the problem is so large that the hydrodynamics can be piggybacked at essentially no additional cost. It is at this step that one can also incorporate non-LTE microphysics (i.e., occupation numbers determined from rate equations) because one now has access to the spectral information required to synthesize radiative rates. Careful thought must then be given to how one structures the equations inasmuch as the radiation field and occupation numbers contain redundant information. In principle, Eddington factors could be supplied to step 2b from a frequency-by-frequency application of step 2a, but one must then overcome the technical problem of transferring information between the adaptive frequency mesh of step 2b and the fixed frequency mesh of step 2a. This procedure may, in fact, not work, and we must, therefore, consider one final step.

In the most general case of strongly nonequilibrium, highly relativistic flow, the angle–frequency coupling of the radiation field becomes utterly inseparable. Hence one is led to step 3, in which one solves the full angle- and frequency-dependent transport problem on adaptive spatial, angle, and frequency meshes concurrently with the hydrodynamics. At this level one can also treat the full microphysics of the radiation–material interaction. In this 3-D problem the number of variables to be computed is immense.

In the program just outlined, step 1 is achievable today with a good deal of hard work. Steps 2a and 2b are not, but are feasible on machines expected to be available in the next few years. To put these steps in perspective, one notes that the level of effort required at each time step is comparable to a major non-LTE model atmosphere calculation using very elaborate physics. Hence, it represents an increase by two to three orders of magnitude over computations that are now fairly routine, though difficult. Step 3 is probably not tractable on machines likely to be available by the end of this decade; machine architectures incorporating massive memories and large numbers of parallel processors are essential for the solution of the problem. Longer word lengths will also be essential to cope with the huge systems we wish to solve.

Parallel programs can also be formulated for problems in two or three spatial dimensions but will, of course, be more difficult to achieve. One instantly recognizes two major difficulties to be faced:

1. An adaptive spatial mesh must be devised that can give the desired resolution while remaining untangled.

2. It remains to be demonstrated that the multifrequency–gray method with tensor Eddington factors will actuallly work in 2-D or 3-D geometries, even for Newtonian flows. If it does not, one is immediately faced with the analogs of our steps 2a and/or 2b simultaneously with the enlarged number of spatial dimensions.

REFERENCES

Axelrod, T. S., Dubois, P. F., and Rhoades, C. E., Jr. (1984). *J. Comput. Phys.* **54**, 205.
Buchler, J. R. (1979). *J. Quant. Spectrosc. Radiat. Transfer* **22**, 293.
Cameron, I. G. (1966). *J. Comput. Phys.* **1**, 1.
Castor, J. I. (1972). *Astrophys. J.* **178**, 779.
Castor, J. I. (1984). In "Astrophysical Radiation Hydrodynamics" (K.-H. A. Winkler and M. L. Norman, eds.). Reidel Publ., Dordrecht, Netherlands.
Castor, J. I., Davis, C. G., and Davison, D. K. (1977). "Dynamical Zoning Within a Lagrangean Mesh by Use of DYN, a Stellar Pulsation Code," Los Alamos Sci. Lab. Rep. LA-6644. University of California, Los Alamos.
Freeman, B. E., Hauser, L. E., Palmer, J. T., Pickard, S. O., Simmons, G. M., Williston, D. G., and Zerkle, J. E. (1968). "The VERA Code: A One-Dimensional Radiative Hydrodynamics Program," Vol. I, Defense Atomic Support Agency Rep. No. 2135. Systems, Science, and Software, Inc., La Jolla, California.
Heaslet, M. A., and Baldwin, B. S. (1963). *Phys. Fluids* **6**, 781.
Landau, L. D., and Lifshitz, E. M. (1959). "Fluid Mechanics," Pergamon Press, Oxford and New York.
Mihalas, D., and Mihalas, B. W. (1984). "Foundations of Radiation Hydrodynamics." Oxford Univ. Press, London and New York.
Mihalas, D., and Weaver, R. P. (1982). "The Dynamics of a Radiating Fluid," Los Alamos Nat. Lab. Rep. LA-UR-82-606. University of California, Los Alamos.
Mihalas, D., Winkler, K.-H. A., and Norman, M. L. (1984). *J. Quant. Spectrosc. Radiat. Transfer* **31**, No. 6, 479.

Noh, W. F. (1976). "Numerical Methods in Hydrodynamics Calculations," Lawrence Livermore Lab. Rep. UCRL-52112. University of California, Livermore.
Pomraning, G. C. (1973). "The Equations of Radiation Hydrodynamics. Pergamon Oxford.
Raizer, Yu. P. (1958). *Sov. Phys—JETP (Engl. Transl.)* **5**, 919.
Sod, G. A. (1978). *J. Comput. Phys.* **27**, 1.
Tscharnuter, W., and Winkler, K.-H. A. (1979). *Comput. Phys. Commun*, **18**, 171.
van Leer, B. (1979). *J. Comput. Phys.* **32**, 101.
von Neumann, J., and Richtmyer, R. D. (1950). *J. Appl. Phys.* **21**, 232.
Winkler, K.-H. A. (1976). A numerical procedure for the calculation of nonsteady spherical shock fronts with radiation. Ph.D. Thesis, University of Göttingen. Issued as technical report MPI-PAE-Astro 90 by the Max-Planck-Institut für Physik und Astrophysik, Munich; English translation: Lawrence Livermore Laboratory Rep. UCRL-Trans-11206. University of California, Livermore, 1977.
Winkler, K.-H. A., and Newman, M. J. (1980). *Astrophys. J.* **238**, 311.
Winkler, K.-H. A., Mihalas, D., and Norman, M. L. (1984a). *Comput. Phys. Commun.* (in press).
Winkler, K.-H. A., Norman, M. L., and Mihalas, D. (1984b). *J. Quant. Spectrosc. Radiat. Transfer* **31**, No. 6, 473.
Winkler, K.-H. A., Norman, M. L., and Newman, M. J. (1984c). *Physica* **12D**, 408.
Woodward, P., and Colella, P. (1983). *J. Comput. Phys.* **54**, 115.
Yorke, H. W. (1979). *Astron. Astrophys.* **80**, 308.
Yorke, H. W. (1980). *Astron. Astrophys.* **85**, 215.
Zel'dovich, Ya. B. (1957). *Sov. Phys.—JETP (Engl. Transl.)* **5**, 919.

7

Multiple Time-Scale Methods in Tokamak Magnetohydrodynamics

STEPHEN C. JARDIN

Plasma Physics Laboratory
Princeton University
Princeton, New Jersey

I. Introduction 186
   A. Scope of Work 186
   B. Fundamental Equations 187
   C. MHD Time Scales 188
   D. General Principles 192
II. Ideal Time-Scale MHD Simulations 193
   A. Introduction 193
   B. Magnetic Flux Coordinates 194
   C. Velocity Decomposition 196
   D. The Transformed Ideal MHD Equations 197
   E. 2-D Dynamic Grid Method 198
   F. 2-D Vacuum Equations 201
   G. Finite-Difference Equations 203
   H. Application 204
III. Resistive Time-Scale MHD Simulations 209
   A. Introduction 209
   B. The Zero-Mass Method 209
   C. The Large-Mass Method 217
IV. Discussion 229
   References 231

I. INTRODUCTION

A. Scope of Work

A high-temperature magnetized plasma is an exceedingly complicated medium with many different resonances and fundamental modes of vibration. Natural time scales for phenomena occurring in a tokamak plasma near reactor parameters range from the subpicosecond period of an electron gyrating about the magnetic field to the several seconds it takes for the toroidal current to penetrate into the discharge. It is evident that to make progress in understanding and describing the behavior of the plasma in a device such as a tokamak, one must be fully aware of the time scales involved and of the nature of the phenomena occurring on each time scale.

The most basic set of equations describing the interaction of an ionized plasma with an electromagnetic field is the combined Maxwell and Vlasov–Fokker–Planck equations for the electric and magnetic fields and for the six-dimensional phase space distribution function. These equations are necessary for describing high-frequency phenomena involving particle orbits, wave-particle resonances, and the evolution of highly non-Maxwellian distribution functions. However, if we restrict our attention to relatively low-frequency macroscopic phenomena that do not depend on the details of the velocity–space distribution function, then, by taking appropriate velocity moments of the kinetic equations, we obtain a closed set of fluid and field equations, the magnetohydrodynamic (MHD) equations. The subject of this chapter is the numerical simulation of phenomena described by these MHD equations. We give particular emphasis to the application of the MHD equations to the description of phenomena occurring in tokamak fusion experiments.

There presently exists a substantial literature describing methods for the numerical simulation of various phenomena occurring in tokamak discharges. The present chapter is restricted further to exclude numerical methods based on the solution of model equations obtained by performing an asymptotic expansion in the geometric parameter a/R, the toroidal aspect ratio. A large amount of literature exists on this subject, and the interested reader is referred to several excellent articles (Strauss, 1976, 1983; Potter, 1976).

The remainder of this introductory section is devoted to describing the MHD equations and the time scales that are present when these equations are applied to describe the physics of the tokamak (Furth, 1975). We then state the general principles, which are developed in the subsequent sections. Section II outlines a method for describing the fastest time scale of interest in MHD, the transverse ideal time scale. In Section III we describe two methods for computing the slowest time scale of interest in MHD, the resistive or transport time scale. The final section briefly discusses ways of treating the intermediate MHD time scales and reviews the literature concerning certain techniques not discussed here.

B. Fundamental Equations

The basic equations describing the time evolution of a magnetically confined plasma comprise field equations (Maxwell's equations ignoring the displacement current),

$$\partial \mathbf{B}/\partial t = -\nabla \times \mathbf{E}, \tag{1a}$$

$$\nabla \times \mathbf{B} = \mu_0 \mathbf{J}, \tag{1b}$$

$$\nabla \cdot \mathbf{B} = 0, \tag{1c}$$

and fluid equations, which are appropriate velocity moments of the Vlasov–Fokker–Planck equations for electrons and ions. The fluid equations are distinguished by their parity in $\mathbf{v}$ (the microscopic particle velocity). The even parity moment equations represent the conservation of particles and energy:

$$\partial n_j/\partial t + \nabla \cdot (n_j \mathbf{u}_j) = S_{nj} \tag{2a}$$

$$(\tfrac{3}{2}) \partial p/\partial t + \nabla \cdot \sum_j (\mathbf{q}_j + (\tfrac{5}{2})p_j \mathbf{u}_j) = \mathbf{J} \cdot \mathbf{E} + S_p + \pi \tag{2b}$$

$$(\tfrac{3}{2}) \partial p_e/\partial t + \nabla \cdot (\mathbf{q}_e + (\tfrac{5}{2})p_e \mathbf{u}_e) = \mathbf{J} \cdot \mathbf{E} + Q_{\Delta e} + S_e + \pi$$
$$\qquad - \mathbf{u}_i \cdot \nabla p_i + \boldsymbol{\pi}_i : \nabla \mathbf{u}_i. \tag{2c}$$

Here, n_j is the particle density, $\mathbf{u}_j$ the conductive heat flux, $p_j = n_j T_j$ the pressure, and $p = p_e + p_i$, S_{nj}, S_p, S_e particle and pressure sources;

$$\pi = -\sum_{e,i} \nabla \cdot (\mathbf{u}_j \cdot \boldsymbol{\pi}_j) \tag{2d}$$

is the viscous heating term,

$$Q_{\Delta e} = 3(m_e/m_i)n_e(T_i - T_e)/\tau_{ei} \tag{2e}$$

the temperature equipartition term; $\boldsymbol{\pi}_j$ the viscous stress tensor, and τ_{ei} the electron–ion collision time. The relation $p_j \gg (\tfrac{1}{2})m_j n_j u_j^2$ has been assumed for the subsonic flows of interest.

The odd parity moments yield the force balance equations for each species:

$$\mathbf{E} + \mathbf{u}_j \times \mathbf{B} = -\mathbf{R}_j/(n_j e_j) + m_j T(u_j)/e_j, \tag{3a}$$

where

$$T(u) = \partial \mathbf{u}/\partial t + \mathbf{u} \cdot \nabla \mathbf{u} \tag{3b}$$

is the inertial term

$$\mathbf{R}_j = -\nabla p_j - \nabla \cdot \boldsymbol{\pi}_j + \mathbf{F}_j \tag{3c}$$

represents the departure from ideal ($\mathbf{E} \times \mathbf{B}$) drift motion,

$$\mathbf{F}_j = \int m_j \mathbf{v} c_j \, d\mathbf{v} \tag{3d}$$

is the collisional friction between electrons and ions, and c_j the collision operator of species j. Equation (3a) for electrons is the general form of Ohm's law. Summing $e_j n_j \times$ Eq. (3a) and using quasi-neutrality, $\sum_{e,i} n_j e_j = 0$, yields the net force balance

$$\rho \mathbf{T}(\mathbf{u}) + \nabla \cdot (\pi_i + \pi_e) = -\nabla p + \mathbf{J} \times \mathbf{B}, \qquad (3a')$$

where $\rho = \sum_{e,i} m_j n_j$ is the mass density, $\mathbf{u} = \sum_{e,i} m_j n_j \mathbf{u}_j / \rho$ the mean mass flow velocity, and $\mathbf{J} = \sum_{e,i} n_j e_j \mathbf{u}_j$ the electrical current density.

To close this set of equations, one must specify the remaining transport quantities $\mathbf{q}_i$, $\mathbf{q}_e$, π_i, π_e, and $\mathbf{F}_e = -\mathbf{F}_i$ that appear in Eqs. (2a)–(3d). The standard collisional transport model is given in an article by Braginskii (1965), in which these are evaluated in the limit of large fields ($\omega_{cj}\tau_j \gg 1$). The collisional friction term is

$$R_e = -R_i = ne(\mathbf{J}_\perp / \sigma_\perp + \mathbf{J}_\| / \sigma_\|) - 0.7 n k_b \nabla_\| T_e - \tfrac{3}{2}(nk_b/\omega_{ce}\tau_e)\hat{b} \times \nabla T_e, \qquad (4a)$$

where $\sigma_\| = 2.0\sigma_\perp = 2.0 ne^2 \tau_e / m_e$ are the electrical conductivities; $\|$ and $\perp$ refer to the direction of the magnetic field; $\mathbf{B} = |\mathbf{B}|\hat{b}$; ω_{ce} and ω_{ci} are the electron and ion gyrofrequencies; and electron and ion collisional times are

$$\tau_j = 3\sqrt{m_j}(k_b T_j)^{3/2}/4\sqrt{2\pi} n_j e^4 \ln \Lambda \qquad (4b)$$

for $j = e, i$, with $\ln \Lambda$ the Coulomb logarithm and k_b Boltzman's constant. The ion and electron heat fluxes are given by

$$\mathbf{q}_i = -K_\|^i \nabla_\| k_b T_i - K_\perp^i \nabla_\perp k_b T_i + K_\wedge^i \hat{b} \times \nabla_\perp k_b T_i, \qquad (4c)$$

$$\mathbf{q}_e = -K_\|^e \nabla_\| k_b T_e - K_\perp^e \nabla_\perp k_b T_e - K_\wedge^e \hat{b} \times \nabla_\perp k_b T_e$$
$$- 0.71(k_b T_e / e)\mathbf{J}_\| + \tfrac{3}{2}(k_b T_e / \omega_{ce}\tau_e e)\mathbf{J} \times \hat{b}, \qquad (4d)$$

with the thermal conductivities given by

$$K_\|^i = 3.9 \frac{nk_b T_i \tau_i}{m_i}, \qquad K_\perp^i = 2 \frac{nk_b T_i}{m_i \omega_{ci}^2 \tau_i}, \qquad K_\wedge^i = \frac{5}{2} \frac{nk_b T_i}{m_i \omega_{ci}},$$
$$K_\|^e = 3.2 \frac{nk_b T_e \tau_e}{m_e}, \qquad K_\perp^e = 4.7 \frac{nk_b T_e}{m_e \omega_{ce}^2 \tau_e}, \qquad K_\wedge^e = \frac{5}{2} \frac{nk_b T_e}{m_e \omega_{ce}}. \qquad (4e)$$

The reader is referred to Braginskii (1965) for the forms of the viscous stress tensors π_e and π_i. The forms of the transport coefficients valid in the more collisionless regimes are found in Hazeltine et al. (1973).

C. MHD Time Scales

We consider the application of the MHD equations (1)–(4) to the toroidal geometry illustrated in Fig. 1. A tokamak has a large externally produced toroidal field $\mathbf{B}_T$ and a smaller poloidal magnetic field $\mathbf{B}_p$, produced primarily by the toroidal plasma current. If R and a are the approximate major and minor radii of the toroidal plasma, we denote the ratio

$$q \simeq B_T a / B_p R$$

7. Tokamak Magnetohydrodynamics

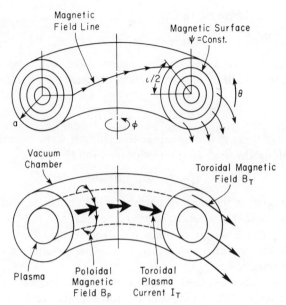

Fig. 1. Axisymmetric toroidal geometry. Magnetic field spirals around two-dimensional magnetic surfaces.

the *safety factor* of the device. (A more precise definition of the safety factor is given in Section II.) Typically, in tokamaks, q is in the range of 2 to 4 at the plasma edge. The inverse of the safety factor, $\iota = 2\pi/q$, measures the number of radians in poloidal angle that a magnetic-field line travels in transiting once the long way around the device.

There are several spatial lengths and directions of importance in the tokamak. In the most idealized case, the magnetic-field lines can be thought to lie in two-dimensional surfaces, called magnetic surfaces. We denote these nested surfaces by a *flux function*, $\psi = $ constant, so that the magnetic field satisfies $\mathbf{B} \cdot \nabla\psi = 0$. Clearly, the relevant spatial length in the normal direction $\hat{n} \equiv \nabla\psi/|\nabla\psi|$ across the magnetic surfaces and perpendicular to the magnetic field is the minor radius a.

For particles or waves traveling parallel to the magnetic field, in the direction $\hat{b} \equiv \mathbf{B}/|\mathbf{B}|$ the periodicity distance in the toroidal angle φ is $2\pi R$, but the periodicity distance in the poloidal angle θ is $2\pi Rq$ (see Fig. 1). It is the latter distance that is physically most relevant, since the equilibrium tokamak is normally symmetric in the φ direction but not in the θ direction.

The MHD equations contain both wavelike (hyperbolic) and diffusionlike (parabolic) characteristics. The time scales associated with the wave phenomena (Friedrichs, 1954) are the time it takes for a fast compressible wave to traverse the minor radius,

$$\tau_f \sim a(\mu_0\rho_0)^{1/2}/B_T, \tag{5a}$$

the time it takes a shear-Alfvén (transverse) wave, which propagates only along magnetic-field lines, to travel a connection length qR along a magnetic surface,

$$\tau_t \sim Rq(\mu_0\rho_0)^{1/2}/B_T, \tag{5b}$$

and the time it takes a slow-compressible (sound) wave, which also propagates only along the field, to travel a connection length,

$$\tau_s \sim Rq(\mu_0\rho_0/\beta)^{1/2}/B_T. \tag{5c}$$

Here $\beta = \mu_0 p_0/B_T^2$ is the ratio of plasma to field pressure, and ρ_0 the mass density. We note that it is the connection length qR and not the minor radius a that enters into Eqs. (5b, c), since both the shear-Alfvén and the slow-compressible waves propagate essentially one-dimensionally along the direction of the magnetic field.

The fundamental cross-field diffusion time scale, the plasma skin time, is the time it takes for the toroidal current (or poloidal field) to penetrate the minor radius a,

$$\tau_R \sim \mu_0 a^2/\eta_\parallel, \tag{5d}$$

where $\eta_\parallel$ is the parallel plasma resistivity, $\eta_\parallel = \sigma_\parallel^{-1}$. Other, related diffusion time scales contained in Eqs. (1)–(4) are the neoclassical (Hazeltine and Hinton, 1973) cross-field heat-conduction time,

$$\begin{aligned}\tau_h &\sim \beta^{-1}(m_e/m_i)^{1/2}(1+2q^2)^{-1}\tau_R \quad \text{high-collisionality regime} \\ &\sim (R/a)^{-3/2}q^{-2}\beta^{-1}(m_e/m_i)^{1/2}\tau_R \quad \text{low-collisionality regime}\end{aligned} \tag{5e}$$

and the neoclassical cross-field particle transport time,

$$\tau_p \sim (m_i/m_e)^{1/2}\tau_h. \tag{5f}$$

The equilibration times parallel to the magnetic field are much shorter. Since both the electron and ion transit times around the device are much shorter than their collision times, the characteristic times for equilibration within the surfaces are given by

$$\tau_{et} \sim qR/v_{te} \quad \text{electrons} \tag{5g}$$

and

$$\tau_{it} \sim qR/v_{ti}, \quad \text{ions.} \tag{5h}$$

We illustrate the multiplicity of time scales present in a present-generation tokamak fusion experiment in Fig. 2, where we have evaluated these expressions for $B_T = 5$ T, $I_p = 4$ MA, $R = 3$ m, $a = 1$ m, $n_0 = 10^{20}$ m^{-3}, and

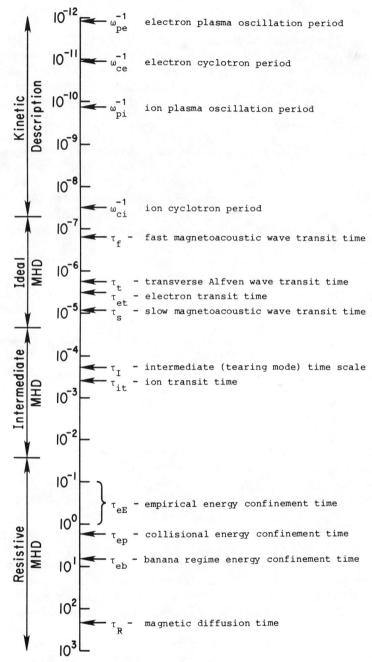

Fig. 2. Some of the fundamental time scales in a tokamak fusion experiment with $B_T = 5$ T, $I_p = 5$ MA, $R = 3$ m, $a = 1$ m, $n_0 = 10^{20}\ m^{-3}$, $T = 5$ keV.

$T = 5$ keV. For comparison, we have also indicated the fundamental kinetic time scales, electron and ion plasma oscillation periods, and cyclotron periods. We see that even neglecting the kinetic phenomena, the MHD phenomena span almost 10 orders of magnitude. Clearly, straightforward time integration of the equations is unfeasible on all but the fastest time scales.

D. General Principles

We discuss here some of the principles that are explored in the subsequent sections of this chapter to enable time integration of the MHD equations on the different time scales. The most obvious but most important principle is this: Only attempt to model transient phenomena accurately on a single time scale in a given simulation. Average over the phenomena occurring on the faster time scales, and ignore the variations occurring on the slower time scales.

Averaging of the fast time-scale phenomena is justified by the fact that there are no unstable modes on this faster time scale, only stable oscillations. If there were unstable modes on a faster time scale, it would make no sense to simulate the slower time scale. For example, we would not try to simulate the diffusion time-scale evolution of a plasma that is unstable to MHD instabilities that grow on the time of the transverse Alfvén wave (kink modes) or the intermediate time scale (tearing modes).

A second principle that we utilize is that natural coordinates aligned with the magnetic field are essential in separating out anisotropic wave motion and anisotropic diffusion. Since the magnetic field, in general, will change its direction during the course of a calculation, this necessitates utilizing a time-dependent coordinate transformation that continually aligns itself with the field. We shall see how this formalism can greatly elucidate the separate physics occurring when modeling both the wave (ideal) and the diffusion (resistive) time scales.

A third technique is to use a combination stream function–potential representation for a two-dimensional velocity field. We shall find that doing this aids in separating out the fast compressible MHD wave from the slower incompressible transverse wave. In fact, in Section II we describe the dynamic grid method that combines this with the previous technique so that the incompressible part of the velocity is treated as Lagrangian and evolves the coordinate transformation or, equivalently, the computational grid. The compressible part of the velocity field is treated as Eulerian or as flow relative to this moving grid.

A final technique that we utilize in Section III is to take some freedom with modifying the phenomena occurring on the faster time scales. If the slower time-scale motion depends only on the long-time asymptotic response of the faster time-scale motion, we are free to modify the dynamics of the faster time-scale phenomena, as long as we do not affect its long-time behavior. In Section III we describe two methods that are able to get at the slow (resistive) time scale by modifying the fast (ideal) time-scale waves in two very different ways. The first method accelerates these waves to infinite velocity and thus changes these characteristics from hyperbolic to elliptic, thereby eliminating the wave time scales from the problem. The second method slows these waves so they can be integrated along with the resistive phenomena.

II. IDEAL TIME-SCALE MHD SIMULATIONS

A. Introduction

In tokamaks there is a large class of MHD instabilities that take place on a time scale short compared with the dissipative times. In describing this motion, we can utilize a simplified fluid description in which the dissipative terms and the differences in the electron and ion properties are ignored. This simplified set of equations describing the evolution of the magnetic field, the density, the velocity, and the pressure is referred to as the ideal MHD equations:

$$\partial \mathbf{B}/\partial t = \nabla \times (\mathbf{v} \times \mathbf{B}), \tag{6a}$$

$$\partial \rho/\partial t + \nabla \cdot (\rho \mathbf{v}) = 0, \tag{6b}$$

$$(\partial/\partial t + \mathbf{v} \cdot \nabla)(p/\rho^\gamma) = 0, \tag{6c}$$

$$\rho \, \partial \mathbf{v}/\partial t + \rho \mathbf{v} \cdot \nabla \mathbf{v} + \nabla p = (\nabla \times \mathbf{B}) \times \mathbf{B}/\mu_0. \tag{6d}$$

We note that neither the electric field $\mathbf{E}$ nor the current density $\mathbf{J}$ explicitly appears in Eqs. (6a–d), although they may be obtained from the infinite conductivity limit of Ohm's law and from Ampere's law, ignoring the displacement current:

$$\mathbf{E} + \mathbf{v} \times \mathbf{B} = 0, \tag{7a}$$

$$\nabla \times \mathbf{B} = \mu_0 \mathbf{J}. \tag{7b}$$

The transient solutions of Eqs. (6a–d) are dominated by the wave solutions. We have seen in Eqs. (5a, b) that the time scale τ_t for the nearly incompressible transverse modes associated with MHD stability phenomena is longer than the time scale for the fast compressible modes by the factor Rq/a, i.e.,

$$\tau_f \sim (Rq/a)\tau_t.$$

This factor Rq/a is typically 10–30 in tokamaks of interest.

There are several different ways of dealing with this disparity. One (Potter, 1976; Strauss, 1976, 1983) is to use the limiting form of the MHD equations as Rq/a approaches infinity. This forces incompressibility and thus annihilates the fast waves altogether, leaving only the transverse time scale in the equations. However, this approximation can be inadequate for Rq/a finite. Another approach (Wesson and Sykes, 1975; Schneider and Bateman, 1975; Bateman et al., 1974; Brackbill, 1976) has been to integrate the primitive compressible equations without taking special care to isolate phenomena occurring on different time scales. This approach is useful when simulating tokamaks with small Rq/a and large $\beta \sim 2\mu_0 p/B^2$ so that the transverse and fast time scales come together. Fully implicit methods have been considered but have not been perceived as practical because of the large number of variables. Alternating-direction implicit methods have not been demonstrated to be useful in such a highly anisotropic medium.

We describe here the dynamic grid method (Jardin et al., 1978), a technique for solving the ideal MHD equations on the transverse time scale by averaging over the fast time-scale phenomena numerically using a partially implicit method. This is made possible by introducing a time-dependent, nonorthogonal, magnetic-flux coordinate transformation that determines the grid used in the computation. The coordinate system is chosen to conform to the field in such a way that the characteristics of the partial differential equations can be represented accurately. In particular, the transverse wave that rotates the magnetic field as it propagates will be represented in the time-dependent coordinate transformation. The fast compressible wave, on the other hand, compresses both the field and the fluid as it propagates. It is removed from the coordinate transformation by forcing the grid velocity to be incompressible, i.e., by making the Jacobian constant in time. This constrains the fast wave motion to appear in the slippage flow; i.e., in the Eulerian motion of the plasma flowing through the grid. We shall see that this leads to a relatively simple scalar equation that can be solved implicitly, effectively averaging over the fast wave phenomena.

B. Magnetic-Flux Coordinates

We introduce the coordinates (Kruskal and Kulsrud, 1958) (ψ, θ, φ), which, along with the time t, serve as the independent variables. The coordinates θ and φ are periodic anglelike variables with period 2π. The angle θ increases when going the short way around the torus (poloidal direction), whereas the angle φ increases when going the long way around the torus (toroidal direction). The coordinate φ is chosen to be constant along a magnetic-field line; i.e., $\mathbf{B} \cdot \nabla \psi = 0$.

A general form for the three-dimensional magnetic field $\mathbf{B}$, a vector function satisfying $\mathbf{B} \cdot \nabla \psi = 0$ and $\nabla \cdot \mathbf{B} = 0$, is given by

$$\mathbf{B} = [\chi_\psi(\psi, t) + h_\varphi(\psi, \theta, \varphi, t)] \nabla \varphi \times \nabla \psi + [\Psi_\psi(\psi, t) - h_\theta(\psi, \theta, \varphi, t)] \nabla \psi \times \nabla \theta, \tag{8}$$

7. Tokamak Magnetohydrodynamics

where here and elsewhere we use subscripts to denote partial differentiation. The two one-dimensional scalar functions $\chi(\psi, t)$ and $\Psi(\psi, t)$ and the three-dimensional scalar function $h(\psi, \theta, \varphi, t)$ completely determine the magnetic field once the coordinate basic vectors $(\nabla\psi, \nabla\theta, \nabla\varphi)$ are known.

We note here that the magnetic field as represented by Eq. (8) can vary in several ways. It can change its magnitude and its orientation within a magnetic surface by changing the functions $\chi(\psi, t)$, $\Psi(\psi, t)$, and $h(\psi, \theta, \varphi, t)$ within the brackets. Alternatively, it can change its orientation by changing the directions of the basis vectors $\nabla\varphi \times \nabla\psi$ and $\nabla\psi \times \nabla\theta$.

It is assumed throughout that the surfaces $\psi = $ constant form nested toroidal surfaces. If this is true initially at time $t = 0$, then it is a property of the ideal MHD equations that this topology will remain invariant for all time. The functions $\chi(\psi, t)$ and $\Psi(\psi, t)$ introduced in Eq. (8) are physically $1/2\pi$ times the amounts of poloidal and toroidal magnetic flux contained within a magnetic surface with label ψ. Thus, by direct computation, the poloidal flux is

$$2\pi\chi(\psi, t) \equiv \frac{1}{(2\pi)} \int \mathbf{B} \cdot \nabla\theta \, d\tau = 2\pi \int_0^\psi \chi_{\psi'}(\psi', t) \, d\psi', \tag{9a}$$

and the toroidal flux is

$$2\pi\Psi(\psi, t) \equiv \frac{1}{(2\pi)} \int \mathbf{B} \cdot \nabla\varphi \, d\tau = 2\pi \int_0^\psi \Psi_{\psi'}(\psi', t) \, d\psi'. \tag{9b}$$

We note here that the safety factor q can be expressed as the ratio $q \equiv d\Psi/d\chi$. The volume element used in Eq. (9) is

$$d\tau \equiv J \, d\psi \, d\theta \, d\varphi, \tag{10a}$$

where

$$J \equiv [\nabla\psi \times \nabla\theta \cdot \nabla\varphi]^{-1} = \left| \frac{\partial \mathbf{x}}{\partial \psi} \times \frac{\partial \mathbf{x}}{\partial \theta} \cdot \frac{\partial \mathbf{x}}{\partial \varphi} \right| \tag{10b}$$

is the Jacobian of the transformation from Cartesian to magnetic-flux coordinates.

It is notationally convenient to introduce the natural basis vectors for the ψ, θ, φ coordinate system,

$$\hat{e}^1 \equiv \nabla\psi, \qquad \hat{e}^2 \equiv \nabla\theta, \qquad \hat{e}^3 \equiv \nabla\varphi, \tag{11}$$

and the reciprocal bases

$$\hat{e}_1 \equiv \partial \mathbf{x}/\partial\psi = x_\psi \hat{x} + y_\psi \hat{y} + z_\psi \hat{z}, \tag{12a}$$

$$\hat{e}_2 \equiv \partial \mathbf{x}/\partial\theta = x_\theta \hat{x} + y_\theta \hat{y} + z_\theta \hat{z}, \tag{12b}$$

$$\hat{e}_3 \equiv \partial \mathbf{x}/\partial\varphi = x_\varphi \hat{x} + y_\varphi \hat{y} + z_\varphi \hat{z}, \tag{12c}$$

or

$$\hat{e}_i = \partial x^j/\partial u^i \, \hat{x}^j. \tag{12d}$$

Here $(x^1 \equiv x, x^2 \equiv y, x^3 \equiv z)$ form a standard Cartesian coordinate system, $(\hat{x}^1 \equiv \hat{x}, \hat{x}^2 \equiv \hat{y}, \hat{x}^3 \equiv \hat{z})$ are unit vectors along the Cartesian axis, and for notational convenience we have let $u^1 \equiv \psi, u^2 \equiv \theta, u^3 \equiv \varphi$. The basis vectors are related by way of the formula

$$\hat{e}^i = (1/2J)\hat{e}_j \times \hat{e}_k \varepsilon_{ijk}, \quad \hat{e}_i = \tfrac{1}{2}J\hat{e}^j \times \hat{e}^k \varepsilon_{ijk}, \quad \hat{e}^i \cdot \hat{e}_j = \delta_{ij},$$

$$\hat{e}^i = g^{ij}\hat{e}_j, \quad \hat{e}_i = g_{ij}\hat{e}^j, \quad g^{ij} = \hat{e}^i \cdot \hat{e}^j, \quad g_{ij} = \hat{e}_i \cdot \hat{e}_j,$$

$$\partial \hat{e}_i/\partial u^j = \{{}^{\,k}_{ij}\}\hat{e}_k, \quad \partial \hat{e}^i/\partial u^j = -\{{}^{\,i}_{jk}\}\hat{e}^k, \tag{13}$$

where (ijk) run from 1 to 3, δ_{ij} is the Kronecker delta, ε_{ijk} is the usual permutation symbol, and summation of repeated indices is implied.

It is readily verified that the metric tensor, the Jacobian, and the Christoffel symbols can be expressed as the following derivatives of the Cartesian coordinates:

$$g_{ij} = (\partial x^k/\partial u^i)(\partial x^k/\partial u^j), \tag{13a}$$

$$g^{ij} = (1/2J^2)\varepsilon_{ikl}\varepsilon_{jmn}g_{km}g_{ln}, \tag{13b}$$

$$J = \varepsilon_{ijk}(\partial x^i/\partial \psi)(\partial x^j/\partial \theta)(\partial x^k/\partial \varphi), \tag{13c}$$

$$\{{}^{\,\alpha}_{\beta\gamma}\} = (1/2J)\varepsilon_{ijk}\varepsilon_{\alpha mn}(\partial^2 x^i/\partial u^\beta \partial u^\gamma)(\partial x^j/\partial u^m)(\partial x^k/\partial u^n). \tag{13d}$$

Thus, for example

$$g_{12} = (x_\psi x_\theta + y_\psi y_\theta + z_\psi z_\theta),$$

$$g^{13} = [(y_\theta z_\varphi - y_\varphi z_\theta)(y_\psi z_\theta - y_\theta z_\psi) + (x_\varphi z_\theta - x_\theta z_\varphi)(x_\theta z_\psi - z_\theta x_\psi) + (x_\theta y_\varphi - y_\theta x_\varphi)(x_\psi y_\theta - y_\psi x_\theta)]/J^2,$$

$$\{{}^{\,1}_{23}\} = [x_{\theta\varphi}(y_\theta z_\varphi - y_\varphi z_\theta) + y_{\theta\varphi}(x_\varphi z_\theta - x_\theta z_\varphi) + z_{\theta\varphi}(y_\varphi x_\theta - y_\theta x_\varphi)]/J.$$

C. Velocity Decomposition

A *coordinate velocity* may be defined by

$$\mathbf{v}_g \equiv \partial \mathbf{x}/\partial t, \tag{14a}$$

where time derivatives, unless explicitly indicated, are taken with (ψ, θ, φ) held fixed. The time rate of change of the Jacobian and the grid velocity are related by the identity (in the flux surface frame)

$$\partial J/\partial t = J\nabla \cdot \mathbf{v}_g. \tag{14b}$$

The coordinate velocity enters naturally when transforming time derivatives from a stationary system to the ψ, θ, φ system. If Ω is an arbitrary scalar quantity, then the chain rule takes the form

$$\partial\Omega/\partial t|_{\psi,\theta,\varphi} = \partial\Omega/\partial t|_{\mathbf{x}} + \mathbf{v}_g \cdot \nabla\Omega. \tag{15}$$

The velocity of a fluid element is thus divided into two parts, the velocity of the coordinates and a velocity relative to the coordinates:

$$\mathbf{v} = \mathbf{v}_g + \mathbf{v}_r. \tag{16}$$

It is clear that only the combination $\mathbf{v}_g + \mathbf{v}_r$ can be determined from the physical evolution equations.

For definiteness, we introduce the six scalar variables (α, β, γ), (u, v, w) so that the grid velocity and the relative velocity are

$$\mathbf{v}_g \cdot \nabla\psi = \alpha(\psi,\theta,\varphi,t), \qquad \mathbf{v}_r \cdot \nabla\psi = u(\psi,\theta,\varphi,t); \tag{16a}$$

$$\mathbf{v}_g \cdot \nabla\theta = \beta(\psi,\theta,\varphi,t), \qquad \mathbf{v}_r \cdot \nabla\theta = v(\psi,\theta,\varphi,t); \tag{16b}$$

$$\mathbf{v}_g \cdot \nabla\varphi = \gamma(\psi,\theta,\varphi,t), \qquad \mathbf{v}_r \cdot \nabla\varphi = w(\psi,\theta,\varphi,t). \tag{16c}$$

D. The Transformed Ideal MHD Equations

By transforming to the coordinate system given by Eqs. (10)–(13) and using the representation for the magnetic field given by Eq. (8) and for the velocity field given by Eqs. (14)–(16), the magnetic-flux conservation equation (6a) becomes

$$(\chi_\psi)_t + (\chi_\psi u)_\psi = 0, \tag{17a}$$

$$(\Psi_\psi)_t + (\Psi_\psi u)_\psi = 0, \tag{17b}$$

$$u_\theta = u_\varphi = 0, \tag{17c}$$

$$h_t + (hu)_\psi + \omega(\chi_\psi + h_\varphi) - v(\Psi_\psi - h_\theta) = 0; \tag{17d}$$

the conservation of mass and entropy equations (6b) and (6c) become

$$(\rho J)_t + (\rho J u)_\psi + (\rho J v)_\theta + (\rho J w)_\varphi = 0 \tag{17e}$$

$$(p^{1/\gamma}J)_t + (p^{1/\gamma}Ju)_\psi + (p^{1/\gamma}Jv)_\theta + (p^{1/\gamma}Jw)_\varphi = 0; \tag{17f}$$

and the grid evolution and momentum equations assume the compact form

$$x_t^i = (\mathbf{v}_g \cdot \nabla u_j) \, \partial x^i / \partial u_j, \tag{17g}$$

$$(\mathbf{v} \cdot \nabla u^i)_t + (1/\rho)(\nabla u_i \cdot \nabla u_j) \, \partial P/\partial u^j + w^i = 0, \tag{17h}$$

where we have defined the total pressure

$$P \equiv p + (1/2\mu_0)B^2, \tag{18a}$$

with

$$B^2 = (\chi_\psi + h_\varphi)^2 g_{22} + 2(\chi_\psi + h_\varphi)(\Psi_\psi - h_\theta)g_{23} + (\Psi_\psi - h_\theta)^2 g_{33} \quad (18b)$$

being the square of the magnetic-field strength and

$$w^i \equiv \mathbf{v} \cdot \nabla(\mathbf{v}_g \cdot \nabla u^i) + \mathbf{v}_r \cdot \nabla(\mathbf{v} \cdot \nabla u^i)$$
$$- (\partial/\partial u^j) |J(\mathbf{B} \cdot \nabla u^i)(\mathbf{B} \cdot \nabla u^j)|/\rho\mu_0 J$$
$$+ |(\mathbf{v} \cdot \nabla u^j)(\mathbf{v} \cdot \nabla u^k) - (\mathbf{B} \cdot \nabla u^j)(\mathbf{B} \cdot \nabla u^k)/\rho\mu_0|\{{}^{\ i}_{jk}\}. \quad (18c)$$

E. 2-D Dynamic Grid Method

To illustrate the dynamic grid method, we consider a 2-D axisymmetric subsystem of the equations presented in Section D obtained by taking

$$x = R(\psi, \theta) \sin\varphi, \quad y = R(\psi, \theta) \cos\varphi \quad z = Z(\psi, \theta), \quad (19)$$

and by letting $R(\psi, \theta)$ and $Z(\psi, \theta)$ be the coordinate functions; see Fig. (3). We consider only motions that have no φ variation other than that explicitly indicated in Eq. (19), i.e., that have $\partial/\partial\varphi \equiv 0$. This corresponds to considering the axisymmetric motion of a toroidal plasma column with no φ variation.

The form of the Jacobian is arbitrary so far, although it is related to the coordinate velocity through Eq. (14b). To force the moving (ψ, θ) mesh to be nearly incompressible and thus free of the fast-time-scale motion, we take the Jacobian to be of the form

$$J = T(t)R^m\psi^n, \quad (20)$$

where m, n are integers to be specified later, and $T(t)$ is a time-dependent normalization constant that we allow to change on the slow time scale. It then follows from Eqs. (14b) and (20) that if $T_t = 0$,

$$\nabla \cdot (R^{-m}\mathbf{v}_g) = 0, \quad (21)$$

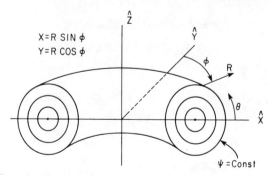

Fig. 3. A two-dimensional axisymmetric subsystem uses (R, Z, φ) as the fixed coordinates and (ψ, θ, φ) as the magnetic-flux coordinates, with φ being ignorable.

7. Tokamak Magnetohydrodynamics

and both the Jacobian and hence the coordinate functions $R(\psi,\theta,t)$ and $Z(\psi,\theta,t)$ are free of the fast-time-scale motion.

From Eq. (21) it follows that a stream function ξ exists for part of the two-dimensional coordinate velocity. For $T_t \neq 0$ we generalize the concept of stream function and write

$$\mathbf{v}_g = R^m \nabla \times (\xi \nabla\varphi) + (T_t/T)(\mathbf{Z} - \mathbf{Z}_0). \tag{22}$$

The second term in Eq. (22) is to accommodate the coordinate system expansion implied by allowing a time-dependent constant $T(t)$ in the expression for the Jacobian, Eq. (20). Thus, in Eq. (16) we have

$$\alpha \equiv \mathbf{v}_g \cdot \nabla\psi = J^{-1}[R^m \xi_\theta - (T_t/T)RR_\theta(Z - Z_0)], \tag{23a}$$

$$\beta \equiv \mathbf{v}_g \cdot \nabla\theta = J^{-1}[-R^m \xi_\psi + (T_t/T)RR_\psi(Z - Z_0)], \tag{23b}$$

$$\gamma \equiv \mathbf{v}_g \cdot \nabla\varphi = 0. \tag{23c}$$

In this 2-D axisymmetric limit, the metric tensor connecting $(\psi,\theta,)$ and (R,Z,φ) simplifies so that $g_{13} = g_{31} = g_{23} = g_{32} = 0$, $g^{13} = g^{31} = g^{23} = g^{32} = 0$, $g_{33} = R^2$, $g^{33} = R^{-2}$. Other nonzero metric terms are given by

$$J \equiv (\nabla\psi \times \nabla\theta \cdot \nabla\varphi)^{-1} = R(R_\psi Z_\theta - Z_\psi R_\theta), \tag{24}$$

$$g^{11} \equiv |\nabla\psi|^2 = (R^2/J^2)(R_\theta^2 + Z_\theta^2),$$

$$g^{22} \equiv |\nabla\theta|^2 = (R^2/J^2)(R_\psi^2 + Z_\psi^2),$$

$$g^{12} = g^{21} \equiv \nabla\theta \cdot \nabla\psi = -\left(\frac{R^2}{J^2}\right)(R_\theta R_\psi + Z_\theta Z_\psi),$$

$$\{{}^{\,1}_{11}\} = (R/J)(Z_\theta R_{\psi\psi} - R_\theta Z_{\psi\psi}), \quad \{{}^{\,1}_{22}\} = (R/J)(Z_\theta R_{\theta\theta} - R_\theta Z_{\theta\theta}),$$

$$\{{}^{\,2}_{11}\} = (R/J)(R_\psi Z_{\psi\psi} - Z_\psi R_{\psi\psi}), \quad \{{}^{\,2}_{22}\} = (R/J)(R_\psi Z_{\theta\theta} - Z_\psi R_{\theta\theta}),$$

$$\{{}^{\,1}_{21}\} = \{{}^{\,1}_{12}\} = (R/J)(Z_\theta R_{\psi\theta} - R_\theta Z_{\psi\theta}),$$

$$\{{}^{\,2}_{21}\} = \{{}^{\,2}_{12}\} = (R/J)(R_\psi Z_{\psi\theta} - Z_\psi R_{\psi\theta}),$$

$$\{{}^{\,3}_{13}\} = \{{}^{\,3}_{31}\} = R_\psi/R, \quad \{{}^{\,3}_{23}\} = \{{}^{\,3}_{32}\} = R_\theta/R,$$

$$\{{}^{\,1}_{33}\} = -R^2 Z_\theta/J, \quad \{{}^{\,2}_{33}\} = R^2 Z_\psi/J.$$

It is natural and convenient to introduce a new variable for the toroidal magnetic field:

$$F(\psi,\theta) \equiv (R^2/J)[\Psi' - h_\theta]$$

so that the axisymmetric magnetic field, a specialization of the general form given by Eq. (8), is given by

$$\mathbf{B} = \chi_\psi(\psi,t) \nabla\varphi \times \nabla\psi + F(\psi,\theta,t) \nabla\varphi. \tag{25}$$

We note that in axisymmetric equilibrium, when $\mathbf{J} \times \mathbf{B} = \nabla p$, it follows from Eq. (25) that $F_\theta = 0$. The analog of the 3-D equations (17) in this 2-D geometry becomes

$$(\chi_\psi)_t + (\chi_\psi u)_\psi = 0, \tag{26a}$$

$$(FJ/R^2)_t + (uFJ/R^2)_\psi + (vFJ/R^2 - \omega\chi_\psi)_\theta = 0, \tag{26b}$$

$$u_\theta = 0, \tag{26c}$$

$$(\rho J)_t + (\rho J u)_\psi + (\rho J v)_\theta = 0, \tag{26d}$$

$$(p^{1/\gamma}J)_t + (p^{1/\gamma}Ju)_\psi + (p^{1/\gamma}Jv)_\theta = 0, \tag{26e}$$

$$R_t = R_\psi \alpha + R_\theta \beta, \tag{26f}$$

$$Z_t = Z_\psi \alpha + Z_\theta \beta, \tag{26g}$$

$$(u + \alpha)_t + \rho^{-1}|\nabla\psi|^2 P_\psi + \rho^{-1}(\nabla\theta \cdot \nabla\psi)P_\theta + w = 0, \tag{26h}$$

$$(v + \beta)_t + \rho^{-1}(\nabla\theta \cdot \nabla\psi)P_\psi + \rho^{-1}|\nabla\theta|^2 P_\theta + S = 0, \tag{26i}$$

$$\omega_t + V = 0. \tag{26j}$$

Here, the total pressure, Eqs. (18a, b) takes the simplified form

$$P = p + (2\mu_0)^{-1}[\chi_\psi^2|\nabla\psi|^2 + F^2]/R^2, \tag{27a}$$

and we have defined

$$w \equiv v(u + \alpha)_\theta + u(u + \alpha)_\psi + (u + \alpha)\alpha_\psi + (v + \beta)\alpha_\theta$$
$$+ (u + \alpha)^2\{{}^1_{11}\} + (v + \beta)^2\{{}^1_{22}\} + 2(u + \alpha)(v + \beta)\{{}^1_{21}\}$$
$$- (\rho\mu_0)^{-1}(\chi_\psi/J)^2\{{}^1_{22}\} + (\omega^2 - F^2/\rho\mu_0)\{{}^1_{33}\}, \tag{27b}$$

$$S \equiv v(v + \beta)_\theta + u(v + \beta)_\psi + (u + \alpha)\beta_\psi + (v + \beta)\beta_\theta$$
$$+ (u + \alpha)^2\{{}^2_{11}\} + (v + \beta)^2\{{}^2_{22}\} + 2(u + \alpha)(v + \beta)\{{}^2_{12}\}$$
$$- (\rho\mu_0)^{-1}(\chi_\psi/J)^2\{{}^2_{22}\} + (\omega^2 - F^2/\rho\mu_0)\{{}^2_{33}\}, \tag{27c}$$

$$U = u\omega_\psi + v\omega_\theta + 2\omega(u + \alpha)\{{}^3_{13}\} + [2\omega(v + \beta) - \chi'F/JR^2\rho\mu_0]\{{}^3_{23}\}$$
$$- (\partial/\partial\theta)[\chi'F/R^2]/\rho\mu_0 J. \tag{27d}$$

We obtain two auxiliary equations for the fast wave by adding the ψ derivative of Eq. (26h) to the θ derivative of Eq. (26i) and by time-differentiating Eq. (27a); thus

$$\Delta_t + [\rho^{-1}|\nabla\psi|^2 P_\psi + \rho^{-1}(\nabla\theta \cdot \nabla\psi)P_\theta]_\psi$$
$$+ [\rho^{-1}(\nabla\theta \cdot \nabla\psi)P_\psi + \rho^{-1}|\nabla\theta|^2 P_\theta]_\theta + Q = 0, \tag{28a}$$

$$P_t + (\gamma p + F^2/\mu_0 R^2)\Delta + N = 0. \tag{28b}$$

Here

$$\Delta \equiv \nabla \cdot v - \frac{1}{J} \mathbf{v} \cdot \nabla J = (u + \alpha)_\psi + (v + \beta)_\theta \tag{29a}$$

is the part of the divergence of the velocity field not associated with spatial variation of the Jacobian,

$$Q \equiv w_\psi + S_\theta, \tag{29b}$$

$$\begin{aligned}
N &\equiv (u + \alpha)[\gamma p J \psi + (1/\mu_0 R) F^2 (J/R)_\psi]/J \\
&+ (v + \beta)[\gamma p J \theta + (1/\mu_0 R) F^2 (J/R)_\theta]/J \\
&+ u[p + \tfrac{1}{2} F^2 / \mu_0 R^2]_\psi + v[p + \tfrac{1}{2} F^2 / \mu_0 R^2]_\theta \\
&- F \omega_\theta \chi_\psi / \mu_0 J + \chi_\psi (|\nabla \psi|^2 / \mu_0 R^2)(u \chi_\psi)_\psi \\
&- \tfrac{1}{2} \chi_\psi^2 (|\nabla \psi|^2 / \mu_0 R^2)_t .
\end{aligned} \tag{29c}$$

F. 2-D Vacuum Equations

The ideal MHD boundary conditions between a plasma region and a vacuum region are for the tangential electric field and the total pressure P [Eq. (27a)] to be continuous. The first condition is equivalent to the requirements that the poloidal flux χ be continuous and that the plasma–vacuum interface remain a flux surface. This is built into the present formulation. The second condition requires the calculation of B^2 on the vacuum side of the plasma–vacuum interface. This is obtained by solving an elliptic magnetostatic problem each time step in the irregularly shaped and changing vacuum region. We describe here a Green's function solution method which allows for the existence of external currents in the vacuum region.

The magnetic field in the vacuum region is represented as

$$\mathbf{B} = (2\pi)^{-1} \nabla \chi \times \nabla \varphi + F_0 \nabla \varphi, \tag{30}$$

where F_0 is a spatial constant related to the total external current flowing through the center (hole) of the torus. We allow for a singular current distribution in the vacuum region owing to M current-carrying rings of negligible cross section located at R_m, Z_m ($m = 1, M$):

$$J = \sum_{m=1}^{M} I_m \delta(\mathbf{x} - \mathbf{x}_m).$$

Away from these singularities, the magnetic field in the vacuum satisfies the equation $\nabla \times \mathbf{B} = 0$ or, from Eq. (30), $R^2 \nabla \cdot R^{-2} \nabla \chi = 0$. The infinite medium toroidal Green's function is given by

$$G(\mathbf{x}; \mathbf{x}') = [(RR')^{1/2}/k][(2 - k^2) K(k^2) - E(k^2)],$$

where

$$k^2 \equiv 4RR'/[(R + R')^2 + (Z - Z')^2]^{1/2}$$

and $K(k^2)$ and $E(k^2)$ are elliptic integrals of the first and second kind. The poloidal flux in the vacuum region then satisfies the integral equation

$$\chi(\mathbf{x}) = \sum_{m=1}^{M} \mu_0 I_m G(\mathbf{x}; \mathbf{x}_m) + \int \frac{dl'}{R'} G(\mathbf{x}; \mathbf{x}') \frac{\partial \chi}{\partial n'}, \qquad (31)$$

where the line integral is evaluated over the plasma–vacuum interface. Equation (31) is solved by approximating the line integral by many line segments Δl_s, over each of which $\partial \chi/\partial n$ is taken as a constant. Evaluating Eq. (30) at each line segment transforms the integral equation into a matrix equation. Thus, for each interface segment Δl_i centered at $\mathbf{x}_i$ with poloidal flux $\chi_i = $ constant, we have

$$\chi_i = \sum_{m=1}^{M} \mu_0 I_m G(\mathbf{x}_i; \mathbf{x}_m) + \sum_{\substack{j=1,N \\ j \neq i}} \frac{\Delta l_j}{R_j} G(\mathbf{x}_i; \mathbf{x}_j) \left(\frac{\partial \chi}{\partial n}\right)_j$$

$$+ \int_{\Delta l_i} \frac{dl'}{R'} G(\mathbf{x}_i; \mathbf{x}') \frac{\partial \chi}{\partial n'}. \qquad (32)$$

Special care is necessary in evaluating the last *self-field* term since $|G_{ij}|$ diverges as $\mathbf{x}_i$ approaches $\mathbf{x}_j$. This is approximated by

$$\int_{\Delta l_i} \frac{dl'}{R'} G(\mathbf{x}_i, \mathbf{x}) \frac{\partial \chi}{\partial n'} \cong \Delta l_i \left[\ln\left(\frac{\Delta l_i}{16 R_i}\right) + 1\right] \left(\frac{\partial \chi}{\partial n}\right)_i. \qquad (33)$$

The integral Eq. (31) is thus reduced to the matrix equation

$$\mathbf{A} \cdot \mathbf{B} = \mathbf{C}, \qquad (34)$$

where

$$c_i = \chi_i - \sum_{m=1}^{M} \mu_0 I_m G_{im}, \qquad (35a)$$

$$b_i = (\partial \chi/\partial n)_i \qquad (35b)$$

$$a_{ij} = (\Delta l_j/R_j) G(\mathbf{x}_i; \mathbf{x}_j), \quad \text{for } i \neq j, \qquad (35c)$$

$$a_{ii} = \Delta l_i \left|\ln(\Delta l_i/16 R_i) + 1\right|. \qquad (35d)$$

Equation (34) is a well-conditioned matrix equation solved by standard techniques. The total pressure on the vacuum side of the plasma boundary is then given by

$$P = (2\mu_0 R^2)^{-1}[(2\pi)^{-2}(\partial \chi/\partial n)^2 + F_0^2]. \qquad (36)$$

G. Finite-Difference Equations

We describe here a partially implicit finite-difference method in which the compressible wave motion is treated *implicitly*, but the incompressible motion and the material convection are treated *explicitly*. The feasibility of this treatment lies in the fact that the only fast-time-scale motion is associated with the fast compressible magnetoacoustic wave, which propagates isotropically and has been isolated in the variables P and Δ through Eqs. (28a, b). The coefficients in Eq. (28) (the metric terms, Q, N, $\gamma p + F^2/R^2$, etc.) can be held constant during the implicit iteration. By time-centering Eqs. (28a, b), we obtain the following equation for the advanced-time total pressure P^{n+1}:

$$P^{n+1} - \tfrac{1}{2} a\, \delta t [(b P^{n+1}_\psi + c P^{n+1}_\theta)_\psi + (c P^{n+1}_\psi + d P^{n+1}_\theta)_\theta] + D = 0. \quad (37)$$

Here we have defined the variables

$$a \equiv \delta t (\gamma P + F^2/\mu_0 R^2)^{n+1/2}, \qquad b \equiv [(1/\rho)|\nabla \psi|^2]^{n+1/2}$$

$$c \equiv [(1/\rho)\nabla\theta \cdot \nabla\psi]^{n+1/2}, \qquad d \equiv [(1/\rho)|\nabla\theta|^2]^{n+1/2}$$

$$D \equiv \delta t\, N^{n+1/2} + 2a\, \Delta^n - P^n - a\, \delta t Q^{n+1/2}$$
$$- \tfrac{1}{2} a\, \delta t [(b P^n_\psi + c P^n_\theta)_\psi + (c P^n_\psi + d P^n_\theta)_\theta]. \quad (38)$$

We now summarize the partially implicit time-advancement scheme to take the solution from time level n to time level $n + 1$:

1. Advance the coordinates to a predicted value at time level $n + 1$, using Eqs. (26f, g) with the velocities at time level n:

$$(\tilde{R}^{n+1} - R^n)/\delta t = (R_\psi \alpha + R_\theta \beta)^n, \qquad (\tilde{Z}^{n+1} - Z^n)/\delta t = (Z_\psi \alpha + Z_\theta \beta)^n.$$

(2). Using the predicted values of the coordinates at time level $n + 1$ to define the plasma–vacuum interface and using Eqs. (35b) and (36), invert the vacuum matrix, Eq. (34), to define the total pressure P^{n+1} on the vacuum side of the interface.

3. Iterate Eq. (37) for P^{n+1} throughout the plasma interior until the convergence criterion is met, using the boundary values from step 2.

4. Advance the velocity equations (26h, i), using P^{n+1} obtained from step 3 to time-center the spatial derivatives. For example,

$$[(u + \alpha)^{n+1} - (u + \alpha)^n]/\delta t + (|\Delta\psi|^2/\rho)^n \tfrac{1}{2}(P^{n+1}_\psi + P^n_\psi)$$
$$+ (\nabla\theta \cdot \nabla\psi/\rho)^n \tfrac{1}{2}(P^{n+1}_\theta + P^n_\theta) + R^{n+1/2}.$$

5. Separate out the compressible velocity components u and v from the coordinate stream function ξ and expansion parameter T to be consistent with Eqs. (16), (23), and (26c). The quantity $(u + \alpha)^{n+1}$ is surface-averaged

to give

$$u(\psi) = \frac{1}{2\pi} \int_0^{2\pi} (u + \alpha) \, d\theta + \frac{\dot{T}}{T^2 \psi''} \frac{1}{2\pi} \int_0^{2\pi} R R_\theta (Z - Z_0)/R^m \, d\theta,$$

which is used to determine $u(\psi)$, except on the last flux surface (plasma–vacuum boundary), where $u(\psi) = 0$ and the surface-averaged equation is used to determine the expansion parameter $\dot{T}(t)$. The surface-varying part of $(u + \alpha)^{n+1}$ gives the coordinate stream function

$$\xi_\theta = T\psi''[(u + \alpha) - u(\psi)] + (\dot{T}/T)[RR_\theta(Z - Z_0)/R^m]$$

and $\xi_\psi = (\partial/\partial\psi) \int_0^\theta \xi_\theta \, d\theta$. The quantities α and β are then obtained from their definitions, Eq. (23).

6. Advance the conservation Eqs. (26a, e), using the velocity variables at level $n + 1$ to time-center the spatial derivatives.

Physically, the compressible wave motion is being treated implicitly, but the incompressible motion and the material convection are being treated explicitly. Since the incompressible motion is still treated by explicit differencing, the Courant stability criteria for the incompressible (transverse) waves

$$\delta t \lesssim (\delta x/c_T)$$

must be obeyed. Since $c_T \sim c_f(Rq/a)^{-1}$, the maximum allowable time step is greatly increased over that for a fully explicit method.

H. Application

As an initial application of the dynamic grid method, we demonstrate the ability of this code to calculate the correct time evolution of a large-aspect-ratio tokamak plasma of circular cross section, initially prepared close to an exact eigenmode, calculated by a linear normal mode or δW code (Chance et al., 1977). In Fig. 4 we show some of the normalized eigenvalues $\tilde{\omega}^2 = \omega^2 a^2 \rho_0/B_0^2$, with poloidal mode number $m = 4$, for a cylindrical equilibrium with minor radius $a = 1$, wall radius $b = 1.1055$, toroidal periodicity length $R = 2\pi$, constant density and current with $\rho_0 = 1$, $q = 2.5$, and pressure profile $\mu_0 p_0 = 0.01[1 - (r/a)^2]B_0^2$. The modes divide into three groupings, associated with the three types of waves in ideal MHD.

Figure 5 shows the velocity vectors at time $t = 0$ and $t = 48\tau_f$ on the equal-area, finite-difference mesh used, initially prepared in the eigenmode labeled A in Fig. 4. There are 18 zones in the radial direction and 60 in the θ direction. The calculation is run for 600 cycles, with the average time step being 10 times larger than that allowed by the Courant criterion based on the fast waves, $0.08\tau_f$.

7. Tokamak Magnetohydrodynamics

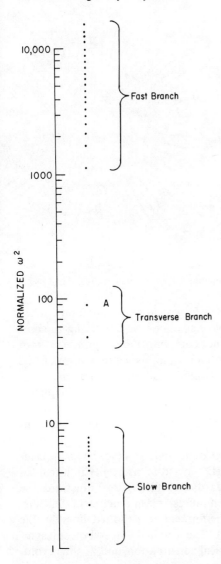

Fig. 4. Spectrum of axisymmetric eigenmodes with azimuthal wave number 4. Mode denoted by A will be examined more closely.

The theoretical period of this oscillation is $\tau = 13.8\tau_f$. Figures 6a, c present an interesting comparison. They represent the time histories of the $\nabla\psi$ component of the velocity, $\nabla\psi \cdot \mathbf{v} = u + \alpha$, the metric tensor component $|\Delta\psi|^2$, the plasma density at a given location, the eighth flux surface from the center, and the fifth θ surface from the horizontal. From Fig. 6a we see that the velocity consists of two components, one rapidly varying and the other slowly

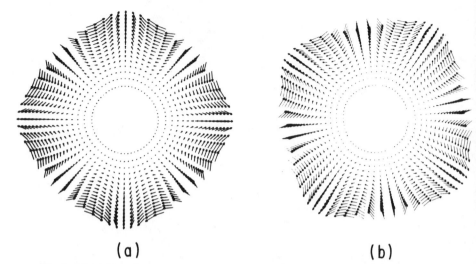

Fig. 5. Transverse eigenmode simulation showing the velocity vectors at times 0 and $48\tau_f$.

varying. The slowly varying component is largest in amplitude and corresponds to the slow, nearly incompressible eigenmode we are trying to compute. The smaller amplitude high-frequency component arises primarily from the inexactness of the finite-difference initial conditions. Examination of Fig. 6b shows that there is no high-frequency component in the metric tensor element $|\nabla\psi|^2$. Figure 6c similarly shows that there is little or no low-frequency component in the time variation of the density. Thus, the separation of the velocity into a slowly varying incompressible part and a rapidly varying compressible part has been successful.

As a second example of the dynamic grid method, we describe its application to the computation of an ideal MHD time-scale instability in a real tokamak experiment (Meade et al., 1975; Jardin, 1978). Figure 7 illustrates the location of the coil currents (pluses and minuses), plasma magnetic-flux surfaces (solid lines), and vacuum magnetic-flux surfaces (dotted lines) in the Princeton Divertor Experiment (PDX). The originally top–bottom symmetric plasma is given a small asymmetrical perturbation, and its time evolution is followed. It is found that the plasma is unstable to an axisymmetric motion attracting it into one of the coils. In Fig. 8 we illustrate the magnetic configuration at time $t = 73\tau_f$, after the instability has had a chance to grow. It is seen that the instability is not merely a rigid displacement but is larger on the inside of the torus and significantly distorts the plasma cross section. Further investigation (Jardin, 1978), not shown, reveals that this instability can be stabilized by the presence of nearby conductors.

7. Tokamak Magnetohydrodynamics

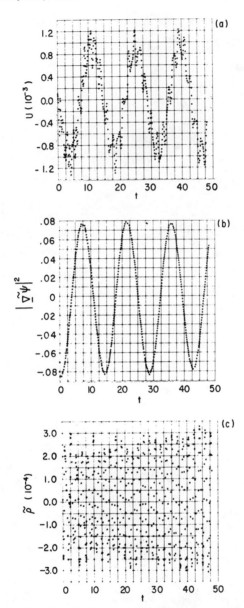

Fig. 6. Some time histories at location $I = 5, J = 8$. (a) $\nabla\psi$ component of velocity; (b) perturbed metric element $|\nabla\psi|^2$; (c) perturbed density.

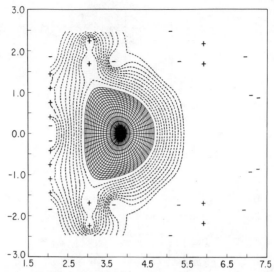

Fig. 7. Magnetic surfaces at time $t = 0$. Dashed surfaces lie in vacuum. Plus and minus indicate polarity of current-carrying coils.

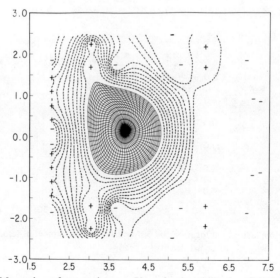

Fig. 8. Magnetic surfaces at time $t = 73\tau_f$, after instability has distorted plasma.

III. RESISTIVE TIME-SCALE MHD SIMULATIONS

A. Introduction

The resistive time scale in a tokamak is that in which the particles, energy, and electrical current penetrate into and diffuse out of the entire discharge. It includes the time it takes for the current to build up to its steady-state value, the plasma heating time, the resistive skin time, and the particle and energy confinement times. The resistive time scale is very slow compared with all three wave transit times τ_f, τ_t, τ_s [Eq. (5)] and to the time it takes for particles and energy to diffuse along (parallel to) the magnetic field.

We consider only axisymmetric configurations, so that a flux function $\chi(R, Z)$ exists, defined by Eq. (25). In this chapter we describe two very different methods for performing calculations on this time scale. The *zero-mass* method (Hirshman and Jardin, 1979) uses an asymptotically valid reduced set of equations obtained by expanding in the small ratios of the Alfvén wave transit time to the resistive diffusion time, and of the perpendicular to parallel mobilities and thermal conductivities. It obtains as solubility constraints one-dimensional time-evolution equations for the surface-averaged thermodynamic variables and magnetic fluxes. The condition that force balance be satisfied, $\nabla p = \mathbf{J} \times \mathbf{B}$, provides a constraint on the allowed motion. A time-dependent coordinate transformation is used to describe the diffusion of plasma quantities through magnetic surfaces of changing shape.

The second method, the *large-mass* method, artificially slows down and damps the Alfvén waves by enhancing the ion mass and viscosity. Since the physical results are known to be asymptotically independent of these parameters, we can increase their values by several orders of magnitude without modifying the physical solution. This allows us to bring together the Alfvén and transport time scales. Of course, one must verify a posteriori that the results obtained are indeed independent of the enhancement factors.

B. The Zero-Mass Method

A formal ordering scheme based on the smallness of the quantity

$$\delta \equiv \tau_A/\tau_R \cong (\rho_0/\mu_0)^{1/2}\eta/aB \tag{39}$$

shows that the inertial terms in the equation of motion, Eq. (4), can be neglected provided that there are no Alfvén time-scale instabilities. This equation is then replaced by the equilibrium constraint

$$\nabla p = \mathbf{J} \times \mathbf{B}. \tag{4'}$$

We again utilize the magnetic-flux coordinate system introduced in Section II.B. The three component equations obtained by operating on Eq. (4') with

$\mathbf{B}\cdot$, $\nabla\psi\times$, and $\nabla\theta\times\nabla\varphi\cdot$ are

$$p(\mathbf{x},t) = p(\psi,t), \qquad (40a)$$

$$F(\mathbf{x},t) = F(\psi,t), \qquad (40b)$$

$$\Delta^*\chi + FF_\chi + \mu_0 R^2 p_\chi = 0. \qquad (40c)$$

Here we have introduced the operator

$$\begin{aligned}\Delta^* a(\mathbf{x}) &\equiv R^2 \nabla\cdot R^{-2}\nabla a \\ &= R^2 J^{-1}\{[(J/R^2)|\nabla\psi|^2 a_\psi + (J/R^2)\nabla\psi\cdot\nabla\theta\, a_\theta]_\psi \\ &\quad + [(J/R^2)\nabla\theta\cdot\nabla\psi\, a_\psi + (J/R^2)|\nabla\theta|^2 a_\theta]_\theta\}.\end{aligned} \qquad (41)$$

Equation (40c) is a two-dimensional, nonlinear, differential constraint, relating the surface quantities $p(\psi,t)$, $F(\psi,t)$, $\chi(\psi,t)$ to the magnetic surface geometry at all times. The incorporation of this constraint into the time evolution equations for the differential particle number, entropy, and magnetic fluxes is the central problem to be addressed in this section. We shall proceed to time-differentiate Eq. (40c) and use the resulting linear equation to determine the velocity of the constant ψ coordinate surfaces and the velocities of the magnetic fluxes relative to the constant ψ surfaces.

1. One-Dimensional Transport Equations

In a confined tokamak plasma, the motion of particles and heat along the magnetic field $\mathbf{B}$ is rapid compared with the resistive motion across magnetic surfaces. This leads to density and temperature profiles that are nearly uniform on a flux surface. Formally, expanding in the small parameter $\Delta \equiv \rho_e/a$, it can be shown that

$$p_j(\mathbf{x},t) = p_j(\psi,t) + \Delta\tilde{p}(\psi,\theta,t), \qquad (42a)$$

$$T_j(\mathbf{x},t) = T_j(\psi,t) + \Delta\tilde{T}(\psi,\theta,t), \qquad (42b)$$

where

$$p_j(\psi,t) \equiv \langle p_j(\mathbf{x},t)\rangle, \qquad (42c)$$

$$T_j(\psi,t) \equiv \langle T_j(\mathbf{x},t)\rangle, \qquad (42d)$$

and we have introduced the flux-surface-average operator

$$\langle a\rangle \equiv \int_0^{2\pi} d\theta\, Ja \bigg/ \int_0^{2\pi} d\theta\, J. \qquad (43)$$

In view of Eqs. (42a) and (42b), the local conservation laws contain more information than is necessary to advance the dominant surface-averaged part

7. Tokamak Magnetohydrodynamics

of the thermodynamic variables. The extraneous information can be annihilated by flux-surface-averaging the conservation equations. The one-dimensional set of transport equations so obtained can be summarized as follows:

$$(N'_j)_t + (N'_j u)_\psi = -(V'\Gamma_j)_\psi + V'\langle S_{nj}\rangle, \tag{44a}$$

$$\sigma'_t + (\sigma' u)_\psi = -\tfrac{2}{5}(\sigma'/p)S, \tag{44b}$$

$$(\sigma'_e)_t + (\sigma' u)_\psi = -\tfrac{2}{5}(\sigma'_e/p_e)S_e, \tag{44c}$$

$$\chi'_t + (\chi' u)_\psi = (E^*_{\|})_\psi, \tag{44d}$$

$$\Psi'_t + (\Psi' u)_\psi = 0. \tag{44e}$$

The variables advanced in time in Eqs. (44a–e) are the differential particle number for each species $N'_j(\psi,t) \equiv n_j(\psi,t)V'$, the differential plasma entropy $\sigma'(\psi,t) \equiv [p(\psi,t)]^{3/5}V'$, the differential electron entropy $\sigma'_e(\psi,t) \equiv [p_e(\psi,t)]^{3/5}V'$, the poloidal flux density per radian [from Eq. (25)]

$$\chi'(\psi,t) = (2\pi)^{-2} \partial\left(\int_0^\psi dx\, \mathbf{B}\cdot\nabla\theta\right)\Big/\partial\psi = J\mathbf{B}\cdot\nabla\theta, \tag{45}$$

and the toroidal flux density per radian

$$\Psi'(\psi,t) \equiv (2\pi)^{-2} \partial\left(\int_0^\psi dx\, \mathbf{B}\cdot\nabla\varphi\right)\Big/\partial\psi = (2\pi)^{-2}F\langle R^{-2}\rangle V', \tag{46}$$

with the differential volume being defined as

$$V'(\psi,t) \equiv \partial\left(\int_0^\psi dx\right)\Big/\partial\psi = 2\pi\int_0^{2\pi} J\,d\theta. \tag{47}$$

Here n_j is the particle density of species j, p_j the pressure of species j, and p the combined electron and ion pressure $p = p_e + p_i$. The surface-averaged entropy and electron entropy terms appearing in Eqs. (44b, c), are defined as

$$S \equiv -\langle \mathbf{J}\cdot\nabla\varphi\rangle E^*_{\|} + \frac{1}{V'}\sum_{j=\mathrm{ie}}[Q_j]_\psi - \langle S_p\rangle, \tag{48a}$$

$$S_e \equiv -\langle \mathbf{J}\cdot\nabla\varphi\rangle E^*_{\|} + \frac{1}{V'}[Q_e]_\psi + (p_i)_\psi \Gamma_i/n_i + \langle u_i\cdot\nabla\cdot\pi_i\rangle - Q_{\Delta e} - \langle S_{pe}\rangle. \tag{48b}$$

Besides the electron–ion equipartition term $Q_{\Delta e}$, there are five transport-model-dependent quantities in Eqs. (44) describing relative slippage in a two-fluid resistive high-temperature plasma; the parallel electric field $\alpha \equiv E^*_{\|} = \langle \mathbf{E}\cdot\mathbf{B}\rangle/\langle \mathbf{B}\cdot\nabla\varphi\rangle$, the relative particle flux $\beta \equiv \Gamma_i/n_i = \Gamma_e/n_e$, the ion heat flux $\gamma \equiv Q_i = V'(\langle \mathbf{q}_i\cdot\nabla\psi\rangle + \tfrac{5}{2}\Gamma_i T_i)$, the electron heat flux $\delta \equiv Q_e = V'(\langle \mathbf{q}_e\cdot$

$\nabla \psi \rangle + \frac{5}{2} \Gamma_e T_e)$, and the viscous heating term $\varepsilon \equiv V' \langle u_i \cdot \nabla \cdot \pi_i \rangle$. It is the objective of transport theory to relate these fluxes to the vector of forces consisting of derivatives of the fundamental thermodynamic variables. It is convenient to define the force vector as

$$\mathbf{F} \equiv [1, (N'/V')_\psi, (\sigma'/V')_\psi, (\sigma'_e/V')_\psi, (\chi' V' \langle |\nabla \psi|^2 / R^2 \rangle / F)_\psi].$$

To specify a particular transport model, it is then sufficient to supply the 25 functions $(\alpha^j, \beta^j, \gamma^j, \delta^j, \varepsilon^j)$, $j = 0, 4$, according to the format

$$\alpha = \sum_{j=0}^{4} \alpha^j F^j, \qquad \beta = \sum_{j=0}^{4} \beta^j F^j, \qquad \text{etc.} \tag{50}$$

These functions are tabulated in Appendix C of Jardin, 1981 for several transport regimes.

2. Incorporation of the Equilibrium Constraint

To complete the specification of the transport problem, we must provide an equation to obtain the one-dimensional toroidal flux velocity $u(\psi, t)$ that appears in Eqs. (44a–e), and to obtain the two-dimensional coordinate velocity $\mathbf{v}_g$, Eqs. (22) and (23), so that the coordinates of the magnetic geometry can be advanced through Eqs. (26f, g). An equation to determine both of these quantities is obtained by time-differentiating the plasma equilibrium equation (40c), and substituting for the time derivatives of σ', χ', and Ψ' from Eq. (44). Using these and the convective derivative Eq. (15), one obtains a linear equation for the normal component of the toroidal flux velocity Ω:

$$\left[\frac{1}{\mu_0 R^2} \chi' \Delta^* + L_0 + L_1 \right] \Omega = 2\mathbf{J} \cdot \nabla \varphi (E_\parallel^*)_\psi + \frac{1}{2\mu_0} B_p^2 [(E_\parallel^*)_\psi / \chi']_\psi - \frac{2}{3} S_\psi. \tag{51}$$

Here $B_p = \chi' |\nabla \psi| / R$ is the magnitude of the poloidal field, $E_\parallel^*$ and S appear in Eqs. (44a–c), and Δ^* is defined in Eq. (41). The new operators appearing in Eq. (51) are defined as

$$L_0(a) \equiv (\mu_0 R^2)^{-1} \{ F^2 (V' \langle R^{-2} \rangle)^{-1} [(V'/\chi') \langle R^{-2} a \rangle]_\psi \}_\psi, \tag{52a}$$

$$L_1(a) \equiv [(p_\psi / \chi')_\psi + (1/\mu_0 R^2)(FF_\psi X')_\psi] a + \frac{5}{3} \{ (p/V')[(V'/\chi') \langle a \rangle]_\psi \}_\psi. \tag{52b}$$

We note that $L_0/L_1 \sim B_t^2/p$ and $\mu_0 L_0 / \chi' \Delta^* \sim B_t^2 / B_p^2$ where $B_t \sim R\mathbf{B} \cdot \nabla \varphi$ is the magnitude of the toroidal field. Thus for a tokamak plasma with $\beta \sim B_p^2/B_t^2 \ll 1$, L_0 is the dominant operator in Eq. (51).

We have used the symbol Ω in Eq. (51) to denote $(\chi'/\Psi') \equiv q^{-1}$ times the absolute time derivative of the toroidal flux function. Thus,

$$\Omega(\psi, \theta, t) \equiv q^{-1} \, d\psi/dt|_x = -\chi'(\psi, t)[u(\psi, t) + \alpha(\psi, \theta, t)]. \tag{53}$$

Here the second equality follows from Eqs. (15), (16a), and (44e). If we impose the natural boundary conditions that $u = 0$ at $\psi = \psi_{max} = 1$, then Eq. (53) can be unraveled to obtain T_t, ξ_θ, ξ_ψ, α, β, and u from Ω by using Eq. (23). Thus, at $\psi = 1$, where $u = 0$, we have

$$T_t/T = \langle R^{-m}\Omega \rangle / \langle RR_\theta(Z - Z_0)/JR^m \rangle \chi'. \tag{54a}$$

With T_t/T given, we compute $u(\psi, t)$, $\xi_\theta(\psi, \theta, t)$, and $\xi_\psi(\psi, \theta, t)$ elsewhere as

$$u(\psi, t) = [\langle R^{-m}\Omega \rangle/\chi' - (T_t/T)\langle RR_\theta(Z - Z_0)/JR^m \rangle]/\langle R^{-m} \rangle, \tag{54b}$$

$$\xi_\theta(\psi, \theta, t) = (T_t/T)RR_\theta(Z - Z_0)/R^m - (u + \Omega/\chi')J/R^m, \tag{54c}$$

$$\xi_\psi(\psi, \theta, t) = \partial/\partial\psi \int_0^\theta \xi_\theta \, d\theta. \tag{54d}$$

Equations (23a, b) may then be used to compute the coordinate velocities $\alpha(\psi, \theta, t)$ and $\beta(\psi, \theta, t)$.

The description employed here is partly Lagrangian in that the (R, Z) coordinates of the ψ, θ grid evolve in time with the velocity $\mathbf{v}_g = (\alpha, \beta)$ according to Eqs. (26f, g). Thus the two-dimensional variable $\Omega(\psi, \theta)$, which is obtained from Eq. (51) and describes how the toroidal flux surfaces change in time, gives several pieces of information. Its flux-surface average, Eqs. (54a, b), gives the uniform expansion of the coordinate system $T_t(t)$ and the one-dimensional velocity $u(\psi, t)$ with respect to a Eulerian grid equally spaced in the geometrical coordinate ψ. Its surface-varying part gives the Lagrangian coordinate stream function ξ, which describes how the constant ψ surfaces change their shape.

Physically we have replaced the equation of motion, Eq. (3a'), which evolves the velocity field in time, with a global constraint equation (51), which determines the toroidal flux velocity as an instantaneous boundary-value problem. This velocity field so determined is free of the fast-time-scale motion associated with wave phenomena but is such that it will evolve one equilibrium configuration into another. The one-dimensional transport equations (44) then describe how the other thermodynamic and field variables evolve relative to the toroidal-flux velocity. Given previously are similar formulations (Byrne and Klein, 1978; Grad, 1970; Nelson and Grad, 1977; Todd and Grimm, 1975; Helton et al., 1977; Hogan, 1978) in which the plasma equilibrium equation is not time-differentiated but is resolved at each time step. Although they are conceptually more straightforward, these methods must solve a nonlinear elliptic equation at each time step, which can be very time consuming.

3. Numerical Method

Once the transport coefficients are given in terms of the plasma variables and their derivatives through Eq. (50), the system of Eqs. (44), (26f, g) and (51), together with the necessary definitions in those sections, provide a closed system of equations, needing only the source functions $\langle S_{nj} \rangle$, $\langle S_p \rangle$, $\langle S_{pe} \rangle$, and the boundary conditions to completely specify a problem. The numerical method utilized is described here. If we denote by **Y** the state vector consisting of the one-dimensional variables being advanced in time in Eq. (44), i.e., N'_j, σ', σ'_e, χ' and Ψ', by $A(\mathbf{Y}, \mathbf{X})$—the vector of the right-hand sides of Eq. (44), by $L(\mathbf{Y}, \mathbf{X})$—the operator on the left-hand side of Eq. (51) operating on Ω, and by $B(\mathbf{Y}, \mathbf{X})$—the right-hand side of Eq. (51), then the evolution equations take the compact form

$$\mathbf{x}_t = \mathbf{v}_g, \tag{55a}$$

$$\mathbf{Y}_t + (\mathbf{Y}u)_\psi = A(\mathbf{Y}, \mathbf{X}), \tag{55b}$$

$$L(\mathbf{Y}, \mathbf{X})\{\Omega\} = B(\mathbf{Y}, \mathbf{X}). \tag{55c}$$

Here, of course, $\mathbf{v}_g$ and u are obtainable from Ω through Eqs. (54) and the definitions, Eqs. (16), and (23). The notation used in Eq. (55) underscores the fact that the operator L is an explicit function of **Y** and **X** and acts on the function Ω.

As discussed in Hirshman and Jardin (1979), the formulation adopted here ensures that the toroidal flux velocity u is "small" compared with the other diffusive velocities; i.e., $u \sim B_p/B_T \ll 1$. This justifies adopting the following prescription for advancing the solution from time level n to time level $(n + 1)$:

$$\mathbf{x}^{(n+1)} - \mathbf{x}^{(n)} = \Delta t \, \mathbf{v}_g^{(n)}, \tag{55a'}$$

$$\mathbf{Y}^{(n+1)} - \mathbf{Y}^{(n)} + \Delta t [\mathbf{Y}^{(n+1/2)} u^{(n)}]_\psi = \Delta t \, \mathbf{A}[\mathbf{Y}^{(n+1/2)}, \mathbf{x}^{(n+1/2)}], \tag{55b'}$$

$$L[\mathbf{Y}^{(n+1)}, \mathbf{X}^{(n+1)}]\{\Omega^{(n+1)}\} = B[\mathbf{Y}^{(n+1)}, \mathbf{X}^{(n+1)}]. \tag{55c'}$$

Since only the old time level of u, $u^{(n)}$, appears in Eq. (55b'), Eqs. (55b', c') decouple, and their finite-difference solution can be considered separately.

Let us consider first the solution of Eq. (55b) for the vector **Y**. Since this equation contains all the transport time scales, a time-centered, implicit, unconditionally stable (Crank–Nicholson) method is called for. Since the right-hand side of Eq. (44e) for Ψ' is zero, this equation splits off from the others and is solved simply by applying centered spatial difference operators to obtain a tridiagonal system. For the remaining four vectors, we introduce the notation that a subscript j corresponds to location $\psi = [j - \frac{1}{2}]\Delta\psi$ and a superscript n corresponds to time level $t = (n - 1)\Delta t$. The finite-difference

7. Tokamak Magnetohydrodynamics

form of the surface-averaged transport equations (44a–d) assumes the form

$$\mathbf{A}_j^n \cdot \mathbf{\Phi}_{j+1}^{n+1} - \mathbf{B}_j^n \cdot \mathbf{\Phi}_j^{n+1} - \mathbf{C}_j^n \cdot \mathbf{\Phi}_{j-1}^{n+1} + \mathbf{D}_j^n = 0. \tag{56}$$

Here $\mathbf{\Phi}_j^n$ is the solution vector at position j and time level n:

$$\mathbf{\Phi}_j^n \equiv \lfloor N_j^{\prime n}, \sigma_j^{\prime n}, \sigma_{ej}^{\prime n}, \chi_j^{\prime n} \rfloor.$$

Each of the tridiagonal elements $\mathbf{A}_j^n$, $\mathbf{B}_j^n$, $\mathbf{C}_j^n$ is a full 4×4 matrix with elements given in terms of the α^i, β^i, etc., of Eq. (50). These elements and those of the four vectors $\mathbf{D}_j^n$ can be found in Appendix A of Jardin (1981). The matrix tridiagonal system Eq. (56) is inverted in the usual way with fixed-value or derivative boundary conditions easily being incorporated at the outer boundary.

We next consider equations of the form

$$(\Omega_\theta a + \Omega_\psi b)_\psi + (\Omega_\theta c + \Omega_\psi a)_\theta + d\Omega + JH[P(V'\langle H\Omega \rangle)_\psi]_\psi \\ + JK[G(V'\langle K\Omega \rangle)_\psi]_\psi + e = 0. \tag{57}$$

Subscripts here denote differentiation and angle brackets denote the usual flux-surface-average operator, defined in Eq. (43). Equation (57) is identical to Eq. (51) or (55c) if we make the identifications

$$a(\psi, \theta) = -(R_\theta R_\psi + Z_\theta Z_\psi)/J, \tag{58a}$$

$$b(\psi, \theta) = (R_\theta^2 + Z_\theta^2)/J, \tag{58b}$$

$$c(\psi, \theta) = (R_\psi^2 + Z_\psi^2)/J, \tag{58c}$$

$$d(\psi, \theta) = J[\mu_0(p_\psi/\chi')_\psi + R^{-2}(FF_\psi/\chi')_\psi]/\chi', \tag{58d}$$

$$e(\psi, \theta) = -\mu_0 J\{2\mathbf{J} \cdot \nabla\varphi(E_{\parallel}^*)_\psi + (\tfrac{1}{2}\mu_0)B_p^2[(E_{\parallel}^*)_\psi/\chi']_\psi - \tfrac{2}{3}S_\psi\}, \tag{58e}$$

$$K(\psi, \theta) = 1/R^2\chi', \tag{58f}$$

$$H(\psi, \theta) = 1/\chi', \tag{58g}$$

$$G(\psi) = F^2/(V'\langle R^{-2} \rangle), \tag{58h}$$

$$P(\psi) = \tfrac{5}{3}\mu_0 p/V'. \tag{58i}$$

Equation (57) is a linear, nonlocal, two-dimensional equation for Ω, to be inverted each time step determining the time evolution of the toroidal-flux surfaces by determining the velocity variables u, α, and β through Eq. (54). Mathematically, it is a generalized differential equation requiring boundary conditions for $\Omega(\psi_{\max}, \theta)$ and a single boundary condition for $\Omega(0, \theta)$. The θ integrals (or flux-surface averages) in Eq. (21) make it nonlocal and thus impractical to solve by iteration techniques. Instead, we solve this equation directly by Fourier decomposing in the angle coordinate θ and by finite differencing in the surface coordinate ψ. Fourier techniques are natural in that the θ integrals and derivatives are treated on an equal footing. The efficiency of this method lies in the fact that for most problems, only 5–10 Fourier harmonics need be kept at each ψ surface to obtain accurate solutions.

For simplicity, we restrict discussion to systems that are symmetric about the plasma midplane. Then Ω has a discrete Fourier representation of the form $\Omega(\psi, \theta) = \Omega^0(\psi) + \sum_{m=1}^{M} \Omega^m(\psi) \cos m\theta$, and the coefficients can likewise be expanded as $b(\psi, \theta) = b^0(\psi) + \sum_{m=1}^{M} b^m(\psi) \cos m\theta$, etc. By substituting these expansions into Eq. (57), evaluating the product terms, and finite-differencing in ψ with the notation $\psi_j = j \, \Delta\psi$, we obtain the canonical system

$$\mathbf{A}_j \cdot \mathbf{\Omega}_{j+1} - \mathbf{B}_j \cdot \mathbf{\Omega}_j + \mathbf{C}_j \cdot \mathbf{\Omega}_{j-1} + \mathbf{D}_j = 0. \tag{59}$$

Here $\Omega_j \equiv \{\Omega_j^0, \Omega_j^2, \ldots, \Omega_m^M\}$ is a vector containing the coefficients of the $M + 1$ Fourier harmonics of Ω at location ψ_j. The elements of the $(M + 1) \times (M + 1)$ matrices A_j, B_j, C_j and the length $(M + 1)$ vectors D_j are defined in terms of the coefficient harmonic amplitudes b^m, c^m, etc., in Appendix B of Jardin, 1981.

The system of Eqs. (59), identical in form to Eq. (56), is solved according to the same algorithm. We note that a single boundary value for Ω_0^0 need be supplied at the origin. As discussed in Jardin (1981), setting $\Omega_0^0 = 0$ corresponds to no toroidal flux being created there. Boundary values for all the $(\Omega_j^m; m = 0, M)$ must be supplied at the outer boundary $j = J$. These are either prescribed or calculated by Green's function techniques to be consistent with currents in external coils.

As an initial condition, the equilibrium (Grad–Shafranov) equation, (40c) is solved at time $t = 0$ using a flux-coordinate equilibrium code (DeLucia et al., 1980). In subsequent time steps, it is, in general, not necessary to solve the equilibrium equation since the solution of Eq. (51) describes how the equilibrium flux surfaces change from one time step to the next. Some truncation error will accumulate in this process so that, as time progresses, the equilibrium equation will not be exactly satisfied. One can constantly monitor this error, and when it builds up to a noticeable value, the flux-coordinate equilibrium code can be called in and run in an adiabatic mode to reduce this error to zero. In practice, this readjustment of the equilibrium is seldom, if ever, needed to limit the accumulated error in the equilibrium equation to a few percent over the course of a complete problem.

4. Application

As an example of the numerical solution procedure described, we analyze the performance of a proposed fusion experiment. A deuterium–tritium tokamak plasma is modeled as it is heated to near ignition conditions by intense neutral particle injection; then it is subjected to major radius compression causing the plasma to ignite. Standard anomalous loss terms, neutral gas refilling models, thermonuclear fusion source terms, and neutral beam source terms have been included to increase the realism of the model (Jardin, 1981).

Using these source and transport models, the calculation proceeds as follows. An initial plasma equilibrium is computed with $R = 200$ cm, $a = 60$ cm, toroidal magnetic field at axis $B_T = 61$ kG, and with plasma current $I_p = 2.47$ MA. The calculation proceeds for 0.5 sec with no external heating source, at which time the 14-MW neutral beam source term is turned on and left on until $t = 1.6$ sec. The major-radius compression takes place in the interval 1.5 sec $< t < 1.6$ sec, and the calculation is halted at $t = 2.0$ sec.

The boundary conditions used in this calculation are as follows. The values of n, T_e, T_i, and χ' were held at fixed values at the plasma–vacuum boundary (pedestal boundary conditions) so that $n_b = 2 \times 10^{13}$ cm^{-3}, $T_{ib} = T_{eb} = 30$ eV, $q_b = 2.68$. The toroidal field in the vacuum is $\mathbf{B}_T = g_v \nabla\varphi$, with g_v a constant that gives the value $|B_T| = 61$ kG at $R = 200$ cm. The surface-averaged part of the toroidal-flux velocity at the boundary u_b is determined self-consistently so that the equilibrium equation remains satisfied across the plasma–vacuum boundary. The surface-varying part of the toroidal-flux velocity is prescribed to be zero, except during the compression phase, at which time it is given a value corresponding to a uniform radially inward velocity.

The results of this calculation are illustrated in Figs. 9–11. Figure 9 shows the magnetic surfaces at $t = 0$ (before compression) and at $t = 2$ sec (after compression). During the compression, the position of the magnetic axis has decreased from $R = 207$ to 140 cm, and the minor radius has decreased from $a = 61$ to 48 cm. Figure 10 shows the midplane values of T_e, T_i, n, and J_φ at various times during the calculation, and Fig. 11 shows the central values of the temperatures and density as a function of the time t.

C. The Large-Mass Method

The method described in Section III.B for integrating the MHD equations on the resistive time scale consisted of taking the asymptotic limit of the equations as the plasma mass goes to zero or, equivalently, as the Alfvén velocity $V_A \sim B/(\mu_0 \rho_0)^{1/2}$ goes to infinity. This has the effect of removing the wave-transit time scales completely from the problem. The penalty paid for doing so is that now an elliptic equation, Eq. (51), must be solved for Ω each time step, reflecting the fact that there is an infinite propagation velocity.

The method described in this section (Jardin and Park, 1981) is the opposite of the zero-mass method in that it artificially increases the plasma mass to slow the Alfvén waves down rather than speed them up. These waves can be slowed down by several orders of magnitude, to almost the transport

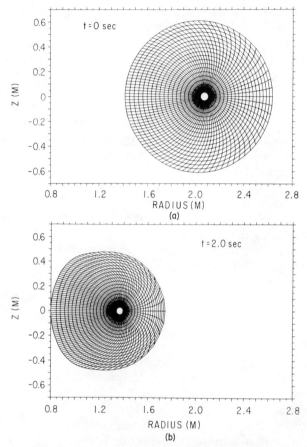

Fig. 9. Constant (ψ, θ) coordinate surfaces at (a) $t = 0$ (before compression) and (b) $t = 2$ sec after compression. The magnetic axis has decreased from $R = 207$ to $R = 140$ cm.

time scale, without affecting the time-averaged physical results. This result follows from the fact that there are two distinct time scales in the equations, ideal (fast) and resistive (slow), with only stable oscillations on the fast time scale and with no intermediate time scales present. The situation becomes quite different when we go to three dimensions and there arise intermediate time scales associated with resistive tearing and interchange modes.

1. Reduced Compressible Magnetohydrodynamic Equations

For simplicity, we restrict consideration to a specific subset of the resistive MHD equations, Eqs. (1)–(4), which describe a single-temperature ($T_e = T_i \equiv T$) collisional regime plasma, so that the viscous heating term $\pi = 0$

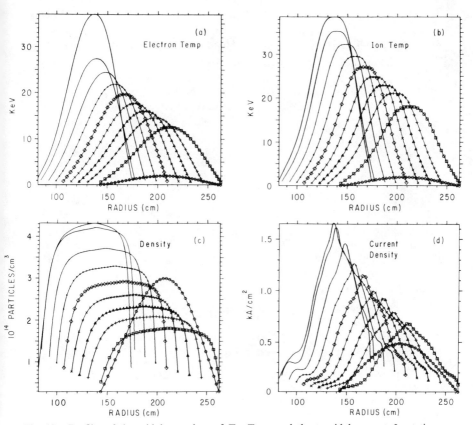

Fig. 10. Profiles of the midplane values of T_e, T_i, n, and the toroidal current J_ψ at times $t = 0.00$, 1.44, 1.51, 1.53, 1.54, 1.56, 1.57, 1.58, 1.60, and 2.0 sec.

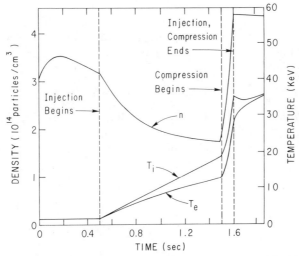

Fig. 11. Time history of the central electron and ion temperatures, $T_e(0)$ and $T_i(1)$ and the central density $n(0)$.

and a simplified Ohm's law applies. These one-fluid equations are

$$(\partial/\partial t)\mathbf{m} + \nabla \cdot (\rho^{-1}\mathbf{mm}) + \nabla p = \mathbf{J} \times \mathbf{B}, \tag{60a}$$

$$\tfrac{3}{2}(\partial/\partial t)p + \nabla \cdot [\mathbf{q} + \tfrac{5}{2}(p/\rho)\mathbf{m}] = \mathbf{J} \cdot \mathbf{E} + S_e, \tag{60b}$$

$$\partial \rho/\partial t + \nabla \cdot \mathbf{m} = S_\rho, \tag{60c}$$

$$\partial \mathbf{B}/\partial t = -\nabla \times \mathbf{E}, \tag{60d}$$

$$\mathbf{E} + \rho^{-1}\mathbf{m} \times \mathbf{B} = \eta \mathbf{J}, \tag{60e}$$

$$\mu_0 \mathbf{J} = \nabla \times \mathbf{B}, \tag{60f}$$

$$k_B T = (M_i/2\rho)p \quad \text{eV}, \tag{60g}$$

$$\eta = 0.002(k_B T)^{-3/2} \quad \text{Ohm m.} \tag{60h}$$

In an axisymmetric geometry with symmetry angle φ, the magnetic field is written, as in Eq. (25), in terms of the poloidal flux per radian χ and the toroidal field function F:

$$\mathbf{B} = \nabla \varphi \times \nabla \chi + F \nabla \varphi. \tag{61}$$

The momentum density $\mathbf{m}$ is written in terms of a stream function A, a toroidal component ω, and a potential Ω; thus,

$$\mathbf{m} = \nabla \varphi \times \nabla A + \omega \nabla \varphi + \nabla \Omega. \tag{62}$$

We are looking for solutions with slowly changing boundary conditions and small mass flow so that the ordering in the reciprocal magnetic Reynolds number

$$(m^2/\rho) \sim \varepsilon^2 B^2/\mu_0 \rho \sim \varepsilon^2 p, \tag{63}$$

with $\varepsilon^2 \ll 1$, is appropriate. Here we have taken the plasma β to be on the order of unity in the ε expansion. The reduced equations are then obtained by using Eq. (63) to justify dropping the convective derivative term in Eq. (60a) and adding a viscous term. Equation (60a) becomes

$$\partial \mathbf{m}/\partial t - \nu \nabla^2 \mathbf{m} = \mathbf{J} \times \mathbf{B} - \nabla p. \tag{60a'}$$

Substituting the definitions Eq. (61) and (62) into Eqs. (60a', b–g), taking the curl and divergence of Eq. (60a), and performing the necessary vector manipulations yield the set of scalar evolution equations solved numerically:

$$\frac{\partial}{\partial t}\nabla^2 \Omega + \nabla \cdot \left[\frac{\Delta^* \chi}{\mu_0 R^2}\nabla \chi + \frac{F}{\mu_0 R^2}\nabla F + \nabla p - \nu \nabla(\nabla^2 \Omega)\right] = 0, \tag{64a}$$

$$\frac{\partial}{\partial t}\Delta^* A + R^2 \nabla \cdot \left[\frac{\Delta^* \chi}{\mu_0 R^2}\nabla \chi \times \nabla \varphi + \frac{F}{\mu_0 R^2}\nabla F \times \nabla \varphi - \frac{\nu}{R^2}\nabla(\Delta^* A)\right] = 0, \tag{64b}$$

7. Tokamak Magnetohydrodynamics

$$\frac{\partial}{\partial t} w + \frac{1}{\mu_0}(\nabla\varphi \times \nabla F) \cdot \nabla\chi - \nu\, \Delta^* w = 0, \tag{64c}$$

$$\frac{\partial \rho}{\partial t} + \nabla^2\Omega + \nabla \cdot \mathbf{\Gamma}_\| = S_\rho, \tag{64d}$$

$$\frac{\partial}{\partial t}\chi + \frac{1}{\rho}(\nabla\varphi \times \nabla A \cdot \nabla\chi + \nabla\Omega \cdot \nabla\chi) = \frac{\eta}{\mu_0}\Delta^*\chi, \tag{64e}$$

$$\frac{\partial}{\partial t} F + R^2\nabla \cdot \left[\frac{F}{\rho R^2}(\nabla\varphi \times \nabla A + \nabla\Omega) - \frac{\omega}{\rho R^2}\nabla\varphi \times \nabla\chi\right] = \frac{R^2}{\mu_0}\nabla \cdot \frac{\eta}{R^2}\nabla F, \tag{64f}$$

$$\frac{\partial}{\partial t} p^{3/5} + \nabla \cdot \left[\frac{p^{3/5}}{\rho}(\nabla\varphi \times \nabla A + \nabla\Omega)\right] = \frac{2}{5} p^{-2/5}(\eta J^2 - \nabla \cdot \mathbf{q} + S_\text{p}). \tag{64g}$$

Here

$$\mathbf{\Gamma}_\| = -(\mathbf{BB}/B^2) \cdot D_\| \eta_\perp \nabla\rho \tag{65a}$$

is the flux of particles parallel to the magnetic field,

$$\mathbf{q} = -(\mathbf{BB}/B^2) \cdot D_\| \eta_\perp \nabla p - (15.2)(p/\mu_0 B^2)\rho\eta_\perp \nabla(p/\rho) \tag{65b}$$

the heat flux vector, and $\Delta^* \equiv R^2 \nabla \cdot R^{-2}\nabla$ the toroidal elliptic operator; $R = |\nabla\varphi|^{-1}$ measures the distance from the axis of symmetry.

If the lengths in Eq. (64) are scaled to a_0, the magnetic field strengths to B_0, the density to ρ_0, the velocity to $v_0 = B_0(\rho_0\mu_0)^{-1/2}$, and the time to $t_0 = a_0 v_0^{-1}$, then the resistivity occurs only in the combination

$$S^{-1} = \eta\rho_0^{1/2}/a_0 B_0 \mu_0^{1/2},$$

where S is the Magnetic Reynolds Number. For parameters typical of a large fusion experiment, we find that $S > 10^6$. The large mass method consists of using a ficticious value of ρ_0 so that the computational S value is only 100–1000, thus eliminating the disparate time scale problem.

The set of Eqs. (64) offers several advantages over more primitive forms. By taking the divergence and the curl of the momentum equation, we have removed the shear Alfvén wave motion from the velocity potential variable Ω and the fast compressible wave motion from the variables A and u. This partial isolation of the different characteristics leads to dramatic improvements in the accuracy, efficiency, and stability of the numerical solutions (see Section 3.).

A physically interesting subset of Eqs. (64) is that obtained by dropping Eqs. (64a, d, g) and setting $\Omega = 0$ and $\rho = \text{const}$ in Eqs. (64b, c, e, f). This is the familiar incompressible approximation. The fact that the incompressible equations are a subset of our full set of equations can be exploited in developing and debugging a numerical solution method for the full set of equations.

2. Physical Interpretation

Here we discuss the replacement of the momentum equation (60a) with the reduced momentum equation (60a') by interpreting the equations using a multiple time-scale formalism. The reduced Eqs. (64) expressed in vector form are

$$\partial \mathbf{m}/\partial t - v\nabla^2 \mathbf{m} = (\mu^0)^{-1}(\nabla \times \mathbf{B}) \times \mathbf{B} - \nabla p, \qquad (66a)$$

$$(\partial/\partial t)\,\mathbf{B} = \nabla \times (\rho^{-1}\mathbf{m} \times \mathbf{B}) - \nabla \times [(\eta/\mu_0)\nabla \times \mathbf{B}], \qquad (66b)$$

$$\partial \rho/\partial t + \nabla \cdot \mathbf{m} + \nabla \cdot \mathbf{\Gamma}_\| = S_\rho, \qquad (66c)$$

$$(\partial/\partial t)\,p^{3/5} + \nabla \cdot (p^{3/5}\mathbf{m}/\rho) = \tfrac{2}{5}p^{-2/5}(\eta J^2 - \nabla \cdot \mathbf{q} + S_\mathrm{p}). \qquad (66d)$$

Equations (66) contain phenomena occurring on two distinct time scales—the ideal time scale:

$$t_0 \sim \tau_\mathrm{f} \sim (a/B)(\mu_0\rho_0)^{1/2};$$

and the resistive time scale:

$$t_1 \sim \tau_\mathrm{R} \sim a^2\mu_0/\eta \sim \varepsilon^{-1}t_0,$$

where $\varepsilon = (\eta/aB)(\rho_0/\mu_0)^{1/2} \ll 1$.

For boundary conditions changing on the (slow) t_1 time scale, we look for an ordering that describes a configuration that is always near ($\sim \varepsilon$) equilibrium, but in which the equilibrium evolves on the t_1 time scale. Such a description is given formally by the multiple-time-scale method with the ordering

$$\eta = \varepsilon\eta, \qquad \mathbf{q} = \varepsilon\mathbf{q}, \qquad \mathbf{\Gamma} = \varepsilon\mathbf{\Gamma}, \qquad (67a)$$

$$\mathbf{B} = \mathbf{B}_0(t_1) + \varepsilon\mathbf{B}_1(t_0,t_1) + \cdots, \qquad (67b)$$

$$\rho = \rho_0(t_1) + \varepsilon\rho_1(t_0,t_1) + \cdots, \qquad (67c)$$

$$p = p_0(t_1) + \varepsilon p_1(t_0,t_1) + \cdots, \qquad (67d)$$

$$\mathbf{m} = \varepsilon\mathbf{m}_1(t_0,t_1) + \cdots, \qquad (67e)$$

$$\partial/\partial t = \partial/\partial t_0 + \varepsilon\,(\partial/\partial t_1) + \cdots, \qquad (67f)$$

with t_0 and t_1 being treated as independent variables. Substitution of the ordering, Eq. (67), into Eq. (66) yields, to order ε^0

$$(\nabla \times \mathbf{B}_0) \times \mathbf{B}_0 = \mu_0 \nabla p_0 \qquad (68)$$

7. Tokamak Magnetohydrodynamics

and to order ε^1

$$\partial \mathbf{m}_1/\partial t_0 - \nu \nabla^2 \mathbf{m}_1 = \mu_0^{-1}[(\nabla \times \mathbf{B}_1) \times \mathbf{B}_0 + (\nabla \times \mathbf{B}_0) \times \mathbf{B}_1] - \nabla p_1. \quad (69)$$

Thus, the zero-order magnetic field and pressure are in equilibrium without zero-order mass flow, and the first-order mass flow evolves on the fast time scale, being driven by the small departure of p and $\mathbf{B}$ from an equilibrium configuration.

Substitution of Eq. (67) into Eq. (66b) gives, to order ε,

$$\frac{\partial \mathbf{B}_0}{\partial t_1} + \frac{\partial \mathbf{B}_1}{\partial t_0} = \nabla \times \left(\frac{\mathbf{m}_1}{\rho_0} \times \mathbf{B}_0 - \frac{\eta}{\mu_0} \nabla \times \mathbf{B}_0 \right). \quad (70)$$

Averaging Eq. (70) over the t_0 time scale and subtracting that average from Eq. (70) gives an equation for the slow time-scale evolution of the equilibrium magnetic field $\mathbf{B}_0$,

$$\frac{\partial \mathbf{B}_0}{\partial t_1} = \nabla \times \left(\frac{\langle \mathbf{m}_1 \rangle}{\rho_0} \times \mathbf{B}_0 - \frac{\eta}{\mu_0} \nabla \times \mathbf{B}_0 \right), \quad (71)$$

where $\langle \mathbf{m}_1 \rangle(t_1) \equiv \tau^{-1} \int m_1(t_0, t_1)\, dt_0$ denotes the average of $\mathbf{m}_1$ over the t_0 time scale, and an equation for the fast-time-scale evolution of $\mathbf{B}_1$

$$\frac{\partial \mathbf{B}_1}{\partial t_0} = \nabla \times \left[\frac{(\mathbf{m}_1 - \langle \mathbf{m}_1 \rangle)}{\rho_0} \times \mathbf{B}_0 \right]. \quad (72)$$

Note that we used the fact that $\langle \mathbf{B}_1 \rangle(t_1) = 0$, which is implied by the two time-scale orderings of Eq. (67). Similar operations on the density and pressure equations (66c, d) yield separate evolution equations for $\rho_0, \rho_1, p_0,$ and p_1:

$$\frac{\partial \rho_0}{\partial t_1} + \nabla \cdot [\langle \mathbf{m}_1 \rangle + \langle \mathbf{\Gamma}_\| \rangle] = 0, \quad (73)$$

$$\frac{\partial \rho_1}{\partial t_0} + \nabla \cdot [\mathbf{m}_1 - \langle \mathbf{m}_1 \rangle] = 0, \quad (74)$$

$$\frac{\partial p_0^{3/5}}{\partial t_1} + \nabla \cdot \left\langle \frac{p_0^{3/5} \mathbf{m}}{\rho_0} \right\rangle = \left\langle \frac{2}{5} p_0^{-2/5} (\eta J^2 - \nabla \cdot \mathbf{q}) \right\rangle, \quad (75)$$

$$\frac{\partial p_1^{3/5}}{\partial t_0} + \nabla \cdot \left[\frac{p_0^{3/5} \mathbf{m}}{\rho_0} - \left\langle \frac{p_0^{3/5} \mathbf{m}}{\rho_0} \right\rangle \right] = 0. \quad (76)$$

A description of the fast t_0 time scale physics is given by Eqs. (69), (72), (74), and (76). Taking the t_0 derivative of Eq. (69) and substituting from Eqs.

(72) and (76) give a dispersion relation for the fast-time-scale mass flow:

$$\frac{\partial^2 \mathbf{m}_1}{\partial t_0^2} - v \nabla^2 \frac{\partial \mathbf{m}_1}{\partial t_0} = \mu_0^{-1}\left\{\nabla \times \left[\nabla \times \left(\frac{\mathbf{m}_1}{\rho_0} \times \mathbf{B}_0\right)\right] \times \mathbf{B}_0 \right.$$
$$\left. + (\nabla \times \mathbf{B}_0) \times \left[\nabla \times \left(\frac{\mathbf{m}_1}{\rho_0} \times \mathbf{B}_0\right)\right]\right\}$$
$$+ \nabla\left(\frac{\mathbf{m}_1}{\rho_0} \cdot \nabla p_0 + \frac{5}{3} p_0 \nabla \cdot \frac{\mathbf{m}_1}{\rho_0}\right). \tag{77}$$

With $v = 0$, Eq. (77) is just the ideal MHD equation of motion, containing both the transverse and the longitudinal wave branches. Since there are no unstable modes in two dimensions, the fast-time-scale behavior is simply that of small amplitude oscillations about the slowly varying equilibrium fields. With $v > 0$, these oscillations are damped. Setting $v = 2aB_0(\mu_0\rho_0)^{-1/2}$ makes the longest wavelength modes critically damped, decaying on the τ_f time scale.

A description of the slow t_1 time-scale physics is given by Eqs. (68), (71), (73), and (75). Taking the t_1 derivative of Eq. (68) and substituting from Eq. (71) and (75) yield the constraint equation determining the slowly varying momentum density $\langle m_1 \rangle \langle t_1 \rangle$:

$$\left\{\nabla \times \left[\nabla \times \left(\frac{\langle \mathbf{m}_1 \rangle}{\rho_0} \times \mathbf{B}_0 - \frac{\eta}{\mu_0} \nabla \times \mathbf{B}_0\right)\right]\right\} \times \mathbf{B}_0 + (\nabla \times \mathbf{B}_0)$$
$$\times \left[\nabla \times \left(\frac{\langle \mathbf{m}_1 \rangle}{\rho_0} \times \mathbf{B}_0 - \frac{\eta}{\mu_0} \nabla \times \mathbf{B}_0\right)\right]$$
$$+ \nabla\left[\frac{\langle \mathbf{m}_1 \rangle}{\rho_0} \cdot \nabla p_0 + \frac{5}{3} p_0 \nabla \cdot \frac{\langle \mathbf{m}_1 \rangle}{\rho_0} + \frac{2}{3}(\nabla \cdot \mathbf{q} - \eta J^2)\right] = 0. \tag{78}$$

Equation (78) is the well-known constraint equation for the velocity field associated with resistive diffusion, the analog of Eq. (51). Other forms of this equation have been derived by Pao (1978), Hazeltine et al. (1973), Turnbull and Storer (1983), and Hirshman and Jardin (1979). The fundamental point to note here is that the resistive component of the velocity $\langle \mathbf{m}_1 \rangle (t_1)$ is not really determined from a time-advancement equation as is the rapidly varying part $\mathbf{m}_1(t_0, t_1)$. It is determined, in fact, from the constraint equation that the system evolves through a series of near-equilibrium states $[B_0(t_1), p_0(t_1)]$ satisfying the equilibrium equation (68). However, this occurs automatically when the system of Eqs. (64) is integrated in time.

3. Numerical Method

We have shown that by artificially enhancing the plasma density, we can reduce the frequency of the Alfvén wave oscillations without affecting the resistive evolution. However, there remain disparate time scales in the equations due to the differences in the velocities of the fast compressible Alfvén waves and the slower shear Alfvén waves, and also due to the differences between the value of the resistivity in the hot plasma and in the cold vacuum regions. We therefore use the technique of sub-cycling to evaluate the diffusive and fast wave terms N times (typically $N = 10$–80) during each time step used by the rest of the problem.

We introduce a variable U for the divergence of the velocity, $U \equiv \nabla^2 \Omega$, and a variable F_0, where F_0/R is the vacuum toroidal field strength far from the plasma. The appropriate forms of equations (64a), (64e), and (64f) to apply subcycling are

$$\frac{\partial}{\partial t} U + \frac{F_0}{\mu_0} \nabla \cdot \frac{1}{R^2} \nabla F + Q = \nabla \cdot v \nabla U, \tag{64a'}$$

$$\frac{\partial}{\partial t} \chi + S = \frac{\eta}{\mu_0} \Delta^* \chi, \tag{64e'}$$

$$\frac{\partial}{\partial t} F + \frac{F_0}{\rho} U + T = \frac{R^2}{\mu_0} \nabla \cdot \frac{\eta}{R^2} \nabla F, \tag{64f'}$$

where the slowly varying Q, S, and T are defined as

$$Q = \nabla \cdot \left[\frac{\Delta^* \chi}{\mu_0 R^2} \nabla \chi + \frac{F - F_0}{\mu_0 R^2} \nabla F + \nabla p \right], \tag{79a}$$

$$S = \frac{1}{\rho} [\nabla \varphi \times \nabla A \cdot \nabla \chi + \nabla \Omega \cdot \nabla \chi], \tag{79b}$$

$$T = R^2 \nabla \cdot \left[\frac{F}{\rho R^2} \nabla \varphi \times \nabla A + \frac{F - F_0}{\rho R^2} \nabla \Omega - \frac{\omega}{\rho R^2} \nabla \varphi \times \nabla \chi \right]$$
$$+ F_0 R^2 \nabla \Omega \cdot \nabla \frac{1}{\rho R^2}. \tag{79c}$$

Thus equations (64a'), (64e'), and (64f') for U, χ and F are updated N times with a time step $\delta t = \Delta t/N$ for each major time step when Q, S, and T are evaluated, $\Delta^* A$, ω, ρ, $p^{3/5}$ are updated, and the elliptic equations for A and Ω are inverted. An explicit time advancement scheme is utilized, the wave and convection terms being differenced using the leapfrog method, and the diffusive terms using the method of Dufort and Frankel (1953). Thus, for example, Eq. (64f′) is differenced as

$$\frac{1}{2\delta t}[F_{i,j}^{n+1} - F_{i,j}^{n-1}] + \frac{F_0}{\rho_0}U^n + T$$

$$= \frac{1}{\mu_0(\Delta R)^2}\left\{\eta_{i+1/2,j}(R_{i,j}/R_{i+1/2,j})\left[F_{i+1,j}^n - \frac{1}{2}(F_{i,j}^{n+1} + F_{i,j}^{n-1})\right]\right.$$

$$\left. - \eta_{i-1/2,j}(R_{i,j}/R_{i-1/2,j})\left[\frac{1}{2}(F_{i,j}^{n+1} + F_{i,j}^{n-1}) - F_{i-1,j}^n\right]\right\}$$

$$+ \frac{1}{\mu_0(\Delta Z)^2}\left\{\eta_{i,j+1/2}\left[F_{i,j+1}^n - \frac{1}{2}(F_{i,j}^{n+1} + F_{i,j}^{n-1})\right]\right.$$

$$\left. - \eta_{i,j-1/2}\left[\frac{1}{2}(F_{i,j}^{n+1} + F_{i,j}^{n-1}) - F_{i,j-1}^n\right]\right\}, \quad (80)$$

where we have used superscript n to denote time (sub) cycle, and subscripts i and j to denote R and Z differences. Equation (80) is solved algebraically for $F_{i,j}^{n+1}$ at each location, with T being recomputed only every N sub-cycles.

The condition for stability of the method is that

$$\delta t \lesssim \Delta R(\mu_0\rho_0)^{1/2}/B_T \quad (81a)$$

and

$$\Delta t = N\delta t \lesssim \Delta R(\mu_0\rho_0)^{1/2}/B_p, \quad (81b)$$

where B_p and B_T denote the poloidal and toroidal field strengths. If we set $N = B_T/B_p$, then the two criterion become identical. The diffusive terms do not impose a time step restriction for stability, but they will lead to inaccuracy if the explicit criterion

$$\delta t \lesssim \mu_0(\Delta R)^2/\eta_{i,j}$$

is greatly exceeded.

4. Application

As an application illustrating the versatility and the usefulness of this method, we shall discuss the computation of the current buildup and cross-sectional shaping phases of a tokamak discharge in the Princeton Beta Experiment (PBX) (Bol et al., 1983). In this calculation, each computational grid point is one of three types: a plasma point, vacuum point, or coil point. For the plasma points, the full set of resistive MHD equations (64) is advanced in time. For the vacuum points, these same equations are solved, except that we add explicit source-sink terms of mass and energy, S_ρ and S_p,

7. Tokamak Magnetohydrodynamics

into Eqs. (64d, g) to force the vacuum density and temperatures to be constants, ρ_v and T_v. Actually, the value of ρ_v is unimportant, but T_v must be chosen to be much less than the plasma temperature ($T_v \ll T_p$) such that the magnetic diffusion time across the vacuum region is short compared with the time scales of physical interest, i.e., the time scales over which the boundary conditions change or the physical instability growth times.

The coil points are treated as vacuum points, except that the velocity is set to zero and an explicit electric-field driving term RE_φ is added to the poloidal-flux equation (64e). Thus, Eq. (64e) takes the form

$$\partial \Phi / \partial t + rI = V, \qquad (82)$$

where $\Phi \equiv -2\pi\chi = LI + \sum_{j \neq i} M_{ij}I_j$ is the actual poloidal flux at grid point i, $r \equiv 2\pi\eta R/(\Delta R \Delta Z)$ the effective resistance, $I = [(\Delta R \Delta Z)/\mu_0 R] \Delta^*\chi$ the current associated with that grid point, and $V = 2\pi RE_\varphi$ the applied voltage. Thus, the actual circuitry of the external coils can be incorporated into the calculation, including active feedback systems.

In Fig. 12 we illustrate the current buildup and bean-shaping stage of the discharge as computed by the time-dependent resistive MHD evolution code STARTUP. Figures 12a–d illustrate the evolution of the magnetic-flux surfaces when perfectly conducting plates are present (nonshaded boxes). The top–bottom plates are connected in antiseries so that they do not couple to the symmetric Ohmic heating (OH) or equilibrium-field (EF) systems but do provide stabilization for up–down motions of the plasma.

The plasma current I_p, equilibrium-field current I_{EF}, and shaping-field current I_{SF} take on the values illustrated in Fig. 13. The plasma size is limited by contact with the midplane limiter at $R = 1.1$ m.

We see from examining Figs. 12a–d that the plasma evolves smoothly from a circular cross section into a strong bean shape without exhibiting any tendency to bifurcate or lose equilibrium. These calculations used classical transport in the plasma, with additional enhanced resistivity inside the $q = 1$ surface to flatten the current density.

To illustrate the necessity of the conducting plates, we show in Figs. 12e–h the same calculation, but with high resistivity in the stabilizing plates. We see that the plasma becomes positionally unstable and, given a small initial downward perturbation, will move downward into the bottom plate on the resistive L/r time scale of the plates.

The validity of the large-mass method rests on the fact that the physical results obtained are insensitive to such variations in the nonphysical parameters as the enhancement factors for the plasma mass and viscosity and to the temperature used in the vacuum region. We have verified this by rerunning the calculation of Figs. 12a–d three additional times, varying each of these

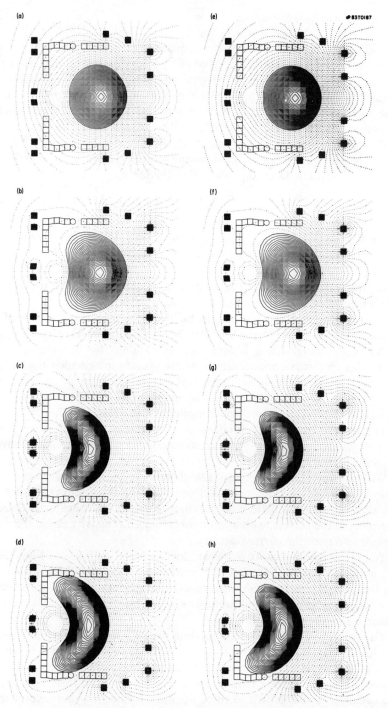

Fig. 12. Two sequences of bean formation calculated by the code STARTUP. The left sequence is totally up–down symmetric. The right sequence has a small perturbation that grows in time.

7. Tokamak Magnetohydrodynamics

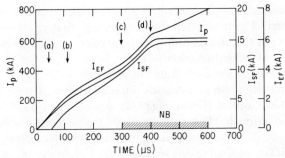

Fig. 13. Time histories of the plasma current I_p, the equilibrium field circuit current I_{EF}, and the shaping field circuit current I_{SF}. The (a), (b), (c), (d) refer to the snapshots in Fig. 12. Neutral beam auxiliary heating is on from 300 to 600 msec.

three quantities by a factor of 2. The results of these calculations look identical to the eye, with the plots of the magnetic axis position versus time differing by at most 2%.

IV. DISCUSSION

The computational methods discussed in the preceding three sections are illustrative of multiple-time-scale techniques used in the simulation of MHD phenomena occurring in tokamak fusion experiments. Needless to say, the specific topics covered are by no means comprehensive. Perhaps the most glaring omission has to do with the discussion of phenomena occurring on the time scales intermediate between ideal and resistive diffusive MHD (see Fig. 2). Resistive instabilities, such as tearing modes, resistive interchange modes, and resistive ballooning modes, grow on these intermediate time scales.

The great majority of simulation work performed to date on resistive instabilities has utilized the "reduced" MHD equations obtained by performing an asymptotic expansion in the geometric inverse aspect ratio of the tokamak, a/R. As mentioned in the introduction, since this expansion effectively annihilates the faster time scales from the equations by analytical means, we have chosen to exclude these methods from discussion.

Recently, some progress has been made by DeLucia (1983) in simulating intermediate-time-scale resistive instabilities using the full (nonreduced) MHD equations. This work utilizes a stream function–potential representation of the magnetic and velocity fields that is a generalization of Eqs. (61) and (62) to arbitrary helical symmetry geometry. Fourier decomposition in the angle and implicit time-differencing of the linear terms is shown to yield a tractable method for addressing the intermediate time scales.

Fully three-dimensional simulations have also not been emphasized in the present chapter. The reason for this is primarily that relatively little work has been done in the area of multiple-time-scale, three-dimensional simulations of tokamak magnetohydrodynamics. However, noteworthy efforts in this area include the work of Brackbill and Pracht (1973), which utilizes a generalized moving mesh and implicit time-integration scheme based on the ICED-ALE technique of Hirt *et al.* (1974), and of Finan (1978), who utilizes alternating-direction implicit techniques in the numerical solution of the time-dependent, three-dimensional, single-fluid, resistive MHD equations.

The complexity of the present subject matter undoubtably makes it difficult for one to distinguish the numerical techniques from the underlying physical processes. There are several central themes, however, that should be evident in the methods described here. One is the use of magnetic coordinates, which implies the use of a time-dependent coordinate transformation since the magnetic field evolves in time.

Aligning one coordinate with the magnetic field is useful in ideal magnetohydrodynamics to represent MHD waves and the plasma–vacuum interfaces accurately. The fluid and the magnetic field are frozen together, and thus one could devise a purely Lagrangian method that utilizes magnetic coordinates, i.e., that has no relative fluid flow crossing the coordinate lines. Such a Lagrangian method would, in general, be compressible, and so the positions of the grid points would exhibit the fast-time-scale motion. Since the equations of motion for the grid points of a Lagrangian grid are very complicated, a fully implicit Lagrangian formulation is impractical. The dynamic grid method described in Section II gets around this by allowing some slippage between the fluid flow and the magnetic coordinate lines so that the fast-time-scale compressible motion of the fluid shows up as Eulerian flow through the moving grid. We remark that this slippage does not represent slippage between the fluid and the magnetic field, which can only occur physically with the introduction of resistivity. Rather, it exploits the freedom inherent in defining a magnetic coordinate system consistent with a given magnetic field to force the coordinate transformation to be free of the fast-time-scale motion.

On the resistive time scale, the fluid and the field are no longer frozen together. In Section III.B we describe a resistive-time-scale method based on the same coordinate transformation as that utilized in Section II. However, here the motivation for using this is somewhat different. Again, the fundamental physics issue is that one coordinate be aligned with the magnetic field, this time to represent the long-time effects of parallel transport accurately. The freedom associated with defining the remaining properties of the magnetic coordinate system are used to divide the physical magnetic-flux velocity into two parts. The incompressible part associated with a change of shape, but not volume, is used to evolve the Lagrangian coordinates of the magnetic surfaces. The remaining compressible part of the velocity field is constant on each magnetic surface. It is represented as the Eulerian flow of toroidal flux across flux surfaces.

The final method in Section III.C evolves the magnetic flux χ in time but does not explicitly use magnetic coordinates. This approach has some advantages in that it is conceptually straightforward and can easily treat situations in which the magnetic topology changes, i.e., in which new minima or maxima appear in the poloidal flux χ.

ACKNOWLEDGMENTS

The author is indebted to his colleagues and collaborators at the Plasma Physics Laboratory. Drs. J. M. Greene, R. C. Grimm, S. P. Hirshman, J. L. Johnson, and W. Park deserve special mention.

This work was supported by the U.S. Department of Energy under contract DE-ACO2-76-CHO-3073.

REFERENCES

Bateman, G., Schneider, W., and Grossmann, W. (1974). *Nucl. Fusion* **14**, 669.
Bol, K., *et al.* (1983). "PBX: The Princeton Beta Experiment," PPPL-2032. Plasma Physics Laboratory, Princeton University, Princeton, New Jersey.
Brackbill, J. V. (1976). *In* "Methods in Computational Physics" (J. Killeen, ed.), Vol. 16, p. 1. Academic Press, New York.
Brackbill, J. U., and Pracht, W. E. (1973). *J. Comput. Phys.* **13**, 455.
Braginskii, S. I. (1965). *Rev. Plasma Phys.* (*Engl. Transl.*) **1**, 205.
Byrne, N., and Klein, H. (1978). *J. Comput. Phys.* **26**, 352.
Chance, M. S., Greene, J. M., Grimm, R. C., and Johnson, J. L. (1977). *Nucl. Fusion* **17**, 65.
DeLucia, J. (1983). Ph.D. Thesis, Princeton University, Princeton, New Jersey.
DeLucia, J., Jardin, S. C., and Todd, A. M. M. (1980). *J. Comput. Phys.* **37**, 183.
Dufort, E. C., and Frankel, S. P. (1953). *Math. Aids Comput.* **7**, 135.
Finan, C. H., III (1978). Ph.D. Thesis, University of California, Davis.
Furth, H. P. (1975). *Nucl. Fusion* **15**, 487.
Grad, H. (1970). *In* "Proceedings of the Congress of Mathematicians, Nice," Vol. 3, p. 105. Gauthier-Villars, Paris.
Grad, H. *et al.* (1958). "Notes on Magnetohydrodynamics," NYU-6486. Courant Institute of Mathematical Sciences, New York University.
Hazeltine, R. D., and Hinton, F. L. (1973). *Phys. Fluids* **16**, 1883.
Hazeltine, R. D., Hinton, F. L., and Rosenbluth, M. N. (1973). *Phys. Fluids* **16**, 1645.
Helton, F. J., Miller, R. L., and Rawls, J. R. (1977). *J. Comput. Phys.* **24**, 117.
Hirshman, S. P., and Jardin, S. C. (1979). *Phys. Fluids* **22**, 731.
Hirt, C. W., Amsden, A. A., and Cook, J. L. (1974). *J. Comput. Phys.* **14**, 227.
Hogan, J. T. (1978). *Oak Ridge Natl. Lab.* [*Rep.*] *ORNL-TM U.S.* **ORNL-TM-6049**.
Ridge Natl. Lab. [*Rep.*] *ORNL-TM U.S.* **ORNL-TM-6049**.
Jardin, S. C. (1978). *Phys. Fluids* **21**, 1851.

Jardin, S. C. (1981). *J. Comput. Phys.* **43**, 31.
Jardin, S. C., and Park, W. (1981). *Phys. Fluids* **24**, 679.
Jardin, S. C., Johnson, J. L., Greene, J. M., and Grimm, R. C. (1978). *J. Comput. Phys.* **29**, 101.
Kruskal, M., and Kulsrud, R. (1958). *Phys. Fluids* **1**, 285.
Meade, D. M., Furth, H. P., Rutherford, P. H., and Seidel, F. G. P. (1975). *In* "Plasma Physics and Controlled Nuclear Fusion Research," Vol. I, p. 605. IAEA, Vienna.
Nelson, D. B., and Grad, H. (1977). *Oak Ridge Natl. Lab.* [*Rep.*] *ORNL-TM* (*U.S.*) **ORNL-TM-6094**.
Pao, Y. (1978). *Phys. Fluids* **21**, 1120.
Potter, D. (1976). *In* "Methods in Computational Physics" (J. Killeen, ed.), Vol. 16, p. 43. Academic Press, New York.
Schneider, W., and Bateman, G. (1975). *In* "Plasma Physics and Controlled Nuclear Fusion Research, 1974," Vol. I, p. 429. IAEA, Vienna.
Strauss, H. R. (1976). *Phys. Fluids* **19**, 134.
Strauss, H. R. (1983). *Nucl. Fusion* **23**, 649.
Todd, A. M. M., and Grimm, R. C. (1975). *Bull. Am. Phys. Soc.* [2] **20**, 1279.
Turnbull, A., and Storer, R. (1983). *J. Comput. Phys.* **50**, 409.
Wesson, J. A., and Sykes, A. (1975). *In* "Plasma Physics and Controlled Nuclear Fusion Research, 1974," Vol. I, p. 449. IAEA, Vienna.

8

Hybrid and Collisional Implicit Plasma Simulation Models

RODNEY J. MASON

Los Alamos National Laboratory
University of California
Los Alamos, New Mexico

    I. Introduction 233
   II. Basic Moment Method 236
       A. General Approach 236
       B. Refinements 239
       C. Relation to the Direct Method 248
  III. Collisional–Hybrid Extensions 249
       A. Governing Equations 250
       B. Moment Method Solution 255
       C. Discussion 262
  IV. Applications 264
   V. Conclusion 267
       References 269

I. INTRODUCTION

Hybrid plasma simulation models treat elements of a plasma as particles, and the remaining elements as one or more fluids. In time-explicit hybrid models the electron mass is often set to zero to eliminate plasma oscillations. An early explicit hybrid model is due to Mason (1969, 1971, 1972). Later examples are those of Chodura (1975), Sgro and Nielson (1976), Hamasaki et al. (1977), Byers et al. (1978), and Hewett and Nielson (1978).

Time-implicit hybrid models were first explored at Los Alamos in the late 1970s to simulate laser-generated electron transport—with electron inertial effects retained. At intensities exceeding 10^{15} W cm^{-2} most of the energy absorbed from CO_2 lasers goes into long-range suprathermal electrons with typical energies of 150 keV and source densities of 10^{19} cm^{-3}. These *hot* electrons transport their energy through a strongly collisional background plasma at characteristic 10^{23}-cm^{-3} electron densities and 1–30-eV temperatures. The long range of the suprathermals and their potentially dominant electromagnetic interactions render them amenable to a collisional (Shanny *et al.*, 1967) particle-in-cell description (Morse, 1970). On the other hand, the extreme density variation and collisionality of the background dictates a fluid treatment. Since both the fluid and particle modeling retain inertia, a time-implicit electric field is demanded to avoid the time-step constraints from plasma oscillations (Langdon, 1979; Crystal *et al.*, 1979).

For the earliest one-dimensional, implicit, hybrid calculations (Mason, 1979b, 1981d) the implicit fields were calculated only crudely. In addition to the particle equations for suprathermals, an auxiliary momentum equation was used to help predict future particle fluxes $j_h^{(m+1)}$, copying the simplicity of the cold momentum equation. For fixed ions, the hot and cold momentum equations were combined with the quasi-neutral condition, $J \equiv j_h^{(m+1)} + j_c^{(m+1)} \simeq 0$, to give the future $E^{(m+1)}$. A computational cycle was completed by advancing the particles and fluids in this future field. Later, in Mason (1980a), an improved field was computed by using the total predicted flux in Ampere's law, $J^{(m+1)} = 1/(4\pi e) \, \partial E/\partial t$. Calculations with these early models, however, lead to a gradual accumulation of charge separation errors, encouraging the serious consideration of an alternate "charge dilation technique" (Mason, 1980a, 1981d). Finally, in Mason (1980b,c, 1981a,b) the separation errors were eliminated by using both momentum and continuity equations to predict a future total source density $N^{(m+1)}$ that was joined with Poisson's equation, i.e., $4\pi e N^{(m+1)} = -\partial^2 \varphi/\partial x^2$, to give $E^{(m+1)} = -\partial \varphi/\partial x$. This final improvement completed the essentials of the implicit moment method.

By this juncture Denavit had developed a very similar collisionless moment model, as on outgrowth of efforts to time-average away the excessive noise in particle codes. In Crystal *et al.* (1979) he had quantified the stabilizing virtues of implicit E fields through a Nyquist stability analysis. Also, the use of continuity and momentum equations for the inversion of a Vlasov plasma representation was recommended. Then, Denavit and Walsh (1980) and Denavit (1981) presented a moment method in which auxiliary moment equations and the time integration of $J^{(*)} = 1/(4\pi e) \, \partial E/\partial t$ with $J^{(*)} \equiv (3J^{(m+1)} + J^{(m-1)})/4$ were used to produce the E-field in which particle electrons and ions were advanced. The $\frac{3}{4}, \frac{1}{4}$ averaging improved the relative suppression of unwanted high-frequency modes. In Denavit (1981) the charge separation errors were eliminated by replacement of the old level-(*m*) field with the field from the solution to Poisson's equation for the old density—essentially by use of the Boris (1970) continuity equation correction.

Generally, the use of implicit fields in these models was found to permit a time step far exceeding the largest local inverse plasma period in a problem, i.e., $\Delta t \gg \omega_p^{-1}$. Still, for stability and accuracy, the time step has remained constrained by an electron Courant condition, $\Delta t < \Delta x/v_T$, with Δx the cell size and v_T the mean thermal speed of the hottest electrons. Since the Debye length is $\lambda_D = v_T/\omega_p$, it follows that $\Delta x \gg \lambda_D$ can prevail in calculations. This can be a clear improvement over explicit particle calculations, which require a cell size no larger than λ_D to avoid anomalous plasma heating from finite-grid instability (Langdon, 1970). The electron Courant time step is, however, much shorter than the ion Courant limit that constrains explicit hybrid codes when electron inertia is suppressed. Consequently, when details of the evolving electron distribution are unimportant, the earlier explicit approach may still be preferred.

The first implicit hybrid models (Mason, 1980a, 1981d) were used to examine the effects of classical and ion-acoustic resistivity on the return currents feeding the hot emission in laser–foil targets. Also, the effects of convective "trough" inhibition were explored. The later moment models (Mason, 1981a; Denavit, 1981) were versatile enough to treat full-particle problems, in which all the hot and cold electrons and ions were treated as particles. The sample problems discussed included electron–electron two-stream instability, ion-acoustic waves, and one- and two-electron temperature ion-acoustic expansions. Relatedly, collisional full-particle implicit moment calculations were carried out in Mason (1981b) to characterize real and apparent thermal inhibition in laser-produced plasmas.

Subsequently, Brackbill and Forslund (1982) (see also Chapter 9, this volume) have extended the implicit moment method to two-dimensional, collisionless problems. They have stressed the value of implicit pressure, as possibly permitting time steps exceeding the Courant limit, and have indicated the need for a minimum time step—for the suppression of finite-grid instability. Sample application has been made to the Weibel instability, and to lateral laser-induced transport on foils (Forslund and Brackbill, 1982). Also, Friedman *et al.* (1981) and Langdon *et al.* (1983) have perfected a direct method that avoids any reference to auxiliary fluid equations. Barnes *et al.* (1983) have applied the direct method to two-dimensional magnetized plasmas, and a merger of orbit-averaging techniques (Cohen *et al.*, 1980) and implicit methods has been accomplished by Cohen *et al.* (1982). Recently, Mason (1983a) has outlined a refined hybrid implicit scheme for the simulation of electron transport through single and multiple foils, and Denavit (1983a,b) has simulated collisional foil expansions and diode dynamics by means of a new, semidirect hybrid scheme.

In this chapter we provide a detailed review of the moment method in Section II and elaborate on the workings of the more complex collisional and hybrid models in Section III. Readers interested principally in hybrid methods will find that Section III is largely self-contained for their purposes.

In Section II we show how auxiliary momentum and continuity equations are used in the collisionless, full-particle limit to predict a future total charge density that is used in Poisson's equation to yield an advanced, implicit E field. Optimal centering of the equations and Newton–Raphson iteration to find the particle position in the advanced field are discussed. A stability analysis of the collisionless differencing is provided, including analysis of maximum and minimum time-step limits and a justification for the spatial differencing employed. The possible virtues of future schemes using implicit pressures are explored. A procedure is outlined for iteration of the moment method field to the exact solution of Poisson's equation at the advanced time. Finally, a connection is drawn between the moment method and the alternate direct method of Friedman *et al.* (1981).

Section III first shows how hybrid implicit schemes are appropriate to laser–matter interaction studies. The governing equations for coupled particles and fluids are then discussed. These are Newton's laws plus emission, drag, and scattering additions for the particles and a set of modified Braginskii equations for the fluids. Next we detail our split-operator approach to the solution of these equations, discussing time-step control, drag and thermal coupling between the components, calculation of the electron–ion scatter-driven coupling of the mean fluid velocities, evaluation of the implicit moment hybrid E field, and computation of the particle and fluid motions through this field. Particular discussion is given to techniques required for the elimination of numerical diffusion in our Eulerian treatment of the hydrodynamics.

The chapter concludes with a sampling of applications, a summary of highlights, and an analysis of future directions.

II. BASIC MOMENT METHOD

A. General Approach

We wish first to solve the one-dimensional, electrostatic, collisionless, Maxwell–Valsov particle-in-cell (PIC) simulated system

$$\frac{\partial E^{(\ddagger)}}{\partial x} = -\frac{\partial^2 \varphi^{(\ddagger)}}{\partial x^2} = 4\pi \sum_\alpha q_\alpha n_\alpha^{(\ddagger)}, \tag{1a}$$

$$u_k^{(m+1+\Gamma)} = u_k^{(m+\Gamma)} + \frac{q_\alpha}{m_\alpha} E^{(*)}(x_k^{(*)}) \, \Delta t, \tag{1b}$$

$$x_k^{(m+1)} = x_k^{(m)} + u_k^{(*)} \, \Delta t, \quad k = 1 \to M, \tag{1c}$$

8. Implicit Plasma Simulation Models

in which $\alpha = e$ (h or c) for electrons (hot or cold) and i for ions, k selects a specific simulation particle from the total M, n_α is the accumulated electron or ion density on the mesh, and $E = -\partial\varphi/\partial x$.

Traditionally, one would use leapfrog velocities, i.e., $\Gamma = -\frac{1}{2}$, and the old $n^{(\ddagger)} = n^{(m)}$, $x_k^{(*)} = x_k^{(m)}$, $E^{(*)} = E^{(m)}$ electric field. However, this can lead to violent instability, if the time step Δt is too large. To see this, consider electrons spreading to the right at the edge of a foil. At $t = 0$ the electrons overlay the ions $E^{(m)} = 0$. Let the ions be fixed. Suppose the time step has been chosen so that on the average the electrons cross a single cell of width Δx in a time step; i.e., $\Delta x \simeq v_e \Delta t$, where $v_e = (kT_e/m_e)^{1/2}$ is their mean thermal speed. From Eq. (1a) this produces a field $E = -4\pi eN \Delta x$, where N is some mean charge separation density established in the cell. In the next cycle the electrons will suffer an acceleration $\Delta v_e = eE/m_e \Delta t = 4\pi e^2 n_e \Delta t^2/m_e(N/n_e)v_e = (\omega_p \Delta t)^2(N/n_e)v_e$ and experience an energy change $\Delta\varepsilon/\varepsilon = 2(\omega_p \Delta t)^2(N/n_e)$, where $\varepsilon = m_e v_e^2/2$. If $N/n_e < \frac{1}{2}$, $\omega_p \Delta t \leq 1$ can be tolerated. For $\omega_p \Delta t \gg 1$ the mean electron energy will change by many multiples upon crossing the cell, leading to extreme divergence of the calculation.

Alternatively, one can anticipate that the excursions of the electrons will be self-regulating in an implicit field, so that only fractional energy changes occur on the average. Until recently, however, the implicit Eq. (1) system of $2M + 1$ equations has had too large and dense a matrix for straightforward inversion. In the implicit moment method we surmount this problem by solving an auxiliary set of momentum and continuity equations with the Poisson equation to provide a predicted field $\tilde{E}^{(m+1)}$. The particles are then advanced in these predicted fields. If desired, an iteration of the field and particle equations can be subsequently performed to establish convergence of $\tilde{E}^{(m+1)}$ to the exact Poisson solution $E^{(m+1)}$ at the advanced particle positions $x_k^{(m+1)}$.

Thus, in place of Eq. (1a) we use

$$\frac{\partial \tilde{E}^{(m+1)}}{\partial x} = -\frac{\partial^2 \tilde{\varphi}^{(m+1)}}{\partial x^2} = 4\pi \sum_\alpha q_\alpha \tilde{n}_\alpha^{(m+1)}, \tag{2a}$$

$$\tilde{j}_\alpha^{(m+1+\Gamma)} = j_\alpha^{(m+\Gamma)} - \frac{1}{m_\alpha}\left[\frac{\partial \Pi_\alpha^{(*)}}{\partial x} - q_\alpha n_\alpha^{(\perp)} \tilde{E}^{(*)}\right]\Delta t, \tag{2b}$$

$$\tilde{n}_\alpha^{(m+1)} = n_\alpha^{(m)} - \frac{\partial \tilde{j}_\alpha^{(*)}}{\partial x}\Delta t, \tag{2c}$$

in which $j_\alpha \equiv n_\alpha U_\alpha$, $\Pi_\alpha \equiv P_\alpha + n_\alpha m_\alpha U_\alpha^2$, and $P_\alpha \equiv n_\alpha kT_\alpha$. Choices for the levels (*) and ($\perp$) are discussed in what follows. The tildes denote predicted values from use of the moment equations.

With a fully implicit choice for $\tilde{E}^{(*)} = \tilde{E}^{(m+1)} = -\partial\tilde{\varphi}^{(m+1)}/\partial x$ and $\Gamma = 0$, the Eqs. (2) combine to yield

$$\frac{\partial}{\partial x}(1+\chi)\frac{\partial \tilde{\varphi}^{(m+1)}}{\partial x} = -4\pi \sum_\alpha q_\alpha \left[\int_0^x n_\alpha^{(m)}\, dx - \frac{\partial}{\partial x}\left(j_\alpha^{(m)} \Delta t - \frac{1}{m_\alpha}\frac{\partial \Pi_\alpha^{(*)}}{\partial x} \Delta t^2 \right) \right], \quad (3)$$

in which $\chi \equiv \omega_p'^2 \Delta t^2$, $\omega_p'^2 \equiv \omega_p^2 (1 + Zm_e/m_i)$ and $\omega_p^2 = 4\pi e^2 n_e^{(\perp)}/m_e$ with $n_e = n_c + n_h$, $q_c = q_h = -e$, $q_i = Ze$, and $Zn_i = n_e$. Integrating for a quiescent or mirror-symmetric left boundary where $j_\alpha(0) = \tilde{E}(0) = 0$, we convert this to

$$\tilde{E}^{(m+1)} = \frac{4\pi \left[\int_0^x \sum_\alpha q_\alpha n_\alpha^{(m)}\, dx - \sum_\alpha q_\alpha j_\alpha^{(m)} \Delta t + \sum_\alpha \frac{q_\alpha}{m_\alpha}\frac{\partial \Pi_\alpha^{(*)}}{\partial x} \Delta t^2 \right]}{(1 + \omega_p'^2 \Delta t^2)}. \quad (4)$$

For the simplest effective particle advancement prescription with Eq. (4), we first assume convergence, i.e., $n = \tilde{n}$, $j = \tilde{j}$ and $E = \tilde{E}$, etc. In practice, we have found no instance when this has lead to identifiable difficulties. Furthermore, we set $\Pi^{(*)} = \Pi^{(m)}$, since the old pressure data are easily accumulated from the particles along with $j^{(m)}$ and $n^{(m)}$. We then accelerate particles using Eq. (1b) for $\Gamma = 0$ in the fully implicit field, $E^{(*)} = E^{(m+1)}(x_k^{(m)})$, readily evaluated at the old particle positions; then we move them by Eq. (1c) with the fully forward velocities $u_k^{(*)} = u_k^{(m+1)}$. The resultant dynamics of problems described with this system are numerically stable for a Courant limited time step set for fastest electrons and generally physically plausible, but disturbances are excessively damped so that refined centering, as described in the next section, is highly desirable.

The Eq. (4) field expression has several interesting limits. Assuming $\omega_p' \Delta t \gg 1$ and using $m_e/m_i \ll 1$, we can look at the quasi-neutral response to charge separation and electron drift and examine the quasi-neutral equilibrium field. Thus, reconsidering the foil charge separation problem examined earlier, we observe that, with old currents and pressure gradients neglected, the first term in Eq. (4) gives for charge separation N over a cell width Δx the field $E = -N \Delta x/(en_e \Delta t^2/m_e)$. The incremental velocity during Δt for electrons situated in this field is $\delta u = -eE \Delta t/m_e = N \Delta x/(n_e \Delta t)$. Therefore, since $\delta n_e = -(n_e \delta u/\Delta x) \Delta t$ through continuity, we can conclude that the incremental density δn_e is just enough to cancel N by the end of the time step. Similarly, for a current flow $j_e^{(m)}$, but negligible charge separation and pressure gradients, the large-time step field becomes $E = m_e j_e^{(m)}/(en_e \Delta t)$. In this case, the incremental velocity $\delta u = -eE \Delta t/m_e$ just cancels j_e. Finally, for the largest $\omega_p' \Delta t \gg 1$ we see that the old densities and currents will be "forgotten," and the Eq. (4) field reduces to

$$E = -(en_e^{(*)})^{-1} \partial \Pi_e^{(*)}/\partial x. \quad (5)$$

Neglecting the electrons' drift speed as compared to their thermal speed, one can conclude that electrons crossing a cell through the Eq. (5) field will experience an energy change of order $\delta\varepsilon = -eE\,\Delta x = n_e^{-1}(\partial P_e/\partial x)\,\Delta x \simeq kT_e$. In Mason (1969, 1971) Eq. (5) was used with $\Pi_e^{(*)} \simeq P_e^{(*)} = n_e^{(*)}kT_e^{(*)}$ and

$$n_e = Zn_i - (4\pi e)^{-1}\,\partial E/\partial x \tag{6}$$

to provide an E field for the earliest ion-acoustic shock studies. Although we see now that this was essentially an implicit field, in Mason (1969) it was used as a level-(m) field for the leapfrog advancement of particle ions in isothermal fluid electrons. The implicit pressure $\Pi_e^{(*)}$ was determined iteratively with $n_e^{(*)}$ and E. Convergence was rapid for cell sizes exceeding a Debye length.

With the assumption $n_e = Zn_i$, Eq. (5) is equivalent to Hewett and Nielson's (1978) hybrid result for one-dimensional and zero B fields. Relatedly, Eq. (3) is equivalent to their inhomogeneous Poisson equation if $1 + \chi$ is changed to χ on the left side of Eq. (3) and the $\sum n_\alpha$ and $\sum j_\alpha$ sources are dropped from the right side. The appearance of the 1 in combination with the susceptibility χ in Eq. (3) and in the denominator of Eq. (4) is due directly to the implicit retention of electron inertia. In low-density regions this 1 serves to automatically avoid singular problems that challenge many magnetohydrodynamic (MHD) (Lindemuth, 1977) and hybrid models (Hewett, 1980; Harned, 1982).

These observations suggest the possible utility of yet another implicit hybrid model that would avoid electron particles and use implicit electron fluid equations retaining inertia. Equation (4) could again give the E field for this model except that the level-(m) data would be acquired from the old electron fluid values, rather than from the latest particle accumulations. Such an implicit pure hybrid model should be the object of future scrutiny. For the present, however, we turn to improvements constrained by the presence of particle electrons and our need to track the evolving electron distribution.

B. Refinements

1. Centering

a. (m) STORAGE. Although a fully implicit treatment provides the greatest stability and the simplest algorithm, in Eqs. (1) and (2) improved centering is required for accuracy. In Mason (1981a) a leapfrog implicit scheme was developed, chiefly for compatibility with earlier leapfrog movers. The scheme presently preferred is an (m)-stored scheme (see Brackbill and Forslund, 1982), in which

$$E^{(*)} = \Theta E^{(m+1)} + (1 - \Theta)E^{(m)}, \tag{7a}$$

$$\Pi^{(*)} = \eta\Pi^{(m+1)} + (1 - \eta)\Pi^{(m)}, \tag{7b}$$

$$U^{(*)} = \psi U^{(m+1)} + (1 - \psi)U^{(m)}, \tag{7c}$$

with

$$u_k^{(*)} = \psi u_k^{(m+1)} + (1 - \psi)u_k^{(m)}, \tag{7d}$$

$$x_k^{(*)} = \Theta x_k^{(m+1)} + (1 - \Theta)x_k^{(m)}, \tag{7e}$$

and

$$n_\alpha^{(\perp)} = \beta n_\alpha^{(m+1)} + (1 - \beta)n_\alpha^{(m)}. \tag{7f}$$

This centering has the straightforward advantage that all the unknowns are stored at the same level (m). The centering parameters Θ to β are individually tunable to adjust the model to physical and practical constraints. When Θ to β are unity, the system is fully implicit; when Θ to $\beta = 0.5$, as we shall see later, the system is stable and undamped. The (m) storage permits a variable Δt — a feature, generally denied in particle codes, but of extreme utility in evolving inhomogeneous problems. Further, for any Θ the system can facilitate improved energy conservation, since a centered Ampere's law works to establish $J^{(m+1/2)} = 0$ is quasi-neutral regions, which moves $U^{(m+1/2)}$ toward zero—dissipating energy in leapfrog schemes, while simply reversing the kinetic drift $U^{(m+1)} = -U^{(m)}$, with (m) storage.

However, in $\omega \Delta t$ both the leapfrog and (m)-stored centerings lead to the first-order damping of oscillations of frequency ω. Crystal *et al.* (1979) have shown that the choice $E^{(*)} \equiv \frac{3}{4}E^{(m+1)} + \frac{1}{4}E^{(m-1)}$ (with Π and J centered similarly) can provide strong damping for high frequencies, whereas the damping of low frequencies is reduced to $O(\omega \Delta t)^3$. Also, Denavit (1981) and Cohen *et al.* (1982) have provided more elaborate centerings with more improved damping characteristics. Since practical advantage from the use of these higher order schemes still warrants further demonstration, and since their use introduces further complexity, especially in hybrid modeling, we avoid these schemes in the present development.

b. *E*-FIELD SAMPLING. The (m)-storage field should be sampled at the $x^{(m+\Theta)}$ position of a particle. To find this, we use $u^{(m+\psi)} = u^{(m)} + (q_\alpha/m_\alpha)E^{(*)}\psi \Delta t$ and $x^{(m+\Theta)} = x^{(m)} + u^{(m+\psi)} \Theta \Delta t$ to construct the function

$$G(x) = x - [x^{(m)} + u^{(m)}\Theta \Delta t + (q_\alpha E^{(*)}(x)/m_\alpha \Theta \psi \Delta t^2]. \tag{8}$$

We seek the solution $G(x^{(m+\Theta)}) = 0$. Since $G(x + \delta x) \cong G(x) + (\partial G/\partial x) \delta x = 0$, an initial guess, $^{(s)}x^{(m+\Theta)}$, is improved by adding

$$^{(s)}\delta x = -\frac{G}{(\partial G/\partial x)} = \frac{G(^{(s)}x^{(m+\Theta)})}{\{1 - (q_\alpha/m_\alpha)[\partial E^{(*)}(^{(s)}x^{(m+\Theta)})/\partial x]\Theta\psi \Delta t^2\}} \tag{9a}$$

to obtain

$$^{(s+1)}x^{(m+\Theta)} = {}^{(s)}x^{(m+\Theta)} + {}^{(s)}\delta x. \tag{9b}$$

8. Implicit Plasma Simulation Models

For time steps satisfying a Courant condition, the convergence of this Newton–Raphson iteration procedure to $\delta x/x < 0.001$ is typically accomplished with three iterations.

A Courant time constraint is needed to avoid singularity of the Eq. (9a) denominator. To see this, consider an e^{ikx} mode of the equilibrium field $E = -(1/en_e)\partial \Pi_e/\partial x = -ik(m_e/e)a_0^2$. For electrons this yields $G'(x) = 1 - (ka_0 \Delta t)^2 \psi \Theta$, so that the centered ($\psi = \Theta = \frac{1}{2}$) particle advancement prescription will diverge unless $ka_0 \Delta t < 2$, or $\Delta t < 0.7\, \Delta x/a_0$. A comparable singularity arises with the $\frac{3}{4}, \frac{1}{4}$ differencing (Denavit, 1981). One can express the denominator in Eq. (9a) in terms of a trapping frequency by noting that the distance separating the actual and free-streaming position for a particle (type α) is $\delta x = x^{(m+1)} - \tilde{x}^{(m+1)} = (q/2m)E\, \Delta t^2$. Singularity is avoided as long as the particle excursions are small enough to satisfy $(k\, \delta x)\psi\Theta \ll 1$, where k denotes the wave number of the electric field. This condition may be expressed as $(\omega_T \Delta t)\psi\Theta \ll 1$, where $\omega_T \equiv (qkE/m)^{1/2}$ denotes the trapping frequency.

At the cost of reduced accuracy, these iterations may be avoided for simplicity by just sampling the field at the old particle positions $x^{(m)}$ and using a sufficiently advanced E field $\Theta \le 1$ to ensure stability. Alternatively, the leapfrog centering in Mason (1981a) naturally avoids the need for this iteration (for $\Theta = 0$) by determining an averaged implicit field for level (m) at the centered position of the particles. In electron two-stream instability simulations, our experience with the leapfrog scheme shows that at least stable and apparently physically plausible results can be obtained with time steps well in excess of ω_T^{-1}.

c. STABILITY ANALYSIS. Linear stability is unaffected by the $n_\alpha^{(\perp)}$ centering choice for β [Eq. (7f)], and experience indicates difficulties (e.g., occasional negative density determinations) in the iterative calculation of a centered $n^{(\perp)}$. Consequently, the convenient choice, $\beta = 0$ for $n^{(\perp)} = n^{(m)}$, is generally employed.

For our stability analysis we assume convergence of the field, i.e., $\tilde{E} = E$. The linear stability of Eqs. (2) can be readily investigated for the simplified case of isothermal electrons oscillating in a uniform motionless ion background, where the mean thermal electron speed is $a_0 = (kT/m)^{1/2}$, and the density is $n_0 = Zn_i$. The linearized equations for e^{ikx} modes become

$$\partial U_k/\partial t = +(\omega_p^2/ikn_0)\bar{\bar{n}}_k - (a_0^2 ik/n_0)\bar{n}_k, \tag{10a}$$

$$\partial n_k/\partial t = -ikn_0 \bar{U}_k. \tag{10b}$$

Here $\omega_p^2 \equiv 4\pi e^2 n_0/m_e$, and $E_k = -4\pi en_k/ik$ has been eliminated. The density $\bar{\bar{n}}$ is associated with the E field, whereas $\bar{n}$ is related to a static pressure $P = \bar{n}kT$. Introducing, for example, the time dependence $n_k = n_0 \exp[i\omega(t - t_0)]$, we obtain $n_k^{(m)} = n_0 \exp(i\omega m\, \Delta t) = n_0 \zeta^m$, with $\zeta = \exp(i\omega\, \Delta t)$ at the discrete levels $m = (t - t_0)/\Delta t$. Then, employing the Eq. (7) expression for the field, pressure, and velocity centerings, such that $\bar{\bar{n}} = \Theta n^{(m+1)} + (1 - \Theta)n^{(m)}$, $\bar{n} = \eta n^{(m+1)} + (1 - \eta)n^{(m)}$ and $\bar{U} = \psi U^{(m+1)} + (1 - \psi)U^{(m)}$, and using, for example, $\partial U_k/\partial t = (U_k^{(m+1)} - U_k^{(m)})/\Delta t$, we derive the dispersion relation

$$(\xi - 1)^2 = -\omega_p^2 \Delta t^2 [\Theta\xi + (1 - \Theta)][\psi\xi + (1 - \psi)] \\ - a_0^2 k^2 \Delta t^2 [\psi\xi + (1 - \psi)][\eta\xi + (1 - \eta)]. \quad (11)$$

With identical centering for the E field, pressure, and velocity, i.e., $\Theta = \eta = \psi$, this simplifies to

$$(\xi - 1)^2 = -c^2 [\Theta\xi + (1 - \Theta)]^2, \quad (12)$$

in which $c^2 \equiv (\omega_p^2 + a_0^2 k^2) \Delta t^2 \equiv \bar{\omega}_p^2$.

Equation (12) has the solution $\xi = [1 \pm ic(1 - \Theta)/(1 \mp ic\Theta)]$, yielding the amplification factor $|\xi| = \{[1 + c^2(1 - \Theta)^2]/(1 + c^2\Theta^2)\}^{1/2}$. Let $\xi \equiv \exp(i\omega_0 \Delta t_e - \gamma \Delta t)$. Then, $|\xi| = \exp(-\gamma \Delta t) = 1/c = 1/\bar{\omega}_p \Delta t$ for $c \gg 1$, and $\gamma = (1/\Delta t) \log(\bar{\omega}_p \Delta t)$. Disturbances decay away on a time scale $T_d = 1/\gamma = \Delta t/\log(\bar{\omega}_p \Delta t)$. Alternatively, in the centered limit $\Theta = \frac{1}{2}$, $|\xi| = 1$, so that for arbitrary c (i.e., all Δt) there is neither damping nor growth. Thus, there is no Courant limit ($a_0 \Delta t < \Delta x$) with centered implicit pressure. Furthermore, for $c \gg 1$, $\xi = -1$; so $e^{i\omega \Delta t} = \cos \omega_0 \Delta t = -1$ and $\omega_0 \Delta t = 2\pi \Delta t/T = \pi$ yield oscillations at period $T = 2 \Delta t$. Also, for $c \ll 1$ and any Θ, $\xi \to 1 \pm ic$; so $e^{i\omega \Delta t} \to 1 + i\omega \Delta t$, yielding $\omega \to \pm(\omega_p^2 + a_0^2 k^2)^{1/2}$. Thus, plasma oscillations are returned for small time steps. Finally, we note that a small degree of damping can be added by taking small deviations $1 \gg \delta > 0$ from the centered condition; i.e., $\Theta = \frac{1}{2} + \delta$. In the $c \gg 1$ limit, the amplification factor is $|\xi| = 1 - 4\delta$; so $\gamma = 4\delta/\Delta t$, whereas for $c \ll 1$, γ/ω_p becomes $\delta(\omega_p \Delta t)$. Correspondingly, for the $\frac{3}{4}$, $\frac{1}{4}$ centering scheme in Denavit (1981), the $c \gg 1$ damping is $\gamma = (\Delta t)^{-1} \log \sqrt{3}$, whereas the low frequency γ/ω is $O(\omega \Delta t)^3$.

Since economy generally dictates the use of the largest possible time steps, of particular interest is the solution to the general dispersion relation, Eq. (11), for $\omega_p \Delta t \gg 1$; i.e.,

$$\xi = -[(1 - \theta) + A(1 - \eta)]/(\Theta + A\eta). \quad (13)$$

Here $A \equiv (a_0 k/\omega_p)^2$. Again, for $\eta = \theta$ this returns the earlier result that stability ensues for $\Theta \geq \frac{1}{2}$. Note that the velocity centering parameter ψ has no influence in this limit. In general, with a fully implicit pressure, $|\xi| = (1 - \Theta)/(\Theta + A)$, so that for large $ka_0 \Delta t$ the required implicitness in the E field is reduced. More important, with a fully explicit pressure ($\eta = 0$), the amplification factor becomes $|\xi| = [(1 - \Theta) + A]/\Theta$. Thus, for stability and $|\xi| \leq 1$, $\Theta \geq (1 + A)/2$ is required. For a minimum-disturbance wavelength $\lambda = 2 \Delta x$, the maximum k value is $\pi/\Delta x$. This means that $ka_0 \Delta t = \pi a_0 \Delta t/\Delta x = \pi C_N$, where $C_N = a_0 \Delta t/\Delta x$ is the Courant number. Consequently, with explicit Π and (m) storage, the stability requirement becomes $\Theta > \{1 + [(\pi C_N)/(\omega_p \Delta t)]^2\}/2$. Thus, for stability in a typical simulation run at $\omega_p \Delta t = 4.0$ and $C_N = 0.5$, the required Θ is $\Theta = 0.58$, instead of the $\Theta = 0.5$ that is sufficient with a centered implicit Π.

In the opposing $\omega_p \Delta t \ll 1$ limit, Eq. (11) describes the purely hydrodynamic, linearized behavior of the Eq. (2) system in nearly field-free regions of the plasma. With identically implicit velocity and pressure, as above, the system is stable for all $ka_0 \Delta t$ when $\psi \equiv \eta \geq \frac{1}{2}$. However, with explicit pressure ($\eta = 0$), instability ensues for *all* $ka_0 \Delta t$ if the velocity is centered ($\psi = \frac{1}{2}$), whereas with a fully forward velocity ($\psi = 1$), $ka_0 \Delta t < 2$ is required for stable calculations. For $k_{max} = \pi/\Delta x$ this sets the maximum stable time step at $0.7 \Delta x/a_0$.

The preceding results are most important, since the complexity of hybrid formulations renders the developmental use of explicit pressures particularly convenient. The limitations of this usage should be well characterized.

d. MINIMUM Δt. Thus far, we have examined the maximum time step accessible with implicit methods. In practice, there may also be a minimum Δt. In explicit calculations, starting with $\Delta x/\lambda_D > 1$, it is generally found that particle aliasing errors to the spatial grid result in a heating of the electrons until their Debye length expands to $\lambda_D \equiv (kT_e/m_e)^{1/2}/\omega_p = 1$ (Langdon, 1970; Okuda, 1972). With the implicit modeling, it is generally possible to increase Θ enough beyond 0.5, so that net cooling of the plasma results (see Mason, 1981a). However, even with a fixed Θ, cooling can be induced by simply increasing Δt. Brackbill and Forslund (1982) have given several numerical examples that quantify these observations. They conclude that $\Delta t > 0.1 \Delta x/v_e$, $v_e = (kT_e/m_e)^{1/2}$, should be obeyed for finite-grid stabilization. Yet, in practical situations, say with 100-keV hot electrons impinging on a 50-eV background, it may be impossible to satisfy this minimum Δt rule for the cold species. For such cases, hybrid modeling with fluid cold electrons is recommended (see Mason, 1983a), as detailed in Section III.

2. Spatial Storage and Differencing

The spatial storage used in our modeling has been selected as a compromise between the sometimes conflicting requirements of particle and fluid simulation. Densities are accumulated, and temperatures and static pressures stored at cell centers j. Currents are accumulated, and velocities and E fields stored at cell boundaries $j + \frac{1}{2}$. An average density $n_{j+1/2} = (n_j + n_{j+1})/2$ is associated at the boundaries with $j \equiv n_{j+1/2}U$ in the moment modeling and with $nE \equiv n_{j+1/2}E$. Dynamic fluid pressures are computed from boundary velocity averages, e.g., $mnU^2 \equiv 0.25mn_j(U_{j-1/2} + U_{j+1/2})^2$. Total particle pressures Π are accumulated at centers. This mixed storage ensures a finite field when, for example, $E = -(en_e)^{-1} \partial P_e/\partial x$, $P = nkT$. If P is nonzero in only cell j or $j + 1$, the boundary n is also nonzero. If, instead, all the variables were stored at centers, $\partial P/\partial x = (P_{j+1} - P_{j-1})/\Delta x$ divided by n_j could be infinite. Similarly, finite new velocities, $U' = U - (mn^{(m)})^{-1} (\partial \Pi/\partial x) \Delta t$, in hydrodynamic flow are guaranteed with the mixed storage.

However, it is well known that mixed storage in particle simulations leads to an erroneous self-force on the particles. This force can be eliminated by temporarily averaging the E field to centers and accelerating the particles in this average field, extrapolated with area weighting to the particle positions (Boris and Lee, 1973).

In the Section III collisional–hybrid extension of the moment method the collision rates ν are stored at cell centers. Boundary average rates are computed harmonically $\nu_{j+1/2} = (1/\nu_j + 1/\nu_{j+1})^{-1}$ to ensure free expansion from collisional regions into a vacuum.

3. Implicit Pressure

a. EVALUATION. A forward static pressure $P^{(m+1)}$ is needed to construct the various centered schemes we have discussed. In practice, $P^{(m+1)}$ has been acquired by iterating the Eq. (2) fluid system (Mason, 1981a; Brackbill and Forslund, 1981) or the related particle equations with the field (Denavit, 1981). In each case, convergence to the forward pressure is restricted to a Courant-like limit. To establish this limit with the fluid equations, we consider again the fully forward differenced Eq. (2) system for stationary ions and a single isothermal electron component. The new electron density after a time step is

$$n^{(m+1)} = n^{(m)} - \frac{\partial}{\partial x}\left[j^{(m)} \Delta t - \frac{kT_e}{m_e}\frac{\partial n^{(*)}}{\partial x} \Delta t^2 - \frac{n_0 eE}{m_e} \Delta t^2\right]. \quad (14a)$$

For an e^{ikx} mode $E_k = -(4\pi en_k)/ik$; so with $a_0^2 = kT_e/m_e$, Eq. (14a) becomes

$$n_k^{(m+1)} = (n_k^{(m)} - ikj_k \Delta t - k^2 a_0^2 \Delta t^2\, n_k^{(*)})/(1 + \omega_p^2 \Delta t^2). \quad (14b)$$

If we choose $^{(s)}n_k^{(m+1)}$ as our first guess for $n_k^{(*)}$, then the differential of Eq. (14b) shows that at each iteration step (s) the errors $^{(s)}\delta n_k^{(m+1)}$ are related by

$$^{(s+1)}\delta n_k^{(m+1)} = -[k^2 a_0^2 \Delta t^2/(1 + \omega_p^2 \Delta t^2)]^{(s)}\delta n_k^{(m+1)}. \quad (14c)$$

The magnitude of the errors diminishes, and the iteration converges to provide a new density and pressure, if $(ka_0 \Delta t)^2/(1 + \omega_p^2 \Delta t^2) < 1$. Thus, convergence to the new pressure is accelerated for large $\omega_p \Delta t$. Alternatively, for large Δt convergence will follow if $k < \omega_p/a_0$ or $\Delta x > \lambda_D$.

Our analysis indicates that convergence to a forward pressure will be rapid in quasi-neutral regions, where the E field forces the electrons to follow the ions. In more tenuous regions, convergence requires adherence to the stricter $ka_0 \Delta t < 1$ Courant condition. Also, in two-electron component systems, where hot electrons can stream through a colder background, the stricter Courant limit must be anticipated to stabilize the counterstreaming fronts.

For more general situations, such as those presented by two-electron fluid counterstreaming, the Eq. (2) system can be solved for any Δt by inversion of a pentadiagonal matrix system. The substitution of Eq. (6) into (5) and the use of $E = -\partial \varphi / \partial x$ produces, for example, a third-order equation in φ; the full Eq. (2) system is fourth order in this variable. Moreover, for the most accurate pressures one might choose to employ a set of energy equations with Eq. (2), so that a $T^{(m+1)}$ could be determined for use in an improved $P^{(*)} = n^{(m+1)} k T^{(m+1)}$. In turn, this energy equation could require closure through a heat–flux equation, for which improvement could be further complicated through a need for flux limiters (see Mason, 1981b). In fact, these issues have been avoided to date, and only the use of iterated pressures has been explored, since successful particle advancement algorithms are still limited to a Courant time step for the hottest electrons.

b. SUBCYCLING. We see by now that the time-step constraint in the moment method stems chiefly from the Courant limit for accuracy in the particle advancement process. Earlier, it was stressed that implicit pressures and fields render the auxiliary fluid–field solution stable for any chosen Δt. When the pressure is made implicit in conventional hydrodynamic schemes (e.g., see Harlow and Amsden, 1971), the time step can be freely picked to resolve significant changes for the variables of interest. A Courant time step would be used to track the passage of the leading edge of an expanding cloud of hot electrons, for example, but subsequent adiabatic adjustments within the electron cloud might be managed with a much larger Δt. A corresponding freedom of Δt choice in PIC simulation may be possible through particle subcycling.

Subcycling was employed in Mason (1980a) to maneuver electrons through intense field regions. Cohen *et al.* (1982) have explored it in the context of orbit averaging, and Adam *et al.* (1982) have proposed its use in conjunction with implicit methods. Cohen makes a detailed review of both orbit averaging and subcycling elsewhere in this volume. Here, we wish to suggest that a combination of subcycling and the moment-method fields computed with implicit pressures may allow for the desired significant increase in Δt (see also Cohen *et al.*, 1982). At the very largest Δt, such pressure-implicit fields would go over to the hybrid-model fields, as exemplified by Eq. (5), which have long been used for modeling on the ion time scales. With particle subcycling, the field in each traversed cell must be sampled so as to guarantee a representation of trapping effects. Thus, computational efficiency would have to come from a reduced frequency for the field calculation and moment accumulations [done $(m_i/m_e)^{1/2}$ times less often] and, possibly, from vectorized coding for particle advancement along fixed orbits during a cycle. The success of related orbit treatments in the hybrid code of Vomvoridis and Denavit (1979) lends support to our speculation that subcycling will prove useful in fully implicit hybrid-like fields.

4. Particle–Field Iterations

a. THE IMPROVED E FIELD. It is possible that the moments accumulated following the particle advancement will differ from those predicted with the implicit solution for the E field. In particular, the accumulated density may differ from its predicted value, implying an error in the E field used to move the particles. Generally, we have found such an error to be of negligible physical significance. However, if desired, it can be eliminated by iteration of the particle and fluid–field equations.

We start with $\partial\,^{(s)}E/\partial x \neq 4\pi \sum q_\alpha\,^{(s)}n_\alpha$ and seek a modified field and density for the next iterate $(s+1)$ such that

$$\frac{\partial\,^{(s+1)}E}{\partial x} = 4\pi \sum_\alpha q_\alpha\,^{(s+1)}n_\alpha. \tag{15}$$

To find this improvement, in the spirit of the convergence analysis of Langdon et al. (1983), we define $^{(s+1)}\delta E \equiv {}^{(s+1)}E - {}^{(s)}E$ and $^{(s+1)}\delta n \equiv {}^{(s+1)}n - {}^{(s)}n$, and we derive

$$^{(s+1)}\delta n_\alpha = -(\partial/\partial x)\left[(q_\alpha/m_\alpha)n_\alpha^{(*)}\,\Delta t^2\,{}^{(s+1)}\delta E^{(*)}\right] \tag{16}$$

from the differentials of Eq. (2) (with explicit $\Pi^{(*)}$). Here, Eq. (16) is viewed as approximate, since the fluid-equation predictions may differ from the actual mean-particle motion. These expressions are then used to convert Eq. (15) to

$$\frac{\partial[(1 + \omega_p'^2\,\Delta t^2)\,{}^{(s+1)}\delta E]}{\partial x} \simeq 4\pi \sum_\alpha q_\alpha\,{}^{(s)}n_\alpha - \frac{\partial\,^{(s)}E}{\partial x} \equiv {}^{(s)}R. \tag{17}$$

For a mirror-symmetric boundary at $x = 0$, this integrates to

$$^{(s+1)}\delta E \simeq \int_0^x {}^{(s)}R\,dx/(1 + \omega_p'^2\,\Delta t^2), \tag{18a}$$

yielding the improved field

$$^{(s+1)}E = {}^{(s)}E + {}^{(s+1)}\delta E. \tag{18b}$$

The iteration cycle is completed by moving the particles in the latest $^{(s+1)}E^{(*)}$ E field from their (m)-level coordinates to the new coordinates $^{(s+1)}u^{(m+1)}$ and $^{(s+1)}x^{(m+1)}$.

b. CONVERGENCE. Convergence properties for this procedure can be assessed through an examination of the ratio of successive integrated residuals, $\int_0^x {}^{(s)}R\, dx$. Since

$$^{(s+1)}R = {}^{(s)}R + 4\pi \sum_\alpha q_\alpha {}^{(s+1)}\delta n_\alpha - \frac{\partial^{(s+1)}\delta E}{\partial x}, \quad (19)$$

the Eq. (17) $^{(s)}R$ definition can be used with Eq. (19) to give

$$\frac{\int_0^x {}^{(s+1)}R\, dx}{\int_0^x {}^{(s)}R\, dx} = \frac{(\omega_p'^2 \Delta t^2)\,{}^{(s+1)}\delta E + \int_0^x (4\pi \sum_\alpha q_\alpha {}^{(s+1)}\delta n_\alpha)\, dx}{(1 + \omega_p'^2 \Delta t^2)\,{}^{(s+1)}\delta E}. \quad (20)$$

The ω_p' terms in Eq. (20) derive from the moment equations. For hybrid extensions we have found it convenient to store the moment method densities at cell centers, and the fields at cell boundaries. Consequently, we difference these terms as

$$\omega_p'^2 \Delta t^2\, \delta E = (\omega_p'^2 \Delta t^2)_{j+1/2}\, \delta E_{j+1/2} = 4\pi \sum_\alpha \left(\frac{q_\alpha}{m_\alpha}\right) \frac{(n_{\alpha j} + n_{\alpha j+1})}{2} \delta E_{j+1/2}. \quad (21)$$

The $\int \delta n_\alpha\, dx$ term in Eq. (20) can be determined with the aid of the "direct method" formalism of Friedman et al. (1981). For this, the density is represented by a sum of "shape functions," one for each particle k; i.e.,

$$n_\alpha = W \sum_k S(x_k - x_j), \quad (22)$$

in which $W \equiv N_\alpha/K$, and N_α is the mean electron or ion number density in a cell with K, the mean number of superparticles per cell used in simulation.

In the absence of an E field, a particle in Δt will advance to the free-streaming position $\hat{x}_k = x_k^{(m)} + u_k^{(m)} \Delta t$. Expanding the shape function about $\hat{x}_k$, one obtains

$$S(x_k^{(m+1)} - x_j) \simeq S(\hat{x}_k - x_j) + (x_k^{(m+1)} - \hat{x}_k)(\partial/\partial x)S(\hat{x}_k - x_j). \quad (23)$$

The presence of an E field will cause a deviation in the fully implicit position

$$x_k^{(m+1)} - \hat{x}_k \simeq \frac{q_\alpha}{m_\alpha} \Delta t^2\, W \sum_i S(\hat{x}_k - x_{i+1/2})E_{i+1/2}^{(m+1)}, \quad (24)$$

with an error on the order of $(\omega_T \Delta t)^2$, due to the evaluation of E at $\hat{x}$ instead of $x^{(m+1)}$. Equations (23) and (24) can now be combined to give the advanced density

$$n_\alpha^{(m+1)} = \hat{n}_\alpha + \delta n_\alpha, \quad (25a)$$

in which

$$\delta n_\alpha = \frac{q_\alpha \Delta t^2}{m_\alpha} W \sum_i \sum_k S(\hat{x}_k - x_{i+1/2})E_{i+1/2} \frac{\partial}{\partial \hat{x}_k} S(\hat{x}_k - x_j). \quad (25b)$$

Finally, integrating Eq. (25b) to $x_{j+1/2}$ and interpreting E as a perturbation δE, we obtain, for the heuristic case of one particle of each type α at each cell center (such that $\bar{n} = (n_j + n_{j+1})/2$), the result

$$\int_0^x (4\pi \sum_\alpha q_\alpha \, \delta n_\alpha) \, dx = -\frac{(\omega_p'^2 \, \Delta t^2)_{j+1/2}}{4} (\delta E_{j-1/2} + 2 \, \delta E_{j+1/2} + \delta E_{j+3/2}). \tag{26}$$

Studying the Eq. (20) residual ratio, one can now see that for long-wavelength disturbances, such that $\delta E_{j-1/2} \simeq \delta E_{j+1/2} \simeq \delta E_{j+3/2}$, the Eq. (21) and (26) terms will nearly cancel, so the convergence of the Eq. (18a) iteration can be extremely rapid. However, for shorter wavelength disturbances, say $\delta E_{j-1/2} = \delta E_{j+3/2} = 0$ with a single $\delta E_{j+1/2}$ spike, the successive δE will only be reduced by a factor of 2 per cycle. For more realistic particle spatial distributions the analysis is more difficult, but experience has shown that in the electron two-stream instability problem of Mason (1981a), for example, convergence to $\delta E/E = 10^{-6}$ is achieved in about 10 iterations. In practice, physically acceptable results have been regularly achieved with the particle–field iterations ignored.

C. Relation to the Direct Method

In the direct implicit method (Friedman *et al.*, 1981; Langdon *et al.*, 1983), Eq. (25a) is combined with Eq. (1a) to produce

$$-\frac{\delta^2 \varphi}{\partial x^2} = 4\pi \sum_\alpha q_\alpha \hat{n}_\alpha + 4\pi \sum_\alpha q_\alpha \, \delta n_\alpha. \tag{27}$$

Use of the Eq. (25b) *strict* spatial differencing for δn_α leads to a pentadiagonal system for φ. Because of this spread of matrix elements with strict differencing, there has been considerable motivation to employ in the direct method "ad hoc" or "simplified" differencing schemes, which reduce the bandwidth of the implicit field equations. This is discussed in Langdon *et al.* (1983). Typically (Friedman *et al.*, 1981), the simpler Eq. (16) choice has been employed, with the changes $\delta E \to E$ and $n^{(*)} \to \hat{n}$, yielding (in the fully implicit limit)

$$-\frac{\partial}{\partial x}(1 + \chi)\frac{\partial \varphi^{(m+1)}}{\partial x} = 4\pi \sum_\alpha q_\alpha \hat{n}_\alpha \tag{28a}$$

with the susceptibility

$$\chi \equiv 4\pi \sum_\alpha \frac{q_\alpha^2}{m_\alpha} \hat{n}_\alpha \, \Delta t^2 \equiv \hat{\omega}_p^2 \, \Delta t^2. \tag{28b}$$

The direct method cycle is concluded by accelerating particles in the new field, $E = -\partial \varphi / \partial x$, and completing their positional advance with Eq. (24).

The use of the field-free extrapolated $\hat{n}_\alpha$ frees the direct field solution from a Courant limit for stability, while this limit persists for accuracy—otherwise, for large Δt, particles might stream well beyond the fields that should trap them. Improved relative efficiency is achieved by avoiding j_α and Π_α accumulations over the particles. The analytic substructure illuminated by the direct approach has proved crucial to the development of consistent spatial smoothing procedures (Langdon et al., 1983) and to the use of particles spanning more than one cell (Barnes et al., 1983).

The direct method Poisson equation (28) converts to the moment method Eq. (3) if the susceptibility χ is calculated with $n^{(*)}$ and the extrapolated particle position density $\hat{n}$ is replaced by the fluid extrapolated density, i.e., if

$$\hat{n}_\alpha \to n_\alpha^{(m)} - \frac{\partial}{\partial x}\left(j_\alpha^{(m)} - \frac{1}{m_\alpha}\frac{\partial \Pi_\alpha^{(*)}}{\partial x}\Delta t \right)\Delta t. \tag{29}$$

In the resultant Eq. (4) E-field expression, the use of an implicit pressure $\Pi^{(*)}$ and subcycling allows, at least in theory, for accuracy in simulations far exceeding the Courant limit. Further, as we shall see, there is a natural relation and compatibility between the moment formalism and hybrid simulation, in which selected plasma components are modeled as fluids.

III. COLLISIONAL–HYBRID EXTENSIONS

The long-range "hot" electrons in laser–matter simulation carry energy away from the laser spot. In the corona they are essentially collisionless. They spread laterally along the surface, directly penetrate the interior, or reflect off the sheath field before penetrating. Inside the target they scatter principally off the ions and are dragged to lower speed by the background electrons, eventually absorbing among them. A collisional PIC treatment is ideal for these hot electrons.

The use of a fluid treatment for the background plasma in laser–matter interaction studies is attractive for several reasons. The electron densities to be modeled range from above 10^{24} cm^{-3} in fully ionized gold, through the critical density at 4×10^{22} cm^{-3} for 0.25-μm KrF laser light (or 10^{19} cm^{-3} for CO_2), and down into the corona to at least 1% of the critical density. If 100 particles per cell are used to represent the background electrons, the minimum describable density in fully ionized gold is 10^{22} cm^{-3}, unless the particles are weighted—with smaller weights initially dispatched to lower density regions. But even with weights, a heavily weighted particle must be divided into many smaller particles upon entering a tenuous region and, optimally, recombined upon returning to a dense region—presenting a most challenging bookkeeping problem. The need to treat this range of densities is particularly evident in studies aimed at modeling the cold-electron return current that runs from the dense interior of a laser target out to the critical density for conversion by the laser to suprathermals (Mason, 1983a).

Strong collisionality presents a second motivation for fluid modeling. With strong collisions the cold-electron distribution is rapidly relaxed to a Maxwellian. Electron transport evolves principally as electron diffusion with density gradients and Ohm's law drift in any E field. An implicit fluid treatment can provide a natural representation of this phenomenology, with computational economy, by allowing a time step well in excess of the cold electron–ion collision frequency, $v_c \Delta t \gg 1$. With a pure particle representation for the background, the time step would be limited to $v_c \Delta t < 1$ by currently available particle collision models (Shanny et al., 1967).

Finally, we note that the development of implicit multifluid hybrid schemes (hot- and cold-electron fluids and *no* particle electrons) is of special interest, since for many problems the effective division of the electron distribution into two Maxwellians may provide sufficient accuracy in modeling and pronounced economy, particularly with the use of implicit pressures, as suggested in Section II.B.2.

A. Governing Equations

1. *Particles*

a. EMISSION. In laser problems a calculational cycle begins with a search of the mesh for the location of the first cell at critical density to be encountered by the laser light. Particles are then emitted from this cell (or its boundary) to represent suprathermal production by resonance absorption of the light. The particles can be emitted in a variety of distributions, e.g., Maxwellian, beamlike, and purely transverse. For resonance absorption we emit them in a drifting Maxwellian confined to a 20° cone about the incident laser direction. The electrons are given velocity component u_h along the laser direction, and a transverse component v_h perpendicular to that direction. Typically, from 2 to 10 particles are emitted each cycle. Their emission temperature is now set from experimental experience, e.g., $T_h \geq 150$ keV for CO_2 illumination, and the emitted density weight associated with each particle is set to carry away the requisite absorbed energy. One might also choose to convert from fluids to particles when the local collisional frequency falls below some prescribed limit. At least two particles must be produced per cycle to conserve momentum. With particle electron emission the background density must be correspondingly reduced to maintain charge neutrality.

b. DRAG. Next, we decrease the speed of each suprathermal electron to model its drag against all the other electrons. The hot-speeds are reduced each cycle by

$$\Delta c = -(4\pi e^4 n_i/m_e k T_e) \Delta t \; G(\xi) \log \lambda, \qquad \xi \equiv (m_e c^2/2kT_e)^{1/2}, \qquad (30a)$$

in which

$$G(\xi) \equiv 0.376\xi/(1 + 0.542\xi - 0.504\xi^2 + 0.752\xi^2) \quad (30b)$$

is a polynomial fit (R. Jones, personal communication, 1982) to the Spitzer (1961) error function combination for particle drag, and $c^2 \equiv u^2 + v^2$.

Thus, for a particle initially moving at an angle ψ relative to the x axis such that

$$\cos \psi = u/c, \quad (31)$$

the new velocities become

$$u' = (c + \Delta c) \cos \psi \quad (32a)$$

and

$$v' = (c + \Delta c) \sin \psi. \quad (32b)$$

The temperature T_e in Eq. (30a) is the temperature of all the electrons when no background fluid is employed (Mason, 1981b) or the thermal electron temperature when a fluid background is present. The drag in Eq. (30a) goes as c^{-2} for speeds well beyond this thermal speed, as in Mason (1980a); it peaks at $\xi = 1$ and goes to zero when $c \to 0$. Lost momentum and energy are accumulated on the mesh during the drag calculation. For this we use the same area weighting prescription as for our cell-centered density and pressure accumulations. Particles with speeds below some specified minimum (usually, the mean thermal speed of the background fluid) are destroyed. Accumulated density, momentum, and energy from the drag deceleration and particle destruction are then added to the corresponding background properties as local sources. The drag is suppressed when $n_h \geq n_c$ to avoid nonphysical heating of low-density thermals. In sufficiently collisional problems there can exist a quasi-equilibrium situation in which the various creation mechanisms for hot particles are balanced by the loss through drag—so that the number of simulation particles needed to represent the suprathermals in flight is relatively constant.

Following particle emission and drag, the intermediate-time properties n'_α, j'_α, and Π'_α are accumulated on the mesh for subsequent use in the field calculation.

c. SCATTER. We assume that electrons crossing a finite thickness of plasma undergo many small angle deflections. Since the successive collisions are independent events, the central limit theorem indicates that, for Rutherford scattering, the distribution of net deflection angles θ from the forward direction of each particle will be approximately Gaussian (Jackson, 1962;

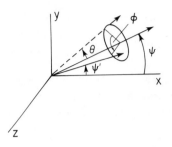

Fig. 1. Relation between the angles involved in a scattering event, as $\psi \to \psi'$ and $\bar{c} \to \bar{c}'$.

Shanny et al., 1967); i.e.,

$$P(\theta) = (\theta/\langle\theta^2\rangle) \exp(-\theta^2/2\langle\theta^2\rangle) \tag{33}$$

with $\langle\theta^2\rangle \simeq 8\pi e^4 m^{-2} c^{-3} \Delta t\, n_i Z(Z+1) \log \lambda$. The $Z+1$ factor replaces the proper ionic Z factor to approximate the generally weaker scatter-off electrons. Here, $c\,\Delta t$ is the straight-line distance crossed between each cumulative scatter.

During each time step new θ values are picked at random from the Eq. (33) Gaussian distribution ($\theta = 0 \to \infty$), and a new azimuthal angle φ, measuring rotation from the plane of θ deflection, is chosen at random from a uniform distribution ($\varphi = 0 \to 2\pi$); see Fig. 1. Thus, electrons still moving at the Eq. (31) angle ψ acquire a new angle ψ', satisfying

$$\cos \psi' = \sin \psi \sin \theta \cos \varphi + \cos \psi \cos \theta \tag{34}$$

and the new velocity values

$$u''_h = c \cos \psi', \tag{35a}$$

$$v''_h = c \sin \psi'. \tag{35b}$$

When the background density is sufficiently low, or when an electron is sufficiently energetic, then $\langle\theta^2\rangle \ll 1$ and the electron is nearly undeflected during the time step. Under these conditions the drag is also minimal, so that the net motion is essentially that of a fully deterministic PIC. On the average this scattering procedure gives a mean effective collision frequency for the hot electrons of $\nu_h = \langle\theta^2\rangle/(2\,\Delta t)$, which should be very nearly equal to the Braginskii value used with the hot fluid in Eq. (36), which follows.

d. *E* ACCELERATION AND TRANSLATION. Following the scattering operations the particle coordinates are simply updated in accordance with Eqs. (1b, c).

However, the foregoing scattering procedure is valid only so long as $\nu_h\,\Delta t \leq 1$. For longer time steps the particle motion should become a random walk, with x-directed displacements significantly reduced below $u_k^{(*)}\,\Delta t$. Phenomenologically, one might expect, for example, that in the presence of a driving

E field the particle translation would go over to (Mason, 1981c)

$$x_k^{(m+1)} = x_k^{(m)} + \frac{u_k''^{(*)} \Delta t}{[1 + v_k(c) \Delta t]^{1/2}} - \frac{eE^{(*)} \Delta t^2}{m_e[1 + v_k(c) \Delta t]}. \tag{36}$$

For $v(c) \Delta t \gg 1$, this establishes the reduced displacements, $\propto c(\Delta t/v)^{1/2}$, expected with Brownian motion, leading on the average to particle diffusion, plus an additional displacement from the drift velocity $eE/(mv)$ established with an Ohm's law.

A pedestrian procedure for avoidance of the $v \Delta t \leq 1$ limit is simply to subcycle the scattering, E field acceleration, and translation of particles, using for each particle a subcycle Δt_s such that $v_k \Delta t_s \leq 1$. More inventively, C. Cranfill, S. Goldman, J. Brackbill, and R. Jones, (private communication, 1983) have made considerable progress toward rigorously establishing an expression like Eq. (36) for a modified single-time-step translation, but with variations guaranteeing accuracy in strong and weak collisional limits. See also Chapter 9 in this volume by Brackbill and Forslund. For the present, we use the simpler sequential drag, scatter, and Eq. (1) updates, absorbing the strongly collisional electrons in a fluid background.

2. Fluid Equations

In hybrid modeling the particle equations for any or all of the components α are replaced by a set of corresponding modified Braginskii (1965) equations:

$$n_\alpha^{(m+1)} = n_\alpha'^{(m)} - \frac{\partial j_\alpha'^{(*)}}{\partial x} \Delta t, \tag{37a}$$

$$j_\alpha^{(m+1)} = j_\alpha'^{(m)} - \frac{1}{m_\alpha}\left(\frac{\partial \Pi_\alpha'^{(*)}}{\partial x} - q_\alpha n_\alpha^{(\perp)} E^{(*)}\right) \Delta t,$$

$$- v_\alpha \Delta t \left(j_\alpha^{(m+1)} - \frac{n_\alpha^{(\perp)} j_i^{(m+1)}}{n_i^{(\perp)}}\right), \quad \alpha = h, c, \tag{37b}$$

$$j_i^{(m+1)} = j_i'^{(m)} - \frac{1}{m_i}\left(\frac{\partial \Pi_i'^{(*)}}{\partial x} - Zen_i^{(\perp)} E^{(*)}\right) \Delta t$$

$$+ \sum_{\alpha=h,c} \frac{m_e}{m_i} v_\alpha \Delta t \left(j_\alpha^{(m+1)} - \frac{n_\alpha^{(\perp)} j_i^{(m+1)}}{n_i^{(\perp)}}\right), \tag{37c}$$

$$I_\alpha^{(m+1)} = I_\alpha'^{(m)} - \left[\frac{\partial(U_\alpha^{(*)} I_\alpha^{(*)})}{\partial x} + \Pi_\alpha'^{(*)} \frac{\partial U_\alpha^{(*)}}{\partial x}\right] \Delta t - \frac{\partial Q_\alpha}{\partial x} \Delta t$$

$$+ v_\alpha \Delta t [m_\alpha n_\alpha (U_\alpha^{(*)} - U_i^{(*)})^2], \quad \alpha \equiv h, c, \tag{37d}$$

$$I_i^{(m+1)} = I_i'^{(m)} - \left[\frac{\partial(U_i^{(*)} I_i^{(*)})}{\partial x} + \Pi_i'^{(*)} \frac{\partial U_i^{(*)}}{\partial x}\right] \Delta t, \tag{37e}$$

in which, as earlier, $j_\alpha \equiv n_\alpha U_\alpha$, but now $\Pi'_\alpha \equiv P'_\alpha + n_\alpha m_\alpha U_\alpha^2$, and $P'_\alpha = n_\alpha k T_\alpha + P_{a\alpha}$, with $P_{a\alpha}$ an artificial viscous pressure. Also, $I_\alpha \equiv \frac{3}{2} n_\alpha m_\alpha k T_\alpha$ is the internal energy. The levels (∗) are chosen according to the Section II.B.1 guidelines. The levels (1) are again set to (m) for convenience. Spatial storage obeys the Section II.B.2 rules. For simplicity, thermoelectric effects that could arise from particle-velocity independent collision rates are suppressed. The missing thermoelectric terms and factors can be readily restored, as accomplished in Mason (1979a, 1980a).

The square-bracketed terms in Eqs. (37d, e) represent the convective transport of internal energy, $\partial(nU3kT/2)/\partial x$, in the various components and hydrodynamic (i.e., PdV) work. The electron–ion scattering rates, $v_\alpha = 1/\tau_\alpha$, are taken from Braginskii (1965); i.e.,

$$v_\alpha = \tfrac{4}{3}(2\pi/m_e)^{1/2}(e^4 Z^2 n_i \log \lambda / T'^{3/2}_\alpha), \tag{38}$$

except that we introduce $T'_\alpha \equiv T_\alpha + m_\alpha U_\alpha^2/3k$, including the U_α^2 term, to approximate the effects of large drift velocities. As for the particle Eq. (31), the replacement $Z^2 \to Z(Z + 1)$ can be used to mock up the effects of electron–electron scatter. The scattering terms in the momentum equations (37b, c) are constructed so as to force $U_\alpha \to U_i$ for $v_\alpha \Delta t \to \infty$. Thermal conduction in the electron fluids is modeled with the classically flux-limited heat flux,

$$Q_\alpha = -K_\alpha/(1 + F_d/F_f)(\partial T_\alpha/\partial x), \qquad \alpha = h, c, \tag{39}$$

in which $K_\alpha \equiv \tfrac{5}{2}(n_\alpha k T_\alpha)/(m_e v_\alpha)$, $F_d \equiv -K_\alpha \partial T_\alpha/\partial x$, $F_f \equiv f n_\alpha v_\alpha T_\alpha$, $f = 0.6$, and $v_\alpha \equiv (kT_\alpha/m_e)^{1/2}$ (Mason, 1981b). Ion thermal conduction is generally much weaker, and is, therefore, neglected here. Collisions between the electrons and ions give rise to the Joule heating terms in Eq. (37d), $v_\alpha \Delta t\, m_\alpha n_\alpha (U_\alpha - U_i)^2$.

The primed properties in Eqs. (37) are defined by

$$n'^{(m)} \equiv n^{(m)} + \dot{n}_\alpha \Delta t, \tag{40a}$$

$$j'^{(m)}_\alpha \equiv j^{(m)}_\alpha + \dot{j}_\alpha \Delta t, \qquad \alpha \equiv h, c, \tag{40b}$$

$$j'^{(m)}_i \equiv j^{(m)}_i, \tag{40c}$$

$$I'^{(m)}_h \equiv I^{(m)}_h + \dot{I}_h \Delta t, \tag{40d}$$

$$I'^{(m)}_c \equiv I^{(m)}_c + \dot{I}_c \Delta t - \Delta H_c, \tag{40e}$$

$$I'^{(m)}_i \equiv I^{(m)}_i + \dot{I}_i \Delta t + \Delta H_c. \tag{40f}$$

These include the density, momentum, and temperature changes (dotted terms) from hot-electron emission and drag, with corresponding changes in the cold-electron properties. When the hot electrons are treated as fluid, the mean drag rate is phenomenologically taken as $v_{dh} \equiv v_h/Z$, referencing Eq. (38). Finally, the electron–ion collisions couple the cold-electron and ion

8. Implicit Plasma Simulation Models

temperatures through (Braginskii, 1965)

$$\Delta H_c = 3 v_c \, \Delta t \, n_c m_e k (T_c - T_i). \tag{41}$$

Since, generally, $v_h n_h \ll v_c n_c$, the direct thermal coupling of hot electrons to ions is neglected.

In summary, the primed properties in Eqs. (40) represent the state of the fluids following hot-electron emission, drag coupling to the thermals, and cold-electron thermal coupling to the ions. The Eqs. (37) then complete the update by including the effects of electron translation as a result of pressure gradients and E fields, as resisted by electron–ion scatter. These equations also include the temperature changes from Joule heating, due to the electron scatter through the ions, flux-limited thermal conduction, convective thermal transport, and shock heating (with artificial viscosity) or expansive cooling from the performance of hydrodynamic work.

3. Fields

In hybrid modeling we encounter the need for large time steps, $\omega_p \Delta t \gg 1$, especially in the dense interior of a laser target or plasma diode; so the requisite electrostatic field equation remains Eq. (1a) with $(\ddagger) = (m + 1)$.

B. Moment Method Solution

Here we detail the procedures followed for time integration of the governing hybrid equations by the implicit moment method.

1. Time-Step Control

First, the level-(m) time step is selected. The (m) storage described in Section II.B.1 facilitates Δt variations. For each component we compute the maximum signal speed

$$a_\alpha = [(\Pi_\alpha + 2P_{a\alpha})/m_\alpha n_\alpha]^{1/2} \tag{42a}$$

over the mesh. As in Eq. (37), Π_α includes the dynamic pressure $n_\alpha m_\alpha U_\alpha^2$, and $P_{a\alpha}$ is the artificial viscous pressure. We then set the time step to

$$\Delta t = 0.7 \min(\Delta x / a_\alpha) \tag{42b}$$

as the minimum value over the various components (usually associated with the hot electrons). To avoid rapid increases as the plasma evolves, we impose the additional constraint

$$\Delta t = \min(\Delta t^{(m)}, 1.2 \, \Delta t^{(m-1)}), \tag{42c}$$

and also fix minimal and maximal acceptable limits

$$\Delta t = \min[\max(\Delta t, \Delta t_{\min}), \Delta t_{\max}]. \tag{42d}$$

Note that with universally strong collisionality the electrons would be everywhere locked to the ions, so that a generally larger time step would be possible. The current Eq. (42) time-step controller does not take advantage of this possibility. Furthermore, with implicit pressures it should be possible to delete the static pressure P_α from Π_α in Eq. (42a), leaving a time step controlled by the fastest mean-fluid-flow rate, rather than by the fastest mean thermal speed (Harlow and Amsden, 1971). This additional improvement should be accessible with fluid modeling of all the components, and, possibly, through subcycling of the particles in hybrid modeling (review Section II.B.2.b).

2. Emission, Drag, and Thermal Coupling

Next in the time integration, we produce any emitted hot component and calculate its absorption due to drag. At present, the model has a "switch" allowing for either hot-particle or hot-fluid emission. The relative merits of these two modes are under comparison for various problems. The treatment of particle emission and drag has already been described.

With fluid electron emission at a fixed temperature (e.g., $T_{\rm eh} = 150$ keV for CO_2 at 10^{15} W cm^{-2}), the Eq. (40a) density added each cycle at critical is

$$\dot{n}_{\rm eh}\,\Delta t = \Delta n_{\rm eh} = \Delta\varepsilon/1.5 k T_{\rm eh}\,\Delta x, \tag{43}$$

in which "eh" denotes emitted hot electrons, and $\Delta\varepsilon$ is the absorbed laser energy per centimeter^2. Correspondingly, $\dot{n}_{\rm c} = -\dot{n}_{\rm h}$. If the electrons in the cell or at its boundaries are moving, their velocities must be adjusted to accommodate the added static density. Similarly, the local $I_{\rm h}$ and $T_{\rm h}$ must be given new values; i.e., $\dot{I}_{\rm eh}\,\Delta t = \Delta\varepsilon$ in Eq. (40d).

For the fluid drag modeling we decrease the hot-electron density each cycle by

$$\Delta n_{\rm dh} = n_{\rm h}^{\prime(m+1)} - n_{\rm h}^{(m)} = -\nu_{\rm dh}\,\Delta t\,n_{\rm h}^{\prime(m+1)}, \tag{44}$$

and, thus, $\Delta n_{\rm dc} = -\Delta n_{\rm dh}$, with "d" denoting changes due to drag. By this implicit differencing, $n_{\rm h}^{\prime(m+1)} \to 0$ for $\nu_{\rm dh}\,\Delta t \gg 1$.

With either particle or fluid hot electrons we complete this stage of the solution by coupling the cold-electron temperature to the ion temperature. From Eqs. (40e,f) and (41) and $I \equiv 3nmkT/2$, we solve the implicit system

$$\tfrac{3}{2} n_{\rm c} m_{\rm e} k\,\partial T_{\rm c}/\partial t = -3\nu_{\rm c} n_{\rm c} m_{\rm e} k (T_{\rm c}^{(m+1)} - T_{\rm i}^{(m+1)}), \tag{45a}$$

$$\tfrac{3}{2} n_{\rm i} m_{\rm i} k\,\partial T_{\rm i}/\partial t = 3\nu_{\rm c} n_{\rm i} m_{\rm e} k (T_{\rm c}^{(m+1)} - T_{\rm i}^{(m+1)}). \tag{45b}$$

The new temperatures are

$$T_c'^{(m+1)} = [(1 + S)T_c^{(m)} + SRT_i^{(m)}]/D \qquad (46a)$$

and

$$T_i'^{(m+1)} = [ST_c^{(m)} + (1 + SR)T_i^{(m)}]/D, \qquad (46b)$$

in which $R \equiv n_i/n_c$, $S \equiv 2(m_e/n_i)v_c \Delta t/R$, and $D \equiv [1 + S(1 + R)]$. When $n_h \to 0$, generally, $R \to Z^{-1}$. For $v_c \Delta t \gg 1$, $S \gg 1$, so $T_i'^{(m+1)} \to T_c'^{(m+1)}$, yielding the required cold-electron and ion thermal equilibration.

The new densities and temperatures n' and T' following the emission, drag, and thermal coupling are now used as input to velocity coupling and implicit E-field calculations. Henceforth, to simplify the notation, the primes associated with emission and drag should be understood and are deleted.

3. Collisional Velocity Coupling

Continuing the moment method solution, we next determine the effects of collisional coupling. Electron–ion scattering tends to lock the component velocities together. After some considerable manipulation the momentum equations (37b, c) rearrange to

$$j_\alpha^{(m+1)} = n_\alpha^{(m)}[A_\alpha - (e\, \Delta t/m_e)B_\alpha E^{(*)}], \qquad (47a)$$

with

$$A_\alpha \equiv C_\alpha(U_\alpha + v_\alpha \Delta t\, A_i), \qquad \alpha \equiv h, c, \qquad (47b)$$

$$A_i \equiv C_i(U_i' + D_h U_h' + D_c U_c'), \qquad (47c)$$

$$B_\alpha \equiv C_\alpha(1 + v_\alpha \Delta t\, B_i), \qquad \alpha \equiv h, c, \qquad (47d)$$

$$B_i \equiv C_i[D_h + D_c - Z(m_e/m_i)], \qquad (47e)$$

$$C_\alpha \equiv (1 + v_\alpha \Delta t)^{-1}, \qquad \alpha \equiv h, c, \qquad (47f)$$

$$C_i \equiv (1 + D_h + D_c)^{-1}, \qquad (47g)$$

$$D_\alpha \equiv (m_e/m_i)(n_\alpha/n_i)[v_\alpha \Delta t/(1 + v_\alpha \Delta t)], \qquad (47h)$$

$$U_\alpha' \equiv U_\alpha^{(m)} - (n_\alpha^{(m)} m_\alpha)^{-1} (\partial \Pi_\alpha'^{(*)}/\partial x)\, \Delta t. \qquad (47i)$$

The Eqs. (47) are valid for all values of $v_\alpha \Delta t$, as a result of the implicit treatment of j_i in the Eqs. (37b,c) collision terms. This rearrangement is due to D. Besnard (Private Communication, 1983). An earlier (Mason, 1983a) and semiexplicit formulation was algebraically simpler, but limited to moderate collisionality; i.e., $(v_\alpha \Delta t)(m_e/m_i) \ll 1$.

The Eq. (47a) coupling solutions go to expected limits. For example, with weak hot-electron collisions, $v_h \to 0$, we get $C_h \to 1$, so $A_h \to U'_h$ and $B_h \to 1$, yielding

$$j_h^{(m+1)} = j_h^{(m)} - m_e^{-1}(\partial \Pi_h'^{(*)}/\partial x)\Delta t - (en_h^{(m)} \Delta t/m_e)E^{(*)}, \quad (48a)$$

since $j_h^{(m)} \equiv n_h^{(m)} U_h^{(m)}$. Alternatively, for strong cold collisions, $v_c \Delta t \gg 1$, with $v_h \to 0$ and $m_e/m_i \ll 1$, it follows that $A_c \to U'_i + U'_c/(v_c \Delta t)$ and $B_c \to (v_c \Delta t)^{-1}$, giving

$$j_c^{(m+1)} = n_c^{(m)} U'_i - (m_e v_c)^{-1}(\partial \Pi_c'^{(*)}/\partial x) - (en_c/m_e v_c)E^{(*)}, \quad (48b)$$

which is a form of Ohm's law.

4. The Hybrid E Field

We next calculate the predicted field $\tilde{E}$ by combining the continuity equation (37a) and the velocity-coupled momentum solution, Eq. (47a), for the fully implicit limit $(*) \equiv (m+1)$, with Poisson's equation (2a), as in Section I.A, to obtain

$$\tilde{E}^{(m+1)} = \frac{4\pi(\sum_\alpha \int_0^x q_\alpha n_\alpha^{(m)} dx - \sum_\alpha q_\alpha n_\alpha^{(m)} A_\alpha \Delta t)}{[1 - \sum_\alpha (4\pi q_\alpha e n_\alpha^{(m)} \Delta t^2/m_e)B_\alpha]}. \quad (49)$$

For negligible collisions this reduces to Eq. (4). Alternatively, for collisionless hot electrons but strongly collisional thermals with initial quasi-neutrality (i.e., $\sum q_\alpha n_\alpha = 0$), Eq. (49) reduces to

$$\tilde{E}^{(m+1)} = -\eta J_h = \eta J_c, \quad (50)$$

the more usual Ohm's law. Here $J \equiv -en^{(m)}U'$ and $\eta_c \equiv 4\pi v_c \omega_p^{-2}(n_c)$. This result also requires negligible ion drift speeds and $n_h \ll n_c/(v_c \Delta t)$.

In the moment method we use the Eq. (49) field as $E^{(m+1)}$ in the acceleration of the various plasma components, even when, for example, the hot electrons and ions are treated as particles, so that the Eq. (47a) Braginskii representations for these components are both redundant and approximate. However, in our applications no identifiable difficulties have been traced to this deficiency, possibly because of the weak collisionality of the hot electrons and the relatively slow evolution of the ions. Favorable results must also stem from the moment reinitializations each cycle, following the particle accumulations and fluid updates. Convergence of the Eq. (49) $\tilde{E}^{(m+1)}$ field to the exact solution $E^{(m+1)}$ of Poisson's equation (1a) can presumably be accomplished through an iteration of the field–particle–fluid equations, as an extension of the collisionless procedure outlined in Section II.A.4.

5. Particle Update

Given the new field, we scatter any electrons and advance any particles as outlined in Sections III.A.1.c, d. We use the fully forward centerings of Section II.B.1 as our most robust choice but must employ $\Theta = \psi \simeq 0.6$ for acceptable energy conservation, when, for example, the particles undergo reflections in a sheath—say at the edges of a foil or diode.

6. Fluids Update

Following modern preference, we divide the hydrodynamics into separate Lagrangian and advection phases. The update is then completed with calculations of the electron temperature changes from Joule heating and thermal conduction.

a. LAGRANGIAN ADVANCEMENT. For this phase of the hydrodynamics, following DeBar (1974), Sutcliffe (1974), and Gentry et al. (1966), we assume that a fixed mass $M_\alpha^{(m)} = m_\alpha \Delta x (n_j^{(m)} + n_{j+1}^{(m)})/2$ is attached to each boundary, so that through the Eq. (47a) action of pressure forces, collisions, and E fields, this mass is accelerated to an intermediate velocity $\hat{U}_\alpha \equiv j_\alpha^{(m+1)}/n_\alpha^{(m)}$. The mnU^2 terms in Π are suppressed during this phase, but the artificial viscous pressure is retained. Here, $P_a \equiv mn_j(U_{j+1/2} - U_{j-1/2})^2$, $U_{j+1/2} < U_{j-1/2}$, and $P_a = 0$, $U_{j+1/2} \leq U_{j-1/2}$.

Next, the cell walls are moved to new Lagrangian positions $\hat{x}_{j+1/2} = x_{j+1/2} + \hat{U} \Delta t$, so that new temporary cell widths, $\Delta \hat{x} = \hat{x}_{j+1/2} - \hat{x}_{j-1/2}$, can be computed, along with new temporary Lagrangian densities, $\hat{n} = n(\Delta x/\Delta \hat{x})$. New Lagrangian temperatures are computed at this juncture from Eqs. (37d, e); i.e., $\hat{T}^{(m+1)} = T^{(m)} - 2/(3n)\Pi (\partial U/\partial x) \Delta t$. When $P_a = 0$, an analytically equivalent temperature can be obtained from the adiabatic relation $\hat{T} = T(\Delta/\Delta \hat{x})^{2/3}$, which conserves the specific entropy

$$s = \log(T^{3/2}/n). \tag{51}$$

Boundary densities below a floor value n_{vac} are considered vacuum densities. We set $\hat{U}_{j+1/2}$ to zero, when $n_{j+1/2} \leq n_{\text{vac}}$. Typically, $n_{\text{vac}} = 10^{12}$ cm^{-3}— seven decades below the CO_2 critical density. The artificial pressures P_a are set to zero in all compressive cells bordering a vacuum.

For improved energy conservation and accuracy the Lagrangian step can be iterated as a predictor–corrector, with the first pass calculated for the reduced time step, $\psi \Delta t$ ($0.5 < \psi \leq 1$), to give approximate properties at the intermediate level $(m + \psi)$, and the second and final velocity update calculated with the intermediate Π and n values. Advection of resultant fluid properties should then proceed with the level-$(m + \psi)$ velocities. For this improvement to have uniform value, the E field for the corrector pass should be sampled at the intermediate position of each boundary, requiring interpolations from the fixed storage positions for the field, as discussed for particles in Section II.B.1. To avoid this complication in the present model, we employ only a single Lagrangian step and advect with the velocities following this step.

b. ADVECTIVE REMAPPING. As one approach to the fluid modeling, we might have tracked each of the fluids on its own moving Lagrangian mesh. This would have introduced some incompatibility, however, since traditional PIC modeling requires fixed cells for the field storage and particle accumulations. Thus, we use a fixed common Eulerian mesh for all the components and remap following the Lagrangian update phase. Remapping leads, however, to numerical diffusion, both internal to the evolving plasma components and at their vacuum–matter interfaces.

In Mason (1983a) the internal numerical diffusion was minimized by performing antidiffusive flux-corrective calculations (Book et al., 1975) in the spirit of Zalesak (1979). We started with first order in space–donor-cell fluxes (Gentry et al., 1966) and then substituted second-order interpolated-donor-cell fluxes (Hirt, 1968) to the extent possible without producing nonphysical maxima or minima in the results. Numerical diffusion at vacuum interfaces was removed through a *gating* operation, by which the flux into the vacuum at a cell boundary was set to zero until the density behind that boundary reached an extrapolation of the interior densities. The Flux Corrected Transport (FCT) corrections tended to produce slight but erroneous discontinuous density steps in expanding flow, and the gating procedure failed to extend readily to higher dimensions. Consequently, improvements were sought for both these procedures.

In the present model second-order-accurate internal flow is approximated through the method of van Leer (1977, 1979), as interpreted by Youngs (1982). Numerical diffusion into the vacuum is diminished by a remap of entropy rather than by the more usual internal energy.

For entropy advection, subsequent to the Lagrangian calculation of viscous heating effects, we use the Eq. (51) definition, following Braginskii (1965), to recast the remaining Eqs. (37d, e) hydrodynamic terms into the form

$$T_\alpha n_\alpha (ds_\alpha/dt) = T_\alpha [\partial n_\alpha s_\alpha/\partial t + \partial(n_\alpha s_\alpha \hat{U}_\alpha)/\partial x] = 0. \tag{52}$$

The left-hand side of Eq. (52) reiterates the isentropic nature of the remaining Lagrangian phase calculations, subsequent to any viscous heating, and the right-hand side provides equations for the subsequent remap. Entropy advection in $P = nkT$ gases has the fortuitous property that $T^{3/2}$ numerically diffuses as n, forcing an apparent adiabatic cooling of diffused density into the vacuum, instead of the more usually encountered erroneous shock heating of the vacuum boundary cells. For a more general interface treatment, capable of managing not only vacuum interfaces in two dimensions but also interfaces

8. Implicit Plasma Simulation Models

that separate different ion components, one should consider the SLIC technique of Noh and Woodward (1976) and the Volume of Fluid (VOF) approach (Hirt and Nichols, 1981), as recently elaborated by Lötstedt (1982).

For the actual advection, following DeBar (1974), we first consider the transport of mass. As each cell wall is returned to its initial position, a mass per unit area (for each component α),

$$\delta M_{j+1/2} \equiv \delta x_{j+1/2}\, m\bar{n}_{j+1/2} = \hat{U}_{j+1/2}\, \Delta t\, m\bar{n}_{j+1/2}, \tag{53a}$$

crosses the wall. Second-order accuracy is approached by using Youngs's (1982) extrapolation,

$$\bar{n}_{j+1/2} = \hat{n}_u + \tfrac{1}{2}(1 - \eta)\, \Delta x\, D_{j+1/2}, \tag{53b}$$

for the density, in which u denotes the donating (upwind) cell and $\eta = \min(|\hat{U}_{j+1/2}|\, \Delta t/\Delta x, 1)$. The term containing $D_{j+1/2}$ allows for the density gradient in the donating cell. The purely second-order choice $D_{j+1/2} = \bar{D}_{j+1/2} \equiv (\hat{n}_{j+1} - \hat{n}_j)/\Delta x$ can give rise to nonphysical maxima and minima in the solution. Youngs (1982) finds that the application of nonlinear limits to $D_{j+1/2}$ ensures that the new density $n_d^{(m+1)}$ will satisfy the monotonicity condition (with d the downwind cell)

$$\min(\hat{n}_u, \hat{n}_d) \leq n_d^{(m+1)} \leq \max(\hat{n}_u, \hat{n}_d), \tag{54}$$

which implies that if the sequence $\ldots, \hat{n}_{j-1}, \hat{n}_j, \hat{n}_{j+1}$ is monotonic, then so is $\ldots, n_{j-1}^{(m+1)}, n_j^{(m+1)}, n_{j+1}^{(m+1)}$. The limited $D_{j+1/2}$ is

$$D_{j+1/2}\, \Delta x = R\, \min(|\bar{D}_{j+1/2}\, \Delta x|, 2|\hat{n}_d - \hat{n}_u|, 2|\hat{n}_u - \hat{n}_r|), \tag{55}$$

in which $R = 0$ if $\mathrm{sgn}(\hat{n}_d - \hat{n}_u) \neq \mathrm{sgn}(\hat{n}_u - \hat{n}_r)$, and $R = \mathrm{sgn}(\hat{n}_d - \hat{n}_u)$ otherwise. Also, $d \equiv j + 1$ with $r \equiv j - 1$, if $\hat{U}_{j+1/2} > 0$, and $d \equiv j$ with $r \equiv j + 2$, if $\hat{U}_{j+1/2} < 0$.

The Eq. (53b) flux manifests similarities to the Book et al. (1975) FCT corrected flux. However, here Youngs's limits are based on the old densities $\hat{n}$ rather than the new $n^{(m+1)}$ and are imposed during the advection process rather than later as an anti-diffusion phase. Closer scrutiny will show that the non-$\bar{D}_{j+1/2}\, \Delta x$ terms in Eq. (55) are a factor η smaller than the corresponding terms under FCT limitation—possibly accounting for the smoother expansion profiles observed with the present scheme.

Thus, with the old cell masses defined as $M_j^{(m)} \equiv mn_j^{(m)}\, \Delta x = m\hat{n}_j\, \Delta \hat{x}$, the new fluid densities following the remap become

$$n_j^{(m+1)} = (M_j^{(m)} - \delta M_{j+1/2} + \delta M_{j-1/2})/(m\, \Delta x), \tag{56}$$

and the new cell masses are $M_j^{(m+1)} = mn_j^{(m+1)}\, \Delta x$.

Similarly, constructing the entropy fluxes $\delta S_{j+1/2} \equiv \delta M_{j+1/2} \bar{s}_{j+1/2}$, one can calculate the new specific entropies

$$s_j^{(m+1)} = (M_j^{(m)} s^{(m)} - \delta S_{j+1/2} + \delta S_{j-1/2})/M_j^{(m+1)} \tag{57}$$

and determine $T_j^{(m+1)}$ with the aid of Eq. (51). In Eq. (57) $\bar{s}$ is obtained by substituting s for n in Eqs. (53b) and (55). Again, the circumflex (^) denotes values at the end of the Lagrangian step. Here, $\eta = \min(|\delta M_{j+1/2}|/M_u, 1)$.

Finally, to advect the velocities, we construct centered mass fluxes $\delta M_j \equiv (\delta M_{j-1/2} + \delta M_{j+1/2})/2$, the momentum fluxes $\delta \Phi_j \equiv \delta M_j \bar{U}$, and boundary-centered masses $M'_{j+1/2} \equiv (M_j + M_{j+1})/2$. Then,

$$U_{j+1/2}^{(m+1)} = (M'^{(m)}_{j+1/2} U_{j+1/2}^{(m)} - \delta\Phi_{j+1} + \delta\Phi_j)/M'^{(m+1)}_{j+1/2}, \tag{58}$$

where we substitute U for n in Eqs. (53b) and (55) to produce $\bar{U}$, and use $\eta = \min(|\delta M_{j+1}|/M'_u, 1)$ (with the d, u and r definitions shifted forward one-half cell). This completes the advection phase.

c. JOULE HEATING AND THERMAL CONDUCTION. Electron temperature increases from the Joule heating terms in Eq. (37d) are calculated for a cell by averaging the $v_{j+1/2} \Delta t (\Delta U_{j+1/2})^2$, $\Delta U = U_e - U_i$, contributions at adjacent boundaries. The resultant consistency with Eq. (37b) guarantees $\Delta U \to 0$ for $v \Delta t \to \infty$ and, therefore, finite heating. No corresponding guarantee exists for the simpler differenced $v_j \Delta t\, 0.25(\Delta U_{j+1/2} + \Delta U_{j-1/2})^2$ Joule heating source. As for the Lagrangian phase velocities, the heating is set to zero in "vacuum" cells for which $n_e < n_{vac}$.

Finally, the electron thermal conduction from Eqs. (37d) and (39) is completed with a fully implicit Gaussian elimination solution (Richtmyer and Morton, 1957). An explicit conduction solution would suffice with the $f = 0.6$ flux limitation of Eq. (39) and our Courant limited time step, but the implicit solution permits solutions with arbitrary alternate f choices. Conduction into $n_e < n_{vac}$ cells is suppressed.

C. Discussion

Figure 2 provides a flowchart of the various procedures that are completed in an implicit moment PIC hybrid time step.

Results might be improved by iterating the particle–fluid–field solutions along the lines suggested in Section II.B.4. We have not found this necessary for the production of physically plausible results. Iteration should correct any discrepancies imposed by the use of centered differences in the Eq. (47) auxiliary fluid calculations leading to the E-field determination and the use of van Leer corrected donor-cell differencing in the actual hydrodynamic advancement of the fluid components. Comparable discrepancies exist between the auxiliary field equations and the equations describing any particle components.

With a fluid description for the background, one avoids the heating derived from finite-grid instability, so that phenomena due to a cold and, perhaps, strongly collisional background can now be explored. Furthermore,

8. Implicit Plasma Simulation Models

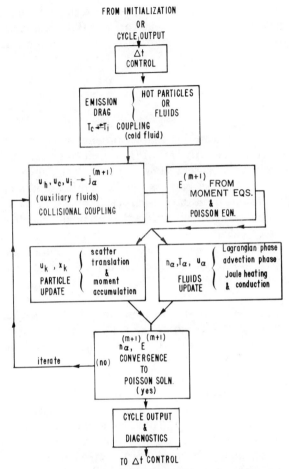

Fig. 2. Flowchart of the implicit moment hybrid calculations.

the use, where acceptable, of fluid descriptions for both the hot- and cold-electron components should allow calculations with time steps approaching the ion Courant limit once the electron pressures have been made implicit. Since particle descriptions are more universally accurate, a more natural direction for improvement may be to generalize the conditions for fluid-to-particle conversion, as has been recently explored by Denavit (1983b), so that particles fill all the weakly collisional cells.

Ultimately, a totally fixed mesh description, such as that used by Book et al. (1975) and Zalesak (1979), may prove superior to the current Lagrangian remap approach, avoiding, for example, the need for field sampling at the

Lagrangian phase positions in any predictor–corrector substeps for improved time centering. Denavit's (1983b) recent work is in this direction. Alternatively, the remap approach may extend more readily to higher dimensions (Mason and Besnard, 1984).

IV. APPLICATIONS

The methods described in this chapter are embodied in the two-dimensional implicit hybrid code ANTHEM (Mason, 1983b, Besnard and Mason 1984). In this section we present sample one-dimensional calculations from ANTHEM in each of its full-particle, hybrid, and full-fluid modes.

Figure 3 gives results from the full-particle implicit calculation of an ion-acoustic diaphragm problem, much like that originally examined in Mason (1971). The earlier calculation was done with isothermal fluid electrons. Here we use particle electrons, but speed the ion evolution by imposing a proton/electron mass ratio of 100. The high density, initially confined by the diaphragm, is 10^{21} electrons cm^{-3}. The initial density jump R is 3:1. Hot electrons are at 430 eV in the dense plasma, and the ions are everywhere at 14 eV; so Θ, the hot-electron–ion temperature ratio, is 30. The low-density electrons to the right of the diaphragm are cold at 1 eV. We used a time step of $\Delta t = 10^{-2}$ psec, so $\omega_p \Delta t = 17.8$ initially in the dense plasma, and we employed 100 cells of width $\Delta x = 2.4 \times 10^{-2}$ μm, so $\Delta x/\lambda_D = 4.9$, with λ_D the dense plasma Debye length.

The ion density profile initially evolves much as the $R = 3$, $\Theta = 30$ case of Mason (1971), with the reflection of an ion precursor foot and a front trailed by ion-acoustic oscillations. However, in the course of the 2.4 psec of evolution described, the hot electrons spread throughout the 2-4-μm test region (the both boundaries being mirror reflective), reducing the maximum temperature by roughly a factor of 4, so that $\Theta \to 7.5$. The late-time post-frontal oscillations are reduced accordingly.

It is interesting that the spreading of the hot electrons results in the "suction" of the cold electrons, as a beam, into the hot region, so that the colds first achieve the hot energy and then mix indistinguishably with the hot dense background. This behavior has been given more detailed analysis in Mason (1981b), which showed that the aggregate effects of the electron transport can be matched by flux-limited diffusive transport with flux limiters ranging from $f = 0.5$ (classical) to 0.03, depending on the degree of collisionality in the plasma and on details of its evolving inhomogeneity. For our 430-eV electrons the mean thermal speed is 8.7 μm psec^{-1} so that in the 2.4 psec shown, there is time for multiple hot-electron crossings through the cold background, consistent with its near isothermalization.

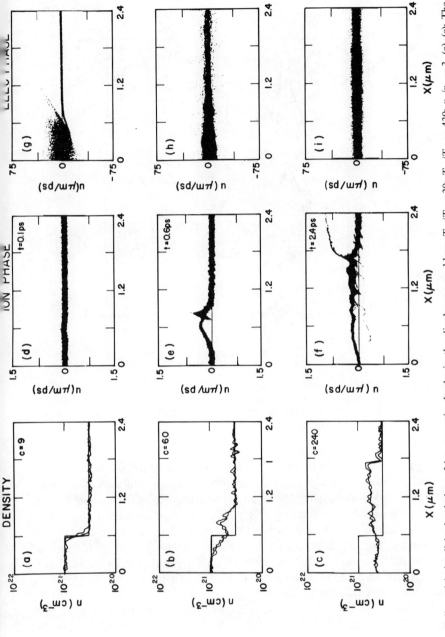

Fig. 3. Full-particle implicit simulation of ion-acoustic shocks in the diaphragm problem: $T_{e1}/T_i = 30$; $T_{e1}/T_{e2} = 430$; $n_1/n_2 = 3$. (a)–(c): The evolving ion and electron densities; (d)–(f): Ion-phase plots displaying the reflection of a characteristic precursor foot. (g)–(i): Electron-phase plots showing rapid isothermalization.

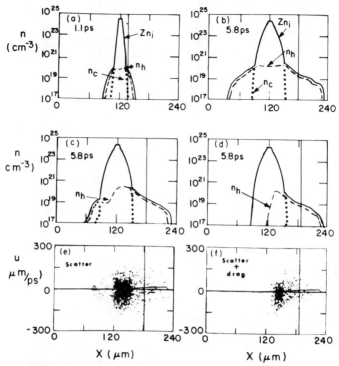

Fig. 4. A foil exposed to 35-keV particle emission. Frames (a) and (b) have the scatter and drag "off," but the E field "on." For (c) scatter has been added. For (d) drag is added with scatter. Frames (e) and (f) display the phase plots corresponding to (c) and (d).

Figure 4 collects hybrid results for laser-generated particle electrons transported through a background of fluid cold electrons and ions. The results are borrowed from Mason (1983a) and were completed with vacuum boundary gating and FCT advection. They are, however, essentially unchanged when recalculated with the currently recommended van Leer entropy advection (see Section III.B.6.b). The figure shows a 9.6-μm gold foil illuminated by 3×10^{15} W cm^{-2} CO_2 light with 35% absorption. We set $T_h = 35$ keV, $T_c = T_i$ and imposed full ionization, $Z = 79$. Again, the proton/electron mass ratio was artificially set to 100.

The frames (a) and (b) are for scattering and drag "off" but with the E field "on." The vertical fiducial line marks the location of the light-absorbing critical density surface. The electron density output has been smoothed by processing through a $0.25(n_{j-1} + 2n_j + n_{j+1})$ filter. By 5.8 psec, frame (b) shows the presence of a fast ion expansion on both sides of the foil. Frame (c) shows the modified evolution when scattering is added—the electron density is an order of magnitude lower on the foil backside. Finally, frame (d) shows a significantly lower backside hot-electron population when drag effects are included as well. The phase plot (f) shows a correspondingly reduced surviving particle density with drag. Here, since the late time deposition is made in a corona populated entirely by hot electrons, the energy was added by reducing the weights of background suprathermals as new particles were added. The need to do this was symptomatic of too low a choice for the emitted suprathermal temperature and can be avoided (Mason, 1983b) by use of an emission temperature exceeding 150 keV.

Finally, Fig. 5 presents results from a full-fluid calculation of 10-keV suprathermal penetration into a plastic ($Z = 4$) background. The computations employed variable time steps, bounded by the minimum hot-electron Courant limit and 50 cells of width $\Delta x = 0.5$ μm. The initial background ion and electron temperatures were 1 eV. The laser delivered 10^{15} W cm^{-2} of 1.06-μm light to the first cell exceeding the critical density 10^{21} cm^{-3}. Cold-electron–ion thermal coupling, hot- and cold-electron thermal conduction (with $f = 0.6$), and Joule heating all were operative. Frame (b) shows how the implicit E field locks the hot-electron and ion densities together in the corona, while excluding the cold electrons. Frames (d)–(f) track the evolving hot-electron, cold-electron, and ion temperatures, as well as $T_{av} \equiv (n_h T_h + n_c T_c)/(n_h + n_c)$. The average temperature drops rapidly with depth into the dense background, possibly implying the need for severe flux limitation ($f = 0.03$; see Mason, 1981b) in a more standard, single-group, limited-diffusion calculation of this phenomenology.

V. CONCLUSION

This chapter has reviewed the implicit moment method and its hybrid extensions. It has demonstrated that use of implicit fields can free the user from the $\omega_p \Delta t < 2$ constraint of older, explicit methods of simulation. Details of the evolving electron distribution can now be tracked with a time step limited only by the shortest electron transit time across a cell. Further, the use of implicit pressures and subcycling may even allow for substantially larger Δt. The moment method has been related to the alternate direct method, which can provide useful variation and important insight into the substructure of implicit particle calculations.

The moment approach has been found to be straightforward and naturally compatible with the fluid modeling of parts of the plasma. The hybrid use of fluid background components has been shown to facilitate the study of long-range electron transport through strongly collisional background plasmas with steep density gradients. Experience is needed to determine the

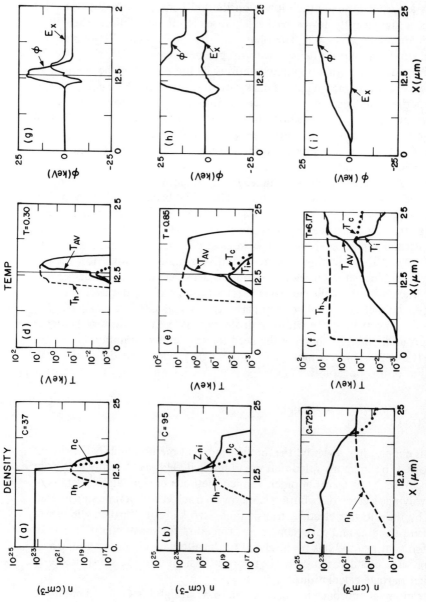

Fig. 5. Full-fluid simulation of 10-keV suprathermal electron penetration into a foil. Frames (a)–(c) show the evolving hot-electron density profile, the exclusion of cold electrons from the corona, and the fast expansion of ions with the hot electrons; Frames (d)–(f) give the various component temperatures, as well as the average electron temperature: Frames (g)–(i) display the electrostatic potential confining the hot electrons, and the E field (arbitrary units) that integrates to that potential.

optimal blend of particle and fluid modeling in problems of interest. Improved procedures for the local conversion from fluid to particle (or particle to fluid) descriptions are in demand.

In the near future we shall see the two-dimensional extension of such hybrid modeling to include B-field effects and the implementation of improved collisional models, permitting a particle treatment, if desired, when $v \Delta t > 1$.

ACKNOWLEDGMENTS

The author wishes to thank J. Brackbill, C. Cranfill, D. Forslund, R. Jones, E. Lindman, and J. Saltzman at Los Alamos; B. Cohen, J. Denavit, A. Friedman, and A. B. Langdon at Livermore; D. Besnard at Limeil; J.-C. Adam at the Ecole Polytechnique; and D. Youngs at Aldermaston for helpful discussions during the course of this work. This work was performed under the auspices of the U.S. Department of Energy.

REFERENCES

Adam, J.-C., Gourdin-Serveniere, A., and Langdon, A. B. (1982). *J. Comput. Phys.* **47**, 229.
Barnes, D. C., Kamimura, T., Leboeuf, J.-N., and Tajima, T. (1983). *J. Comput. Phys.* **52**, 480.
Besnard, D., and Mason, R. (1984). Lemeil Lab. [Rep.] CEA-N-2413.
Book, D. L., Boris, J. P., and Hain, K. (1975). *J. Comput. Phys.* **18**, 248.
Boris, J. P. (1970). *Proc. Conf. Numer. Simul. Plasmas, 4th, 1970,* p. 3.
Boris, J. P., and Lee, R. (1973). *J. Comput. Phys.* **12**, 131.
Brackbill, J. U., and Forslund, D. W. (1982). *J. Comput. Phys.* **46**, 271.
Braginskii, S. I. (1965). *Rev. Plasma Phys.* **1**, 205.
Byers, J. A., Cohen, B. I., Condit, W. C., and Hanson, J. D. (1978). *J. Comput. Phys.* **27**, 363.
Chodura, R. (1975). *Nucl. Fusion* **15**, 55.
Cohen, B. I., Brengle, T. A., Conley, D. B., and Freis, R. P. (1980). *J. Comput. Phys.* **38**, 45.
Cohen, B. I., Fries, R. P., and Thomas, V. (1982). *J. Comput. Phys.* **45**, 345.
Crystal, T. L., Denavit, J., and Rathmann, C. E. (1979). *Comments Plasma Phys. Controlled Fusion* **5**, 17.
DeBar, R. B. (1974). *Lawrence Livermore Lab.* [Rep.] *UCID* **UCID-19683**.
Denavit, J. (1981). *J. Comput. Phys.* **42**, 337.
Denavit, J. (1983a). *Proc. 10th Annu. Conf. Numer. Simul. Plasmas, 1983.*
Denavit, J. (1983b). *Bull. Am. Phys. Soc.* [2] **28**, 1125.
Denavit, J., and Walsh, J. M. (1980). *Proc. Conf. Numer. Simul. Plasmas, 9th, 1980* Paper C3.
Denavit, J., Crystal, T. L., Rathmann, J. L., and Vomvoridis, J. L. (1979). In "Proceedings of the Informal Conference on Particle and Hybrid Codes for Fusion, Napa, California," Memo No. UCB/ERL M79/79, Pap. No. 1. Electronics Research Laboratory, University of California, Berkeley.
Forslund, D. W. and Brackbill, J. U. (1982). *Phys. Rev. Lett.* **48**, 1614.
Friedman, A., Sudan, R. N., and Denavit, J. (1978a). *Proc. Conf. Numer. Simul. Plasmas, 8th, 1978.*
Friedman, A., Sudan, R. N., and Denavit, J. (1978b). *Bull. Am. Phys. Soc.* [2] **23**, 842.
Friedman, A., Langdon, A. B., and Cohen, B. I. (1981). *Comments Plasma Phys.* **6**, 225.
Gentry, R. A., Martin, R. E., and Daly, B. J. (1966). *J. Comput. Phys.* **1**, 87.

Hamasaki, S., Krall, N. A., Wagner, C. E., and Byrne, R. N. (1977). *Phys. Fluids* **20**, 65.
Harlow, F. H., and Amsden, A. A. (1971). *J. Comput. Phys.* **8**, 197.
Harned, D. S. (1982). *J. Comput. Phys.* **47**, 452.
Hewett, D. W. (1980). *J. Comput. Phys.* **38**, 378.
Hewett, D. W., and Nielson, C. W. (1978). *J. Comput. Phys.* **29**, 219.
Hirt, C. W. (1968). *J. Comput. Phys.* **2**, 339.
Hirt, C. W., and Nichols, B. D. (1981). *J. Comput. Phys.* **39**, 201.
Jackson, J. D. (1962). "Classical Electrodynamics." Wiley, New York.
Langdon, A. B. (1970). *J. Comput. Phys.* **6**, 247.
Langdon, A. B. (1979). *J. Comput. Phys.* **30**, 202.
Langdon, A. B., Cohen, B. I., and Friedman, A. (1983). *J. Comput. Phys.* **51**, 1983.
Lindemuth, I. R. (1977). *J. Comput. Phys.* **25**, 104.
Lötstedt, P. (1982). *J. Comput. Phys.* **47**, 211.
Mason, R. J. (1969). *Bull. Am. Phys. Soc.* [2] **14**, 1043.
Mason, R. J. (1971). *Phys. Fluids* **14**, 1943.
Mason, R. J. (1972). *Phys. Fluids* **15**, 845.
Mason, R. J. (1979a). *Phys. Rev. Lett.* **43**, 1795.
Mason, R. J. (1979b). *In* "Proceedings of the Informal Conference on Particle and Hybrid Codes for Fusion, Napa, California" Memo No. UCB/ERL M79/79, Pap. No. 23. Electronics Research Laboratory, University of California, Berkeley.
Mason, R. J. (1980a). *Phys. Fluids* **23**, 2204.
Mason, R. J. (1980b). *Proc. 10th Annu. Anomalous Absorption Conf., 1980* Paper A10.
Mason, R. J. (1980c). *Bull. Am. Phys. Soc.* [2] **25**, 926.
Mason, R. J. (1981a). *J. Comput. Phys.* **41**, 233.
Mason, R. J. (1981b). *Phys. Rev. Lett.* **47**, 652.
Mason, R. J. (1981c). *Proc. Am. Nucl. Soc. Int. Top. Meet., Adv. Math. Methods Nucl. Eng. Probl., 1981* Vol. 2, p. 407.
Mason, R. J. (1981d). *In* "Laser Interaction and Related Phenomena," Vol. 5, p. 743. Plenum, New York.
Mason, R. J. (1983a). *J. Comput. Phys.* **51**, 484.
Mason, R. J. (1983b). Los Alamos Lab. [Rep.] LA-10112-C.
Mason, R. J., and Besnard, D. (1986) To be published.
Morse, R. L. (1970). *Methods Comput. Phys.* **9**, 213.
Noh, W. F., and Woodward, P. (1976). *In* "Proceedings of the Fifth International Conference on Numerical Methods in Fluid Dynamics" (A. I. van de Vooren and P. J. Zandbergen, eds.), p. 330. Springer-Verlag, Berlin and New York.
Okuda, H. (1972). J. *Comput. Phys.* **10**, 475.
Richtmyer, R. D., and Morton, K. W. (1957). "Difference Methods for Initial Value Problems," pp. 198–201. Wiley (Interscience), New York.
Sgro, A. G., and Nielson, C. W. (1976). *Phys. Fluids* **19**, 126.
Shanny, R., Dawson, J. M., and Greene, J. M. (1967). *Phys. Fluids* **10**, 1281.
Spitzer, L. G. (1961). "Physics of Fully Ionized Gases." Wiley, New York.
Sutcliffe, W. G. (1974). *Lawrence Livermore Lab.* [*Rep.*] *UCID* **UCID-17013**.
van Leer, B. (1977). *J. Comput. Phys.* **23**, 276.
van Leer, B. (1979). *J. Comput. Phys.* **32**, 101.
Vomvoridis, J. L., and Denavit, J. (1978). *Phys. Fluids* **22**, 367.
Youngs, D. L. (1982). *In* "Numerical Methods for Fluid Dynamics" (K. W. Morton and M. J. Baines, eds.), p. 273. Academic Press, 1982.
Zalesak, S. T. (1979). *J. Comput. Phys.* **31**, 335.

9

Simulation of Low-Frequency, Electromagnetic Phenomena in Plasmas

J. U. BRACKBILL and D. W. FORSLUND

Los Alamos National Laboratory
University of California
Los Alamos, New Mexico

I. Introduction.	272
II. Implicit Plasma Simulation	274
A. The Implicit Moment Method.	274
III. Implicit Formulation of the Dynamic Equations	280
A. The Particle Description.	280
B. The Moment Description.	287
C. The Self-Consistent Fields	291
D. Spatial Differencing.	292
IV. The Algorithm for the Implicit Moment Method	293
A. Solving the Potential Equations.	293
V. Properties of the Implicit Moment Method	295
A. Dispersion Analysis of the Moment Equations.	296
B. The Energy Integral.	299
VI. Computational Examples.	300
A. Laser-Induced Ion Emission.	301
B. Collisionless Shocks.	305
VII. Conclusions.	308
References.	309

I. INTRODUCTION

The dynamic equations of collisionless plasma physics are well known. They are mathematically interesting, describe phenomena observed in space plasmas and on the sun, and guide us in the development of power from controlled fusion. However, the equations are difficult to solve. They describe complex nonlinear interactions among large numbers of charged particles, waves propagating at the speeds of light and sound, sheaths and boundary layers a gyroradius thick, and solar flares many times the size of the earth. They describe all of these simultaneously on time scales varying over many orders of magnitude.

Because of the complexity of the equations, especially for collisionless plasmas, it has been fruitful to simulate plasmas on a computer. In the simulations, plasmas are modeled by as many as a million finite-sized particles interacting through their self-consistent electric and magnetic fields. There have been many successful applications of plasma simulations in magnetic confinement, in inertial confinement fusion, and in space plasma physics. Methods are described in several excellent review articles, including Birdsall et al. (1970), Langdon and Lasinski (1976), and Dawson (1983).

However, the most interesting problems are still out of reach, even with recent improvements in computers. The problems we want to do are always larger than the problems we have already done. They require more particles or run for a longer time than our computer resources will allow.

To make progress faster than computers advance, we must improve our plasma simulation methods however we can, through improved data management, asymptotic analysis, or algorithm development. In this chapter, we discuss a new algorithm for modeling low-frequency plasma phenomena using an implicit formulation that allows large time steps. This leaves unchanged the restrictions on the range of time scales that can be encompassed in a single calculation, but it allows us greater freedom to choose the time step to resolve the time scale of interest.

Implicit methods are only one type among the many that can be used for modeling low-frequency phenomena. For example, one can solve the reduced equations of the Darwin model and eliminate light waves (Nielson and Lewis, 1976). One can represent electrons as a collisional, sometimes massless, fluid and eliminate electron oscillations (Dickman et al., 1969; Busnardo-Neto et al., 1977; Hewett and Nielson, 1978; Byers et al., 1978; Hewett, 1980). One can average over the particle orbits (Cohen, Chapter 10). One can even use the magnetohydrodynamic approximation and eliminate particle effects entirely (Chase et al., 1973; Colombant et al., 1976; Pert, 1981). Even with all of these possibilities, one still can make a strong argument for solving the full system of dynamic equations. To answer questions that cannot be answered by fluid models, such as the role of ion kinetics in the magnetohydrodynamic stability of magnetically confined plasmas or the origin of dissipation in collisionless shocks, it may be necessary to retain ion and electron kinetic and inertial terms. To verify phenomenological modeling, as in calculating energy transport in laser target plasmas, it is necessary to calculate from first principles. Thus implicit methods, with their potential for reducing the cost of numerical calculations, are potentially very useful.

Implicit methods for plasma simulation are not new. For example, Nielson and Lindman (1974) describe an implicit treatment of light waves that eliminates the Courant condition related to signals propagating with the speed of light. However, their use in plasma simulation has been inhibited by the cost of solving implicitly formulated particle and field equations. Langdon (1979) points out that solving coupled field and particle equations require iterations that are expensive and, possibly, not convergent.

Recently, two new techniques have been developed for solving the implicit equations; these techniques reduce the cost by reducing the number of equations that must be iterated. In their simplest forms, the direct (or kinematic) method, as described in Friedman *et al.* (1981), extrapolates the particle motion forward in time along the unperturbed particle orbits, and the implicit moment (or fluid dynamic) method approximates the solution using fluid equations (Mason, 1981; Denavit, 1981). [Implicit methods are reviewed briefly in Cohen (1982).] The use of either of these techniques allows us to apply implicit methods to the simulation of low-frequency plasma phenomena.

This chapter describes our use of the implicit moment method to model low-frequency, nonrelativistic, electromagnetic waves in two-dimensional plasmas. Specifically, we exploit the special properties of implicit equations, including spectral compression and selective damping, to construct interpolating approximations that match existing asymptotic equations in both the high- and low-frequency limits. We illustrate this approach by applying it to collisionless plasmas in the magnetohydrodynamic limit, to guiding center plasmas, and finally to collisional plasmas.

We organize the development as follows: In Section II, we motivate the use of implicit equations by comparing the plasma dispersion function for implicit and explicit difference equations. We also briefly outline the implicit moment method. In Section III we develop the finite-difference equations for models describing collisionless plasmas, guiding center motion, and collisions. We also address the choice of approximations in the moment equations and how they affect the maximum time step. In Section IV we describe the overall algorithm, as well as methods for solving the potential equations. In Section V, we analyze the properties of the implicit equations and estimate the numerical error. In Section VI, we present results from several calculations to illustrate the application of the implicit method to modeling plasma phenomena.

II. IMPLICIT PLASMA SIMULATION

A. The Implicit Moment Method

In their earlier papers, Mason and Denavit demonstrated that moment and particle descriptions of the plasma could be used cooperatively in formulating a stable algorithm for very large time steps (Mason, 1981; Denavit, 1981). We shall review briefly the fundamentals of plasma simulation and describe how the implicit moment method works.

1. The Fundamental Equations of Plasma Physics

The positions and velocities of all the particles describe the plasma completely. The evolution of this information in time is determined by solving the equations of motion for the particles,

$$d\mathbf{x}/dt = \mathbf{u} \tag{1}$$

$$d\mathbf{u}/dt = (q/m)(\mathbf{E} + \mathbf{u} \times \mathbf{B}/c), \tag{2}$$

where $\mathbf{x}$ and $\mathbf{u}$ are the particle position and velocity, q/m the particle charge to mass ratio, and $\mathbf{E}$ and $\mathbf{B}$ the electric and magnetic fields.

The self-consistent fields, $\mathbf{E}$ and $\mathbf{B}$, are calculated from the Maxwell relations: Faraday's law,

$$\partial \mathbf{B}/\partial t - c(\nabla \times \mathbf{E}) = 0, \tag{3}$$

the solenoidal condition,

$$\nabla \cdot \mathbf{B} = 0, \tag{4}$$

Ampere's law,

$$-\partial \mathbf{E}/\partial t + c(\nabla \times \mathbf{B}) = 4\pi \mathbf{j}, \tag{5}$$

and Gauss's law,

$$\nabla \cdot \mathbf{E} = 4\pi n. \tag{6}$$

Note that $\mathbf{E}$ and $\mathbf{B}$ are completely determined by $n(\mathbf{x}, t)$ and $\mathbf{j}(\mathbf{x}, t)$ and the boundary conditions.

The velocity moments of the particle distribution in position and velocity space from which **E** and **B** are calculated are defined by summations over the particles,

$$\mathbf{M}^0 = n(\mathbf{x}) = \sum_p q_p \delta(\mathbf{x} - \mathbf{x}_p), \tag{7}$$

$$\mathbf{M}^1 = \mathbf{j}(x) = \sum_p q_p \mathbf{u}_p \delta(\mathbf{x} - \mathbf{x}_p) \tag{8}$$

$$\mathbf{M}^2 = \mathbf{\Pi}(\mathbf{x}) = \sum_p q_p \mathbf{u}_p \mathbf{u}_p \delta(\mathbf{x} - \mathbf{x}_p), \tag{9}$$

where p is an index labeling each particle, and $n(\mathbf{x})$, $\mathbf{j}(\mathbf{x})$, and $\mathbf{\Pi}(\mathbf{x})$ the charge density, current density, and stress density, respectively.

The self-consistent solution of the Maxwell relations, together with the particle equations of motion, completely describes the evolution of the plasma.

2. The Properties of Implicit Differencing

First, we consider implicit and explicit finite-difference approximations to the particle equations of motion. In the implicit formulation the position of the particle is given by

$$\mathbf{x}_{ps}^1 = \mathbf{x}_{ps}^0 + \bar{\mathbf{u}}_{ps} \Delta t, \tag{10}$$

where $\mathbf{x}_{ps}^1$ is the position of particle p of specie s at time $(n + 1) \Delta t$, and $\mathbf{x}_{ps}^0$ its position at $n \Delta t$. The particle's average velocity over the interval Δt is defined by $\bar{\mathbf{u}} \equiv (\mathbf{u}^1 + \mathbf{u}^0)/2$. The velocity in electric and magnetic fields is given by

$$\mathbf{u}_{ps}^1 = \mathbf{u}_{ps}^0 + (q_s/m_s)(\mathbf{E}^\theta + \bar{\mathbf{u}}_{ps} \times \mathbf{B}^0/c) \Delta t \tag{11}$$

where q_s/m_s is the charge to mass ratio for specie s, and $\mathbf{E}^\theta$ an intermediate value of the electric field given by interpolation: $\mathbf{E}^\theta = \theta \mathbf{E}^1 + (1 - \theta)\mathbf{E}^0$, $\frac{1}{2} < \theta < 1$. In the momentum equation, $\mathbf{E}^\theta$ and $\mathbf{B}^0$ are evaluated at $\bar{\mathbf{x}} = (\mathbf{x}^1 + \mathbf{x}^0)/2$. Thus, when $\theta = \frac{1}{2}$ the method is time-centered and second order accurate in Δt.

The electric field $\mathbf{E}^1$ is calculated from Gauss's law [Eq. (6)]:

$$\nabla \cdot \mathbf{E}^1 = 4\pi n^1.$$

The explicit formulation, the *leapfrog* scheme, is written

$$\mathbf{x}_{ps}^1 = \mathbf{x}_{ps}^0 + \mathbf{u}_{ps}^{1/2} \Delta t, \tag{12}$$

where $\mathbf{u}_{ps}^{1/2}$ is calculated from the momentum equation

$$\mathbf{u}_{ps}^{1/2} = \mathbf{u}_{ps}^{-1/2} + (q_s/m_s)(\mathbf{E}^0 + \bar{\mathbf{u}}_{ps}^0 \times \mathbf{B}^0/c) \Delta t, \tag{13}$$

where $\mathbf{E}^0 = \mathbf{E}(\mathbf{x}^0, t)$ and $\bar{\mathbf{u}}^0_{ps} = (\mathbf{u}^{1/2}_{ps} + \mathbf{u}^{-1/2}_{ps})/2$. In the leapfrog method, the position and velocity are advanced alternately. The method is time-centered and second order accurate.

The implicit and explicit methods have similar properties for small Δt, but for large Δt the explicit method is unstable with exponentially growing solutions, and the implicit method is stable for all Δt. The difference in stability is clearly illustrated by the numerical plasma dispersion function (Lindman, 1970; Langdon, 1970).

Using the method outlined by Langdon (1979), we calculate the dispersion relation directly from the particle equations of motion for an unmagnetized plasma of electrons with a fixed charge-neutralizing background of ions and variation only in the x direction. In a time-varying electric field, the particles deviate from their unperturbed orbits $x_{(0)}$ by an amount $x_{(1)}$. The unperturbed orbits are given by

$$x^1_{(0)} = x^0_{(0)} + u_{(0)} \Delta t, \tag{14}$$

where $u_{(0)}$ is the unperturbed particle velocity, and the perturbed orbits are given by eliminating u between Eqs. (10) and (11):

$$x^1_{(1)} - 2x^0_{(1)} + x^{-1}_{(1)} = (q/m)[E^\theta_{(1)}(x^{1/2}) + E^{\theta-1}_{(1)}(x^{-1/2})] \Delta t^2/2. \tag{15}$$

Following the sequence of steps outlined by Langdon (1979), we linearize the equation, replace x by $x_{(0)}$ on the right-hand side (RHS), assume the explicit time dependence $\exp(i\omega t)$, Fourier transform in x, and, finally, calculate the perturbed charge density $n_{(1)}$ from the continuity equation and the definition of the charge density to obtain the numerical dispersion relation

$$0 = \varepsilon_1(k, \omega) \equiv 1 - \exp[i(\theta - 1/2)\Delta t](\omega_e \Delta t/2)^2$$
$$\times \int du [g(u) \cos(\omega - ku) \Delta t/2]/\sin^2(\omega - ku) \Delta t/2, \tag{16}$$

where $g(u)$ is the unperturbed probability distribution and ω_e^2 the electron plasma frequency: $\omega_e^2 = 4\pi n_e q^2/m_e$.

The corresponding dispersion relation for the explicit leapfrog scheme is derived by Langdon (1979):

$$0 = \varepsilon_E(k, \omega) \equiv 1 - (\omega_e \Delta t/2)^2 \int du\, g(u)/\sin^2(\omega - ku) \Delta t/2. \tag{17}$$

Equations (16) and (17) are similar except for an exponential factor in front and the cosine in the integral in Eq. (17). These small differences, however, account for the large differences in the stability of explicit and implicit schemes.

Consider a cold plasma with $g(u) = \delta(u)$. For the explicit scheme, the dispersion relation is

$$\sin^2(\omega \Delta t/2) = (\omega_e \Delta t/2)^2.$$

9. Electromagnetic Phenomena in Plasmas

When $\omega_e \Delta t < 2$, the roots are real and the equation is stable, but when $\omega_e \Delta t > 2$, the roots are complex and the equation is exponentially unstable. For the implicit scheme, the dispersion relation (with $\theta = \frac{1}{2}$) is

$$\tan(\omega \Delta t/2) \sin(\omega \Delta t/2) = (\omega_e \Delta t/2)^2.$$

For small Δt, the roots of the dispersion relation for the implicit scheme are indistinguishable from those for the explicit scheme, but for large Δt, the roots lie in the interval $0 < \omega \Delta t < \pi$ and are purely real; in this interval the tangent function can assume any positive value. As $\omega_e \Delta t$ increases, ω is bounded by the Nyquist frequency $\pi/\Delta t$, and the spectrum is compressed into a smaller and smaller interval. Thus, implicit differencing avoids instability by restricting the range of frequencies in the solution.

When θ is different from $\frac{1}{2}$ in the implicit scheme, the roots of the dispersion relation are complex. When $\theta < \frac{1}{2}$, the imaginary part corresponds to an exponentially growing solution, and when $\theta > \frac{1}{2}$, to an exponentially decaying solution.

Numerical solutions of the dispersion function are needed to calculate the roots for more complex distributions, for which purpose the dispersion relations are expressed as continued fractions. We can express ε_l this way by integrating by parts and substituting a series expansion for the cosecant function to obtain

$$\varepsilon = 1 + \exp[i(\theta - \tfrac{1}{2})\omega \Delta t]\omega_e^2/k \int du(dg/du) \sum_q (-1)^q/(\omega - ku - 2\pi q/\Delta t). \tag{18}$$

Aside from the exponential factor, the only difference from the expression for an explicit scheme (Langdon, 1979) is the alternating sign of the terms in the summation.

As described by Langdon (1979), we can include finite spatial grid effects in the dispersion relation above to examine the effect of implicit differencing on the finite grid instability (Lindman, 1970). Because of this instability, $\text{Im}(\omega)$ is greater than zero for some values of k, corresponding to an exponentially growing instability. When Δt and θ both increase, ω eventually becomes complex, causing damping of the plasma oscillations. The damping reduces the growth rate of the finite grid instability, as shown in Fig. 1, where the growth rate of the instability is decreased a hundredfold with $\Delta t = 1$ and $\theta = 0.6$ from its value with $\Delta t = 0$.

The need to suppress the finite grid instability imposes a lower limit on the time step. Empirically, when Δt satisfies the inequality $\omega_s \Delta t (\lambda_D/\Delta x) > 0.1$, the finite grid instability is suppressed. In a way, this is a consistency requirement that forces us to the quasi-neutral time scale $\omega_s \Delta t > 1$ as we calculate longer wavelength phenomena, $\lambda_D/\Delta x < 1$.

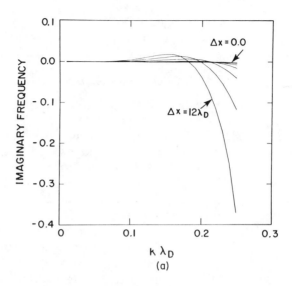

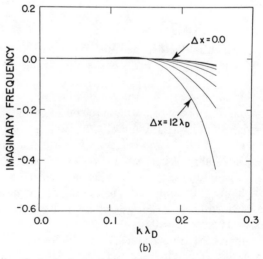

Fig. 1. For wavelengths in the interval $0.1 < k\lambda_D < 0.2$, particle aliasing on the grid causes a numerical instability with growth rates on the order of the electron plasma frequency. With $\Delta t = 0$, the imaginary part of the frequency is positive, as shown in (a), but is essentially zero with $\omega_e \Delta t = 1$ as shown in (b).

3. Field Estimates from the Moment Equations

The stability of the implicit equations for large Δt is useful only if we can solve them. As Langdon (1979) points out, it is expensive, perhaps impossible, to iterate the full particle and field equations to solve the equations. Rather, we solve simplified equations to estimate the self-consistent fields in which the particles move.

The method for estimating the fields is based on the implicit moment method (Mason, 1981; Denavit, 1981). Using the implicit moment algorithm, we march difference equations time step (Δt) by time step. At the beginning of each time step, we calculate $n(\mathbf{x})$ and $\mathbf{j}(\mathbf{x})$ from the particle data using Eqs. (7) and (8). We then calculate $\partial n/\partial t$, $\partial \mathbf{j}/\partial t$, $\partial \mathbf{E}/\partial t$, and $\partial \mathbf{B}/\partial t$ self-consistently from difference approximations to Eqs. (3–6) to provide estimates of $\mathbf{E}(\mathbf{x}, t)$ and $\mathbf{B}(\mathbf{x}, t)$ over the interval Δt. Once we have obtained these estimates, we advance the particle orbits using Eqs. (10) and (11). We repeat this sequence of steps each time step.

Using the moments allows us to estimate the fields at some future time. However, we must recreate the moments from the particle data each time step to obtain information that would otherwise have to be calculated from an energy equation and other higher moment equations. The particles carry detailed information about the plasma from one time step to the next; the moment equations contain only enough information to estimate the fields.

Of course, using the moments also simplifies the equations for the self-consistent fields. The coupled moment and field equations comprise a much smaller system of equations than would the particle and field equations, making the implicit time advancement of the fields easier and cheaper.

Evolution equations of motion for the moments are easily derived by differentiation. For example, the continuity equation results from differentiating $n(\mathbf{x})$ with respect to time:

$$\frac{\partial n(\mathbf{x})}{\partial t} = \sum_p - q_p \left[\frac{d\mathbf{x}_p}{dt} \cdot \nabla \delta(\mathbf{x} - \mathbf{x}_p) \right] = -\nabla \cdot \mathbf{j}(\mathbf{x}). \tag{19}$$

Similarly, the momentum equation results from differentiating $\mathbf{j}(\mathbf{x})$:

$$\frac{\partial \mathbf{j}(\mathbf{x})}{\partial t} = \sum_p \frac{d\mathbf{u}_p}{dt} \delta(\mathbf{x} - \mathbf{x}_p) - \mathbf{u}_p \frac{d\mathbf{x}_p}{dt} \cdot \nabla \delta(\mathbf{x} - \mathbf{x}_p).$$

By substituting from the particle equations of motion, the momentum equation can be rewritten in the familiar form,

$$\partial \mathbf{j}(\mathbf{x})/\partial t = (q/m)[n(\mathbf{x})\mathbf{E} + \mathbf{j}(\mathbf{x}) \times \mathbf{B}/c] - \nabla \cdot \pi. \tag{20}$$

Evolution equations for any moment can be derived in similar fashion. Over a brief time compared with the particle transit time across the length scale of interest, the first few moment evolution equations give accurate estimates of the evolution of the self-consistent fields because the contributions of the higher moments are small. In fact, we need only $n(\mathbf{x}, t)$ and $\mathbf{j}(\mathbf{x}, t)$ to determine $\mathbf{E}$ and $\mathbf{B}$, and these are computed using $\partial n/\partial t$ and $\partial \mathbf{j}/\partial t$. Solving these equations gives us values for $n(\mathbf{x}, t)$ and $\mathbf{j}(\mathbf{x}, t)$ that are very good approximations to the values we would have obtained by solving the individual particle equations of motion, but at a fraction of the cost. Over a longer interval, the correspondence between the solution of a truncated set of moment equations and the particle solutions is poor because the contributions of the higher moments become dominant. However, the time during which the simple moment description suffices is usually long compared with the transit time for light waves or the frequency of electrostatic oscillations.

The lower bound on the time step imposed by the need to suppress the finite grid instability and the upper bound imposed by the convergence of the moment expansion restrict Δt to the interval

$$O(0.1) < v_{\text{th}} \Delta t/\Delta x = (\lambda_D/\Delta x)(\omega_s \Delta t) < O(1.0). \tag{21}$$

In this interval the numerical calculations are stable and accurate. Satisfying the inequality has not been restrictive in practice, because the conjunction of long time and space scales is typical of many plasma problems.

The actual implementation of the implicit moment method for the simulation of plasma phenomena in two dimensions in electromagnetic fields will require considerable discussion. However, we hope that the details will not distract the reader from the underlying simple structure of the method.

III. IMPLICIT FORMULATION OF THE DYNAMIC EQUATIONS

A. The Particle Description

Grad (1969) argues that much of the physics of magnetically confined plasmas is described by the motion of a single particle in specified electric and magnetic fields. This argument is true for magnetically confined plasmas because they are often collisionless (Freidberg, 1982), but it also accounts for the success of particle simulations in modeling collisionless plasma phenomena generally. The importance of particle trapping or wave–particle resonances to the physics of the plasma as a whole makes it natural to concentrate first on the particle equations of motion in developing a self-consistent formulation for the plasma dynamic equations.

Because we also wish to apply the simulations to very-low-frequency waves in inhomogeneous plasmas, we cannot ignore collisions. In a collisional plasma the single-particle motion includes a large random component, and the motion of a single particle tells very little about the physics of the plasma as a whole. Thus, it is better to use a fluid model (Braginskii, 1965) than a particle simulation when collisions are dominant. However, by including collisions in a particle simulation, we can make connections between particle simulations and fluid calculations that are useful but difficult to make otherwise. For example, energy transport in laser target plasmas is often calculated using fluid models with flux limiters (Malone *et al.*, 1975) to prevent divergence of the moment expansions when collisions are infrequent. Using particle simulations with collisions allows us to understand flux inhibition on a more fundamental level and, perhaps, to verify the results of the fluid calculations.

Formulating the particle equations of motion implicitly imposes some special requirements. More than in explicit methods, we must understand the consequences of finite time steps and exploit them to model low-frequency phenomena. For example, in developing a model for collisional or strongly magnetized plasmas, we must consider cases in which Δt is long compared with the mean free time between collisions or with the period of gyration. To do this, we develop an interpolating approximation to the equation of motion for a single particle [Eq. (32)] that describes collisionless or collisional, weakly or strongly magnetized plasmas. (In Section III.B we derive the corresponding moment equations to complete the formulation.)

We define an interpolating approximation as one that gives us the correct limits for both small and large time steps. We require that the solutions to the finite-difference equations converge to the solution of the differential equations when Δt is small, and to the time asymptotic solutions when Δt is large. In between, where we have no solutions to match, we require that the numerical solutions behave smoothly. Thus, we have the ability to check the importance of any given time scale to the results by reducing or increasing the time step.

The interpolating approximation is central to our formulation of the finite-difference equations. To illustrate its application, we analyze equations that completely resolve the particle motion when Δt is small but approach the zero gyroradius or magnetohydrodynamic limit when Δt is large, extend the equations to model a strongly magnetized plasma by adding finite gyroradius corrections, and, finally, model electron scattering in a Lorentz gas using a Monte Carlo method.

1. The Particle Momentum Equation in a Collisionless Plasma

We solve the particle momentum equation for $\bar{\mathbf{u}}_{ps}$ by calculating $\bar{\mathbf{u}}_{ps} \cdot \mathbf{B}^0$ and $\bar{\mathbf{u}}_{ps} \times \mathbf{B}^0$, and substituting the results into Eq. (11). For convenience, we define an intermediate velocity,

$$\tilde{\mathbf{u}}_{ps} = \mathbf{u}_{ps}^0 + (q_s/m_s)\mathbf{E}^\theta \Delta t/2 \qquad (22)$$

and a cyclotron frequency,

$$\Omega_s \equiv (q_s/m_s)\mathbf{B}^0/c,$$

and write the equation for $\bar{\mathbf{u}}_{ps}$,

$$\bar{\mathbf{u}}_{ps} = [\tilde{\mathbf{u}}_{ps} + (\tilde{\mathbf{u}}_{ps} \times \boldsymbol{\Omega}_s)\,\Delta t/2 + (\tilde{\mathbf{u}}_{ps} \cdot \boldsymbol{\Omega}_s)\boldsymbol{\Omega}_s\,\Delta t^2/4]/(1 + \Omega_s^2\,\Delta t^2/4). \quad (23)$$

Equations (10) and (23) are coupled because the fields are evaluated at the advanced particle positions $\bar{\mathbf{x}}_{ps}$.

We assume that these fields are representative of the fields acting on the particles over the time interval Δt, that is, that the plasma is collisionless. If the plasma were collisional, close encounters between particles would result in very large and rapidly varying forces on the particle to which the value of the electric field $\mathbf{E}^\theta$, interpolated between $\mathbf{E}^1$ and $\mathbf{E}^0$, bears no relation.

To reduce collisions to a level that makes this assumption valid, we may either increase the number of particles we use to represent the plasma, or give the particles a finite size (Birdsall and Fuss, 1969; Morse and Nielson, 1969). As has been described previously, replacing the Dirac δ functions $\delta(\mathbf{x} - \mathbf{x}_p)$, used to define the velocity moments in Eqs. (7)–(9), by interpolation functions $S(\mathbf{x} - \mathbf{x}_p)$ with finite width reduces the maximum value of the force acting between particles and thus the collision frequency (Birdsall et al., 1970). In one reference, collisions are found to be proportional to the ratio of the Debye length λ_D to the width of S, typically Δx (Dawson, 1983). In explicit calculations, with $\lambda_D/\Delta x = O(1)$, the collision frequency is one hundred times smaller because the particles have finite size. In implicit calculations, $\lambda_D/\Delta x = O(10^{-1} - 10^{-3})$, and the collision frequency is correspondingly reduced. (To simplify this discussion, however, we continue to treat the particles as point particles with the understanding that the particles have finite size in actual applications.)

Consider the solutions of the equations of motion. In static fields, the particle energy is constant for all values of Δt. The energy change from time step to time step is [from Eq. (11)] $(\mathbf{u}^1)^2 - (\mathbf{u}^0)^2 = 2\bar{\mathbf{u}} \cdot (\mathbf{u} \times \boldsymbol{\Omega})\,\Delta t \equiv 0$.

For large time steps or strong magnetic fields (i.e., $\Omega \Delta t \gg 1$), the particle motion corresponds to the zero gyroradius limit. In this limit, the mean velocity is calculated from Eq. (23),

$$\mathbf{u}_{ps} = (\mathbf{E}^\theta \times \mathbf{B}^0)/B^{02} + (\mathbf{u}_{ps}^0 + (q_s/m_s)\mathbf{E}^\theta\,\Delta t) \cdot \mathbf{B}^0\mathbf{B}^0/B^{02}, \quad (24)$$

and describes the $\mathbf{E} \times \mathbf{B}$ drift of the particle perpendicular to the field and its drift and acceleration along the magnetic field. The terms, however, that describe the gyro motion of a particle about a magnetic-field line become negligible with increasing $\Omega\,\Delta t$. Thus, ∇B drifts, which depend on finite gyro orbits, are absent (Chandrasekhar, 1960). As we further increase $\mathbf{B}^0$, we find that $\mathbf{u}_{ps}$ perpendicular to $\mathbf{B}^0$ tends to zero, even as the motion along magnetic-field lines is unrestricted. The behavior of the equations as Δt increases corresponds to the behavior of equations that are asymptotic in m_e/m_i when $\Omega_e \Delta t \gg 1$ and $\Omega_i \Delta t < 1$, as in the two-fluid, magnetohydrodynamic (MHD) limit (Freidberg, 1982).

2. Strongly Magnetized Plasmas

As discussed earlier by Barnes and Kamimura (1982), we can describe the ∇B drifts in a strongly magnetized plasma using the guiding center approximation (Northrup, 1961, 1963). Including only the lowest order terms, the effective electric field becomes

$$(q_s/m_s)\mathbf{E}_{\text{eff}} = (q_s/m_s)\mathbf{E}_\| - \mu_{ps} \nabla B \qquad (25)$$

where μ_{ps} is the magnetic moment;

$$\mu_{ps} \equiv (\mathbf{u}_{ps} - \mathbf{B}(\mathbf{u}_{ps} \cdot \mathbf{B})/B^2)^2/2B.$$

[We assume that μ_{ps} does not change very much over a time step, which is a less restrictive condition than the usual $\partial \mu_{ps}/\partial t \ll \mu_{ps}\Omega_s$. Numerically, it may be possible to simulate cases such as that described in Kim and Cary (1983) where μ is not a good invariant.] Equation (25) describes the motion of a particle only when $\Omega_s \Delta t \gg 1$.

To write an interpolating approximation that is valid for all values of $\Omega_s \Delta t$, we exploit the special properties of the implicit finite-difference equation. Consider a magnetic moment $\tilde{\mu}$ defined by

$$\tilde{\mu} \equiv \{\tfrac{1}{2}[(\mathbf{u}_\perp^1)^2 + (\mathbf{u}_\perp^0)^2] - \bar{\mathbf{u}}_\perp^2\}/B = (\mathbf{u}_\perp^1 - \mathbf{u}_\perp^0)^2/8B, \qquad (26)$$

where $\mathbf{u}_\perp \equiv \mathbf{u} - \mathbf{B}(\mathbf{u} \cdot \mathbf{B})/B^2$. We have chosen this definition of $\tilde{\mu}$ because it is negligible for small $\Omega_s \Delta t$ and equal to the magnetic moment for large values of $\Omega_s \Delta t$. The reader can easily verify from Eq. (23) that $(\mathbf{u}_\perp^1 - \mathbf{u}_\perp^0)^2$ is given by

$$\tilde{\mu} = \tilde{\mathbf{u}}_\perp^2 \Omega_s^2 \Delta t^2/[8(1 + \Omega_s^2 \Delta t^2/4)B].$$

When $\Omega_s \Delta t \ll 1$ and $\tilde{\mu} = O(\Omega_s^2 \Delta t^2)$, the contribution of the ∇B term to the acceleration is $O(\Delta t^2)$ compared with the other terms, which are $O(1)$. When $\Omega_s \Delta t \gg 1$, $\tilde{\mu} \cong \mathbf{u}_\perp^2/B$. Thus, we only must replace $\tilde{\mathbf{u}}_{ps}$ [defined by Eq. (22)] by:

$$\tilde{\mathbf{u}}_{ps} = \mathbf{u}_{ps}^0 + [(q_s/m_s)\mathbf{E}^\theta - \tilde{\mu}_{ps} \nabla B^0] \Delta t. \qquad (27)$$

Now, even for $\Omega_s \Delta t \gg 1$, the momentum equation includes finite gyroradius effects;

$$\mathbf{u}_{ps}^1 = [(\mathbf{E}^\theta - \tilde{\mu}_{ps} \nabla B) \times \mathbf{B}^0]/B^{02} + [\mathbf{u}_{ps}^0 + (q_s/m_s)(\mathbf{E}^\theta - \tilde{\mu}_{ps} \nabla B) \Delta t] \cdot \mathbf{B}^0/B^{02}. \qquad (28)$$

The ∇B drift perpendicular to the field is explicitly present, as is the mirror force that accelerates a particle along magnetic-field lines toward the region of weaker field.

Several numerical solutions for a strongly magnetized plasma are illustrated in Figs. 2 and 3. In Fig. 2, we depict a particle moving in a mirror field. For all values of Δt the particle remains trapped in the magnetic mirror, even as the gyroradius decreases to zero. In Fig. 3, we depict the motion of a particle in fields such that $\mathbf{E}^\theta - \tilde{\mu}_{ps} \nabla B = 0$, and $\mathbf{E} \times \mathbf{B}$ and ∇B drifts balance. The particle remains motionless for large values of $\Omega_s \Delta t$ but moves in a retrograde fashion for $\Omega_s \Delta t < 1$, indicating that higher order corrections are important. Nevertheless, to the order of approximation of the equations, $\tilde{\mu}_{ps}$ gives the correct intermediate behavior, as well as the correct limits.

3. Collisional Plasmas

In dense plasmas, the accumulated effect of many collisions may result in very large deflections of the particle over a time step. When the collision frequency is large, the terms describing their effect may impose a constraint on the time step that increases the cost of the calculation significantly. Here,

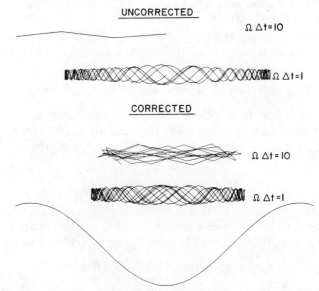

Fig. 2. When the time step is large, finite gyroradius corrections are necessary to calculate the particle motion in a magnetic mirror correctly. From the bottom, the magnetic field (mirror ratio = 2), particle motion with corrections for $\Omega \Delta t = 1$ and 10, and without corrections for the same time steps. Without corrections, the particle motion is not confined by the mirror when $\Omega \Delta t = 10$, even though it should be physically.

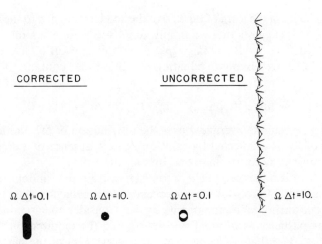

Fig. 3. Without finite gyroradius corrections, the particle may drift when it should not. In electric and magnetic fields such that the gradient B drift downward and the $E \times B$ drift upward balance to lowest order, the corrected motion is retrograde for small Δt (because of higher order terms), as shown on the left, or stationary for large Δt, as shown by the orbit second from the left. Without corrections, the particle drifts upward at the $E \times B$ speed, as shown on the right.

we describe an implicitly differenced model of collisions that allows us to treat an arbitrarily collisional plasma without reducing the time step (Cranfill et al., 1984).

For simplicity, we consider the elastic scattering of electrons due to collisions with massive ions, the so-called high Z approximation. Such collisions result in deflections of the electron's direction of travel without loss of energy. This problem has been treated by many authors; the most nearly similar method is described by Mason (1983).

Starting with a model developed by Rechester and Rosenbluth (1978) to describe electron transport in stochastic magnetic fields, we calculate scattering as though it were due to randomly directed magnetic fields, except that the fields are constrained to give the correct collisional scattering even in the presence of dc magnetic fields. As is well known, stochastic fields cause scattering perpendicular to magnetic drift surfaces only. The constraint causes the scattering due to collisions to have no preferential direction.

In a random magnetic field, the particle velocity is rotated:

$$d\mathbf{u}/dt = (q_s/m_s)(\mathbf{u} \times \delta\mathbf{B}/c), \tag{29}$$

where $\delta\mathbf{B}$ is the random-field component, and the particle energy is preserved. When $\delta\mathbf{B}$ is a vector constant in magnitude but random in direction, the average result of many rotations models the average effect of many collisions.

The similarity of the form of Eq. (29) to the acceleration due to the magnetic field in Eq. (11) suggests that we simply add $\delta \mathbf{B}$ to $\mathbf{B}$ to give a total rotation vector $\mathbf{R}_{ps}$, defined by $\mathbf{R}_{ps} \equiv \mathbf{\Omega}_{ps} + \delta\mathbf{\Omega}_{ps}$, with $\delta\mathbf{\Omega}_{ps} = \delta\mathbf{B}\, q_s/m_s c$. However, when we do so, we obtain an equation of motion that contains an inherent anisotropy,

$$\bar{\mathbf{u}}_{ps} = [\mathbf{u}_{ps}^0 + (\mathbf{u}_{ps}^0 \times \mathbf{R}_{ps}\, \Delta t/2) + (\mathbf{u}_{ps}^0 \cdot \mathbf{R}_{ps})\mathbf{R}_{ps}\, \Delta t^2/4]/(1 + \mathbf{R}_{ps}^2\, \Delta t^2/4),$$

for when we ensemble average $\bar{u}$ over the distribution of $\delta\Omega$, motion in the direction of $\mathbf{R}_{ps}$ is unaffected by collisions. In the absence of a dc magnetic field, we must thus constrain $\delta\Omega$ so that $\mathbf{u}^0 \cdot \delta\mathbf{\Omega} = 0$.

When there is a magnetic field, we must constrain the random-field component differently to recover the correct asymptotic behavior when many collisions occur during each time step: $v_s\, \Delta t \gg 1$. Consider an ensemble average over many particles with initial velocity $\mathbf{u}_{ps}^0$ and the requirement that $\delta\mathbf{\Omega}_{ps}$ be constructed so that collisions slow the average drift of the particles:

$$-v_{ps}\langle \bar{\mathbf{u}}_{ps}\rangle = \langle \bar{\mathbf{u}}_{ps} \times \delta\mathbf{\Omega}_{ps}\rangle. \tag{30}$$

For $\delta\mathbf{\Omega}_{ps}$ satisfying this condition, we can average Eq. (11),

$$\langle \mathbf{u}_{ps}^1 \rangle = \mathbf{u}_{ps}^0 + [(q_s/m_s)\mathbf{E}^\theta + \langle \bar{\mathbf{u}}_{ps}\rangle \times \mathbf{\Omega}_s - v_{ps}\langle \bar{\mathbf{u}}_{ps}\rangle]\, \Delta t, \tag{31}$$

where $\mathbf{u}_{ps}^0$, $\mathbf{E}^\theta$, and $\mathbf{\Omega}_s$ are independent of the average over $\delta\mathbf{\Omega}_{ps}$. We can then solve Eq. (31) for $\langle \bar{\mathbf{u}}_{ps}\rangle$:

$$\langle \bar{\mathbf{u}}_{ps}\rangle = \frac{\tilde{\mathbf{u}}_{ps}' + \tilde{\mathbf{u}}_{ps}' \times \mathbf{\Omega}_{ps}'\, \Delta t/2 + (\tilde{\mathbf{u}}_{ps}' \cdot \mathbf{\Omega}_{ps}')\mathbf{\Omega}_{ps}'\, \Delta t^2/4}{1 + \mathbf{\Omega}_{ps}'^2\, \Delta t^2/4}, \tag{32}$$

where $\tilde{\mathbf{u}}_{ps}' \equiv \tilde{\mathbf{u}}_{ps}/(1 + v_{ps}\, \Delta t/2)$ and $\mathbf{\Omega}_{ps}' = \mathbf{\Omega}_s/(1 + v_{ps}\, \Delta t/2)$. When $v_{ps}\, \Delta t \gg 1$, $\Omega_{ps}'\, \Delta t/2 \sim \Omega_s \tau_{ps}$.

A random field $\delta\mathbf{\Omega}_{ps}$, which obeys the constraint given by Eq. (30), has a component parallel to $\tilde{\mathbf{u}}_{ps}$ equal to

$$(\delta\Omega)_{\|} = -\Omega_{\|}',$$

and a component in a random direction perpendicular to $\tilde{\mathbf{u}}_{ps}$ given by

$$(\delta\Omega)_\perp = (2v_{ps}\, \Delta t\{1 + [\Omega_{\|}^2 + \Omega_\perp^2(1 + v\, \Delta t/2)]\, \Delta t^2/4\})^{1/2} \tag{33}$$

($\Omega_\perp'$ and $\Omega_\|'$ are components perpendicular and parallel to $\tilde{\mathbf{u}}_{ps}$, respectively). The derivation of the random field is given by Cranfill *et al.* (1984).

For this random field we can analyze the particle motion as $v\, \Delta t$ increases, especially as the plasma becomes more collisional. When Ω is zero and $v\, \Delta t \gg 1$, the particle experiences a drift,

$$\langle \mathbf{u}_{ps}\rangle \simeq (q_s/m_s)\mathbf{E}^\theta \tau_{ps},$$

independent of $\mathbf{u}_{ps}^0$ and equal to the velocity a particle acquires when accelerated by $\mathbf{E}^\theta$ over the mean free time between collisions. In the absence of an electric field, and for arbitrary $v \Delta t$, the mean square displacement is

$$(\mathbf{u}_{ps} \Delta t)^2 \simeq \langle \tilde{\mathbf{u}}_{ps}^2 \rangle (1/\Delta t + 1/\tau_{ps})^{-1} \Delta t. \tag{34}$$

When τ_{ps} is large, the distance a particle travels increases as Δt. When τ_{ps} is small, the distance increases as $\Delta t^{1/2}$, as it should for a particle executing a random walk (Chandrasekhar, 1954). When an electric field is acting on the particles, the energy increases linearly in time proportional to v, the collision frequency.

Although we obtain the correct Ohm's law, which depends on the average velocity, we do not obtain the correct equipartition of energy when $\Omega'_s \Delta t > 1$ because the rotation due to collisions depends on $\bar{\mathbf{u}}$, whose value perpendicular to the field decreases to zero as $\Omega'_s \Delta t$ increases. Thus, scattering out of the perpendicular direction is decreased more than scattering into it, and the parallel energy decreases as $\Omega'_s \Delta t$ increases even as $\Omega_s \tau_s$ remains constant. This property of the model presently limits its application to cases in which $\Omega'_s \Delta t \gtrsim 2$, unless collisions are unimportant.

B. The Moment Description

We derived the differential form of the evolution equations (19) and (20) from the definitions of the moments by differentiating them with respect to time. Now we compute the change in the moments over the finite interval Δt by substituting the finite-difference approximations to the particle equations of motion into the definitions of the moments and expanding them in powers of the particle displacement over a time step.

1. The Local Approximation

Consider the evolution of the charge density. At $t = (n + 1) \Delta t$, the density is given by

$$\bar{n}_s(\mathbf{x}) = \sum_p q_s \, \delta(\mathbf{x} - \bar{\mathbf{x}}_{ps}). \tag{35}$$

This expression can be expanded about $\mathbf{x}_{ps}^0$, the time when the particle data are known:

$$\bar{n}_s(\mathbf{x}) = q_s \sum_p \left[\delta(\mathbf{x} - \mathbf{x}_{ps}^0) - \bar{\mathbf{u}}_{ps} \Delta t/2 \cdot \nabla \delta(\mathbf{x} - \bar{\mathbf{x}}_{ps}) \right.$$
$$\left. + \tfrac{1}{2}(\bar{\mathbf{u}}_{ps}\bar{\mathbf{u}}_{ps}) \Delta t^2/4 : \nabla \nabla \delta(\mathbf{x} - \mathbf{x}_{ps}^0) + \cdots \right]. \tag{36}$$

The sequence of terms will converge only when the average particle displacement in a time step is less than the length scale of variation of the data, $k v_{\text{th}} \Delta t < 1$. Using the definitions above and writing the expansion in terms of the moments,

$$n_s^1(\mathbf{x}) = n_s^0(\mathbf{x}) - \nabla \cdot \bar{\mathbf{j}}_s \Delta t + \tfrac{1}{2} \nabla \nabla : \Pi \, \Delta t^2/2 + \cdots, \tag{37}$$

we observe that we may neglect the $O(\Delta t^2)$ term when $kv_{\text{th}} \Delta t < 1$ so that Eq. (37) is approximately the continuity equation. This we call the local approximation.

As is discussed in an earlier paper (Brackbill and Forslund, 1982b) the current equation is expanded in a similar way:

$$\bar{\mathbf{j}}_s(\mathbf{x}) = q_s \sum_p \{u_{ps}^0[\delta(\mathbf{x} - \mathbf{x}_{ps}^0) - \bar{u}_{ps} \Delta t/2 \cdot \nabla\delta(\mathbf{x} - \bar{\mathbf{x}}_{ps})]$$
$$+ (\bar{\mathbf{u}}_{ps} - \mathbf{u}_{ps}^0) \delta(\mathbf{x} - \bar{\mathbf{x}}_{ps})\} + \cdots. \qquad (38)$$

We evaluate the first two terms directly from the definitions of the moments. We evaluate the last term by substituting from the equation of motion for individual particles, including both finite gyroradius corrections and collisional effects, Eq. (32). After substitution into the current equation, we obtain

$$\bar{\mathbf{j}}_s(\mathbf{x}) = \mathbf{j}_s^0(\mathbf{x}) + [(q_s/m_s)(\bar{n}_s \mathbf{E}^\theta + \bar{\mathbf{j}}_s \times \mathbf{B}^0/c)$$
$$- M_s \nabla B^0 - \nabla \cdot \Pi - v_s \bar{\mathbf{j}}_s] \Delta t/2, \qquad (39)$$

where v_s, the effective collision frequency, is defined by

$$v_s \equiv \sum_p \delta(\mathbf{x} - \mathbf{x}_{ps})/\sum_\rho \delta(\mathbf{x} - \mathbf{x}_{ps})/v_{ps},$$

and M_s, the magnetization, is defined by

$$M_s \equiv \sum_p \mu_{ps} \delta(\mathbf{x} - \mathbf{x}_{ps}).$$

We now can solve for $\bar{\mathbf{j}}_s$ with the result,

$$\bar{\mathbf{j}}_s = \frac{\tilde{\mathbf{j}}_s + (\tilde{\mathbf{j}}_s \times \Omega') \Delta t/2 + (\tilde{\mathbf{j}}_s \cdot \Omega')\Omega' \Delta t^2/4}{(1 + \Omega'^2 \Delta t^2/4)}, \qquad (40)$$

where $\tilde{\mathbf{j}}_s$ is defined by

$$\tilde{\mathbf{j}}_s \equiv [\mathbf{j}_s^0 + (\omega_s^2/4\pi \mathbf{E}^\theta - \nabla \cdot \Pi - M_s \nabla B^0)] \Delta t/2/(1 + v_s \Delta t/2), \qquad (41)$$

$\Omega' \equiv \Omega^0/(1 + v_s \Delta t)$, and $\omega_s^2 = 4\pi n_s q_s/m_s$ is the plasma frequency for species s. (Note that $\Omega' \Delta t/2$ approaches $\Omega\tau$ as $v \Delta t$ increases.)

Equation (40) for the current can be evaluated in various limits. Consider a collisional, magnetized plasma with Δt large so that $\Omega \Delta t > v \Delta t \gg 1$ resulting in $\Omega' \Delta t \cong \Omega\tau > 1$. The current $\tilde{\mathbf{j}}_s$ is then given approximately by

$$\tilde{\mathbf{j}}_s \cong [\omega_s^2/4\pi \mathbf{E}^\theta - \nabla \cdot \Pi - M_s \nabla B^0]\tau_s.$$

To cast the equation in a form we can compare with standard references, we neglect the contributions of the pressure and the magnetization current and substitute $\tilde{\mathbf{j}}_s$ into Eq. (40), resulting in an Ohm's law,

$$\bar{\mathbf{j}}_s = (\omega_s^2 \tau_s/4\pi)[\mathbf{E}^\theta + \mathbf{E}^\theta \times \Omega_s\tau_s + (\mathbf{E}^\theta \cdot \Omega_s)\Omega_s\tau_s^2]/(1 + \Omega_s^2\tau_s^2), \qquad (42)$$

that corresponds exactly to the conductivity tensor given by Jackson (1962).

2. The Method of Extrapolation

Previously, we described an implicit approximation to the pressure (Brackbill and Forslund, 1982b). Since then we have found that an implicit approximation to the pressure sometimes allows a slightly increased time step but more often leads to failure of the field calculation because it introduces nonlinear terms into Poisson's equation (54).

To replace the implicit pressure in future applications, we plan to use the extrapolation techniques developed for the direct method, as described in Friedman *et al.* (1981). In a manner similar to the direct method, we derive the moment equations by expanding in powers of the displacement from extrapolations along the unperturbed orbits. By doing so, we obtain moment equations that approximate the particle data more accurately than the moment equations above for longer time intervals. We eliminate the pressure and replace the time-step restriction determined by the thermal velocity by one determined by the drift velocity or trapping frequency.

The midpoint of the unperturbed orbit is calculated from

$$\hat{\mathbf{x}}_{ps} = \mathbf{x}_{ps}^0 + \hat{\mathbf{u}}_{ps} \Delta t/2, \tag{43}$$

where $\hat{\mathbf{u}}_{ps}$ is calculated from Eq. (32) by setting $\mathbf{E}^\theta$ to zero and evaluating Ω'_s at $\hat{\mathbf{x}}_{ps}$. [The latter approximation linearizes Eq. (32).] When the acceleration or the drifts due to the electric field are small compared with the streaming velocity $|\bar{\mathbf{u}}_{ps} - \hat{\mathbf{u}}_{ps}| < |\bar{\mathbf{u}}_{ps}|$, $\hat{\mathbf{x}}_{ps}$ is a good approximation to $\bar{\mathbf{x}}_{ps}$. Thus, when we repeat the sequence of steps leading to the continuity equation (36), this time by expanding about $\hat{\mathbf{x}}_{ps}$, we find that the higher order terms are smaller:

$$\bar{n}_s(\mathbf{x}) = \hat{n}_s(\mathbf{x}) - \nabla \cdot (\bar{\mathbf{j}}_s - \hat{\mathbf{j}}_s) \Delta t/2 \\ - \sum_p (\bar{\mathbf{x}}_p - \hat{\mathbf{x}}_p)(\bar{\mathbf{x}}_p - \hat{\mathbf{x}}_p) : \nabla \nabla \delta(\mathbf{x} - \hat{\mathbf{x}}_{ps}) \cdots. \tag{44}$$

Compared with the last term in Eq. (36), the last term in Eq. (44) is smaller and can be neglected. The extrapolated density $\hat{n}_s(\mathbf{x})$ is defined by

$$\hat{n}_s(\mathbf{x}) \equiv q_s \sum_p \delta(\mathbf{x} - \hat{\mathbf{x}}_{ps}). \tag{45}$$

Similarly, when we repeat the steps leading to the momentum equation (38), this time, expanding about $\hat{\mathbf{x}}_{ps}$, we find

$$\bar{\mathbf{j}}_s(x) = \sum_p \hat{\mathbf{u}}_{ps}[\delta(\mathbf{x} - \hat{\mathbf{x}}_{ps}) - (\bar{\mathbf{u}}_{ps} - \hat{\mathbf{u}}_{ps}) \cdot \nabla \delta(\mathbf{x} - \hat{\mathbf{x}}_{ps})] \\ + (\bar{\mathbf{u}}_p - \hat{\mathbf{u}}_{ps}) \delta(\mathbf{x} - \bar{\mathbf{x}}_{ps}). \tag{46}$$

Because $(\bar{\mathbf{u}}_{ps} - \hat{\mathbf{u}}_{ps})$ depends only on fields that have the same value for every particle at a given position, the second term in Eq. (46) contains only a mean flow contribution to the momentum equation. The last term must be evaluated from the particle momentum equation. Evaluating $\mathbf{\Omega}'_s$ at $\hat{\mathbf{x}}_{ps}$ makes it possible to write

$$\bar{\mathbf{u}}_{ps} = \hat{\mathbf{u}}_{ps} + \frac{(q_s \Delta t/2m_s)[\mathbf{E}^\theta + (\mathbf{E}^\theta \times \mathbf{\Omega}'_s)\,\Delta t/2 + (\mathbf{E}^\theta \cdot \mathbf{\Omega}'_s)\mathbf{\Omega}'_s\,\Delta t^2/4]}{1 + \Omega'^2_s \Delta t^2/4} \quad (47)$$

where $\hat{\mathbf{u}}$ depends entirely on the data at time level 0, $\mathbf{E}^\theta$ is evaluated at $\bar{\mathbf{x}}_{ps}$ as before, and $\mathbf{\Omega}'_s$ is evaluated at $\hat{\mathbf{x}}_{ps}$. (We note that energy of gyromotion is conserved as before.) We can now rewrite Eq. (40) in the standard form:

$$\bar{\mathbf{j}}_s(\mathbf{x}) = \hat{\mathbf{j}}_s(\mathbf{x}) - \nabla \cdot [(\bar{\mathbf{j}}_s - \hat{\mathbf{j}}_s)\hat{\mathbf{j}}_s/\hat{n}_s]\frac{\Delta t}{2}$$
$$+ \frac{(q_s \Delta t/2m_s)[\mathbf{E}^\theta + (\mathbf{E}^\theta \times \mathbf{\Omega}'_s)\,\Delta t/2 + (\mathbf{E}^\theta \cdot \mathbf{\Omega}'_s)\mathbf{\Omega}'_s\,\Delta t^2/4]}{1 + \Omega'^2_s \Delta t^2/4} \quad (48)$$

where $\hat{\mathbf{j}}_s(\mathbf{x})$ is defined by

$$\hat{\mathbf{j}}_s(\mathbf{x}) = q_s \sum_p \hat{\mathbf{u}}_p\, \delta(\mathbf{x} - \hat{\mathbf{x}}_{ps}). \quad (49)$$

Compared with the formulation given by Barnes *et al.* (1983), the momentum equation (48) contains the convective derivative explicitly. This may eliminate the instability noted by them in a drifting plasma.

By eliminating the contribution of the peculiar velocities (i.e., those that contribute to the pressure) from the moment equations, we eliminate the stability constraint imposed by the Courant condition on particle thermal speed and replace it by one on the drift speed. When the plasma is warm, the terms we have neglected in order to eliminate the pressure are small. When the plasma is cold ($0 < kv_{\text{th}} < \omega_s$), the explicit pressure is a better approximation than extrapolation.

Extrapolating the orbits requires that particles be moved in two steps. At the end of each cycle, $\bar{\mathbf{u}}_{ps}$ must be calculated from Eq. (47), and $\mathbf{x}^1_{ps}$ is calculated from

$$\mathbf{x}^1_{ps} = \hat{\mathbf{x}}_{ps} + (\bar{\mathbf{u}}_{ps} - \hat{\mathbf{u}}_{ps})\,\Delta t. \quad (50)$$

This calculation must be followed by a calculation of $\hat{\mathbf{u}}_{ps}$ for the next cycle from Eq. (32) with $\mathbf{E}^\theta = 0$ and $\mathbf{\Omega}'_{ps}$ evaluated at $\hat{\mathbf{x}}_{ps}$, which is calculated from Eq. (43). The position $\hat{\mathbf{x}}_{ps}$ and velocity $\hat{\mathbf{u}}_{ps}$ are saved until the next cycle. The moments $\hat{n}_s$ and $\hat{\mathbf{j}}_s$ are calculated from Eqs. (45) and (49) and saved. These are used in the moment evaluation equations (48) and (44) to estimate the fields $\mathbf{E}^\theta$ and $\mathbf{B}^1$, as outlined in Section III.C.

The advantages of this method over the one outlined in the previous section are the larger time steps for warm plasmas and the elimination of the pressure. The disadvantage is that the particle information and the moments calculated from it are based on estimated positions and velocities, and

9. Electromagnetic Phenomena in Plasmas

diagnostics derived from this information are based on approximate particle positions and velocities.

C. The Self-Consistent Fields

The implicit formulation of the Maxwell equations is written

$$(\mathbf{B}^1 - \mathbf{B}^0) + c(\nabla \times \mathbf{E}^\theta)\,\Delta t = 0, \tag{51}$$

$$\nabla \cdot \mathbf{B} = 0, \tag{52}$$

$$(\mathbf{E}^1 - \mathbf{E}^0) - c(\nabla \times \mathbf{B}^\theta)\,\Delta t = -4\pi c\bar{\mathbf{j}}\,\Delta t \tag{53}$$

and

$$\nabla \cdot \mathbf{E} = 4\pi n^\theta. \tag{54}$$

The moment equations derived in the previous section are

$$n_s^1 - n_s^0 = -\nabla \cdot \bar{\mathbf{j}}_s\,\Delta t, \tag{55}$$

$$\mathbf{j}_s^1 - \mathbf{j}_s^0 = -\nabla \cdot \mathbf{\Pi}^0\,\Delta t + (q_s/m_s)(\bar{n}_s\mathbf{E}^\theta + \bar{\mathbf{j}}_s \times \mathbf{B}^0/c)\,\Delta t. \tag{56}$$

[We have found that linearizing Eq. (56) by replacing $\bar{n}_s$ by n_s^0 has no perceptable effect on either the stability or the accuracy of the solutions.]

1. The Potentials

The electric and magnetic fields, **E** and **B**, are replaced as dependent variables in Maxwell's equations by the scalar and vector potentials, Φ and **A**:

$$\mathbf{E} = -\nabla\Phi - (\partial\mathbf{A}/\partial t)/c, \tag{57}$$

$$\mathbf{B} = \nabla \times \mathbf{A}, \tag{58}$$

with the Coulomb gauge condition, $\nabla \cdot \mathbf{A} = 0$.

With these substitutions, the homogeneous Maxwell equations [Faraday's law and the solenoidal condition on **B**, Eqs. (51) and (52)] are automatically satisfied. The inhomogeneous equations [Ampere's law and Poisson's equation (53) and (54)] depend on n and **j**:

$$-(\partial^2\mathbf{A}/\partial t^2)/c^2 + \nabla^2 A = [-4\pi\mathbf{j} + \nabla\,(\partial\Phi/\partial t)]/c, \tag{59}$$

$$\nabla^2\Phi = -4\pi n. \tag{60}$$

2. Consistent Differencing in Time

We have already given the time differencing of the equations in terms of **E** and **B**. Consistent with these choices, Poisson's equation and Ampere's law must be written

$$\nabla^2 \Phi^1 = -4\pi n^1, \tag{61}$$

$$-(\mathbf{A}^1 - 2\mathbf{A}^0 + \mathbf{A}^{-1})/c^2 \, \Delta t^2 + \nabla^2 \mathbf{A}^\theta = [-4\pi \bar{\mathbf{j}} + (\nabla \Phi^1 - \nabla \Phi^0)/\Delta t]/c, \tag{62}$$

where we have defined $\nabla \times \mathbf{A}^0 = \mathbf{B}^0$ and $(\partial \mathbf{A}/\partial t)^\theta = (\mathbf{A}^1 - \mathbf{A}^0)/\Delta t$. There is an inconsistency in our choices for the source terms, n and $\mathbf{j}$. Consider Ampere's law. As described in Nielson and Lewis (1976), the Darwin approximation, which neglects retardation effects, yields an Ampere's law written

$$\nabla^2 \mathbf{A} = -4\pi \mathbf{j}_t/c, \tag{63}$$

where $\mathbf{j}_t$ is defined by $\mathbf{j}_t \equiv \mathbf{j} - \nabla(\partial \Phi/\partial t)/4\pi$. We will consider, as did they, charged fluid motion transverse to the x direction. All quantities vary only in the x direction, and $\mathbf{A}$ and $\mathbf{j}$ are transverse to the x direction. The density n is constant. The equations of motion are then $\mathbf{j}^1 - \mathbf{j}^0 = (q/m)\mathbf{E}^\theta \Delta t$, $\mathbf{E}^\theta = -(\mathbf{A}^1 - \mathbf{A}^0)/c \, \Delta t$, and $\nabla^2 \mathbf{A}^\theta = -(4\pi/c)\bar{\mathbf{j}}$. Solving these equations for $\mathbf{A}^1 - \mathbf{A}^0$ yields a Helmholtz equation,

$$\nabla^2(\mathbf{A}^1 - \mathbf{A}^0) = (1/2\theta)(\omega_e^2/c^2)(\mathbf{A}^1 - \mathbf{A}^0),$$

with a scale length equal to the collisionless skin depth c/ω_e. In the case considered by Nielson and Lewis (1976), the solution is unstable because $\theta < 0$. In our case, the solution is stable because $\frac{1}{2} < \theta < 1$. There is an error, however, because the collisionless skin depth is multiplied by $(2\theta)^{1/2}$, which can vary between 1 and $2^{1/2}$. If we were to evaluate $\mathbf{j}$ at $(n + \theta) \Delta t$, we could eliminate this error. However, we would also introduce dissipation, which would rapidly damp the particle motion.

The choice of $\bar{\mathbf{j}}$ in Ampere's law forces us to write the continuity equation as in Eq. (37):

$$n^1 - n^0 + \nabla \cdot \bar{\mathbf{j}} \, \Delta t = 0,$$

for when we take the divergence of Ampere's law, we find

$$-4\pi(\nabla \cdot \bar{\mathbf{j}}) + \nabla^2(\Phi^1 - \Phi^0)/c \, \Delta t = 0.$$

From Poisson's equation, we find that $\nabla^2(\Phi^1 - \Phi^0) = -4\pi(n^1 - n^0)$. Thus, any choice other than $\bar{\mathbf{j}}$ yields an inconsistency in Ampere's law that manifests itself as $\nabla \cdot \mathbf{A} \neq 0$.

D. Spatial Differencing

Space does not permit us to give much emphasis to the approximation of spatial derivatives. Fortunately, the methods discussed in many references can be used with the implicit formulation. For example, Dawson (1983) describes a Fourier transform method for solving the field equations. Langdon

9. Electromagnetic Phenomena in Plasmas

and Lasinski (1976) describe finite-difference approximations to the derivatives appearing in the equations for **E** and **B**, and Morse and Nielson (1969) describe difference equations for the potential equations that give essentially similar results.

One can identify many errors and make many improvements in the approximation of spatial derivatives. However, we have found that the resulting improvements in accuracy of the implicit moment method are marginal, which agrees generally with the analysis given for the direct method (Barnes and Kamimura, 1982; Cohen et al., 1984; Langdon et al., 1983). Thus, we are motivated to look beyond standard methods for further improvements.

We have learned from the implicit code VENUS that, as we extend the time scales of our calculations, we also extend the spatial domain of dependence of the solution. Our computational domain is no longer small compared with the physical domain, or far from any boundary. Therefore, we must represent the geometry of the domain, whether it is a limiter in a tokamak or the support stalk in a laser target, if we are to model the physics accurately. Further, as we move our boundaries to greater distances, we must still be able to resolve small-scale structures in the domain.

Body-fitted coordinates (Thompson et al., 1974) provide variable resolution and can be applied to plasma simulation using an adaptive-grid generator to give control over the distribution of nodes in the interior of the domain (Brackbill and Saltzman, 1982). Much of the formulation is similar to particle-in-cell (PIC) calculations of fluid flow (Brackbill and Ruppel, 1984).

IV. THE ALGORITHM FOR THE IMPLICIT MOMENT METHOD

We must solve the finite-difference approximations to the continuity and current equations simultaneously with Maxwell's equations to estimate the fields at $t = (n + 1) \Delta t$ using the particle data at $t = n \Delta t$. The equations are coupled and nonlinear, and we must solve them iteratively.

A. Solving the Potential Equations

In the iteration, we solve the potential equations for φ^θ and $\mathbf{A}^1$, calculate from these the electric field $\mathbf{E}^\theta$, and, by back substitution, calculate the resulting charge and current density, n^1 and $\bar{\mathbf{j}}$. To explain how the potential equations are solved, we shall describe the algorithm for the solution of Poisson's equation in detail.

If we simply solve Poisson's equation [Eq. (54)] as written, the iteration will diverge; the implicit dependence of n^θ on φ^θ in the source term will dominate when Δt is large. We must first transpose all dependence of φ^θ to the left-hand side (LHS).

We can identify the dependence on φ^θ using the continuity and momentum equations. The charge density n^θ is given by the continuity equation (37) and the relation $n^\theta = \theta(n^1 - n^0)$. The dependence of the current density on $\mathbf{E}^\theta$ is explicitly identified in Eq. (48). Combining these equations to evaluate n^θ yields the expression

$$n^\theta = \sum_s \{4\pi[n_s^0 - \nabla \cdot \hat{j}_s]\theta \, \Delta t + (\mathbf{A}^1 - \mathbf{A}^0) \cdot \nabla \cdot \mathbf{k}_s/c \, \Delta t + \nabla \cdot \mathbf{k}_s \cdot \nabla \varphi^\theta\}, \quad (64)$$

where $\mathbf{k}_s$ is defined by

$$\mathbf{k}_s \cdot \mathbf{E}^\theta \equiv \beta_s[\mathbf{E}^\theta + (\mathbf{E}^\theta \times \mathbf{\Omega}_s') \, \Delta t/2 + (\mathbf{E}^\theta \cdot \mathbf{\Omega}_s')\mathbf{\Omega}_s' \, \Delta t^2/4], \quad (65)$$

and β_s is defined by

$$\beta_s \equiv (\omega_s^2 \theta \, \Delta t^2/2)/(1 + \Omega_s'^2 \, \Delta t/4). \quad (66)$$

Noting that $4\pi n^\theta - \nabla \cdot \mathbf{k} \nabla \varphi^\theta$ does not depend on φ^θ, we can rewrite Poisson's equation in a form that is suitable for iteration:

$$-\nabla \cdot (\mathbf{k} + \mathbf{I}) \nabla \varphi^\theta = 4\pi n^\theta - \nabla \cdot \mathbf{k} \nabla \varphi^\theta \quad (67)$$

where $\mathbf{I}$ is the unit tensor.

In an inhomogeneous, magnetized plasma, $\mathbf{k}$ is a spatially varying tensor and it is not possible to solve Eq. (67) as written using standard, fast Poisson solvers. Instead, one can use one of a number of different methods such as incomplete factorization (Kershaw, 1978) or multi-grid iteration (Brandt, 1977). We have not used these, because in our experience the method described by Concus and Golub (1973) and Buzbee et al. (1970) to accelerate convergence of a Poisson iteration has been satisfactory.

One can easily show that $\mathbf{k}/|\mathbf{k}|$, which simply rotates $\nabla \varphi$ without changing its magnitude, can be replaced by a scalar for both large and small values of Δt.

We first examine the components of $\mathbf{k}$, which we write

$$\hat{x} \cdot \mathbf{k} \cdot \nabla \varphi = k_{11}(\partial \varphi/\partial x) + k_{12}(\partial \varphi/\partial y),$$

$$\hat{y} \cdot \mathbf{k} \cdot \nabla \varphi = k_{21}(\partial \varphi/\partial x) + k_{22}(\partial \varphi/\partial y),$$

where the coefficients are given by,

$$k_{11} = \sum_s \beta_s(1 + \Omega_{1s}'^2 \, \Delta t^2/4),$$

$$k_{12} = \sum_s \beta_s(\Omega_{3s}' + \Omega_{1s}'\Omega_{2s}' \, \Delta t/2) \, \Delta t/2,$$

$$k_{21} = \sum_s \beta_s(-\Omega_{3s}' + \Omega_{1s}'\Omega_{2s}' \, \Delta t/2) \, \Delta t/2,$$

and

$$k_{22} = \sum_s \beta_s(1 + \Omega_{1s}'^2 \, \Delta t^2/4),$$

where the subscripts 1–3 denote the x, y, z components, respectively. The determinant of **k**, from Eq. (66), $(K - 1)^2 \equiv \det(\mathbf{k}) = \sum_s (\omega_s \theta \, \Delta t^2/2)^2$, is independent of $\Omega' \, \Delta t$. Furthermore, for both large and small values of Δt, k_{11} and $k_{22} \simeq K - 1$, but k_{12} and $k_{21} \ll 1$. Thus, we can replace Eq. (67) by the equation,

$$-\nabla \cdot K \nabla \varphi^\theta = 4\pi n^\theta - \nabla \cdot (K - 1) \nabla \varphi^\theta, \tag{68}$$

where K is now a scalar. In this equation, the residual dependence of the right hand side on φ^θ is small and the iteration will converge.

We may now make the change of variables suggested by Concus and Golub (1973) to accelerate the convergence of the iteration. Substituting $\psi = K^{1/2} \varphi$ into Eq. (68) and writing the resulting equation in a form suitable for iteration yields,

$$-\nabla^2 \psi^{(l+1)} + \bar{\Lambda} \psi^{(l+1)} = s - \nabla \cdot (K - 1) \nabla \varphi^{(l)} / K^{1/2} + (\bar{\Lambda} - \Lambda) \psi^{(l)}, \tag{69}$$

where $\Lambda \equiv \nabla^2 K^{1/2} / K^{1/2}$, $s \equiv 4\pi n^\theta / K^{1/2}$, and $\bar{\Lambda} \equiv (\Lambda_{\max} + \Lambda_{\min})/2$. The superscripts refer to the iterate level.

Space does not permit our describing the solution of the vector potential equation (59) in similar detail. The method is similar to the one above and is described in Brackbill and Forslund (1982b).

V. PROPERTIES OF THE IMPLICIT MOMENT METHOD

We have discretized the particle equations of motion, the moment equations, and the Maxwell relations and now must examine the numerical dispersion for finite time steps. First, we must establish the numerical stability of the equations. Only for stable solutions is it meaningful to ask that the numerical solutions faithfully reproduce the solution to the partial differential equations. Second, we must understand the effect of finite time steps on the dispersion so that we can relate the numerical and physical solutions. Third, we must estimate the errors due to finite time steps so that we can establish the validity of the results of a particular calculation.

To examine the numerical dispersion, we have used two standard methods, linear dispersion analysis and the energy method. In Section II.A, we have applied the numerical dispersion function to the finite grid instability. In Section V.A, we linearize the moment equations to examine the stability of equations used to estimate the fields. Finally, in Section V.B, we calculate the energy integral in a form that allows us to identify the sources of error and to monitor the accuracy of numerical computations.

A. Dispersion Analysis of the Moment Equations

Consider the moment equations. Their properties accurately reflect the properties of the plasma in certain limits, but not all. The moment and particle descriptions are equivalent in the following ways. If the number of particles is finite, then a large but finite number of moment equations will completely describe the particle data. In particular, when there are so few particles that the average velocity of the particles at each point is equal to the velocity of the particle at that point, all moments can be calculated from the zeroth and first moments. For example, the pressure tensor, $\Pi(\mathbf{x})$, may be expressed in terms of the current and charge densities, $\Pi(\mathbf{x}) = q \sum_p \mathbf{u}_p \mathbf{u}_p \, \delta(\mathbf{x} - \mathbf{x}_p) = \mathbf{j}(\mathbf{x})\mathbf{j}(\mathbf{x})/n(\mathbf{x})$. All higher moments are similarly expressible in terms of the zeroth and first moments.

Further, the moment evolution equations describe the evolution of the particle distribution more and more accurately as $v_{\text{th}}\tau'/L$, the ratio of the mean free path to the gradient scale length, becomes smaller, where τ'_s is the effective mean free time between collisions equal to the harmonic mean of the physical collision time τ_s and the time step Δt, as given by Eq. (34). Thus, either a more collisional plasma or a shorter time step will result in a closer correspondence between the evolution of the moments and the evolution of the particle distribution.

Thus, we can apply meaningfully the results of the dispersion analysis of the moment equations to the evolution of the particle distribution in a conditional way. We recognize that the correspondence between moment and particle solutions is inexact in general, and that stability of the moment equations does not guarantee stability of the particle equations. However, instability of the moment equations certainly causes instability of the overall system.

1. Spectral Compression and Damping

The equations used to estimate the fields, Eqs. (37), (40), and (51)–(54) are time-centered when $\theta = \frac{1}{2}$ but are used, usually with $\theta > \frac{1}{2}$, to introduce dissipation of light waves and electrostatic oscillations. From the linearized equations derived by writing the dependent variables in the equations of motion as the sum of constant and fluctuating parts, linearizing the dependence on the fluctuating part, and Fourier-analyzing the resulting equations,

$$0 = n_k^1 - n_k^0 + i\mathbf{k} \cdot \bar{\mathbf{j}}_k \, \Delta t, \tag{70}$$

$$0 = \mathbf{j}_k^1 - \mathbf{j}_k^0 - (q/m)(n_0 \mathbf{E}_k^\theta + \mathbf{j}_k \times \mathbf{B}_0/c) \, \Delta t + i\mathbf{k} \cdot \Pi_k \, \Delta t, \tag{71}$$

$$0 = \mathbf{B}_k^1 - \mathbf{B}_k^0 + ic(\mathbf{k} \times \mathbf{E}^\theta) \, \Delta t, \tag{72}$$

$$0 = \mathbf{E}_k^1 - \mathbf{E}_k^0 - ic(\mathbf{k} \times \mathbf{B}_k^\theta) \, \Delta t + 4\pi \bar{\mathbf{j}}_k \, \Delta t, \tag{73}$$

we obtain the characteristic equation. When solutions of the form $\lambda n_k^0 = n_k^1$ are substituted into these equations and the determinant of coefficients set to zero, the solutions of the characteristic equation are the normal modes of the system. The solutions are stable, that is, have no exponentially growing solutions, when $\lambda \lambda^* \leq 1$.

When there is no plasma, the Maxwell equations describe the propagation of electromagnetic waves in a vacuum. The amplification is given then by

$$\lambda \lambda^* = \{1 + [(1 - \theta)ck\,\Delta t]^2\}/[1 + \theta ck\,\Delta t)^2]. \tag{74}$$

Only for $\theta \geq \frac{1}{2}$ are the difference equations stable with $\lambda\lambda^* \leq 1$. When $\theta > \frac{1}{2}$, $\lambda\lambda^* < 1$, and modes with $k > 0$ decay exponentially.

The dispersion properties of the entire system are similar. Numerical differencing with $\theta \geq \frac{1}{2}$ will result in a stable solution, with increasing dissipation as Δt increases. The dispersion properties of the coupled field and moment equations are illustrated in Figs. 4–6. When Δt is small, the variation of ω, the real part of the frequency, with k, the wave number, reproduces the physical dispersion as shown in Fig. 4. However, when we increase the time step to $\Delta t = \pi\omega_e^{-1}$, the numerical dispersion compresses the entire spectrum into the interval $-1 < \omega/\omega_e < 1$, as shown in Fig. 5. The lowest frequencies are reproduced accurately, but the highest frequencies are replaced by the

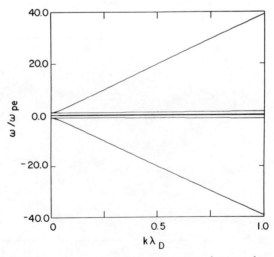

Fig. 4. The numerical dispersion curves for the moment equations are shown for wave propagation in the direction of the magnetic field. From the top, the curves represent light waves, electron plasma oscillations, whistler and ion-acoustic waves. With $\Delta t = 0.01\omega_e^{-1}$, physical frequencies are accurately reproduced. [From Brackbill and Forslund (1982b).]

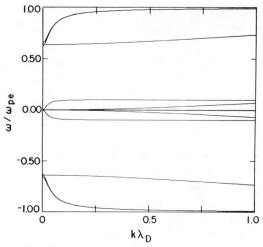

Fig. 5. For the same case as that of Fig. 4, but with $\Delta t = \pi \omega_e^{-1}$, the numerical frequencies are compressed into the interval $-1 \leq \omega/\omega_e \leq 1$. The curves lie in the same order as in Fig. 4. The whistler and ion-acoustic frequencies are distinguishable and accurately reproduce the physical values, even though the frequencies of the light waves and plasma oscillations are much smaller than their physical values. [From Brackbill and Forslund (1982b).]

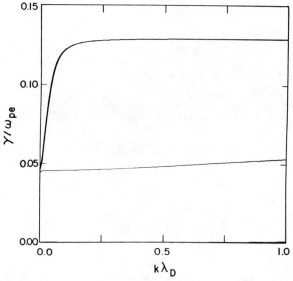

Fig. 6. The damping for $\Delta t = \pi \omega_e^{-1}$, $\theta = 0.6$ is shown. The curves are in the same order as in Fig. 5, showing that damping increases with $\omega \Delta t$. [From Brackbill and Forslund (1982b).]

9. Electromagnetic Phenomena in Plasmas

Nyquist frequency so that the amplitude alternates in sign from one cycle to the next. In Fig. 6, where γ, the imaginary part of the frequency is plotted, we see that unresolved waves with $\omega \Delta t \sim 1$ are strongly damped.

It is precisely this spectral compression of the high frequencies and the selective damping of unresolved modes that allows us to model embedded time-scale phenomena. The unresolved high-frequency waves are eliminated by this damping, but can be recovered by decreasing the time step. By contrast, they are also eliminated from the Darwin approximation but cannot be recovered no matter how small the time step. There are cases in which this can matter. For example, low-frequency electromagnetic waves can be excited by plasma oscillations in a cavity and driven to large amplitudes.

B. The Energy Integral

Accounting for the energy exchange between various modes of a system is often central to solving nonlinear problems. For example, identifying the source of dissipation is essential to understanding the saturation of the lower hybrid drift instability, the level of turbulence to which that corresponds, and, hence, to the effect of the instability on transport (Drake and Lee, 1981). To apply the implicit moment method to such a problem, we must identify the sources of error in energy conservation and estimate their size.

The change in total energy in a time step is calculated by summing the changes in the field and particle energies given by the finite-difference approximations to the equations of motion. The change in the field energy ΔW_F is

$$\Delta W_F = \frac{1}{4}\pi \int [\bar{\mathbf{E}} \cdot (\mathbf{E}^1 - \mathbf{E}^0) + \bar{\mathbf{B}} \cdot (\mathbf{B}^1 - \mathbf{B}^0)] \, dV. \tag{75}$$

Substituting from Eqs. (51) and (53) and noting that $\mathbf{E}^\theta = \bar{\mathbf{E}} + (\theta - \frac{1}{2})(\mathbf{E}^1 - \mathbf{E}^0)$, and similarly for $\mathbf{B}^\theta$, we may write

$$\Delta W_F = (c\,\Delta t/4\pi) \int \nabla \cdot (\mathbf{E}^\theta \times \mathbf{B}^\theta) \, dV - \Delta t \int (\mathbf{E}^\theta \cdot \bar{\mathbf{j}}_F) \, dV$$
$$- \left(\theta - \frac{1}{2}\right) \int [(\mathbf{E}^1 - \mathbf{E}^0)^2 + (\mathbf{B}^1 - \mathbf{B}^0)^2] \, dV. \tag{76}$$

The first term in Eq. (76) is the integral of the divergence of the Poynting flux, and its value is equal to the net flux of field energy across the boundary of the domain. For an isolated system, its value is zero. The second term is the work done by the fields on the particles and is proportional to $\bar{\mathbf{j}}$, the current at $(n + \frac{1}{2})\,\Delta t$ estimated from the moment equations. The third term is the dissipation of the field energy with $\theta > \frac{1}{2}$, which we have deliberately introduced to damp high-frequency electromagnetic and electrostatic oscillations.

The change in the particle kinetic energy in a time step ΔW_P is given by

$$\Delta W_P = \sum_s \sum_p m_s \bar{\mathbf{u}}_{ps} \cdot (\mathbf{u}_{ps}^1 - \mathbf{u}_{ps}^0). \tag{77}$$

The motion of each individual particle satisfies an equation of motion whose ensemble average is given by Eq. (31), and whose scalar product with $\bar{\mathbf{u}}_{ps}$ is given by

$$\Delta W_p = \sum_s q_s \sum_p \bar{\mathbf{u}}_{ps} \cdot \mathbf{E}^\theta \, \Delta t = \Delta t \int \bar{\mathbf{j}}_p \cdot \mathbf{E}^\theta \, dV. \tag{78}$$

The energy exchange between the particles and the field may result in a change in the total energy because our estimate of the current from the fluid equations does not correspond exactly to the actual particle motion. We have made certain approximations in summing the contributions of the particles. For example, the τ_s we use to estimate the fields is an ensemble average value, not the actual value used to move the particles.

In our previous discussion of the energy analysis (Brackbill and Forslund, 1982b), we explored the relationship between energy conservation and the closure assumption. Our subsequent experience with numerical calculations has shown the principal value of the calculation of changes in the total energy is in diagnosing the overall accuracy of a particular numerical calculation. Typically, an increase in the total energy indicates numerical instability, and a decrease of a few percent over many hundreds of time steps indicates a successful calculation. The most common reason for an increase in field energy is the presence of a finite grid instability.

VI. COMPUTATIONAL EXAMPLES

The implicit formulation for a weakly magnetized collisionless plasma described in Section III.A is embodied in the code VENUS. Among the published studies using VENUS are the Weibel instability (Brackbill and Forslund, 1982b) and the related heat-flow-driven filamentation instability in laser plasmas (Brackbill and Forslund, 1982a), collisionless shocks (Quest et al., 1983; Forslund et al., 1984), lateral electron energy transport in laser targets (Forslund and Brackbill, 1982), and the nonlinear evolution of the lower hybrid drift instability (Brackbill et al., 1984). In all of these studies, the greater length and time scales accessible to the implicit formulation have been essential.

We review two of these studies to illustrate the usefulness of the implicit method and to characterize its behavior. In all of these studies, the important physical length scale is the collisionless skin depth c/ω_e, which corresponds to the quasi-neutral limit when the electron velocities are non-

9. Electromagnetic Phenomena in Plasmas

relativistic, $u_e/c \ll 1$. In this limit, the collisionless skin depth corresponds to many Debye lengths $\lambda_D/(c/\omega_e) = u_e/c \ll 1$, and choosing a grid spacing the order of c/ω_e scale lengths causes the mesh spacing to correspond to many Debye lengths $\lambda_D/\Delta x < 1$.

Since the time step is chosen to resolve the motion of the electron through the grid, $0 < u_e \Delta t/\Delta x < 1$, $\omega_e \Delta t > 1$. Thus, long space scales relative to the screening distance λ_D correspond to long times relative to the plasma frequency.

With an explicit formulation, a calculation on the c/ω_e length scale is very expensive. For stability, we must require that $\omega_e \Delta t \gtrsim O(1)$ and $\lambda_D/\Delta x = O(1)$, resulting in $(c/\omega_e)/\Delta x \gg 1$ when the plasma is nonrelativistic. Thus, with the explicit formulation, we must use a finer mesh and a smaller time step at much greater cost.

We now describe the results of several calculations.

A. Laser-Induced Ion Emission

Light energy is converted to hydrodynamic energy in a laser target. The conversion occurs in several steps. Laser energy is absorbed first by electrons. Especially in long-wavelength lasers, such as the CO_2 laser with $\lambda = 10.6$ μm, comparatively few electrons are heated to very high temperatures: $T_e = 50$–500 keV. The hot electrons transport energy away from the laser spot and into the target, where it is eventually converted to hydrodynamic motion.

Details of the conversion, such as the partitioning of energy between ablation in the corona and compression of the target, are determined by details of the absorption and transport on the electron time scale. Plasma simulations are useful in modeling laser absorption and have been successful in matching experimental data. The understanding we have gained through simulation enables us to model absorption phenomenologically. Fluid modeling of electron energy transport in laser targets also has been useful, and the results match some experimental data. There are some discrepancies, however, and the fluid models are not really applicable because the electrons are collisionless. Flux limiters are generally used to prevent divergences in the fluid equations, but these have an ad hoc character that is not entirely satisfactory (Malone et al., 1975; Brackbill and Goldman, 1983).

In modeling electron transport with VENUS, we have found that convective transport is more important than thermal diffusion in collisionless plasmas with focused laser beams. As is well known, the lateral temperature gradient at the edge of the laser spot and the normal density gradient sustained by a rarefaction shock at the critical surface interact to generate a strong magnetic field. When the plasma is weakly collisional, $\Omega_e \tau_{ei} > 1$, the magnetic field and hot electrons convect laterally to great distances from the laser spot. The hot electrons and magnetic field together form a sheath in which much of the electron energy is stored until the electrostatic field due to the trapped electrons extracts ions from the target surface.

In the simulation, the laser energy absorption is modeled by a localized heating of electrons in a spot 60 μm wide from a background temperature of 2.5 to 25 keV, corresponding to a laser intensity of 5×10^{13} W cm^{-2}. The density of the target increases from zero to twice the critical density in a distance equal to c/ω_e.

After nearly 10^4 plasma periods, the density profile is altered, as shown in Fig. 7. The laser beam enters from the left. Behind the spot, and for some distance on either side, the target is compressed. In the spot region, the compression is a reaction to the formation of a plasma of accelerated ions. To the sides, the compression is due to the self-generated magnetic fields depicted in Fig. 8. The height of the plot indicates a magnetic field intensity that reverses sign as one moves laterally through the spot and normally through the surface. The first reversal is due to symmetry, the second to the

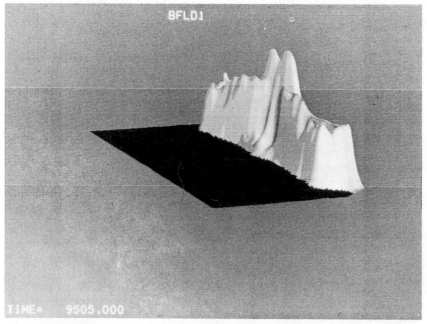

Fig. 7. The density profile in a laser target plasma is altered by the self-generated field. In the illustration, the height of the figure represents the density, which is driven above its initial value behind the laser spot by the reaction to ion blow-off, and to the sides of the spot by the magnetic field.

9. Electromagnetic Phenomena in Plasmas 303

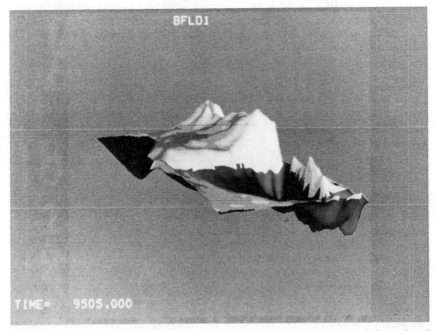

Fig. 8. A magnetic field is generated in a laser target plasma when the laser is focused and the plasma is collisionless. The laser enters from the left and is absorbed by dense plasma on the right. The height of the figure represents the intensity of the magnetic field, whose maximum value exceeds 10^6 G.

structure of the magnetic shock at the surface as discussed in B. Bezzerides and R. Jones (private communication, 1983).

Experimental evidence for lateral transport is given by Yates et al. (1982), where photographs of x-ray emission characterizes the deposition of electron energy into the target. The dark regions correspond to areas where there is no deposition because of magnetic insulation.

Other evidence comes from measurements of fluences of accelerated ions at some distance from the target. In Fig. 9, the contours of the ion current show that most of the accelerated ions are generated in or near the laser spot. Some ions are accelerated from outside the spot but represent only a small fraction of the energy. Evidently, the magnetic field collimates the ion plume in the spot region, for without a magnetic field the plume diverges rapidly as it leaves the surface.

We have compared the results of VENUS with those from a magnetohydrodynamic (MHD) model (Brackbill and Goldman, 1983). Comparisons with MHD results indicate that fluid models cannot describe lateral transport in detail but can reproduce some of its features. We must impose a flux

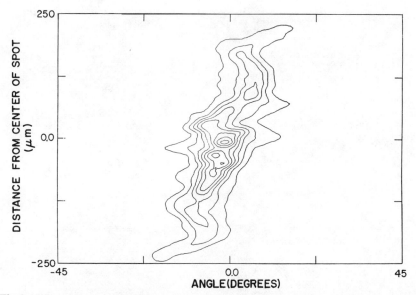

Fig. 9. Ions are accelerated in a laser plasma primarily in the spot and normal to the surface, as shown by the variation of the contours of the constant ion fluence with the direction of travel relative to the surface normal and point of origin on the target. The fluence is peaked at 0 angle (normal to the surface) and 0 position (center of the spot).

limiter similar to Eq. (34) when the mean free path approaches the characteristic scale lengths to prevent divergences in the higher order terms in the fluid equations. We either must include some finite electron mass effects to introduce the (c/ω_e) scale length in the sheath that occurs physically (Jones, 1983) or else impose the scale length phenomenologically using the mesh spacing or the resistive diffusion, either of which can limit the gradient scale lengths in the sheath. We must treat the electrons as a multicomponent fluid when the electron–electron collisions are infrequent and hot and cold components behave independently. All these modifications add to the cost and complexity of the fluid model.

The value of the particle simulation lies in its identification of the physics in a limit that is inaccessible to the fluid model without considerable modification. Being able to compare the simulation results, we can make these modifications with detailed knowledge of the desired results.

The simulation cannot, at present, replace the fluid model. Details of the target, such as material properties and collisional effects, require extensions to VENUS that have not yet been made. Further, for shorter wavelength lasers, where the hot-electron temperature is lower and the background density is higher, collisions are frequent and fluid models are more appropriate.

B. Collisionless Shocks

On the sunward side of the earth, a collisionless bow shock forms where the solar wind and the earth's magnetosphere collide. The shock transition is accompanied by turbulence, plasma heating, and heat flow. The character of the shock depends strongly on the orientation of the magnetic field with respect to the shock normal, which can assume different values at different positions in the shock. In a simulation, however, we consider only one orientation at a time.

More data exist for the bow shock than for most terrestrial plasma experiments. The satellites that collect the data are small compared with the plasma Debye length and are capable of measuring not only electron and ion temperatures, magnetic-field intensities, and flow speeds, but also details of the velocity distribution functions. The data provide an excellent test of theoretical calculations.

Much of the theoretical modeling of collisionless shocks has been done using one-dimensional particle simulations (Biskamp and Welter, 1972; Auer et al., 1971) or one-dimensional hybrid simulations (Leroy et al., 1981, 1982). In one-dimensional simulations, current-driven instabilities are suppressed by the lack of a second dimension. Their absence reduces the resistivity and increases the fraction of reflected ions above what is observed. In the hybrid simulations, the resistive heating is treated phenomenologically, and although many of the observed features of the shock are reproduced, the source of the resistivity is not identified.

To investigate the dissipation in the shock, we have modeled a quasi-perpendicular shock (Forslund et al., 1984) using VENUS with $\beta_e = 2.0$, $\beta_i = 0.2$, and an Alfvén Mach number of 4.5. According to the usual classification scheme, the angle between the magnetic field, 78.7° is close enough to 90° to model a perpendicular shock but still allow heat flow through the shock front. Since the scale length in the shock transition is $O(c/\omega_i)$, the ion collisionless skin depth, the ratio m_e/m_i is set at a larger than physical value of 0.01 to compress the range of time scales.

We have made comparisons between calculations with the diamagnetic current into and out of the plane of calculation to identify the source of the dissipation. By rotating the magnetic field about the shock normal, we can cause the current to flow into and out of the plane. When the current is in the plane, current-driven instabilities can and do occur, as is shown in Figs. 10 and 11, where the magnetic-field and density profiles exhibit considerable turbulence. When the current is out of the plane, no current-driven instability is possible, and we observe none in the simulation results shown in Figs. 12 and 13, which consequently look very much like those of Biskamp and Welter (1972). The turbulence reduces the number of reflected ions, causes

TIME= 8532.000

Fig. 10. A turbulent magnetic field is generated in a collisionless shock wave when the current is in the plane. The height of the plot represents the magnetic-field intensity.

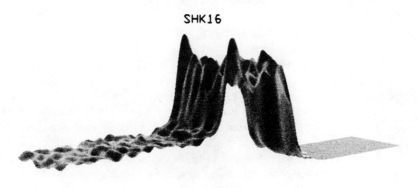

TIME= 8532.000

Fig. 11. The density fluctuations in a collisionless shock are large when the current is in the plane. The height of the plot represents the plasma density.

TIME=   8716.000

Fig. 12. Compared with Fig. 10, the magnetic field is much less turbulent when the current is out of the plane.

TIME=   8716.000

Fig. 13. Compared with Fig. 11, the density fluctuations are much smaller when the current is out of the plane.

heating of ions and electrons, decreases the thickness of the shock, alters the overshoot height and thickness, and enhances the upstream field-aligned electron heat flow. All of these features are observed.

By allowing us to treat larger time and space scales, using VENUS makes it possible to model collisionless shocks in two dimensions while retaining the electron inertial effects. It does not answer all questions. For example, when the Mach number is lower and current-driven instabilities weaker, is the ion-acoustic instability the source of turbulence? To answer this question requires resolving the length scale c/ω_e, which requires heroic expenditures of computer time.

VII. CONCLUSIONS

The implicit formulation of the plasma dynamic equations has enlarged the scope of plasma simulation significantly. We now can calculate phenomena with VENUS on a time scale that is long compared with the electron plasma frequency, while retaining electron kinetic and inertia effects. As the examples illustrate, there is a lot of physics on longer time scales.

It is difficult to compare the implicit moment method with the direct method, since they are not only at different stages of their development but also have developed differently. However, the heuristic approach adopted in developing the moment method and the more formal approach adopted in developing the direct method have both shown that many approximations can be made without sacrificing accuracy and stability. So far, we have found the implicit moment method to be satisfactory. It is especially easy to modify and add to, as in adding collisions. However, in the future we expect to incorporate features of the direct method into our implicit moment formulation (as in the method of extrapolation described earlier) when it is clear they will increase the speed or accuracy of our calculations.

There are many possible variations on the basic method. We have modified VENUS for cylindrical geometry (Wallace et al., 1984) and have reported some results with the collisional model (Goldman et al., 1983). We have also found it useful to compare fluid hybrid and full kinetic calculations with VENUS to understand the source of dissipation in the lower hybrid drift instability (Brackbill et al., 1984). Other versions of the implicit moment method are planned. An adaptively zoned particle simulation code is being developed, and new ways of solving the potential equations are being explored.

These improvements should be quite useful and will further extend the scope of plasma simulation. One might ask, however, whether the new methods are best employed simulating problems in the old way (but bigger and faster), or whether there are more fruitful directions for future development. Certainly doing large problems is a useful role for the implicit codes to fill, but we believe the most valuable results and insights will come through bridging the gap between collisionless plasma simulation and the elaborate fluid transport models that have been developed for phenomena on the MHD time scale.

ACKNOWLEDGMENTS

We gratefully acknowledge many useful discussions with D. Barnes, B. Cohen, C. Cranfill, R. Goldman, A. B. Langdon, R. Mason, and J. Wallace, and the support provided by the U.S. Department of Energy.

REFERENCES

Auer, P. L., Kilb, R. W., and Crevier, F. (1971). *J. Geophys. Res.* **76**, 1971.
Barnes, D. C., and Kamimura, T. (1982). *Res. Rep.—Nagoya Univ., Inst. Plasma Phys.* **IPPJ-570**.
Barnes, D. C., Kamimura, T., Leboeuf, J.-N., and Tajima, T. (1983). *J. Comput. Phys.* **52**, 480.
Bezzerides, B., Forslund, D. W., and Lindman, E. L. (1978). *Phys. Fluids* **21**, 2179.
Birdsall, C. K., and Fuss, D. (1969). *J. Comput. Phys.* **3**, 494.
Birdsall, C. K., Langdon, A. B., and Okuda, H. (1970). *Methods Comput. Phys.* **9**, 241.
Biskamp, D., and Welter, H. (1972). *Nucl. Fusion* **12**, 663.
Brackbill, J. U., and Forslund, D. W. (1982a). *Bull. Am. Phys. Soc.* **27**, 990.
Brackbill, J. U., and Forslund, D. W. (1982b). *J. Comput. Phys.* **46**, 271.
Brackbill, J. U., and Goldman, S. R. (1983). *Commun. Pure Appl. Math.* **36**, 415.
Brackbill, J. U., and Ruppel, H. M. (1984). *J. Comput. Phys.* (submitted for publication).
Brackbill, J. U., and Saltzman, J. S. (1982). *J. Comput. Phys.* **46**, 342.
Brackbill, J. U., Forslund, D. W., Quest, K. B., and Winske, D. (1984). *Phys. Fluids* **27**, 2682.
Braginskii, S. I. (1965). *Rev. Plasma Phys.* **1**, 205.
Brandt, A. (1977). *Math. Comput.* **31**, 333.
Busnardo-Neto, J., Pritchett, P. L., Lin, A. T., and Dawson, J. M. (1977). *J. Comput. Phys.* **23**, 300.
Buzbee, B. L., Golub, G. H., and Nielson, C. W. (1970). *SIAM J. Numer. Anal.* **7**, 62.
Byers, J. A., Cohen, B. J., Condit, W. C., and Hanson, J. D. (1978). *J. Comput. Phys.* **27**, 363.
Chandrasekhar, S. (1954). *In* "Selected Papers on Noise and Stochastic Processes" (N. Wax, ed.), p. 3. Dover, New York.
Chandrasekhar, S. (1960). "Plasma Physics." Univ. of Chicago Press, Chicago, Illinois.
Chase, J. B., LeBlanc, J. M., and Wilson, J. R. (1973). *Phys. Fluids* **16**, 1142.
Cohen, B. I. (1983). *In* "Energy Modeling and Simulation Proceedings," (A. S. Kydes, ed.), p. 383. North-Holland, Amsterdam.
Cohen, B. I., Langdon, A. B., and Friedman, A. (1984). *J. Comput. Phys.* **56**, 51.
Colombant, D. G., Davis, J., Tidman, D. A., Whitney, K. G., and Winsor, N. K. (1976). *Phys. Fluids* **18**, 1687.
Concus, P., and Golub, G. H. (1973). *SIAM J. Numer. Anal.* **10**, 1103.
Cranfill, C., Goldman, R., and Brackbill, J. U. (1984). *J. Comput. Phys.* (in preparation).
Dawson, J. (1983). *Rev. Mod. Phys.* **55**, 403.
Denavit, J. (1981). *J. Comput. Phys.* **42**, 337.
Dickman, D., Morse. R. L., and Nielson, C. U. (1969). *Phys. Fluids* **12**, 1708.
Drake, J. F., and Lee, T. T. (1981). *Phys. Fluids* **24**, 1115.

Forslund, D. W., and Brackbill, J. U. (1982). *Phys. Rev. Lett.* **48**, 1614.
Forslund, D. W., Quest, K. B., Brackbill, J. U., and Lee, K. (1984). *JGR, J. Geophys. Res.* **89**, 2142.
Freidberg, J. P. (1982). *Rev. Mod. Phys.* **54**, 801.
Friedman, A., Langdon, A. B., and Cohen, B. I. (1981). *Comments Plasma Phys. Controlled Fusion* **6**, 225.
Goldman, S. R., Brackbill, J. U., Forslund, D. W., and Wallace, J. M. (1983). *Bull. Am. Phys. Soc.* [2] **28**, 1145.
Grad, H. (1969). *Phys. Today* Dec., p. 34.
Hewett, D. W. (1980), *J. Comput. Phys.* **38**, 378.
Hewett, D. W., and Nielson, C. W. (1978). *J. Comput. Phys.* **29**, 219.
Jackson, J. D. (1962). "Classical Electrodynamics," p. 345. Wiley, New York.
Jones, R. D. (1983). *Phys. Rev. Lett.* **51**, 1269.
Kershaw, D. S. (1978). *J. Comput. Phys.* **26**, 43.
Kim, J. S., and Cary, J. R. (1983). *Phys. Fluids* **26**, 2167.
Langdon, A. B. (1970). *J. Comput. Phys.* **6**, 247.
Langdon, A. B. (1979). *J. Comput. Phys.* **30**, 202.
Langdon, A. B., and Lasinski, B. F. (1976). *Methods Comput. Phys.* **16**, 327.
Langdon, A. B., Cohen, B. I., and Friedman, A. (1983). *J. Comput. Phys.* **51**, 107.
Leroy, M. M., Goodrich, C. C., Winske, D., Wu, C. S., and Papadopoulos, K. (1981). *Geophys. Res. Lett.* **8**, 1269.
Leroy, M. M., Winske, D., Goodrich, C. C., Wu, C. S., and Papadopoulos, K. (1982) *JGR. J. Geophys. Res.* **87**, 5081.
Lindman, E. L. (1970). *J. Comput. Phys.* **5**, 13.
Malone, R. C., McCrory, R. L., and Morse, R. L. (1975). *Phys. Rev. Lett.* **34**, 721.
Mason, R. J. (1981). *J. Comput. Phys.* **42**, 233.
Mason, R. J. (1983). *J. Comput. Phys.* **51**, 484.
Morse, R. L., and Nielson, C. W. (1969). *Phys. Fluids* **12**, 2418.
Nielson, C. W., and Lewis, H. R. (1976). *Methods Comput. Phys.* **16**, 367.
Nielson, C. W., and Lindman, E. L. (1974). *Proc. Conf. Numer. Simul. Plasmas, 6th, 197* Paper E3.
Northrup, T. G. (1961). *Ann. Phys. (N.Y.)* **15**, 79.
Northrup, T. G. (1963) "The Adiabatic Motion of Charged Particles." Wiley (Interscience), New York.
Pert, G. J. (1981). *J. Comput. Phys.* **43**, 111.
Quest, K. B., Forslund, D. W., Brackbill, J. U., and Lee, K. (1983). *Geophys. Res. Lett.* **10**, 471.
Rechester, A. B., and Rosenbluth, M. N. (1978). *Phys. Rev. Lett.* **40**, 38.
Thompson, J. F., Thames, F. C., and Mastin, C. W. (1974). *J. Comput. Phys.* **15**, 299.
Wallace, J. M. Brackbill, J. U., and Forslund, D. W. (1984). *J. Comput. Phys.* (to be published).
Yates, M. A., van Hulsteyn, D. B. Rutkowski, H., Kyrala, G., and Brackbill, J. U. (1982). *Phys. Rev. Lett.* **49**, 1702.

10

Orbit Averaging and Subcycling in Particle Simulation of Plasmas

BRUCE I. COHEN

Lawrence Livermore National Laboratory
University of California
Livermore, California

I. Introduction. 311
   A. Multiple Time Scales in Plasmas. 311
   B. Multiple-Time-Scale Methods in Particle Simulation . . 313
II. Electron Subcycling. 316
   A. Explicit Electrostatic Algorithm 316
   B. Linear Dispersion Relation: Stability, Filtering, and Damping 317
III. Orbit Averaging 320
   A. Explicit Magnetoinductive Model 320
   B. Implicit Methods with Orbit Averaging. 323
   C. Applications of Orbit Averaging 327
IV. Discussion. 329
   References. 332

I. INTRODUCTION

A. Multiple Time Scales in Plasma Physics

Plasma physics phenomena occur over an enormous range of space and time scales. The motion of charged particles in inhomogeneous and time-varying electric and magnetic fields with collisions and collective effects is very complicated. A plasma contains a multitude of distinct time and space

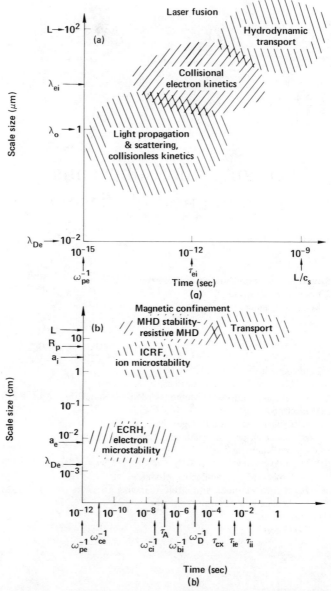

Fig. 1. Space and time scales for plasma phenomena. (a) Schematic of plasma phenomena for a plasma generated by a neodymium–glass laser with wavelength λ_0, electron mean free path for scattering by ions λ_{ei}, electron Debye length λ_{De}, plasma dimension L, plasma frequency ω_{pe}, electron–ion mean collision time τ_{ei}, and sound speed c_s. (b) Schematic for a deuterium plasma in a magnetic mirror with electron and ion Larmor radii a_e and a_i, plasma radius R_p, plasma length L, electron and ion cyclotron frequency ω_{ce} and ω_{ci}, Alfvén-wave transit time τ_A, ion-bounce and drift frequencies ω_{bi} and ω_D, neutral-beam charge-exchange time τ_{cx}, mean ion–electron and ion–ion scattering times τ_{ie} and τ_{ii}.

scales that span many orders of magnitude. Figure 1 gives examples with a range of 4 orders of magnitude in spatial scales and a range of 6 to 12 orders of magnitude in time scales! This makes realistic simulation of time-dependent plasma phenomena generally very difficult.

The particle simulation methods reviewed here successfully resolve disparate time scales. These methods complement the implicit particle simulation methods described in this volume—the implicit moment method (Mason, 1981; Denavit, 1981; Brackbill and Forslund, 1982) and the direct implicit method (Friedman *et al.*, 1981; Langdon *et al.*, 1983). The two multiple-time-scale methods discussed here are electron subcycling and orbit averaging. Both achieve significant gains in efficiency over traditional particle simulation approaches for appropriate applications.

B. Multiple-Time-Scale Methods in Particle Simulation

In a conventional particle simulation, the Newton–Lorentz equations of motion are advanced for the particles to obtain velocity and position data from which charge and current densities are collected on a spatial mesh. These densities are then used as sources in Maxwell's equations, which determine the self-consistent electric and magnetic fields. A common time step is used for all equations, and the differencing schemes are usually accurate to $\mathcal{O}(\Delta t^2)$. Particle simulations can reproduce the most complicated collective plasma effects while retaining kinetic details. However, the computer requirements may be stringent if a high degree of realism is desired in the simulation.

In real plasmas, the electrons and ions can stream along magnetic-field lines, gyrate around a magnetic-field line, bounce back and forth in either an electric or magnetic trap, drift and precess across the magnetic field, and collide with other particles. Each of these motions usually possesses its own characteristic frequency, and the large difference in ion and electron masses increases the range of time scales. The inclusion of self-consistent electromagnetic fields adds additional time scales for the collective oscillations of the plasma. Figure 1 gives dramatic evidence of how disparate plasma time scales are.

From the point of view of numerical analysis, stiff problems commonly arise in plasma simulation because the phenomena of interest develop on relatively slow time scales, but the systems support high-frequency normal modes. The goal of using the largest possible time step to resolve the interesting physics and to minimize the computer cost, while preserving numerical stability and accuracy, makes stringent demands on the numerical methods.

Subcycling and orbit averaging successfully interleave modules for advancing the field and particle equations with different time steps. These methods differ significantly from "reduced" methods in which high-frequency modes are removed from the governing equations, for example, by setting a time derivative or the electron mass equal to zero. Darwin algorithms (Nielson and Lewis, 1976), which eliminate the transverse displacement and radiation fields, and electrostatic algorithms, which solve Poisson's equation and omit electromagnetic fields altogether, are popular examples of reduced methods. These models traditionally do not separate particle and field time steps.

An early example of subcycling is given by Boris (1970), who multistepped the field equations in between successive particle advances. The integration was explicit, reversible, time-centered, and second order accurate in Δt. Boris's scheme was stable for sufficiently small time steps, and a simplified dispersion analysis indicated linear stability for

$$1/\Delta t^2 \geq c^2/\Delta x^2 + c^2/\Delta y^2 + \omega_{pe}^2/4 \qquad (1)$$

in a cold plasma, where c is the speed of light, ω_{pe} the plasma frequency, Δx and Δy the grid spacings, and Δt the time step. The economy achieved here by subcycling derives from the longer time step used for the particle advance and, hence, the fewer particle pushes. This can be an appreciable savings because particle pushing accounts for the major fraction of the calculational burden in a particle simulation. The applicability of subcycling in this electromagnetic model depends on the wave propagation speed c being significantly greater than the typical plasma velocities.

Langdon and Lasinski (1976) subsequently applied subcycling in their work on two-dimensional electromagnetic simulation. In some applications they found a serious numerical instability when the light-wave frequency equals the Nyquist frequency for the particles $\pi/\Delta t$. A similar instability occurs for subcycling in an electrostatic model (Adam *et al.*, 1982) and is described in Section II.

Godfrey (1974) has studied numerical Cherenkov instability in subcycled electromagnetic algorithms. This instability involves a resonance between a spurious beaming mode and the electromagnetic normal modes. For particles advanced every N time steps, stability is ensured for

$$N \, \Delta t \leq \Delta x/v, \qquad (2)$$

where v is the velocity of a single cold beam. This result is appropriate for difference schemes like that in CYLRAD (Boris, 1970) or ZOHAR (Langdon and Lasinski, 1976). Solution of Maxwell's equations using a Fourier transform technique (Godfrey, 1974) or the Langdon–Dawson advective algorithm (Godfrey and Langdon, 1976) avoids the numerical Cherenkov instability. In recent work, Adam *et al.* (1982) introduced subcycling to the particle simulation of a multispecies plasma. They exploit the difference in ion and electron inertia to devise a scheme in which the cost of advancing the ions is negligible compared with that for the electrons. In a subcycled electrostatic algorithm, Poisson's equation is solved and the electrons are advanced on the same time step Δt_e; ions are advanced after N steps with time step $N \Delta t_e$, $N \gg 1$. This scheme is explicit and centered. It can be made numerically stable with the addition of weak damping, but the stability of electron plasma waves requires $\omega_{pe} \Delta t_e < \mathcal{O}(1)$. An implicit subcycled electrostatic algorithm has been suggested by Adam et al. (1982); Section II describes subcycling in more detail.

The method of orbit averaging takes another approach to the problem of multiple time scales in plasmas. An instability or plasma transport often occurs over very long time scales compared with those for particle-transit, cyclotron, bounce, or drift motions. It is then advantageous to follow the particle motion on a fine time scale to resolve kinetic detail accurately and then temporally average the charge and current densities to retain only the slow time variations. The field equations are solved on the slow time scale, and the slowly varying fields are used to advance the particles. Temporal averaging of the particle data allows fewer particles to be used than in a conventional simulation and leads to a cost savings.

Analytical orbit averaging has been used to derive drift-kinetic (Rutherford and Frieman, 1968) and gyrokinetic (Frieman, 1970; Catto, 1978) formalisms. These formalisms yield reduced kinetic equations for the evolution of the phase-space distribution function, in which all dependence on the gyrophase has been annihilated by a formal averaging procedure. There is invoked an ordering that presumes that the frequencies of the perturbed quantities are small compared with the cyclotron frequency and that the gyroradius is small compared with the system scale lengths. These methods can be applied to plasma simulation and result in the complete elimination of the cyclotron time scale from the calculations (Lee, 1983).

Orbit averaging as employed by Cohen et al. (1980, 1982b; Cohen and Freis, 1982) accommodates arbitrarily large gyroradius or banana-radius effects. This is accomplished by retaining all the fast orbital time scales in the particle trajectory calculations. Fields are calculated on a much longer time scale, and an iteration of the computational cycle is performed to improve the centering. Stability analyses and simulation experience have indicated that orbit averaging of an explicit magnetoinductive model is stable with a large time step (Cohen and Freis, 1982). However, if charge separation electric fields are included, then an implicit field solution is required for stability with a large time step (Cohen et al., 1982b). Section III reviews orbit averaging in detail. A discussion of the applicability of subcycling and orbit averaging is presented in Section IV.

II. ELECTRON SUBCYCLING

A. Explicit Electrostatic Algorithm

Electron subcycling improves the efficiency of multiple-species particle simulation by making the cost of advancing the ions negligible compared with that of the electrons. In the simplest explicit subcycling scheme, the electron plasma frequency sets a constraint on the time step for advancing the electrons and the electric field, $\omega_{pe} \Delta t_e < \mathcal{O}(1)$. However, contrary to implicit algorithms that relax this constraint (Mason, 1981; Friedman et al., 1981), there is no restriction to wavelengths larger than the Debye length or to weak field gradients.

In the basic subcycling algorithm formulated by Adam et al. (1982), a standard leapfrog scheme is used to advance ions and electrons. The electron equations of motion and Poisson's equation are solved on each time step Δt_e. The ion time step Δt_i is an integer multiple N of Δt_e. The electric field felt by the ions is a filtered average of the electric fields seen by the electrons; a simple average works well (Adam et al., 1982).

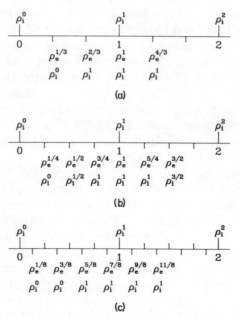

Fig. 2. Schematics of time levels for odd and even versions of the electrostatic subcycling algorithm. The leapfrog advance of the ion velocities and positions allows calculation of the ion density above the time line before it is needed below the line in the electric-field calculation. (a) Odd-N scheme; (b) even-N scheme; (c) even-N scheme.

The subcycling algorithm differs slightly, depending on whether N is even or odd (Fig. 2). In Fig. 2 ion densities are shown above the time line and electron densities below. The ion density used in conjunction with the electron density in Poisson's equation is shown below the electron densities. The superscripts designate the time levels in units of the ion time step. The half-integer superscripts on the ion densities in Fig. 2b are simple averages of the ion densities at adjacent integer time levels. The ion and electron densities are never known at once in Fig. 2c. The even-N schemes are less convenient than the odd-N scheme, and Adam et al. (1982) favor the latter. Both odd- and even-N schemes are time-centered, reversible, and second order accurate in time.

B. Linear Dispersion Relation: Stability, Filtering, and Damping

The electron subcycling algorithm can be unstable when $\omega_{pe} \Delta t_i \approx l\pi$, where l is an integer. This is a temporal aliasing phenomenon caused by the coupling of the two electron plasma normal modes $\omega = \pm \omega_{pe}$ by the ions. Moreover, the manner in which the electric field seen by the ions is averaged and filtered can lead to a strong instability for $\omega_{pe} \Delta t_i \approx 2\pi$. Damping in the particle equations can stabilize these instabilities at the expense of numerical cooling of the electrons (Cohen et al., 1982a).

Adam et al. (1982) made a detailed study of the stability and dispersion characteristics of electron subcycling. Spatial grid effects do not affect the dispersion any differently than in a conventional particle code. In the limit $\Delta x = 0$, the dispersion relation for electrostatic modes in an odd-N scheme is

$$1 + \chi_i^{-1}(k, \omega) = \sum_{q=-M}^{M} \frac{H_1(\omega_q) H_2(\omega_q)}{1 + \chi_e^{-1}(k, \omega_q)}, \tag{3}$$

where $\omega_q \equiv \omega - 2\pi q/\Delta t_i$, $2M + 1 = N = \Delta t_i/\Delta t_e$, and χ_e and χ_i are the Fourier transforms of the usual electron and ion linear susceptibilities relating the charge densities to the potential, including the effects of finite Δt (Langdon, 1979a). Note that $H_1(\omega)$ arises from replicating the ion density at successive electron time steps:

$$H_1(\omega) = \frac{1}{N} \sum_{n'=-M}^{M} \exp(i\omega n' \Delta t_e) = \frac{\sin(\omega \Delta t_i/2)}{N \sin(\omega \Delta t_e/2)}; \tag{4}$$

$H_2(\omega_q)$ is the transfer function for the filter $\{a_{n'}\}$ introduced by averaging the electric field seen by the ions, $E_i = \sum_{n'=-M}^{M} a_{n'} E_e^{n'}$:

$$H_2(\omega) = \sum_{n'=-M}^{M} a_{n'} \exp(-i\omega n' \Delta t_e). \tag{5}$$

With cold ions and warm electrons,

$$\chi_i = -(\omega_{pi} \Delta t_i)^2 / [2 \sin(\omega \Delta t_i/2)]^2 \qquad (6)$$

and

$$\chi_e(k, \omega) = \frac{\omega_{pe}^2}{k^2} \int dv \, k \frac{\partial f_0}{\partial v} \frac{\Delta t_e}{2} \cot\left[(\omega - kv)\frac{\Delta t_e}{2}\right]$$

$$\approx \chi_{e0}(k, \omega) - \frac{1}{12}(\omega_{pe} \Delta t_e)^2 + \cdots \qquad (7)$$

for $\omega_{pe} \Delta t_e < 1$. The χ_{e0} is the continuous result and can be expressed in terms of the Fried-Conte function for a Maxwellian electron distribution. For $\{a_{n'}\}$ a simple average $a_{n'} = 1/N$ and $H_2 = H_1$. With $N = 1$, $H_1 H_2 = 1$ and Eq. (3) give $1 + \chi_e + \chi_i = 0$, the correct result in the limit of no subcycling.

Equation (3) resembles the dispersion relation for parametric instability. The ions couple electron modes $\omega = \pm \omega_{pe}$ to give instability via temporal aliasing. With $\omega_{pe} \approx \pi/\Delta t_i$ and $H_2 = H_1$, there are roots $\omega \approx \omega_{pe}$ in a cold plasma, and the maximum growth rate is given by

$$\text{Im}(\omega/\omega_{pe}) = \tfrac{1}{2}(\omega_{pi}/\omega_{pe})^2. \qquad (8)$$

The next band of instability occurs when $\omega_{pe} \Delta t_i \approx 2\pi$. Here the ion-acoustic wave is destabilized by coupling to two Langmuir waves. For cold ions and warm electrons, there is an unstable mode with $\omega \sim \omega_{pi}$ and a maximum growth rate approximately equal to that in Eq. (8). Detailed solution of the dispersion relation in Eq. (3) indicates that instability corresponds more precisely to the condition that the product of the Bohm-Gross frequency and the ion time step be slightly less than $l\pi$. The growth rates for $\omega_{pe} \Delta t_i \approx 2\pi$ can be significantly larger than Eq. (8) if $H_2(2\pi/\Delta t_i) > 0$, instead of $H_2 = 0$ when $H_2 = H_1$; the growth rates can be as large as Im $\omega \sim \omega_{pi}(H_2)^{1/2}$. Thus, the electric-field filter H_2 influences stability and cannot be chosen arbitrarily. Figure 3 exhibits growth rates determined by numerical solutions of Eq. (3) and from simulations when $\omega_{pe} \Delta t_i \approx \pi$ and 2π. The mass ratio here was $m_i/m_e = 100$ in Fig. 3a and $m_i/m_e = 400$ in Fig. 3b; $\Delta t_i/\Delta t_e = 15$ and $k\lambda_{De} = 0.0982$ in both cases. A second unstable branch emerged in Fig. 3a, which was not predicted by the theory, but this branch did not appear for $m_i/m_e \geq 400$. In general, the maximum growth rates, the domain of instability, and the scaling with parameters agree with theory. For mass ratios exceeding 900, the subcycling algorithm was stable, which is probably because of collisional damping.

The bands of instability are fairly narrow (Fig. 3), and the stability of uniform plasma simulations can be guaranteed by choosing $\omega_{pe} \Delta t_i \neq l\pi$. This

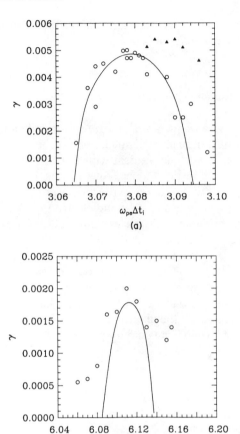

Fig. 3. Comparison of the theory and results of particle simulation for $\Delta t_i/\Delta t_e = 15$ and $k\lambda_{De} = 0.0982$. In (a) $m_i/m_e = 100$ and $\omega_{pe}\Delta t_i \approx \pi$. The circles are data points that fit the theory, and the triangles are data indicating a branch not predicted by the theory. In (b) $m_i/m_e = 400$ and $\omega_{pe}\Delta t_i \approx 2\pi$.

may be difficult to achieve in a nonuniform plasma. Furthermore, the instability is nonresonant and saturates at a very large amplitude; therefore, complete stabilization is desirable. Adam *et al.* (1982) show that numerical damping has a strong stabilizing influence.

Numerical damping can be added to the electron equations of motion (Langdon, 1979a; Cohen *et al.*, 1982a), such that

$$v_{n+1/2} = v_{n-1/2} + a_n \Delta t, \tag{9}$$

$$x_{n+1} = x_n + v_{n+1/2} \Delta t + c_1 \Delta t^2 (a_n - a_{n-1}). \tag{10}$$

This introduces dissipation and a phase shift in the electron plasma waves; the complex frequency shift $\delta\omega$ is given by

$$\mathrm{Re}(\delta\omega/\omega_0) = \tfrac{1}{2}(\omega_0 \Delta t)^2(\tfrac{1}{12} - c_1) + \mathcal{O}(\Delta t^3), \tag{11}$$

$$\mathrm{Im}(\delta\omega/\omega_0) = -(c_1/2)(\omega_0 \Delta t)^3 + \cdots, \tag{12}$$

where ω_0 is the plasma frequency. For $\omega_{pe}\Delta t_i \approx \pi$ and damping rates in Eq. (12) less than the maximum growth rate in Eq. (8), the net growth rates in the simulations were reduced in agreement with Eq. (12). Stability was achieved when the damping rate in Eq. (12) exceeded the maximum growth rate. For $\omega_{pe}\Delta t_i \approx 2\pi$, Eq. (12) with $\omega_0 \sim \omega_{pi}$ overestimates the stabilizing influence of the dissipation and indicates that the electron dissipation has a much weaker influence on ion-acoustic waves than on electron plasma waves.

Adam et al. (1982) commented that stabilization achieved with numerical dissipation would be accompanied by cooling of the plasma (Cohen et al., 1982a). For m_i/m_e large, the value of c_1 required for stabilization need not be large. With a damping scheme that is third order in Δt, like Eqs. (9) and (10), and a moderate or small value of c_1, the cooling may be small enough to be innocuous. However, the algorithm is stable with no damping if $\Delta t_e \ll \Delta t_i < \pi/\omega_{pe}$, and most of the maximum possible savings in computations is realized.

III. ORBIT AVERAGING

A. Explicit Magnetoinductive Model

Orbit averaging has been successfully applied to a magnetoinductive model of a plasma (Cohen et al., 1980; Cohen and Freis, 1982). In this case, charge separation effects are presumed to be negligible. A particle species carries a current, which in turn supports a magnetic field. The time derivative of the magnetic field generates an inductive electric field that is determined by solution of Faraday's law. Strict charge neutrality is assumed. Ambipolar electric fields are unimportant in affecting the dynamics of energetic magnetically confined particles if the ambipolar potential energy is much smaller than the typical particle energies. This scenario is appropriate for a number of laboratory plasma experiments, including, for example, neutral-beam-injected mirror experiments in the 2XIIB facility at Lawrence Livermore National Laboratory (Cohen et al., 1980; Cohen and Freis, 1982). The generalization to include charge separation and ambipolar electric fields is described later in this chapter.

1. Magnetoinductive Algorithm

In the orbit-averaged magnetoinductive scheme, particles are advanced with a small time step,

$$x^{n+1} = x^n + v^{n+1/2} \Delta t, \tag{13}$$

$$v^{n+1/2} = v^{n-1/2} + q\frac{\Delta t}{m}\sum_j S(x^n - X_j)\left[E_j^* + \frac{1}{2c}(v^{n+1/2} + v^{n-1/2}) \times B_j^*\right], \tag{14}$$

where $S(x^n - X_j)$ is the particle spline interpolating the fields defined on the grid at position X_j to the position x^n where the particle is accelerated,

$$(E, B)^* = W_1(M, n')(E, B)^M + [1 - W_1(M, n')](E, B)^{M+1}, \tag{15}$$

and $W_1(M, n')$ is an interpolation function with $0 \le n' \le N - 1$, $N = \Delta T/\Delta t$, and $n = MN + n'$. Maxwell's equations are used to calculate the fields:

$$\nabla \times \left[\alpha B_j^{M+1} + (1-\alpha) B_j^M\right] = (4\pi/c)\langle J_j\rangle^{M+1/2} \tag{16}$$

$$\nabla \times \left[\beta E_j^{M+1} + (1-\beta) E_j^M\right] = -\left(B_j^{M+1} - B_j^M\right) c\, \Delta T, \tag{17}$$

where

$$\langle J_j\rangle^{M+1/2} = \frac{1}{2}\sum_{n'=0}^{N-1} W_2(n') \sum_i q\left[S(x_i^n - X_j) + S(x_i^{n+1} - X_j)\right] v_i^{n+1/2}, \tag{18}$$

α and β are centering parameters controlling dissipation with $0 \le \alpha, \beta \le 1$, and $W_2(n')$ is a digital filter with $\sum_{n'=0}^{N-1} W_2(n') = 1$. Figure 4 gives a schematic

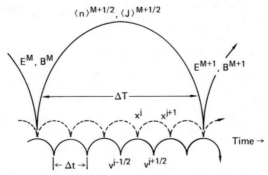

Fig. 4. Schematic of time levels in generic orbit averaging. The quantities $\langle n\rangle$ and $\langle J\rangle$ are averaged in time over the number and current densities accumulated at each small time step.

of orbit averaging. To approximately center $(\mathbf{E}, \mathbf{B})^*$, a predictor–corrector iteration is performed: $W_1(M, n') = 1$ and $(\mathbf{E}, \mathbf{B})^* = (\mathbf{E}, \mathbf{B})^M$ on the predictor step, and $W_1(M, n') = 1 - (n'/N)$ or $W_1(M, n') = 0$ on the corrector step with $(\mathbf{E}, \mathbf{B})^{M+1}$ used in $(\mathbf{E}, \mathbf{B})^*$ calculated earlier on the predictor step. Equations (15–17) are formally only first order accurate in ΔT, but with $(\alpha, \beta) = \frac{1}{2} + \varepsilon$, $\varepsilon \ll 1$, and $W_1 = 1 - (n'/N)$ on the corrector iteration, the algorithm is effectively second order accurate.

2. Magnetoinductive Linear Dispersion Relation

Cohen and Freis (1982) derived a linear dispersion relation for Eqs. (13)–(18) in a cold uniform plasma and with $\Delta x = 0$. As with subcycling, spatial grid effects and orbit averaging do not lead to any new unusual behavior. A quartic dispersion relation in ω was obtained for a one-dimensional slab model with wave propagation perpendicular to a uniform magnetic field. In the limit $\omega_c^2 \Delta T^2 \ll 1$, two of the four roots are compressional Alfvén waves,

$$\omega^2 = k^2 v_A^2/(1 + k^2 c^2/\omega_p^2), \tag{19}$$

where v_A is the Alfvén speed $v_A = c\omega_c/\omega_p$, ω_c the cyclotron frequency, ω_p the plasma frequency, and k the wave number. The other two roots are unphysical.

With $W_1(M, n') = 0$ on the corrector iteration and $\beta = 1$, the Alfvén waves are neutrally stable for $\omega_p^2/k^2 c^2 \ll 1$ and all values of α. The Alfvén waves acquire damping for finite $\omega_p^2/k^2 c^2$. One of the unphysical modes is an odd–even (Nyquist) oscillation that is neutrally stable for $\alpha = \frac{1}{2}$ and all values of

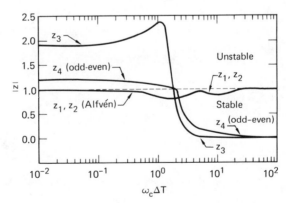

Fig. 5. The amplification factor $z = \exp(-i\omega \Delta T)$ versus $\omega_c \Delta T$ for cold-plasma normal modes in an explicit orbit-averaged magnetoinductive algorithm for $(\omega_p/k_c)^2 = 3$, $\beta = \alpha = 1$ in Eqs. (16) and (17), and $\Delta T/\Delta t = 10^3$. The four roots of the dispersion relation are all stable for $\omega_c \Delta T \gg 1$.

ω_p^2/k^2c^2, whereas the fourth normal mode is purely growing for $\omega_c^2 \Delta T^2 \ll 1$ and $\omega_p^2/k^2c^2 > 1$ and is heavily damped for $\omega_p^2/k^2c^2 \ll 1$. Numerical solution of the dispersion relation is shown in Fig. 5 for $\alpha = \beta = 1$ and $\omega_p^2/k^2c^2 = 3$; note that there is stability for $\omega_c \Delta T > 2$ and strong instability for $\omega_c \Delta T < 1$. For $\omega_c^2 \Delta T^2 \gg 1$ and $\alpha > \frac{1}{2}$, all modes are stable. The Alfvén waves are weakly unstable if $\omega_p^2/k^2c^2 > 1$ with $\omega_c^2 \Delta T^2 \gg 1$ and $\alpha = \frac{1}{2}$. The two unphysical modes are both damped Nyquist oscillations for $\alpha > \frac{1}{2}$, $\omega_c^2 \Delta T^2 \gg 1$, and all values of ω_p^2/k^2c^2. The Alfvén-wave frequencies approach $\pm \mathcal{O}(\pi/\Delta T)$ for $\omega_c^2 \Delta T^2 \gg 1$, and the modes acquire some damping for $\alpha > \frac{1}{2}$. Thus, no instability is encountered even though $kv_A \Delta T \gg 1$. Simulation experience with orbit averaging has confirmed stability for $\omega_c^2 \Delta T^2 > 1$ and instability at small time steps (Cohen et al., 1980; Cohen and Freis, 1982).

B. Implicit Methods with Orbit Averaging

The inclusion of space-charge fields significantly alters the physics model and requires a different approach if stability is to be ensured with a large time step. Langdon has given a proof of the necessity for implicit time differencing to achieve stability for $\omega_p^2 \Delta t^2 \to \infty$ when solving Poisson's equation and particle equations of motion (Cohen et al., 1982a). Cohen et al., (1982b) combined orbit averaging with the implicit moment method (Mason, 1981) to produce a stable algorithm for $\omega_p^2 \Delta T^2 \gg 1$ and operated with fewer particles than conventional explicit and implicit moment algorithms. Orbit-averaged *explicit* electrostatic algorithms were unstable for $\omega_p^2 \Delta T^2 > \mathcal{O}(1)$, which was consistent with Langdon's general considerations.

1. Implicit Moment Method with Orbit Averaging

Fluid equations are introduced in the implicit moment method as intermediaries between the particle and field equations. The two lowest moment equations for mass and momentum conservation allow one to exhibit the dependence of the number and current densities on the electric field algebraically, whose evaluation can then be made implicitly:

$$(\partial/\partial t)n_s + \nabla \cdot \frac{\mathbf{J}_s}{q_s} = 0, \qquad (20)$$

$$(\partial/\partial t)\mathbf{J}_s = (q_s/m_s)[-\nabla \cdot \mathbf{P}_s^\dagger + q_s n_s \mathbf{E} + (\mathbf{J}_s \times \mathbf{B})/c], \qquad (21)$$

where $\mathbf{P}_s^\dagger = \sum_i m_s \mathbf{v}_i \mathbf{v}_i$ is the kinetic stress tensor for species s and is summed over all the particles. Equations (20) and (21) are used to supply slowly varying charge and current densities for use in Maxwell's equations. Because $\mathbf{E}$ and $\mathbf{B}$ are presumed to have only slow temporal variations, time-averaging Eqs. (20) and (21) involves no convolution integrals and is trivial.

Suitable difference representations of Eqs. (20) and (21) in time are

$$n_s^{M+1} = n_s^M - (\Delta T/q_s) \nabla \cdot \mathbf{J}_s^{M+1/2} \tag{22}$$

$$\mathbf{J}^{M+1/2} = \langle \mathbf{J} \rangle^{M-1/2} + (q_s \Delta T/m_s)[-\nabla \cdot \langle \mathbf{P}_s \rangle^M + q_s n_s^M \mathbf{E} \\ + (\mathbf{J}_s^{M+1/2} + \mathbf{J}_s^{M-1/2}) \times \mathbf{B}^M/2c], \tag{23}$$

where $\mathbf{E} = \theta \mathbf{E}^{M+1} + (1 - \theta)(\mathbf{E}^{m+1} + 2\mathbf{E}^M + \mathbf{E}^{M-1})/4$, $0 \le \theta \le 1$, and the orbit-averaged particle current density and kinetic stress tensor are

$$\langle \mathbf{J}_s \rangle_j^{M-1/2} = \sum_{n=0}^{N-1} \sum_i q_s \mathbf{v}_i^{n'+1/2}[S(\mathbf{x}_i^{n'} - \mathbf{X}_j) + S(\mathbf{x}_i^{n'+1} - \mathbf{X}_j)]/2N \tag{24a}$$

for $M \Delta T \le t = M \Delta T + n' \Delta t \le (M + 1) \Delta T$ and

$$\langle \mathbf{P}_s^\dagger \rangle_j^M = \sum_{n'=0}^{N-1} \sum_i q_s \mathbf{v}_i^{n'+1/2} \mathbf{v}_i^{n'+1/2}[S(\mathbf{x}_i^{n'} - \mathbf{X}_j) + S(\mathbf{x}_i^{n'+1} - \mathbf{X}_j)]/2N \tag{24b}$$

for $(M - \frac{1}{2}) \Delta T \le t = M \Delta T + n' \Delta t \le (M + \frac{1}{2}) \Delta T$. Equations (22) and (23) are effectively second order accurate in ΔT for $\theta \ll 1$. The particle equations of motion are used to advance x_i and v_i. In an electrostatic model Gauss's law,

$$\nabla \cdot \mathbf{E}^{M+1} = 4\pi \sum_s q_s n_s^{M+1} \tag{25}$$

determines the electric field. An implicit electromagnetic model could be formulated in the manner of Brackbill and Forslund (1982) or Langdon (1983a,b).

Figure 6 illustrates the time advance of the fluid and particle quantities for Eqs. (13), (14), and (22)–(25). A time-centered $\langle \mathbf{P}_s^\dagger \rangle^M$ requires particle velocity and position data from $M \Delta T$ to $(M + \frac{1}{2}) \Delta T$ that are not yet computed. This difficulty exists in the basic implicit moment method (Mason, 1981; Denavit, 1981; Brackbill and Forslund, 1982) and is overcome by making a prediction or extrapolation of $\langle \mathbf{P}_s^\dagger \rangle^M$, solving the field equation, advancing the particles to $t = (M + 1) \Delta T$, and calculating an iterative correction to $\langle \mathbf{P}_s^\dagger \rangle^M$.

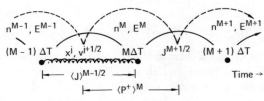

Fig. 6. Time advance of fluid and particle variables in the orbit-averaged implicit moment equation method.

A stability analysis for an electron plasma oscillation in the orbit-averaged, implicit moment equation algorithm was presented by Cohen et al. (1982b). The corrector iteration is superfluous in this analysis because the kinetic stress tensor is negligible for a linear plasma oscillation in a cold, nondrifting plasma. In terms of $z \equiv \exp(-i\omega \Delta T)$, the linear dispersion relation for a cold, uniform, unmagnetized plasma is a quartic:

$$z^2(z-1)^2 + \omega_{pe}^2 \Delta T^2(z^2 - z/2 + \tfrac{1}{2})[\theta z^2 + \tfrac{1}{4}(1-\theta)(z^2 + 2z + 1)] = 0. \quad (26)$$

For $\omega_{pe}^2 \Delta T^2 \ll 1$, there are plasma oscillations

$$\omega \approx \pm\omega_{pe}[1 \pm i(1-2\theta)\omega_{pe}\Delta T/4], \quad (27a)$$

which are damped if $\theta > \tfrac{1}{2}$, and heavily damped solutions

$$z = \pm i(1-\theta)\omega_{pe}\Delta T/8. \quad (27b)$$

For $\omega_{pe}^2 \Delta T^2 \gg 1$, the plasma oscillation becomes significantly damped:

$$z = (1 \pm i\sqrt{7})/4; \quad (28a)$$

the remaining solutions are

$$z = (-1 + \theta \pm i2[\theta(1-\theta)]^{1/2})/(1+3\theta), \quad (28b)$$

which are damped for $0 \le \theta \le 1$.

Figure 7 gives a plot of the solutions of Eq. (26) and some simulation data points that are accurately predicted by the theory. This particular algorithm is stable for all values of $\omega_{pe}^2 \Delta T^2$ when $\theta > \tfrac{1}{2}$. As with subcycling, stability is achieved at the expense of cooling. Cohen et al. (1982b) have synthesized

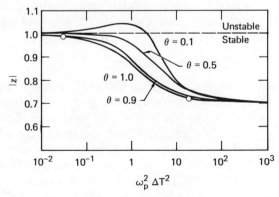

Fig. 7. Absolute value of the amplification factor z for electron plasma oscillations determined by Eq. (26) as a function of $\omega_{pe}^2 \Delta T^2$ for the orbit-averaged implicit-moment equation method. Two simulation data points agreeing with the theory are superposed.

and analyzed several modifications to this algorithm and have found instability when they tried to couple the fluid number density closer to the corresponding orbit-averaged particle data, for example, by using the orbit-averaged particle number density on the right side of Eq. (22).

2. Direct Implicit Method with Orbit Averaging

Cohen et al. (1982b) have applied orbit averaging to the direct implicit method (Friedman et al., 1981) to obtain an electrostatic scheme that is conditionally stable for all values of $\omega_{pe}^2 \Delta T^2$. In one dimension and with suitable boundary conditions, Gauss's law, Eq. (25), is equivalent to

$$\frac{\partial E}{\partial t} + 4\pi \sum_s J_s = 0.$$

This can be differenced as

$$\frac{E^{M+1} - E^M}{\Delta T} + 4\pi \sum_s \langle J_s \rangle^{M+1/2} = 0, \tag{29}$$

where E^* in the equations of motion (13) and (14) is given by $E^* = \alpha E^{M+1} + (1 - \alpha)E^M$.

An implicit $\langle J_s \rangle^{M+1/2}$ can be obtained by linearizing the particle equations of motion with respect to a small increment in the trajectory produced by the advanced electric field E^{M+1}. The orbit-averaged current becomes

$$\langle J_s \rangle^{M+1/2} = \langle J_s^e \rangle^{M+1/2} + \langle \partial J/\partial E^{M+1} \rangle^{M+1/2} E^{M+1}, \tag{30}$$

where $\langle J_s^e \rangle^{M+1/2}$ is the explicit contribution to the orbit-averaged current and $\langle \partial J/\partial E^{M+1} \rangle$ is an orbit-averaged conductivity; both are calculated from Eq. (24a) and the particle equations of motion with $E^{M+1} = 0$ when evaluated in E^*. The validity of the Taylor-series expansion requires

$$|q_s \Delta t^2 (\partial E/\partial x)/m_s| < 1.$$

With the addition of a spatial grid, Eqs. (29) and (30) lead to a matrix equation for E^{M+1}. This matrix is not necessarily sparse; the sparseness depends on the typical particle excursion through the mesh over the time interval $M \Delta T$ to $(M + 1) \Delta T$. This contrasts with the matrix equation obtained in the orbit-averaged implicit moment algorithm, which is very sparse and banded. Such a difference between the direct implicit and implicit moment methods does *not exist* in the absence of orbit averaging (Langdon et al., 1983).

Cohen et al. (1982b) have derived a linear dispersion relation for an electron plasma oscillation supported by this algorithm in a cold, uniform, stationary, unmagnetized plasma:

$$(z - 1)^2 + (\omega_{pe}^2 \Delta T^2/2)(z + 1)(\alpha z + 1 - \alpha) = 0. \tag{31}$$

For $\omega_{pe}^2 \Delta T^2 \ll 1$, the solutions are plasma oscillations

$$\omega = \pm \omega_{pe}[1 \mp i(2\alpha + 1)\omega_{pe} \Delta T/4], \tag{32}$$

which corrects Eq. (23a) of Cohen et al. (1982b). For $\omega_{pe}^2 \Delta T^2 \gg 1$,

$$z = (\alpha - 1)/\alpha, -1 + [8/(2\alpha - 1)\omega_{pe}^2 \Delta T^2]. \tag{33}$$

All solutions are damped for $\alpha > \frac{1}{2}$.

C. Applications of Orbit Averaging

One- and two-dimensional orbit-averaged magnetoinductive and electrostatic algorithms have been implemented. Code results have confirmed the linear dispersion theory (Cohen et al., 1982b; Cohen and Freis, 1982). Use of the orbit-averaged implicit moment algorithm so far has demonstrated a reduction in the number of particles and stability for $\omega_{pe} \Delta T \gg 1$ (Cohen et al., 1982b). An orbit-averaged implicit moment simulation of an ion-acoustic wave in an unmagnetized and initially uniform plasma is shown in Fig. 8. In the simulation, $\omega_{pe} \Delta T = 4$, $k\lambda_{De} = 0.5 \times 10^{-2}$ for the principal Fourier mode, $\Delta T/\Delta t = 20$, and $\Delta t_e = \Delta t_i$; there were 512 ions and electrons, $m_i/m_e = 25$, and there were 32 grid cells. One corrector iteration was performed. The relative agreement between the orbit-averaged particle and fluid density perturbations is quite good for the coarseness of the mesh used and the small number of particles.

A discouraging aspect of the orbit-averaged implicit moment simulations as described in Eqs. (22)–(25) was the apparent divergence of the algorithm for $kv_e \Delta T \gtrsim 0.1$, where k is the largest wave number retained. Use of $kv_e \Delta T \approx 0.1$ is a factor at most 2–4 times more restrictive than that required to control artificial cooling and to ensure the accuracy of the plasma dielectric response (Langdon et al., 1983). In this algorithm, orbit averaging reduces the number of particles and the computer memory requirement but does *not* decrease the total number of operations. Cohen et al. (1982b) suggested modifications of the algorithm to relax the constraint on $kv_e \Delta T$—the most promising of which is an implicit prediction of the kinetic stress tensor—but have not reported any successful results as yet.

Contrary to the experience with orbit averaging in an unmagnetized plasma, orbit-averaged magnetoinductive simulations of plasmas with an externally applied magnetic field have demonstrated a great improvement in computational efficiency (Cohen et al., 1980; Cohen and Freis, 1982). The particle orbits were accurately resolved for $\omega_{ci} \Delta t < 1$ in these applications,

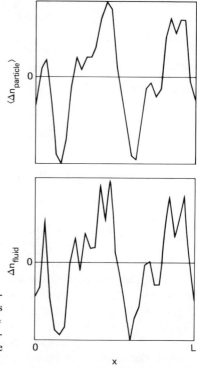

Fig. 8. Orbit-averaged and fluid perturbed number densities $\Delta n \equiv n_i - n_e$ ($\Delta n/n_0 \leq 0.1$) as functions of position x at $kc_s t = 4\pi$ corresponding to $t = 3000\,\Delta T$ with $\Delta T/\Delta t_e = 20$ and $\Delta t_e = \Delta t_i$ for a spatially periodic simulation of an ion acoustic wave with the orbit-averaged implicit moment method.

and the field equations were advanced with $kv_A\,\Delta T$, $\omega_{ci}\,\Delta T \gg 1$ to follow quasi-static changes. Cohen *et al.* (1980) compared one- and two-dimensional magnetoinductive simulations of the formation of field-reversing ion rings with and without orbit averaging. The physics content of these simulations was unchanged by orbit averaging. The one-dimensional simulations of an infinite cylinder of axis-encircling ions possessed a high degree of phase-space order; there was a minimal statistical requirement, and orbit averaging did *not* lead to a significant reduction in the number of particles. However, there was a large reduction in electric-field noise originating from the discreteness of the particle injection after applying orbit averaging.

Two-dimensional orbit-averaged simulations modeling diffuse injection in the 2XIIB mirror experiment at the Lawrence Livermore National Laboratory required substantially fewer particles than did previous conventional simulations (Cohen and Freis, 1982). These simulations studied attempts to achieve reversal of the magnetic field by means of neutral-beam injection across the vacuum magnetic field. Low-temperature electrons (energy $<$ 100 eV) were frozen on field lines and large-orbit ions (energy $\sim$ 13 keV) carried the diamagnetic current. The orbit-averaged simulations (SUPER-AVERAGE) simultaneously resolved the vacuum ion cyclotron frequency $\omega_{ci} = 2.8 \times 10^7 \text{ sec}^{-1}$ and the ion–electron drag rate $v_s^{i/e} = 3 \times 10^2 \text{ sec}^{-1}$ without artificial compression of the disparity in time scales. Steady states were achieved at times corresponding to several milliseconds in 2XIIB with 300–400 A of neutral-beam current maintaining the plasma against losses due to drag, quasi-linear diffusion due to ion cyclotron waves, and loss of adiabaticity (nonconstancy of the magnetic moment).

The use of realistic injection and loss rates allowed the ions to complete enough axial bounces for loss of adiabaticity to play a more significant role in shortening the ion confinement lifetimes than was true in earlier SUPERLAYER simulations, which had accelerated the drag and diffusion rates by more than 10^2 (Byers, 1977). The SUPERAVERAGE simulations required 500–1000 ions rather than the 20,000 previously needed, used correspondingly less memory, and were able to cycle through 10–100 more time steps. The principal physical consequences of the increased axial losses in these simulations were a downward revision of the field-reversal values $\Delta B_z/B_z$ predicted by SUPERLAYER (Turner et al., 1979) and a lengthening of the hot plasma. Figure 9 exhibits some of these results. These particle simulations of 2XIIB with realistic ion parameters were possible only with the enormous improvements in computational efficiency achieved with orbit averaging.

IV. DISCUSSION

Both subcycling and orbit averaging use multiple-time-scale methods to reduce the number of particle calculations in a simulation. Subcycling in an explicit algorithm is appropriate when both ions and electrons must be represented as particles and when short-wavelength $k\lambda_{De} = \mathcal{O}(1)$ space-charge electric fields are relevant. The simulation of ion acoustic turbulence is an important example where subcycling should be efficacious. Subcycling with an explicit field solution can be applied to a wide class of problems but is restricted to small time steps, $\omega_{pe} \Delta t_e \leq \mathcal{O}(1)$. Adam et al. (1982) have suggested an implicit version of subcycling that would implicitly solve for the electric field on the slower ion time scale and advance electrons with a smaller time step. This yields an algorithm that differs minimally from the implicit version of orbit averaging.

The improvement in computational efficiency achieved by orbit averaging depends on the physics application and model. Orbit averaging allows a reduction in the particle statistical requirement when a natural separation exists between the time scales for advancing the fields and particles. The

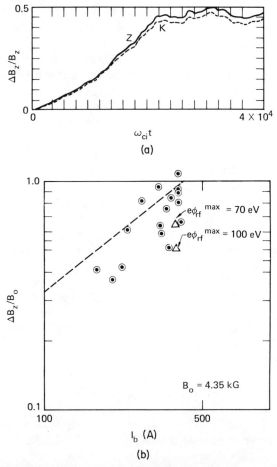

Fig. 9. (a) SUPERAVERAGE simulations of 2XIIB field reversal attempts, showing $Z = 1 - (B_z)_{min}/(B_z)_{max}$ and $K = 1 - (B_z)_{min}/B_0$ at the mirror midplane ($z = 0$), maximized with respect to radius. (b) Comparison of $\Delta B_z/B_0$ from SUPERLAYER and SUPERAVERAGE simulations and 2XIIB experiments on April 14, 1977. The vacuum magnetic field $B_0 = 4350$ G, and $e\varphi_{rf}^{max}$ indicates the peak wave amplitude used in the quasi-linear model of ion cyclotron turbulence: ---, SUPERLAYER; Δ, SUPERAVERAGE data; ●, 2XIIB data (4/14/1977).

presence of an applied magnetic field, which introduces cyclotron, bounce, and drift time scales, lends itself to the application of orbit averaging. In the orbit-averaged electrostatic model with an explicit kinetic stress tensor reviewed here, the convergence requirement $kv_e \Delta T \leq 0.1$ on the field solution required a large time step ΔT, which would have been desirable as the small time step Δt; this was an impediment to successful orbit averaging. Nevertheless, implicit orbit averaging reduced the number of particles and the size of the code, allowed $\omega_{pe}\Delta t > 1$, and effectively filtered high-frequency electric-field noise.

The success of orbit averaging in reducing the number of particles in applications where $\Delta t_i, \Delta t_e \ll \Delta T$ can be understood from the following simple statistical argument. In a uniform plasma in thermal equilibrium, the electric-field energy in a single Fourier mode is

$$\langle E^2/8\pi \rangle_k V = \tfrac{1}{2} Tg, \tag{34}$$

where V is the volume, $\langle\,\rangle$ denotes an ensemble average, T is the temperature, and $g = g(k^2\lambda_{\text{De}}^2, k^2c^2/\omega_{pe}^2, k_\perp^2 v_\perp^2/\omega_{ce}^2, T/m_ec^2, \ldots)$ is a dimensionless form factor depending on the type of physics model; e.g., $g = (1 + k^2\lambda_{\text{De}}^2)^{-1}$ in an electrostatic model with thermal electrons. The derivation of Eq. (34) for a gridded simulation plasma is given by Langdon (1979b).

We divide both sides of Eq. (34) by the number of particles N, identify the particle density $n = N/V$, and obtain

$$\langle E^2/8\pi \rangle_k / nT = g/2N. \tag{35}$$

If each particle is split into M independent smaller particles, $n \to nM$, $m_s \to m_s/M$, $T \to T/M$, $q_s \to q_s/M$, etc., then g and the algebraic form of the left side of Eq. (35) are both unchanged; but $N \to NM$, and the value of the relative fluctuation level given in Eq. (35) is decreased. If the ΔT used in the orbit average is less than the period of the particle orbits and if the field solution suppresses high-frequency oscillations ($\omega \geq \pi/\Delta T$), then the microscopic current and number densities collected on successive small time steps Δt_s before averaging [see Eq. (18) and Fig. 5] will be statistically independent of one another in the field calculation. Hence, the effective number of particles is increased by orbit averaging from N to $N\,\Delta T/\Delta t$. If the orbital period τ is less than ΔT and if the accumulated particle data are replicated, there is no improvement of the statistics with $\Delta T > \tau$; and the effective number of particles is then only $N\tau/\Delta t$. When ΔT and τ are incommensurate, then the effective number of particles remains $N\,\Delta T/\Delta t$. Orbit averaging has been most successful when $(\tau, \Delta T) \gg \Delta t$ and when there is considerable disorder in the particle phase space.

An interesting variation of the orbit-averaged algorithm has recently emerged. At the end of each large time step ΔT, fields are calculated with time derivatives suppressed and the particles are then rearranged randomly to sample phase space adequately. Subsequent field calculations correspond to iterations, which converge to the equilibrium electric and magnetic fields. Such a scheme provides an efficient self-consistent kinetic calculation of equilibria and transport. Algorithms based on the same principle have been used successfully for a long time to calculate electrostatic fields and beam trajectories (Freis, 1973).

Space limitations have not permitted discussion of a number of important issues that were given attention in the papers cited on orbit averaging and subcycling. For example, accuracy considerations in implicit orbit averaging are described in detail by Cohen et al. (1982b). This review has emphasized instead the basic structure of subcycling and orbit-averaged algorithms and their linear dispersion and stability properties. Some of the applications of these new methods reported so far have given dramatic evidence of significant gains in computational efficiency, but much more code experience is needed to define their practical worth and the limits of their applicability.

ACKNOWLEDGMENTS

The author is very grateful to his collaborators, T. A. Brengle, R. P. Freis, A. B. Langdon, and V. Thomas for their many contributions to this work. Thanks also go to J. C. Adam, C. K. Birdsall, J. A. Byers, W. C. Condit, A. Friedman, and A. Gourdin Heron for their numerous suggestions and enthusiastic support, and to M. Donohue for her assistance in editing the manuscript.

This work was performed under the auspices of the U.S. Department of Energy by the Lawrence Livermore National Laboratory under contract No. W-7405-ENG-48.

REFERENCES

Adam, J. C., Gourdin-Serveniere, A., and Langdon, A. B. (1982). *J. Comput. Phys.* **47**, 229.
Boris, J. P. (1971). *Proc. Conf. Numer. Simul. Plasmas, 4th, 1970* Stock 085100059, pp. 3–67.
Brackbill, J. U., and Forslund, D. W. (1982). *J. Comput. Phys.* **46**, 271.
Byers, J. A. (1977). *Phys. Rev. Lett.* **39**, 1476.
Catto, P. J. (1978). *Plasma Phys.* **20**, 719.
Cohen, B. I., and Freis, R. P. (1982). *J. Comput. Phys.* **45**, 367.
Cohen, B. I., Brengle, T. A., Conley, D. B., and Freis, R. P. (1980). *J. Comput. Phys.* **38**, 45.
Cohen, B. I., Langdon, A. B., and Friedman, A. (1982a). *J. Comput. Phys.* **45**, 15.
Cohen, B. I., Freis, R. P., and Thomas, V. (1982b). *J. Comput. Phys.* **45**, 345.
Denavit, J. (1981). *J. Comput. Phys.* **42**, 337.
Freis, R. P. (1973). *Nucl. Fusion* **13**, 247.
Friedman, A., Langdon, A. B., and Cohen, B. I. (1981). *Comments Plasma Phys. Controlled Fusion* **6**, 225.
Frieman, E. A. (1970). *Phys. Fluids* **13**, 490.
Godfrey, B. B. (1974). *J. Comput. Phys.* **15**, 504.
Godfrey, B. B., and Langdon, A. B. (1976). *J. Comput. Phys.* **20**, 251.
Langdon, A. B. (1979a). *J. Comput. Phys.* **30**, 202.
Langdon, A. B. (1979b). *Phys. Fluids* **22**, 169.
Langdon, A. B. (1983a). *Proc. Conf. Numer. Simul. Plasmas, 10th, 19* ■ Paper 2A2.
Langdon, A. B. (1983b). "Electromagnetic Direct Implicit Algorithm," Rep. UCRL-88371. Lawrence Livermore Nat. Lab., Livermore, California.
Langdon, A. B., and Lasinski, B. F. (1976). *In* "Methods in Computational Physics" (J. Killeen, eds.), Vol. 16, pp. 327–366. Academic, New York.

Langdon, A. B., Cohen, B. I., and Friedman, A. (1983). *J. Comput. Phys.* **51**, 107.
Lee, W. W. (1983). *Phys. Fluids* **26**, 556.
Mason, R. J. (1981). *J. Comput. Phys.* **41**, 233.
Nielson, C. W., and Lewis, H. R. (1976). *In* "Methods in Computational Physics" (J. Killeen, eds.), Vol. 16, pp. 367–388. Academic Press, New York.
Rutherford, P. H., and Frieman, E. A. (1968). *Phys. Fluids* **11**, 569.
Turner, W. C., Clauser, J. F., Coensgen, F. H., Correll, D. L., Cummins, W. F., Freis, R. P., Goodman, R. K., Hunt, A. L., Kaiser, T. B., Melin, G. M., Nexsen, W. E., Simonen, T. C., and Stallard, B. W. (1979). *Nucl. Fusion* **19**, 1011.

11

Direct Implicit Plasma Simulation

A. BRUCE LANGDON
Lawrence Livermore Laboratory
University of California
Livermore, California

D. C. BARNES*
Institute for Fusion Studies
The University of Texas at Austin
Austin, Texas

I. Introduction	336
A. Characteristic Time Scales in Weakly Collisional Plasma	337
B. Particle-in-Cell Electrostatic Force Calculation	337
C. Explicit Time Differencing of the Particle Equations of Motion.	339
D. Implicit Time Differencing of the Particle Equations of Motion.	339
II. Direct Method with Electrostatic Fields	341
A. Solution of the Implicit Equations	341
B. Representative Results.	348
C. Properties of Implicit Particle Codes	354
III. Gyroaveraged Particle Simulation	361
A. Individual Particle Motion.	362
B. The Direct Method for Electrostatic Gyroaveraged Particle Simulation	364
IV. Electromagnetic Direct Implicit Method	366
A. Darwin or Magnetoinductive Fields.	367
B. Implicit Electromagnetic Fields.	367

* Present address: Science Applications, Inc., Austin, Texas 78746.

C. The Direct Method for Implicit Particles
   and Fields. 369
D. Comparison of Darwin and Implicit Codes 372
V. Concluding Remarks. 373
   References 374

I. INTRODUCTION

Characteristic time scales for collective phenomena in plasmas encompass many orders of magnitude. Where kinetic effects are crucial, i.e., where fluid descriptions are inadequate, computer simulation methods have been applied very successfully to studies of the nonlinear evolution of plasma phenomena on the faster time scales. For both applications and basic studies, there is increasing interest in extending simulation techniques to kinetic phenomena on much longer time scales.

One approach to modeling long-time-scale behavior in such systems is to alter the governing equations to eliminate uninteresting high-frequency modes. Examples include the electrostatic- and Darwin-field approximations in plasma simulation, and incompressible hydrodynamics. Other approaches, described here and in other chapters in this volume, are subcycling, orbit averaging, and implicit time integration.

The most adaptable and reliable tools for the study of complex kinetic plasma behavior are the *particle-in-cell* (PIC) codes, in which the plasma evolution is modeled by 10^3–10^7 simulation "particles," each representing a large number of plasma particles and moving according to the classical Newton–Lorentz equations of motion in fields governed by Maxwell's equations. In plasmas, many instability, dissipation, and nonlinear saturation mechanisms are kinetic in nature. Because particle codes have been very successful in studying such phenomena, improving their efficiency for long-time-scale simulation is of great value.

This chapter describes a method for implementing implicit time differencing in PIC plasma codes, in which the equations for the time-advanced quantities are constructed directly from the particle equations of motion by linearization, rather than by introducing fluid (velocity moment) equations. This "direct" method is outlined in Section II.A.1; a simple but practical implementation is in Section II.A.3.

The divisions of this chapter are as follows: The remainder of this section defines the explicit electrostatic PIC algorithm, introduces the notation to be used throughout the chapter, and outlines the properties of explicit differencing and the implicit scheme we use. Section II presents the direct

11. Direct Implicit Plasma Simulation

method as applied to the electrostatic-field case, along with some results, and discusses remaining limitations on time step, conservation properties, and linear stability theory. For strongly magnetized plasma, the gyroaveraged algorithm in Section III removes the unwanted large cyclotron frequency. In Section IV the direct method is generalized to include the full electromagnetic field.

A. Characteristic Time Scales in Weakly Collisional Plasma

The highest frequencies are associated with the small mass of electrons and the high speed of light. The Langmuir frequency ω_{pe}, also called simply the plasma frequency, characterizes charge separation oscillations. Others include the cyclotron frequency ω_{ce} and the transit time for electrons or light to cross a characteristic distance.

In contrast, long time scales can be set by ion inertia, electromagnetic effects, and large spatial scale lengths. The ratios of electron to ion plasma and cyclotron frequencies, and of hydrodynamic to electron transit times, are determined by the small number Zm_e/m_i, where Z is the ionic charge state. Where the dominant forces are from magnetic fields due to currents in the plasma itself, the frequencies are reduced relative to ω_{pe} by at least the ratio c/v_e, where c and v_e are light and electron speeds.

B. Particle-in-Cell Electrostatic Force Calculation

The plasma simulation models treated here use a spatial grid to mediate the particle interactions. Originally developed at Stanford, this approach is used in some form in almost all modern plasma work. Instead of interactions being summed over particle pairs, a charge density is formed from the particle positions onto a spatial grid. By using partial difference equations on this grid, an electric field is found. Then the particles are individually advanced in time using classical equations of motion, with the acceleration found by interpolation from the electric field on the grid. This method is not only faster than summing over particle interactions, it avoids very large accelerations of closely spaced particles, which create complications irrelevant to the simulation of collective effects in weakly collisional plasma.

In the description and analysis of these algorithms, it is helpful to consider the charge of the particle as a diffuse "cloud," with $qS(\mathbf{x})$ the charge density of a particle whose center is at the origin. The charge density of a plasma with particle number density $n(\mathbf{x}, t)$ is

$$\rho_c(\mathbf{x}, t) = S(\mathbf{x}) * qn(\mathbf{x}, t) = \int d\mathbf{x}'\, S(\mathbf{x} - \mathbf{x}')qn(\mathbf{x}', t), \tag{1}$$

where the asterisk denotes convolution. This charge density is sampled at the lattice points of a regular grid in space. To avoid clutter, we temporarily restrict the discussion to one dimension. The charge density associated with grid point j is taken to be $\rho_j(t) = \rho_c(X_j, t)$, where $X_j \equiv j\,\Delta x$ and Δx is the grid spacing. In practice, the contribution of each simulation particle is added to ρ_j:

$$\rho_j = \sum_i q_i S(X_j - x_i), \tag{2}$$

where i is the particle index, q_i the charge, and S the particle–grid interpolation spline.

The form of the weighting function $S(x)$ is designed to provide good numerical properties. As developed originally at Stanford, these models used "nearest grid-point" weighting between particle and grid quantities (Hockney, 1965; Burger et al., 1965; Yu et al., 1965; Boris and Roberts, 1969). Most present work is done with linear weighting (Birdsall and Fuss, 1969; Morse and Nielson, 1969). In one dimension, S is the linear spline, a tent shape extending over two cells and zero elsewhere.

From ρ a potential and an electric field are derived on the same mesh. These satisfy relations such as

$$E_j = -(\varphi_{j+1} - \varphi_{j-1})/2\,\Delta x \equiv -\nabla_j \varphi, \tag{3}$$

$$(\varphi_{j+1} - 2\varphi_j + \varphi_{j-1})/\Delta x^2 = -\rho_j, \tag{4}$$

in rationalized cgs units (Panofsky and Phillips, 1962; Jackson, 1962).

Additional smoothing is often applied to ρ_j or φ_j. Consider this to be a convolution with a function $\hat{S}$.

The particle force field is obtained by an interpolation of the form

$$F_i = q_i\,\Delta x \sum_j E_j S(X_j - x_i), \tag{5}$$

where the sum is over grid points j and S is the same function used in Eq. (2).

In three dimensions the locations of grid points are given by

$$\mathbf{X_j} = (j_x\,\Delta x, j_y\,\Delta y, j_z\,\Delta z) = \mathbf{j} \cdot \mathbf{\Delta x}, \tag{6}$$

where $\mathbf{j}$ is a vector with integer components, and $\mathbf{\Delta x}$ a tensor. In the simplest case of a cubic mesh with spacing Δx, $\mathbf{X_j}$ is just $\mathbf{j}\,\Delta x$. In Eq. (5) Δx is replaced by $|\mathbf{\Delta x}| = \det \mathbf{\Delta x}$, the volume of one grid cell.

We shall maintain a convention in which n and $\mathbf{F}$ refer to particle quantities defined on a spatial continuum, whereas ρ, φ, and $\mathbf{E}$ are defined on the spatial grid.

C. Explicit Time Differencing of the Particle Equations of Motion

For the integration of the particle equations of motion, numerical time-differencing algorithms of the elegant type discussed by Buneman (1967) are almost universally used.

$$(\mathbf{x}_{n+1} - \mathbf{x}_n)/\Delta t = \mathbf{v}_{n+1/2}; \quad (7a)$$

$$(\mathbf{v}_{n+1/2} - \mathbf{v}_{n-1/2})/\Delta t = \mathbf{a}_n + \tfrac{1}{2}(\mathbf{v}_{n+1/2} + \mathbf{v}_{n-1/2}) \times (q\mathbf{B}_n/mc), \quad (7b)$$

where

$$\mathbf{a}_n = (q/m)\mathbf{E}_n(\mathbf{x}_n), \quad (7c)$$

and subscript n denotes time level $t_n = n\,\Delta t$. This is usually called the centered leapfrog scheme.

Equations (2)–(5) and (7a, b) outline the time cycle of an explicit PIC simulation.

In the simplest, unmagnetized case, longitudinal oscillations in a uniform cold plasma consist simply of harmonic oscillations. The properties of the leapfrog scheme are illustrated by considering a single particle with a linear restoring force, $a_n = -\omega_0^2 x_n$, with no magnetic-field term. Setting $x_n = Xz^n = X\exp(-i\omega n\,\Delta t)$, where $z \equiv \exp(-i\omega\,\Delta t)$ is the (complex) change in an amplitude per time step, we find

$$(\omega_0\,\Delta t)^2 = -(z-1)^2/z = 4\sin^2(\omega\,\Delta t/2). \quad (8)$$

For $\omega_0\,\Delta t < 2$, the roots z lie on the unit circle, i.e., ω is real; the oscillations are neither growing nor damped. This property is valuable in plasmas, where oscillatory behavior is ubiquitous and the distinction between stable and growing oscillations is crucial to many studies.

For $\omega_0\,\Delta t > 2$, one root z lies *outside* the unit circle (Im $\omega > 0$); this numerical instability arises for *any* explicit scheme for $\omega_0\,\Delta t$ above some threshold of order unity (Cohen et al., 1982b). Instability can be avoided through the use of implicit time integration, at the expense of increased complexity.

D. Implicit Time Differencing of the Particle Equations of Motion

The first major issue is the choice of finite-differenced equations of motion for the particles that have the necessary stability at large time steps and are accurate enough for the low-frequency phenomena to be simulated. We choose not to consider backward-biased schemes with relative errors of order

Δt. It is not expensive to achieve relative error of order Δt^2, with error Δt^3 in Im ω, the growth–decay rate.

Several suitable schemes for time-differencing the particles have been analyzed and applied (Cohen et al., 1982b). Here, we shall discuss only the "D_1" scheme, also called the $\bar{1}$ scheme (Barnes et al., 1983b), which can be written

$$(\mathbf{x}_{n+1} - \mathbf{x}_n)/\Delta t = \mathbf{v}_{n+1/2}; \tag{9a}$$

$$(\mathbf{v}_{n+1/2} - \mathbf{v}_{n-1/2})/\Delta t = \bar{\mathbf{a}}_n + \tfrac{1}{2}(\mathbf{v}_{n+1/2} + \mathbf{v}_{n-1/2}) \times q\mathbf{B}_n/mc, \tag{9b}$$

where

$$\bar{\mathbf{a}}_n = \tfrac{1}{2}[\bar{\mathbf{a}}_{n-1} + (q/m)\mathbf{E}_{n+1}(\mathbf{x}_{n+1})]. \tag{10}$$

Or, if the recursive filter is applied to the fields rather than to the particles (Barnes et al., 1983b), we write

$$\bar{\mathbf{a}}_n = (q/m)\bar{\mathbf{E}}_n(\mathbf{x}_n) = \tfrac{1}{2}(q/m)[\bar{\mathbf{E}}_{n-1}(\mathbf{x}_n) + \mathbf{E}_{n+1}(\mathbf{x}_n)], \tag{11}$$

where

$$\bar{\mathbf{E}}_n = \tfrac{1}{2}[\bar{\mathbf{E}}_{n-1} + \mathbf{E}_{n+1}]. \tag{12}$$

This choice saves storage of one vector quantity per particle, relative to Eq. (10). In fact, the particle mover coding is exactly the same as for the explicit leapfrog method [Eq. (7)]! The only difference is that the electric acceleration is from $\bar{\mathbf{E}}_n$ instead of $\mathbf{E}_n$. Note that $\mathbf{E}_{n+1}$ is evaluated at *different* positions in Eqs. (10) and (11). This creates side effects, which we discuss in Sections II.A, II.C.2, and II.C.4.

To check the accuracy of this scheme, we can derive and solve a dispersion relation for harmonic oscillations, analogous to Eq. (8):

$$(\omega_0 \Delta t)^2 + (2/z - 1/z^2)(z - 1)^2/z = 0. \tag{13}$$

For $\omega_0 \Delta t \lesssim 1$, we find (Cohen et al., 1982b)

$$\pm \text{Re}\, \omega/\omega_0 = 1 - \tfrac{11}{24}(\omega_0 \Delta t)^2 + \cdots, \qquad \text{Im}\, \omega/\omega_0 = -(\omega_0 \Delta t)^3/2 + \cdots,$$

and an extraneous damped mode with $|z| \to \tfrac{1}{2}$. For $\omega_0 \Delta t \gg 1$, the modes are heavily damped, $|z| \to (\omega_0 \Delta t)^{-2/3}$.

Equation (9b) can be solved exactly for $\mathbf{v}_{n+1/2}$ by adding $\tfrac{1}{2}\bar{\mathbf{a}}_n \Delta t$ to $\mathbf{v}_{n-1/2}$, doing a rotation, and again adding $\tfrac{1}{2}\bar{\mathbf{a}}_n \Delta t$. The result is

$$\mathbf{v}_{n+1/2} = \tfrac{1}{2}\bar{\mathbf{a}}_n \Delta t + \mathbf{R} \cdot [\mathbf{v}_{n-1/2} + \tfrac{1}{2}\bar{\mathbf{a}}_n \Delta t], \tag{14}$$

where the operator $\mathbf{R}$ effects a rotation through angle $-2\tan^{-1}(\Omega \Delta t/2)$ (where $\boldsymbol{\Omega} \equiv q\mathbf{B}_n/mc$) and can be written

$$(1 + \theta^2)\mathbf{R} = (1 - \theta^2)\mathbf{I} + 2\boldsymbol{\theta}\boldsymbol{\theta} - 2\boldsymbol{\theta} \times \mathbf{I}, \tag{15}$$

where $\theta \equiv \Omega \, \Delta t/2$ and $\mathbf{I}$ is the unit tensor. For small $\Omega \, \Delta t$, $\mathbf{R} \cong \mathbf{I} - \Omega \, \Delta t \times \mathbf{I}$. For large $\Omega \, \Delta t$, $\mathbf{R} \cong -\mathbf{I} + 2\Omega^{-2}\Omega\Omega - (4/\Omega^2 \, \Delta t)\Omega \times \mathbf{I}$.

The optimum design of these time-difference equations is the first, but simpler, issue in practical implementation of large-time-step methods.

With explicit differencing, the time cycle is split between advancing particles and fields; these calculations alternate and proceed independently. A price we pay for implicit differencing is that time-cycle splitting is more complicated. An implicit code must solve the coupled set of Eqs. (2)–(5), with (9) and (10) or (11) and (12). This is the second issue in implementation.

II. DIRECT METHOD WITH ELECTROSTATIC FIELDS

In all implicit schemes the future positions $\mathbf{x}_{n+1}$ depend on the accelerations $\mathbf{a}_{n+1}$ due to the electric field $\mathbf{E}_{n+1}$. But this field is not yet known, because it depends on the density ρ_{n+1} of particle positions $\{\mathbf{x}_{n+1}\}$. The solution of this large system of nonlinear coupled particle and field equations is the other major implementation issue.

A. Solution of the Implicit Equations

In the first method implemented for this solution, the fields at the new time level are predicted by solving coupled field and fluid equations in which the kinetic stress tensor is approximately evaluated from particle velocities known at the earlier time. After the fields are known, the particles are advanced to the new time level, and, if desired, an improved stress tensor is calculated and the process iterated. This approach has been described in detail by Denavit (1981), and Mason (1981).

It is also practical to predict the future electric field $\mathbf{E}_{n+1}$ quite directly by means of a linearization of the particle–field equations. One form of this method, its implementation, and some examples verifying its performance have been outlined by Friedman *et al.* (1981). Another form is described by Barnes *et al.* (1983b); compare Eq. (10) with Eqs. (11) and (12), and see Section II.C.4. Langdon *et al.* (1983) explore the algorithm with great generality and consider many important details, such as spatial differencing and filtering, and iterative solution of the implicit equations.

1. Outline of the "Direct Method"

The essence of the "direct" method is that we work *directly* with the particle equations of motion and the particle–field coupling equations. These are linearized about an estimate (extrapolation) for their values at the new time level $n + 1$. The future values of $\{\mathbf{x}, \mathbf{v}\}$ are divided into two parts:

(a) increments $\{\delta\mathbf{x}, \delta\mathbf{v}\}$, which depend on the (unknown) fields at the future time level $n + 1$;

(b) extrapolations $\{\mathbf{x}_{n+1}^{(0)}, \mathbf{v}_{n+1/2}^{(0)}\}$, which incorporate *all* other contributions to the equation of motion.

The charge density $\rho_{n+1}^{(0)}$, corresponding to positions $\{\mathbf{x}_{n+1}^{(0)}\}$, is collected, as are the coefficients in an expression for the difference $\delta\rho(\{\delta\mathbf{E}\}) = \rho_{n+1} - \rho_{n+1}^{(0)}$ between the densities obtained after integration with $\mathbf{E}_{n+1}^{(0)}$ and with the corrected field $\mathbf{E}_{n+1} = \delta\mathbf{E} + \mathbf{E}_{n+1}^{(0)}$. These comprise the source term in Gauss's law:

$$\nabla \cdot \mathbf{E}_{n+1} = \delta\rho(\{\delta\mathbf{E}\}) + \rho_{n+1}^{(0)}. \tag{16}$$

This becomes a linear elliptic equation, for $\delta\varphi$ or φ_{n+1}, with nonconstant coefficients.

The care with which we express the increment $\{\delta\rho\}$ is a compromise between complexity and strong convergence (Langdon *et al.*, 1983; Barnes *et al.*, 1983b). If necessary, $\delta\rho$ may be evaluated rigorously as derivatives of Eq. (2) [*strict differencing* (Langdon *et al.*, 1983, Section 4)] or as simplified difference representations (Langdon *et al.*, 1983, Section 3.4; Barnes *et al.*, 1983b) of Eq. (25) for each species. In the following subsections, we consider these two cases in one dimension, generalize to higher dimensions and to include a magnetic field, and briefly discuss an iterative solution of the field equation in two dimensions.

2. A One-Dimensional Realization

The direct implicit method is illustrated in the following one-dimensional, unmagnetized, electrostatic example. The position $\mathbf{x}_{n+1}$ of a particle at time level t_{n+1}, as given by an implicit time-integration scheme, can be written as

$$x_{n+1} = \beta \, \Delta t^2 \, a_{n+1} + \tilde{x}_{n+1}, \tag{17}$$

where $\beta > 0$ is a parameter controlling implicitness and equal to $\frac{1}{2}$ for the D_1 scheme; $\tilde{x}_{n+1}$, the position obtained from the equation of motion with the acceleration a_{n+1} omitted, is known in terms of positions, velocities, and accelerations at times t_n and earlier. Eliminating $v_{n+1/2}$ between Eqs. (9) and (10), we find

$$\tilde{x}_{n+1} = x_n + v_{n-1/2} \, \Delta t + \tfrac{1}{2} \bar{a}_{n-1} \, \Delta t^2. \tag{18}$$

In its most obvious form, which we adopt for this example, the direct im-

11. Direct Implicit Plasma Simulation

plicit algorithm is derived by linearization of the particle positions relative to $\tilde{x}_{n+1}$; that is, $E_{n+1}^{(0)} = 0$ and, therefore, $x_{n+1}^{(0)} = \tilde{x}_{n+1}$.

At the grid point located at $X_j \equiv j \, \Delta x$, the charge density $\tilde{\rho}_{j,n+1}$ is formed as in (2) by adding the contribution of the simulation particles at positions $\{\tilde{x}_{i,n+1}\}$:

$$\tilde{\rho}_{j,n+1} = \sum_i q_i S(X_j - \tilde{x}_{i,n+1}). \tag{19}$$

If we expand S in Eq. (2) with respect to position, then

$$\delta\rho_{j,n+1} = -\sum_i q_i \, \delta x_i \, S'(X_j - \tilde{x}_{i,n+1}), \tag{20}$$

with $\delta x_i = x_{i,n+1} - \tilde{x}_{i,n+1}$ and $S'(X) = dS/dX$. In terms of E_{n+1}, the particle acceleration is obtained from Eq. (5) evaluated at $\tilde{x}_{n+1}$:

$$m_i a_{i,n+1} = q_i \, \Delta x \sum_j E_{j,n+1} S(X_j - \tilde{x}_{i,n+1}). \tag{21}$$

From the particle equation of motion in the form of Eq. (17),

$$\begin{aligned}\delta x_i &= \beta \, \Delta t^2 \, a_{i,n+1} \\ &= \beta \, \Delta t^2 (q_i/m_i) \, \Delta x \sum_j S(X_j - \tilde{x}_{i,n+1}) E_{j,n+1}. \end{aligned} \tag{22}$$

The densities $\tilde{\rho}_{n+1}$ and $\delta\rho_{n+1}$ are inserted into the field equations (3) and (4). With summation over species understood, Poisson's equation in rationalized cgs units becomes

$$\begin{aligned} -\tilde{\rho}_{j,n+1} &= \frac{\varphi_{j+1,n+1} - 2\varphi_{j,n+1} + \varphi_{j-1,n+1}}{\Delta x^2} \\ &\quad + \Delta x \sum_{i,k} \beta q_i^2 \frac{\Delta t^2}{m_i} S'(X_j - \tilde{x}_{i,n+1}) S(X_k - \tilde{x}_{i,n+1}) \\ &\quad \times \frac{\varphi_{k+1,n+1} - \varphi_{k-1,n+1}}{2 \, \Delta x}. \end{aligned} \tag{23}$$

For linear splines, $\Delta x^2 \, S'(x) = \pm 1$ or 0, there is no contribution for $|j - k| > 1$, and, therefore, the field equation is pentadiagonal. After solution of Eq. (23), E is formed using Eq. (3); then the particles can be brought serially to their new positions using Eq. (17) or its equivalent.

The derivation of Eq. (23) respects the actual field interpolation and charge weighting used with the particles; because of this and its linear stability properties, we call this a *strict* implementation of the direct method. Further discussion is found in Friedman et al. (1981), Section 2 of Langdon et al. (1983), Cohen et al. (1984), and Section II.C.2 of this chapter.

3. Simplified Differencing

It is convenient to implement a simpler field equation that Eq. (23), while retaining benefits of the direct method. Writing Eq. (20) as

$$\delta\rho = -[\nabla \cdot \sum q\, \delta\mathbf{x}\, S(\mathbf{x} - \tilde{\mathbf{x}}_{n+1})]_{\mathbf{x}=\mathbf{x}_j}, \qquad (24)$$

we see that Eqs. (20) and (23) are finite-element representations of

$$\delta\rho = -\nabla \cdot [\tilde{\rho}\, \delta\mathbf{x}], \qquad (25)$$

and the elliptic partial difference equation

$$-\tilde{\rho} = \nabla \cdot [1 + \chi(\mathbf{x})] \nabla\varphi, \qquad (26)$$

where $\chi(\mathbf{x}) = \beta\tilde{\rho}(\mathbf{x})(q/m)\Delta t^2$ summed over species; i.e., $\chi = \beta(\omega_p \Delta t)^2$. Because of the similarity of Eq. (26) to the field equation in dielectric media, we call χ the implicit susceptibility. Where $\omega_p \Delta t$, is large, which is the regime we wish to access, note that $\chi \gg 1$ is dominant in the right-hand side of Eq. (26).

With the extrapolated charge density $\tilde{\rho}_{j,n+1}$ and a reasonable finite-difference representation of the linearized implicit contribution $\delta\rho = -\partial(\chi E)/\partial x$, the field equation in one dimension is

$$\tilde{\rho}_{j,n+1} = [(1 + \chi_{j+1/2})E_{j+1/2,n+1} - (1 + \chi_{j-1/2})E_{j-1/2,n+1}]/\Delta x. \qquad (27)$$

Two representation of $\chi_{j+1/2}$ used here are [Langdon et al., 1983, Eqs. (28a, b)]

$$\chi_{j+1/2} = \tfrac{1}{2}(\chi_j + \chi_{j+1}) \qquad (28a)$$

or

$$\chi_{j+1/2} = \max(\chi_j, \chi_{j+1}), \qquad (28b)$$

where

$$\chi_j = \Delta t^2 \sum_s (\beta\tilde{\rho}_{j,n+1} q/m)_s \qquad (29)$$

is a sum over species index s. In both Eqs. (27) and (29), $\tilde{\rho}_{j,n+1}$ is given by Eq. (19).

In terms of the field

$$E_{j,n+1} = \tfrac{1}{2}(E_{j-1/2,n+1} + E_{j+1/2,n+1}) \qquad (30)$$

formed from E at half-integer positions, the particle acceleration is evaluated at $\tilde{x}_{n+1}$ using Eq. (21). This algorithm is the shortest implicit PIC scheme we have seen and was the most robust in the test problem of Section II.B.1.

4. General Electrostatic Case

We return to the multidimensional case, possibly including a magnetic field imposed by external currents, showing the calculational steps to be performed in the two cases resulting from the choice of (10) or (11). We begin by restating the method in a more general form.

EVALUATION OF THE EXTRAPOLATED DENSITIES. The extrapolated charge density $\rho_{n+1}^{(0)}$ is evaluated as in (2), but from positions $\{\mathbf{x}_{n+1}^{(0)}\}$ obtained from the equation of motion with $\mathbf{a}_{n+1}$ given by $\mathbf{E}_{n+1}^{(0)}$, which is a guess for $\mathbf{E}_{n+1}$. This charge density does *not* correctly correspond to the field $\mathbf{E}_{n+1}^{(0)}$; that is, $\nabla \cdot \mathbf{E}_{n+1}^{(0)} \neq \rho_{n+1}^{(0)}$. We wish to calculate an improved field $\mathbf{E}_{n+1}$ with which the particles are reintegrated to positions $\{\mathbf{x}_{n+1}\}$, whose charge density ρ_{n+1} does satisfy

$$\nabla \cdot \mathbf{E}_{n+1} = \rho_{n+1}. \qquad (31)$$

To this end we rewrite Eq. (31) as

$$\nabla \cdot \delta\mathbf{E}_{n+1} - \delta\rho_{n+1} = \nabla \cdot \mathbf{E}_{n+1}^{(0)} - \rho_{n+1}^{(0)}, \qquad (32)$$

where $\mathbf{E}_{n+1} = \mathbf{E}_{n+1}^{(0)} + \delta\mathbf{E}_{n+1}$, and similarly for ρ_{n+1}; $\delta\rho_{n+1}$ is due to the increments $\{\delta\mathbf{x}\}$ in the particle positions, which in turn are due to the difference between $\mathbf{E}_{n+1}$ and $\mathbf{E}_{n+1}^{(0)}$.

EVALUATION OF THE INCREMENTS DUE TO FUTURE FIELDS. Using Eq. (25) and the equation of motion, we express $\delta\rho_{n+1}$ as a linear functional of $\delta\mathbf{E}_{n+1}$. In the general case, the increments $\{\delta\mathbf{x}, \delta\mathbf{v}\}$ are evaluated by linearization of each equation of motion (Langdon et al., 1983; Barnes et al., 1983b) about position $\mathbf{x}_{n+1}^{(0)}$; here, we have

$$\delta\mathbf{x}_{n+1} = \delta\mathbf{v}_{n+1/2}\, \Delta t,$$

or
$$\delta\mathbf{v}_{n+1/2} = (q\, \Delta t/2m)\, \delta\mathbf{E}_{n+1}(\mathbf{x}_{n+1}^{(0)}) \qquad \text{unmagnetized}, \qquad (33)$$
$$\delta\mathbf{v}_{n+1/2} = \mathbf{T} \cdot (q\, \Delta t/2m)\, \delta\mathbf{E}_{n+1}(\mathbf{x}_{n+1}^{(0)}) \qquad \text{magnetized},$$

where $\mathbf{T} \equiv \tfrac{1}{2}[\mathbf{I} + \mathbf{R}_n(\mathbf{x}_{n+1}^{(0)})]$, which follows from (29).

With Eq. (33), the implicit term $\delta\rho = -\nabla \cdot (\rho\, \delta\mathbf{x}_{n+1}^{(0)})$ in Eq. (32) is seen to be

$$\delta\rho = -\nabla \cdot (\rho_{n+1}^{(0)}\, \delta\mathbf{x}) = -\nabla \cdot \left[\sum_s (\rho_{n+1,s}^{(0)} q_s\, \Delta t^2/2m_s)\mathbf{T}_s\right] \cdot \delta\mathbf{E}$$
$$= -\nabla \cdot (\boldsymbol{\chi} \cdot \delta\mathbf{E}). \qquad (34)$$

The $[\cdots]$ is a sum over *species* s, not each particle. If only the electrons are implicit, only they appear in Eq. (34). In this case, the terms $[\cdots]$ require only a knowledge of the electrons' ρ [in addition to the *net* ρ used on the right side of Eq. (36)]. In general, it is sufficient to accumulate $\rho_{n+1,s}^{(0)}$ separately from species with differing q/m. This requires more storage, but *no more computation* than for an explicit code.

The implicit susceptibility

$$\chi = \left[\sum_s (\rho^{(0)}_{n+1,s} q_s \, \Delta t^2 / 2 m_s) \mathbf{T}_s \right] \tag{35}$$

is a *tensor* due to the rotation **R** induced by **B**. The more general expression

$$\chi = \sum_s [\omega_P^2 \, (\partial \mathbf{x}_{n+1}/\partial \mathbf{a}_{n+1})]_s$$

includes equations of motion altered, for example, to include collisions. Elastic collisions of electrons on ions may be modeled by adding to the equation of motion a random rotation in the Galilean frame in which the mean ion velocity is zero. The momentum change must be added to the ions.

We now have everything needed to write an equation for $\delta \mathbf{E} = -\nabla \delta \varphi$. On substituting our expressions for $\rho^{(0)}_{n+1}$ and $\delta \rho$ into the field equation (32), we have our electrostatic implicit field equation

$$\nabla \cdot [1 + \chi] \cdot \nabla \, \delta \varphi_{n+1} = \nabla \cdot \mathbf{E}^{(0)}_{n+1} - \rho^{(0)}_{n+1}. \tag{36}$$

This is an elliptic field equation whose coefficients depend directly on particle data accumulated on the spatial grid in the form of an effective linear susceptibility. The rank of the matrix equation is determined by the number of field quantities defined on the zones; it is independent of the number of particles and is normally much smaller.

The field corrector equation (36) can also be expressed in terms of time-filtered quantities (Barnes *et al.*, 1983b). In this representation, the time-filtered $\bar{\varphi}^{(0)}$ and $\bar{\rho}^{(0)}$ appear on the right and the adjustment $\delta\bar{\varphi}$ to $\bar{\varphi}$ on the left.

This formalism guides successful implementation of spatial smoothing (Langdon *et al.*, 1983; Barnes *et al.*, 1983b). If spatial smoothing, denoted by the operator $\hat{S}$, is to be applied to ρ and φ on the grid, then $\hat{S}$ must be included in χ if the field solution is to take this into account. In some applications, this has been essential (Sections II.B.2 and II.B.3). Inconsistent smoothing has consequences to linear stability (Section II.C.4).

ADVANCING THE PARTICLES. The field $\mathbf{E}_{n+1}$ is evaluated at positions $\{\mathbf{x}^{(0)}_{n+1}\}$ in (5) in integrating the particles to their final positions $\{\mathbf{x}_{n+1}\}$. The error resulting from this approximation and from the linearization of $\delta\rho$ introduces a possible limitation on Δt that depends on field and density gradients; see Section II.C.1.

After advancing each particle to position $\mathbf{x}_{n+1}$, one can immediately calculate its $\tilde{\mathbf{x}}_{n+2}$ and its contribution to $\tilde{\rho}_{n+2}$. In this way, only one pass through the particle list is required per time step, an advantage when the particles are stored on a slower memory device, such as a rotating magnetic disk.

5. Iterative Solution of Corrector Equation in Two Dimensions with Magnetic Field

Except for one-dimensional systems, the use of direct inversion for the solution of the field corrector equation (36) requires an impractically large amount of computer time and memory. Global iterative methods (Concus and Golub, 1973; Nielson and Lewis, 1976; Busnardo-Neto et al., 1977) are effective for the inversion of variable-coefficient elliptic operators of type (36). Other successful methods include preconditioned conjugate gradient for asymmetric operators (Kershaw, 1978, 1980; Petravic and Kuo-Petravic, 1979), "dynamic ADI" (alternating direction implicit) (Doss and Miller, 1979), and "multigrid adaptive" methods (Brandt, 1977); see also Brackbill and Forslund (Chapter 9).

The variable-coefficient operator is approximated by a simpler operator whose inverse may be obtained directly. In global iteration techniques, this approximate inverse is applied to the residual (difference between right- and left-hand members) of the full elliptic equation. The adjustment to the solution is used to change the residual. If the approximate operator is chosen judiciously, convergence of the residual to an acceptably small value will occur in a few iterations.

The simplest such scheme for Eq. (36) is obtained by replacing the variable χ by a constant tensor χ_0 (Barnes and Kamimura, 1982; Tajima and Leboeuf, 1981). The resulting constant coefficient operator may be directly inverted using fast-Fourier-transform techniques. It is natural to choose χ_0 to be the average value of χ over the domain; however, this choice leads to divergence of the iteration for certain density profiles. A more reliable choice, suggested by Concus and Golub (1973), is to choose χ_0 corresponding to the average of the maximum and the minimum densities. Thus, χ_0 is taken as

$$\chi_0 = \tfrac{1}{2}(\rho_{e,\max} + \rho_{e,\min}) \sum_j \chi_j \Big/ \sum_j \rho_{e,j}. \tag{37}$$

The global iteration then proceeds by defining the residual as

$$\varepsilon = \nabla^2 \varphi^{(0)} + \rho^{(0)} + \nabla \cdot [1 + \chi(\mathbf{x})] \cdot \nabla \, \delta\varphi_m, \tag{38}$$

where m is the iteration index. The iteration equation is obtained using the approximate operator as

$$-\nabla \cdot [1 + \chi_0] \cdot \nabla[\delta\varphi_{m+1} - \delta\varphi_m] = \varepsilon_m. \tag{39}$$

Equation (39) is easily inverted by Fourier transforming. If $\mathbf{k}$ is the Fourier-transform variable, the iteration may be represented as

$$\delta\varphi_{m+1}(\mathbf{k}) = \delta\varphi_m + \varepsilon_m(\mathbf{k})/[k^2 + \mathbf{k} \cdot \chi_0 \cdot \mathbf{k}]. \tag{40}$$

After $\delta\varphi_m$ is replaced by $\delta\varphi_{m+1}$, the residual is recomputed and convergence is tested by computing the maximum of ε_{m+1} over the mesh. It is convenient to use a combination of **x** and **k** space to compute ε_{m+1}. The constant coefficient part, $\nabla^2 \delta\varphi_{m+1}$, may be evaluated in **k** space. The nonconstant coefficient part is evaluated by transforming $\delta\varphi_{m+1}$ to **x** space, multiplying by χ, and transforming back to **k** space. If χ contains a smoothing operator $\hat{S}$, this is applied in **k** space and also included in χ_0.

The global iteration described here has been shown to be convergent and efficient in a number of applications (Barnes et al., 1983a,b). Some applications are described in Section II.B.

6. Comparison with the "Moment" Method

The field predictor (26) has some features in common with the field predictor in the moment method. This is not surprising since both methods attempt an approximate solution of similar equations. The relation of terms arising in the implicit moment and direct implicit viewpoints is discussed by Mason (Chapter 8) and Langdon et al. (1983); here we summarize and comment briefly.

The counterpart to χ in the moment method is approximated from the density ρ_n instead of being formed from $\tilde{\rho}_{n+1}$ or $\rho_{n+1}^{(0)}$. The source term $\tilde{\rho}_{n+1}$ in Eq. (26) is replaced by ρ_n plus the divergence of current and stress tensor terms. If there could be no finite-difference errors, this combination would equal $\tilde{\rho}_{n+1}$. These approximations result in a stability constraint $kv_t \Delta t \lesssim 1$ or $k\bar{v} \Delta t \lesssim 1$ (Section II.C.1) that does not arise in the direct method, as well as requiring more computation. We believe that the particle equations themselves are a better guide to the zero-order state than are the moment equations.

As moment and direct codes are borrowing features from each other, the distinction becomes more one of viewpoint in deriving algorithms and less in the resulting codes themselves. Experience with moment codes using ad hoc spatial differencing encouraged experimentation with simpler differencing in direct codes. Insight gained from the direct method shows how to eliminate the stress tensor, which contributes to the $kv_t \Delta t$ constraint in moment codes (Brackbill and Forslund, Chapter 9). Design of optimal codes requires understanding the fruits of both viewpoints.

B. Representative Results

1. Free Expansion of a Plasma Slab

Denavit (1981) used the one-dimensional expansion of a plasma slab as one check on his early moment-method code. Featuring sharp gradients initially and a large range of densities, this problem is also used to check and improve variations of direct-method algorithms. Friedman et al. (1981) emphasize best performance in a one-dimensional electrostatic implementation. Langdon et al. (1984), as outlined here, sought to isolate those forms most suitable for extension to a two-dimensional and electromagnetic code. Without resort to spatial smoothing, adequate accuracy and robust behavior were obtained. Smoothing makes the field equation more expensive to solve, and the resulting loss of resolution may be too costly in two dimensions. With $n \Delta x = 64$ particles per cell in the slab initially, and values of $\omega_{pe} \Delta t$ as large as 120 (much larger than have been reported previously for unmagnetized plasma), they find that two parameters measure the stress on the algorithm.

The more important parameter is $\chi_1 = \beta q^2 \Delta t^2 / m |\Delta x|$, where q and m are the particle charge and mass, and $|\Delta x|$ the zone volume. This is a worst-case measure of the validity of linearization in the field prediction, as stressed by short-wavelength sampling fluctuations in the charge density. Using the simplified algorithm of Section II.A.3 with either Eq. (28a) or (28b), they obtain reasonable results with χ_1 well over 100. Most other variants, including the momentum-conserving algorithm (Section II.C.2), suffer nonlinear numerical instability at the edge of the expansion when $\chi_1 \gtrsim 1$.

With χ_1 written as $\beta(\omega_{pe} \Delta t)^2 / N_c$, where N_c is the number of particles per cell ($n \Delta x$ in one dimension), the value of being able to run with $\chi_1 \gg 1$ is clarified. Many applications of two-dimensional explicit codes require only $N_c \gtrsim 10$. If we were restricted to $\chi_1 \lesssim 1$, we would need a number of particles per cell exceeding $(\omega_{pe} \Delta t)^2$, which would be a severe limitation.

The second parameter is $v_t \Delta t / \Delta x$, the ratio of thermal electron transit distance per cycle to the zone size. With $v_t \Delta t / \Delta x > 1$, energy conservation is degraded in the absence of spatial smoothing. The ability of a direct implicit code to remain stable is useful with nonuniform meshes, where some cell dimensions may be much smaller, and helps evade failures due to uninteresting transients or small regions.

2. Ion-Acoustic Fluctuations of a Nonequilibrium Plasma

The ion-acoustic fluctuations of a uniform, thermal, unmagnetized, two-temperature, one-dimensional plasma are examined (Barnes et al., 1983b). Since the ion-acoustic fluctuations represent an extremely small part of the total fluctuation energy of a thermal plasma and are strongly affected by electron Landau damping in the parameter range studied, these results represent a severe test of the applicability of the model.

Plasma parameters for the case shown here were electron Debye length $\lambda_D = 0.05$, mass ratio $m_i/m_e = 100$, and temperature ratio $T_e/T_i = 20$; initial electron and ion densities were equal and uniform. The numerical parameters used were cell size $\Delta x = 1$, system length $L_x = 128$, and number of particles of each species $N_0 = 9216$. Periodic boundary conditions were taken and the particle shape, introduced as a smoothing of grid quantities, was given by a Gaussian $\hat{S}(x) = \exp(-r^2/2a^2)/a(2\pi)^{1/2}$, with $a = 3\,\Delta x$. The time step was fixed at $\omega_{pe}\,\Delta t = 10$, a factor of 50–100 over that possible for an explicit algorithm.

Simulation results are summarized in Figs. 1 and 2. The time evolution of the total energy normalized to its initial value is shown in Fig. 1. A very slow cooling of the plasma is observed. The total energy loss is less than 10% over the time interval $\omega_{pe}t = 0.0\text{–}1.0 \times 10^5$. The physical cooling rate of the electrons onto the ions is much larger for the simulation parameters than the numerical rate (Barnes et al., 1983b).

The collective behavior of the plasma at frequencies $\omega \ll \omega_{pe}$ is displayed in Fig. 2. The time-averaged electrostatic energy per degree of freedom $k_B T_e/2$ (k_B is Boltzmann's constant), or fluctuation spectrum, is shown in the figure.

A stringent test of the electron response at low frequencies is afforded by comparison of the simulation results with theory. If the resonant electron response is retained by the direct method, the fluctuation spectrum will be given by the upper curve in Fig. 2. If the low-frequency electron response is

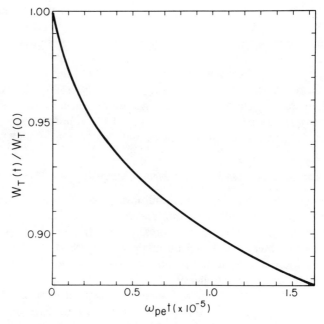

Fig. 1. Total energy normalized to its $t = 0$ value, as a function of time.

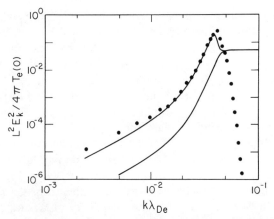

Fig. 2. Fluctuation spectrum for thermal plasma. Normalized electrostatic field energy is shown as a function of wave number. Upper curve is theory including electron Landau damping; lower curve is theory neglecting electron Landau damping; points are simulation results.

only adiabatic, the lower curve is predicted. The ratio of the former to the latter is approximately T_e/T_i. The simulation spectrum indicated by dots closely follows the prediction that includes resonant electron response. Note also that the wave numbers with maximum energy observed in the simulation and predicted by theory agree exactly. At larger wave numbers, the theory is questionable because of approximations made in the derivation.

In summary, the fluctuation spectrum indicates that the resonant electron response at low frequencies is described accurately by the implicit method. It is also clear from these results that fluctuations at mode frequencies higher than the ion-acoustic range have been suppressed; otherwise, the spectrum would have been nearly flat in this frequency range.

3. *Gravitational Interchange Instability of a Magnetized Plasma*

Using the direct method, and treating both the electrons and ions as zero gyroradius particles using the method described in Section III (with gyroradius effects neglected), a two-dimensional magnetized plasma is examined. An unstable gravitational interchange is studied for an inhomogeneous plasma. This calculation is carried out for system size $L_x = L_y = 32\,\Delta x$, number of particles $N_0 = 4608$, mass ratio $m_i/m_e = 100$, electron cyclotron frequency $\Omega_e = \omega_{pe}$, and $a = 3\,\Delta x$ in the Gaussian shape factor $\hat{S}$. The magnetic field is normal to the simulation plane.

Electrons and ions are loaded initially with their guiding center velocity $V_\perp = 0$ in such a way that the distribution of particles is uniform in the left half of the simulation domain. No particles are loaded in the right half of the domain. A gravitational acceleration to the right drives an unstable interchange localized near the interface at the middle of the domain.

The effect of the polarization motion of the ions is very important for this high-density plasma ($\omega_{pi}/\Omega_i \gg 1$). If the polarization motion is neglected, the growth rate becomes unphysically large for high density. For the simulation parameters considered here, the physical growth rate is one-seventh the growth rate found neglecting polarization.

In the simulation the global iteration method described above and an extremely large time step of $\omega_{pe} \Delta t = 10^3$ are used. Simulation results for an appropriate gravitational acceleration are summarized in Figs. 3 and 4. In this case the plasma is initially perturbed with the longest y wavelength so that the growth rate of a single mode may be more accurately measured. In Fig. 3 the electrostatic-field energy is shown in a semilog plot as a function of time. The observed growth rate is within 15% of the theoretically predicted value, verifying the correct modeling of ion polarization motion.

Figure 4 shows three different snapshots of the ion density contours. The positions of roughly 10^3 of the ions are also shown as points in the figure. As can be seen, the unstable interface near the middle of the simulation domain evolves through a linear growth stage, toward a nonlinear "spike and bubble" stage. The only saturation mechanism is provided by the finite size of the periodic simulation domain. Thus, plasma leaving on the right (left) reenters on the left (right). In this way, the configuration evolves toward a state of nearly steady flow with constant potential driving stationary vortices.

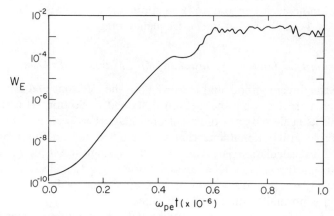

Fig. 3. Electrostatic field energy (normalized) as a function of time for unstable gravitational interchange.

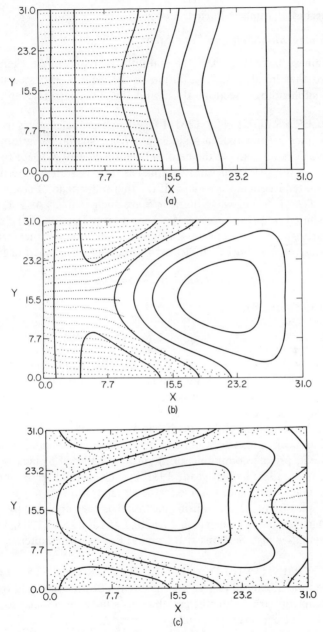

Fig. 4. Snapshots of finite-size ion-density contours are shown, along with point plots of ion positions at three times near saturation of gravitational interchange. (a) $\omega_{pe}\Delta t = 3 \times 10^5$; (b) $\omega_{pe}\Delta t = 4 \times 10^5$; (c) $\omega_{pe}\Delta t = 5 \times 10^5$.

C. Properties of Implicit Particle Codes

1. Remaining Limitations on Time Step

Although we have overcome the stability limit on $\omega_{pe} \Delta t$, there remain restrictions that involve Δt. In addition to the aspects discussed under the following subheadings, Sections II.C.3 and II.C.4 are also relevant.

DOPPLER FREQUENCY LIMIT. When a particle moves in one time step a distance greater than a scale length L for spatial variation of fields, the particle does not sample the field sufficiently closely in space to respond accurately to the field structure. In analysis, this is stated in terms of the Doppler-shifted frequency $\omega - \mathbf{k} \cdot \mathbf{v}$. If the thermal spread of electron velocities $v_t = (T_e/m_e)^{1/2}$, and wave number k are such that $kv_t \Delta t \gtrsim 1$, then the Doppler-shifted frequency exceeds Δt^{-1} for a large fraction of the electrons. It is not surprising that the collective response is qualitatively incorrect, showing an excess of shielding, as indicated by the dielectric function becoming

$$\varepsilon \cong 1 + \beta(\omega_{pe} \Delta t)^2 + \text{ion response} \tag{41a}$$

instead of the correct result,

$$\varepsilon \cong 1 + (k\lambda_D)^{-2} + \text{ion response.} \tag{41b}$$

Violation of the restriction $kv_t \Delta t \lesssim 1$ in direct-method electrostatic codes does not, in general, result in instability. In many applications, the sharp field gradients may arise only during time intervals or in spatial regions such that the inaccuracy does not interfere with the purpose of the simulation. In contrast, strong instability, which has been observed with the moment method, has been traced to a constraint on $kv_t \Delta t$ (Sakagami *et al.*, 1981; J. D. Denavit, private communication) and is attributed to lack of convergence (Denavit, 1981; Cohen *et al.*, 1982a). Such instability, as opposed to innocuous inaccuracy, is disruptive to the simulation.

The Doppler-frequency or transit-time limit has been analyzed by Langdon (1979a), Denavit (1981), Cohen *et al.* (1982b), Langdon *et al.* (1983), and references therein. See also Sections II.B.1 and II.C.4 of this chapter.

LIMITS DUE TO FIELD AND DENSITY GRADIENTS. Linearization of the time-advanced charge density breaks down when field or density gradients are too large. Nonetheless, the codes have been made to function successfully, but convergence is not guaranteed.

Evaluation of $\delta \mathbf{x}$ in terms of $\mathbf{E}_{n+1}$ at $\mathbf{x}_{n+1}^{(0)}$ rather than at $\mathbf{x}_{n+1}$ creates an error $\cong \delta \mathbf{x} \cdot \nabla \mathbf{E}$. Denavit (1981) interprets this term in terms of the frequency of oscillation of a particle trapped in a local potential minimum. The relative error is $\beta(\omega_{\text{trap}} \Delta t)^2$. To compare the severity of this and the transit-time limitations, note that

$$\beta(\omega_{\text{trap}} \Delta t)^2 = \beta(kv_e \Delta t)^2 q\varphi/T_e \qquad (42)$$

Thus, if we take $kv_e \Delta t \lesssim 1$ as given, the linearization is justified if $q\varphi \lesssim T$. Put another way, if $kv \Delta t \lesssim 1$, then $(\omega_{\text{trap}} \Delta t)^2 = ka \Delta t^2$ is small if $a \Delta t$, the impulse in one time step, is small compared to v. In grossly nonneutral regions, where the net charge $\Delta\rho$ is comparable to the electron charge ρ_e,

$$\beta(\omega_{\text{trap}} \Delta t)^2 \cong (\omega_{\text{pe}} \Delta t)^2 \Delta\rho/\rho_e \qquad (43)$$

will not be small unless $\omega_{\text{pe}} \Delta t$ is small in that region.

A similar limit due to the magnetic-field gradient can be interpreted as $(\omega_\beta \Delta t)^2 \lesssim 1$, where ω_β is the "betatron frequency."

The expression $\delta\rho = -\nabla \cdot (\rho \, \delta\mathbf{x})$ is valid when $\delta\mathbf{x}$ is smaller than density gradient scale lengths.

The validity of linearization has been treated by Denavit (1981), Langdon et al. (1983), Cohen et al. (1984), and references therein, and in Section II.B.1 of this chapter.

2. Momentum Conservation

Properties of the physical system can often be reproduced exactly in the simulation, even though the governing equations are not solved exactly. Momentum conservation is easily arranged for in explicit electrostatic simulations. Here, we show how to carry this property over to implicit simulation.

Referring to Section II.A, we see from Eqs. (19) and (21) that the net particle acceleration can be written

$$\sum_i m_i a_{i,n+1} = \Delta x \sum_j \tilde{\rho}_{j,n+1} E_{j,n+1}. \qquad (44)$$

In explicit codes, in the analogous sum $\Delta x \sum \rho E$, E is the solution with ρ as source. The question of momentum conservation is then confined to the symmetry properties of the partial differential equations and the boundary conditions. Momentum conservation is easily obtained where expected physically, at least with Cartesian meshes.

The same would be true in an implicit code if we solved the field and particle equations exactly, but we do not. In Eq. (44), $E_{j,n+1}$ corresponds not to $\tilde{\rho}_{j,n+1}$ but to

$$-(\nabla^2 \varphi)_{j,n+1} = \tilde{\rho}_{j,n+1} + \delta\rho_j = \sum_i q_i [S(X_j - \tilde{x}_{i,n+1}) - \delta x \, S'(X_j - \tilde{x}_{i,n+1})], \qquad (45)$$

with δx given by (22). So, if we revise Eq. (21) to

$$m_i a_{i,n+1} = \Delta x \, q_i \sum_j E_{j,n+1}[S(X_j - \tilde{x}_{i,n+1}) - \delta x \, S'(X_j - \tilde{x}_{i,n+1})], \quad (46)$$

where $[\cdots]$ is the same as in (45), and if the Poisson equation (23) is solved exactly, then ρ and E correspond, and momentum is conserved when appropriate physically.

Let us interpret (46) as two steps in the particle mover. First, evaluate an interim position,

$$x'_{i,n+1} = \delta x_i + \tilde{x}_{i,n+1}$$

$$= \beta \, \Delta t^2 \frac{q_i}{m_i} \Delta x \sum_j S(X_j - \tilde{x}_{i,n+1}) E_{j,n+1} + \tilde{x}_{i,n+1}, \quad (47)$$

using (22) for δx. Then, evaluate a at position x' using an expansion of (21) about $\tilde{x}_{n+1}$, which is (46). This $a_{i,n+1}$ is used for the final values of $v_{n+1/2}$ and x_{n+1}. The source term for Poisson's equation, $\tilde{\rho} + \delta\rho$, is a representation of the density of positions $\{x'\}$ at which the field is evaluated; this makes momentum conservation possible.

We can retain momentum conservation while simplifying the field equation by using a simpler expression for δx in forming the field equation (45) and in moving the particles [Eq. 47]. For example, we can replace (22) by

$$\delta x_i = -\beta \, \Delta t^2 \frac{q_i}{m_i} \Delta x \sum_j S_0(X_{j+1/2} - \tilde{x}_{i,n+1}) \frac{\varphi_{j+1,n+1} - \varphi_{j,n+1}}{\Delta x}, \quad (48)$$

where S_0 is the nearest grid-point (NGP) weighting function. If the function S used in (45) and (46) is the linear spline S_1, then (45) becomes a tridiagonal field equation for φ, instead of the pentadiagonal equation found with strict differencing. When integrating the particles, the interim position [Eq. (47)] is also modified to use S_0. The field equation can be written

$$\tilde{\rho}_{j,n+1} = \frac{(1 + \chi_{j+1/2}) E_{j+1/2,n+1} - (1 + \chi_{j-1/2}) E_{j-1/2,n+1}}{\Delta x}, \quad (49)$$

with

$$E_{j+1/2,n+1} = -\frac{\varphi_{j+1,n+1} - \varphi_{j,n+1}}{\Delta x} \quad (50)$$

and

$$\chi_{j+1/2} = \sum_i \beta \, \Delta t^2 \frac{q_i^2}{m_i} S_0(X_{j+1/2} - \tilde{x}_{i,n+1}) \quad (51)$$

from which

$$E_{j,n+1} = \tfrac{1}{2}(E_{j-1/2,n+1} + E_{j+1/2,n+1}). \quad (52)$$

[Equations (48)–(52) are the same in one dimension as for the Hamiltonian algorithm (Langdon et al., 1983, Section 4.2)]. This scheme is almost as simple as that in Section II.A.3.

This topic was first treated by Friedman et al. (1983), who extended the treatment to include spatial smoothing.

Note that momentum conservation cannot be contrived for the formulation of Barnes et al. (1983b), because time filtering of mesh variables destroys Galilean invariance.

3. Energy Conservation and Artificial Cooling

To understand better the origins and control of nonphysical cooling [a manifestation of error in codes using damped equations of motion (Adam et al., 1982; Barnes et al., 1983b)], the Lenard–Balescu collision operator, corresponding to implicit time integration, is examined. We find two spurious terms due to phase errors associated with damping. One is a nonresonant contribution to the polarization drag. The other is a spurious nonresonant contribution to dynamical friction and corresponds to the drag calculated by Cohen et al. (1982b) but with the field given by the thermal fluctuation spectrum.

Quantitative calculations of cooling rates based on the kinetic theory described here have not yet been carried out. However, cooling rates observed in simulations of thermal plasmas (Barnes et al., 1983b) indicate that cooling is likely to be dominated by damping of a broad band of thermal fluctuations. Thus, the reduction in cooling for schemes with third-order damping compared with those with first-order damping is much less than the reduction in the damping of low-frequency plasma modes. The cooling rate is reduced when the number of modes is increased, decreasing the fluctuation energy per mode. Further, the most effective means of reducing cooling is to cut the width of the fluctuation spectrum by cutting off in wave number (spatial smoothing) or in frequency (by increasing Δt). Finally, acceptably low cooling rates may be obtained in a third-order damping scheme with a moderate amount of spatial smoothing.

KINETIC THEORY. The kinetic theory is based on Birdsall and Langdon (1985), Chapters 9 and 12. Here we neglect the effects of the spatial grid. In the Fokker–Planck collision operator (Balescu–Lenard equation), corresponding to implicit time integration, the velocity diffusion is not altered in any interesting way, but the velocity drag terms are.

POLARIZATION DRAG. One part of the drag, due to anisotropic polarization of an unmagnetized plasma by the test particle, in a Galilean-invariant code [e.g., using Eqs. (9)–(10)], is

$$\langle \mathbf{a} \rangle_{\text{pol}} = \frac{q^2}{m} \int \frac{d\mathbf{k}}{(2\pi)^3} \frac{\mathbf{k}}{k^2} \operatorname{Im} \frac{1}{\varepsilon(\mathbf{k}, \mathbf{k} \cdot \mathbf{v})}$$

$$= -n_0 \frac{q^4}{m^2} \int \frac{d\mathbf{k}}{(2\pi)^3} \frac{\mathbf{k}}{k^2} \int d\mathbf{v}' \frac{f(\mathbf{v}')}{|\varepsilon(\mathbf{k}, \mathbf{k} \cdot \mathbf{v})|^2} \operatorname{Im} \left(\frac{\mathbf{X}}{\mathbf{A}} \right)_{\mathbf{k} \cdot (\mathbf{v} - \mathbf{v}') + i0} \quad (53)$$

(in rationalized cgs units), in which

$$\varepsilon(\mathbf{k}, \omega) = 1 + \omega_p^2 \int d\mathbf{v} \, f_0(\mathbf{v}) \left(\frac{\mathbf{X}}{\mathbf{A}} \right)_{\omega - \mathbf{k} \cdot \mathbf{v} + i0}, \quad (54)$$

where $(\mathbf{X}/\mathbf{A})_{\omega - \mathbf{k} \cdot \mathbf{v} + i0}$ is the ratio of the Fourier amplitudes of $\mathbf{x}^{(1)}$ and $\mathbf{a}^{(1)}$ resulting from the finite-difference equation of motion [and would be $-(\omega - \mathbf{k} \cdot \mathbf{v} + i0)^{-2}$ for exact integration], and $\varepsilon(\mathbf{k}, \omega)$ the corresponding dielectric function. The meaning of the term "$+i0$" is that $(\cdots)_{\omega + i0}$ is understood to mean the limit of $(\cdots)_{\omega + i\gamma}$ as γ approaches zero through *positive* values. Normally, $\operatorname{Im}(\mathbf{X}/\mathbf{A})_{\omega - \mathbf{k} \cdot \mathbf{v} + i0} = -\pi \delta'(\omega - \mathbf{k} \cdot \mathbf{v})$, which leads to the resonant (Landau) contribution, but here $\operatorname{Im}(\mathbf{X}/\mathbf{A})_{\omega - \mathbf{k} \cdot \mathbf{v} + i0}$ is *also* nonzero for $\omega - \mathbf{k} \cdot \mathbf{v} \neq 0$ owing to phase errors associated with numerical damping.

"DYNAMICAL FRICTION". The other part of the drag, the "dynamical friction" can also be expressed in terms of $\operatorname{Im}(\mathbf{X}/\mathbf{A})$:

$$\langle \mathbf{a} \rangle_{\text{fluct}} = \frac{q^2}{m^2} \int \frac{d\mathbf{k}}{(2\pi)^3} \frac{d\omega}{2\pi} \mathbf{k} \cdot (\mathbf{EE})_{\mathbf{k}, \omega} \operatorname{Im} \left(\frac{\mathbf{X}}{\mathbf{A}} \right)_{\omega - \mathbf{k} \cdot \mathbf{v} + i0}, \quad (55)$$

where $(\mathbf{EE})_{\mathbf{k},\omega}$ is the fluctuation spectrum. This result corresponds to that in Section 5 of Cohen et al. (1982b), in which the field consisted of a single wave rather than a spectrum. Here we use the thermal fluctuation spectrum (Langdon, 1979b; Birdsall and Langdon, 1985)

$$(\mathbf{EE})_{\mathbf{k},\omega} = \frac{\kappa \kappa}{K^2} (\rho^2)_{\mathbf{k},\omega}, \quad (56)$$

$$(\rho^2)_{\mathbf{k},\omega} = \frac{2\pi \rho_0 q}{|\varepsilon(\mathbf{k}, \omega)|^2} \int d\mathbf{v} \, f_0(\mathbf{v}) \sum_q \delta(\omega - \mathbf{k} \cdot \mathbf{v} - q\omega_g), \quad (57)$$

where $\omega_g \equiv 2\pi/\Delta t$, and κ, K^2 express the following ratios between the quantities ρ, φ, and $\mathbf{E}$ defined on the grid: $\mathbf{E} = -i\kappa\varphi$, $\rho = K^2 \varphi$ (in rationalized cgs units). As a check on the theory, we can verify that the expressions (53) and (55), with the thermal spectrum, together conserve momentum of the overall distribution of particles.

ENERGY LOSS (COOLING). The Fokker–Planck equation describing the evolution of the velocity distribution function $f(\mathbf{v})$ is

$$\frac{\partial f}{\partial t} = \frac{\partial}{\partial \mathbf{v}} \cdot \left[-f\langle \mathbf{a}\rangle_{\text{pol}} - f\langle \mathbf{a}\rangle_{\text{fluct}} + \frac{\partial}{\partial \mathbf{v}} \cdot f\mathbf{D} \right], \qquad (58)$$

where $\mathbf{D}(\mathbf{v})$ is the diffusion tensor. Because the resonant parts of (53) and (55) cancel, Eq. (58) conserves energy with continuous time. We can compute the rate of cooling due to the nonresonant part of numerical origin:

$$\frac{d}{dt}\text{K.E} = \int d\mathbf{v}\,\mathbf{v} \cdot \langle \mathbf{a}\rangle f(\mathbf{v}) = -\frac{1}{2}n_0 \frac{q^4}{m^2} \int \frac{d\mathbf{k}}{(2\pi)^3} \frac{1}{k^2} \int d\mathbf{v}\,d\mathbf{v}'\, f(\mathbf{v})f(\mathbf{v}')$$
$$\times \left[\frac{1}{|\varepsilon(\mathbf{k},\mathbf{k}\cdot\mathbf{v})|^2} + \frac{1}{|\varepsilon(\mathbf{k},\mathbf{k}\cdot\mathbf{v}')|^2}\right] \mathbf{k}\cdot(\mathbf{v}-\mathbf{v}')\,\text{Im}\left(\frac{X}{A}\right)_{\mathbf{k}\cdot(\mathbf{v}-\mathbf{v}')}, \quad (59)$$

where it is now to be understood that the resonant part of $\text{Im}(X/A)$ is dropped. For the C_1 equation-of-motion scheme, $\text{Im}(X/A)_\omega = c_1 \Delta t^2 \sin \omega \Delta t$, whereas for the D_1 scheme, $\text{Im}(X/A)_\omega = \Delta t^2 \sin \omega \Delta t/(5 - 4\cos \omega \Delta t)$ (Cohen et al., 1982b). In both cases, if the spread in particle velocities is less than $\pi/k_{\max} \Delta t$, then the integrand is always positive; so only cooling results. This is quite generally true for damped time integration, including explicit integration (Adam et al., 1982).

For these schemes with third-order damping, the last two factors in the integrand together are proportional to $[\mathbf{k}\cdot(\mathbf{v}-\mathbf{v}')\Delta t]^2$ for small values. With first-order damping, this approaches a nonzero constant instead (Cohen et al., 1982b). Other implementations of third-order schemes, for example, Barnes et al. (1983b), produce different phase errors and hence different cooling rates.

We hope that application of this analysis will lead to insight into control of errors in implicit simulation.

4. Effects of Imperfect Field Solution on Linear Stability and Dispersion

Analysis of linear stability and dispersion in uniform plasma, as described by dispersion relations, provides guidance to permissible simplifications of the field solution. Generically, these relations are of the form

$$K^2(\mathbf{k})\varepsilon(\mathbf{k},\omega)\varphi = ([\nabla \cdot \chi\nabla]_{\text{code}} - [\nabla \cdot \chi\nabla]_{\text{strict}})_{\mathbf{k}}\,\delta\varphi, \qquad (60)$$

where $[\cdots]$ are the Fourier transforms of the $\nabla \cdot \chi\nabla$ operator in uniform plasma; "strict" and "code" refer to the rigorous representation of Section II.A.2 and to the representation used in the code; K^2 is the ratio ρ/φ, $\delta\varphi$ the transform of $\varphi_{n+1} - \varphi_{n+1}^{(0)}$, and $\varepsilon(\mathbf{k},\omega) = 0$ the dispersion relation applicable to *exact* solution of the implicit finite-differenced particle and field equations.

Expressions for ε for warm plasma, including exactly the effects of Δx and Δt, have been published (Langdon, 1979a). It is assumed that iteration over the particles is not in use. Here we show some contributors to the right-hand side of Eq. (60) and some qualitative consequences, but we do not attempt an encyclopedic enumeration of the possibilities.

Far from being relevant only to direct-method codes, similar, but more complex, methods and results arise with moment-method codes (Denavit, 1981).

NONSTRICT DIFFERENCING AND INCONSISTENT FILTERING. For the one-dimensional simplified algorithm (Section II.A.3),

$$\varepsilon(k,\omega)\varphi = -\beta(\omega_p \Delta t)^2 \sin^2(k\,\Delta x/2)\,\delta\varphi. \tag{61}$$

If smoothing of ρ or φ is used without being included in the susceptibility term, we find, now ignoring the spatial grid,

$$\varepsilon(k,\omega)\varphi = -\beta(\omega_p \Delta t)^2 (1 - \hat{S}^2)\,\delta\varphi. \tag{62}$$

Spatial differencing and/or smoothing also modify ε.

With $\varphi_{n+1}^{(0)} = 0$, $\delta\varphi = \varphi$. The $\delta\varphi$ terms do not upset linear stability (assuming $\hat{S} \leq 1$). On the other hand, with $\varphi_{n+1}^{(0)} = \varphi_n$, the phase of $\delta\varphi/\varphi = (1 - z^{-1})$ is destabilizing in the above examples. Such analysis provides guidance among the many possibilities (Cohen et al., 1984).

NONLOCAL SUSCEPTIBILITY. In addition to the effects of simplified differencing, Barnes et al. (1983b) acquire an additional contribution due to evaluation of $\mathbf{E}_{n+1}$ at $\mathbf{x}_n$ rather than at $\mathbf{x}_{n+1}$ [Eq. (11) versus Eq. (10)] without taking this into account in the susceptibility. When spatial grid effects are ignored their $\delta\bar{\rho}$ is

$$\delta\bar{\rho}(\mathbf{x}) = \nabla \cdot \left[\tfrac{1}{2}\rho_{n+1}^{(0)}(\mathbf{x})(q/m)\,\Delta t^2\, \nabla\,\delta\bar{\varphi}(\mathbf{x} - \mathbf{v}^{(0)}\,\Delta t)\right] \tag{63}$$

weighted over zero-order particle velocities $\mathbf{v}^{(0)}$. In practice, the $\mathbf{v}^{(0)}\,\Delta t$ term is dropped, a local approximation. With no iteration over the particle list and field prediction, the dispersion relation becomes

$$\begin{aligned}
\varepsilon(\mathbf{k},\omega)\bar{\varphi} &= \frac{1}{2}(\omega_p\,\Delta t)^2 \left[1 - \int d\mathbf{v}\, f_0(\mathbf{v})\exp(i\mathbf{k}\cdot\mathbf{v}\,\Delta t)\right]\delta\bar{\varphi} \\
&= \frac{1}{2}(\omega_p\,\Delta t)^2 \left\{1 - \exp\left[i\mathbf{k}\cdot\bar{\mathbf{v}}\,\Delta t - \frac{1}{2}(kv_t\,\Delta t)^2\right]\right\}\delta\bar{\varphi},
\end{aligned} \tag{64}$$

where f_0 is the particle velocity distribution function and the second line applies to a drifting Maxwellian. Owing to the destabilizing effect of the right-hand side, this formulation is not recommended for application to drifting plasma.

The linear dispersion errors that are responsible for this destabilization may be corrected without requiring additional particle quantities, if this is desirable. As the analysis above shows, the destabilizing phase error comes from the neglected displacement between the time n particle positions, where the acceleration of Eq. (11) is applied, and the time $n + 1$ particle positions, which produce the time $n + 1$ density. If the acceleration (26) is redefined as

$$\bar{\mathbf{a}}_n = (q/2m)[\bar{\mathbf{E}}_{n-1}(\mathbf{x}_n - \Delta t\, \mathbf{v}_{n-1/2}) + \mathbf{E}_{n+1}(\mathbf{x}_n + \Delta t\, \mathbf{v}_{n+1/2})], \qquad (65)$$

this displacement vanishes. When the dispersion relation of this scheme, with no iteration over the particle list and field predictor, is investigated for a cold drifting beam, no linear instability is found for any value of the drift speed or time step.

III. GYROAVERAGED PARTICLE SIMULATION

In this section a useful technique is developed for studying magnetized plasma phenomena, which are much slower than the period of cyclotron oscillation for either or both species. In such applications, it is neither desirable nor efficient to follow the rapid gyration of individual particles about the magnetic field. Rather, the primitive equations of motion should be replaced by equations that contain only the slower time scales of interest, that is, the transit and drift time scales.

The approach adopted here is to replace the nearly point particles, which represent the fastest time scales, by the gyroaverage of such a charge. Thus, in both the charge and force calculations, the shape factor is chosen to represent a ring of charge corresponding to the motion of a point charge during a single gyroperiod.

Such gyroaveraging leads to equations of motion that still contain the terms driving high-frequency motion. Implicit differencing of these gyroaveraged equations removes the vestiges of this unwanted high-frequency branch. The resulting model selects the solution that evolves only on the slower transit and drift time scales.

An alternative approach (Lee, 1983; Dubin *et al.*, 1983) replaces the primitive physical fields with fields obtained by a several-term adiabatic Hamiltonian theory. It seems difficult to include effects of spatially inhomogeneous magnetic geometries in this alternate approach. The techniques described in this section may be applied to arbitrary geometries if the gyroradius is taken small but finite (Barnes and Kamimura, 1982). The case of strong inhomogeneity and large $k_\perp r_L$ (where $k_\perp$ is the wave number perpendicular to the magnetic field and r_L a typical gyroradius) seems difficult to treat satisfactorily. Some comments on these difficulties are given at the end of this section.

The properties of gyroaveraged particle simulation (GAPS) are developed for the simplest case: The magnetic field **B** is assumed uniform and the electric field **E** is assumed electrostatic ($\nabla \times \mathbf{E} = 0$). First, the motion of individual simulation particles, representing guiding-center positions, are considered. Then the implementation of GAPS is shown to depend on the direct method of the previous section. Some implementation details are given. Extensions to nonuniform **B** and to electromagnetic **E** are briefly indicated.

A. Individual Particle Motion

An intuitively appealing technique for removing the fast cyclotron time scale from particle motion is the replacement of the instantaneous Lorentz force on a particle by its average over the fast time scale. Northrup (1961) has shown that such a procedure describes drift motion in an electromagnetic field under appropriate conditions. Such a procedure is based on the existence of an adiabatic invariant associated with the wide separation of the time scales for gyration and for variation of the electromagnetic field.

When such an assumption is appropriate, as is supposed in the case of interest here, there exists an invariant μ given to lowest order by the magnetic moment:

$$\mu = mv_\perp^2/2B. \tag{66}$$

Thus, the particle orbit is given to lowest order by

$$\mathbf{x} = \boldsymbol{\xi} + \mathbf{g}(\theta), \tag{67}$$

where $\boldsymbol{\xi}$ is the (slowly varying) guiding-center position, **g** the (rapidly varying) instantaneous gyrovector, and θ the gyrophase. The magnitude of **g** is the gyroradius $r_L = \sqrt{2\mu/q\Omega}$, and its direction rotates about **B** with θ. If **i** is any unit vector normal to **B**,

$$\mathbf{g} = r_L[\cos\theta\, \mathbf{i} + \sin\theta(\mathbf{i} \times \mathbf{B}/B]. \tag{68}$$

With the decomposition of Eq. (67), the Lorentz force is replaced by its gyroaverage. Thus, the equation of motion of a simulation particle becomes

$$m_i \frac{d\mathbf{V}_i}{dt} = \frac{1}{2\pi}\int_0^{2\pi} d\theta\, \mathbf{F}[\boldsymbol{\xi}_i + \mathbf{g}(\theta)] + q_i \mathbf{V}_i \times \mathbf{B}, \tag{69}$$

where the **B** field has been assumed uniform, and $\mathbf{V}_i = d\boldsymbol{\xi}_i/dt$ is the guiding-center velocity.

The gyroaveraged force may be simplified using Eq. (5):

$$\mathbf{F}_{\mu i} = \frac{1}{2\pi}\int_0^{2\pi} d\theta\, \mathbf{F}[\boldsymbol{\xi} + \mathbf{g}(\theta)] = q_i|\Delta\mathbf{x}|\sum_\mathbf{j} \mathbf{E}_{\mu \mathbf{j}} S(\mathbf{X}_\mathbf{j} - \boldsymbol{\xi}_i), \tag{70}$$

where
$$\mathbf{E}_{\mu\mathbf{j}} = -\nabla_{\mathbf{j}}\varphi_{\mu},\tag{71}$$
with
$$\varphi_{\mu j} = \frac{1}{2\pi}\int_0^{2\pi} d\theta\, \varphi(\mathbf{X_j} + \mathbf{g}).\tag{72}$$

The gyroaveraged equation of motion (69) describes drift and transit motion if the proper initial conditions are chosen (Northrup, 1961). These initial conditions associate a unique perpendicular velocity $\mathbf{V}_\perp = \mathbf{U}_\mu(\xi, V_\parallel, 0)$, with a guiding center whose initial position is ξ and whose initial parallel velocity is $V_\parallel$. The general solution consists of the desired drift branch, a rapid perpendicular gyration about the magnetic field, and the nonlinear interaction of these two branches.

To select the desired drift solution in a difference approximation to (69), the initial $\mathbf{V}_\perp$ assigned to each guiding center is chosen as close to $\mathbf{U}_\mu$ as practical. Unwanted gyrations are avoided by the proper differencing of the last term on the right of (69). If the differencing of this term is modified slightly from that used in explicit plasma simulation [Eq. (7a, b)], gyrations are weakly damped and the desired drift motion is recovered.

The simplest treatment is to decenter the leapfrog difference equations slightly. This leads to the difference equations

$$(\xi_{n+1} - \xi_n)/\Delta t = \mathbf{V}_{n+1/2},\tag{73}$$

$$(\mathbf{V}_{n+1/2} - \mathbf{V}_{n-1/2})/\Delta t = \mathbf{a}_{\mu n} + [(\tfrac{1}{2}+\gamma)\mathbf{V}_{n+1/2} + (\tfrac{1}{2}-\gamma)\mathbf{V}_{n-1/2}] \times (q\mathbf{B}/mc),\tag{74}$$

where $\mathbf{a}_{\mu n} = \mathbf{F}_{\mu n}/m$ and $\gamma > 0$ is a small decentering parameter. The effects of the dissipation and phase error introduced by decentering the $\mathbf{V} \times \mathbf{B}$ term are very small, since only the motion perpendicular to $\mathbf{B}$ is affected. Since gyroaveraging effectively removes cyclotron noise, perpendicular motion is little modified by decentering.

To obtain the contribution of a simulation particle to the charge density, recall that the same particle cloud contributes to the charge as samples the force. Thus, $\mathbf{F}_{\mu i}$ may be written as

$$\mathbf{F}_{\mu i} = q_i \int d\mathbf{x}\, \mathbf{E}(\mathbf{x}) S_\mu(\mathbf{x} - \xi_i),\tag{75}$$

where
$$S_\mu(\delta\mathbf{x}) = \frac{1}{2\pi}\int_0^{2\pi} d\theta\, S(\delta\mathbf{x} + \mathbf{g}).\tag{76}$$

The charge density associated with a plasma of gyroaveraged particles is thus

$$\rho_c(\mathbf{x}, t) = \sum_\mu S_\mu * q n_\mu = \sum_\mu \int d\mathbf{x}' \, S_\mu(\mathbf{x} - \mathbf{x}') q n_\mu(\mathbf{x}', t), \tag{77}$$

where n_μ is the number density of magnetic moments μ.

B. The Direct Method for Electrostatic Gyroaveraged Particle Simulation

In this section, the GAPS method is combined with the direct method to give a long-time-step method in which there is no restriction on the time step from either the plasma oscillation period or the cyclotron period of either species.

For small gyroradius particles, the gyroaveraged acceleration $\mathbf{a}_{\mu n}$ is nearly $\mathbf{a}_n$. In that case, Eqs. (73) and (74) are the same as the earlier equations of motion (7a, b) for $\gamma = 0$. For $\gamma > 0$, the susceptibility is slightly modified.

As in Section II.A, the equations of motion are now linearized about a prediction $\{\xi_{n+1}^{(0)}, \mathbf{V}_{n+1/2}^{(0)}\}$. Linearization of Eqs. (73) and (74) gives

$$\delta \xi = \Delta t \, \delta \mathbf{V}, \tag{78}$$

$$\delta \mathbf{V} = \mathbf{T}_\gamma \cdot (q \, \Delta t / 2m) S * \delta \mathbf{E}, \tag{79}$$

where the tensor $\mathbf{T}_\gamma$ is given by

$$(1 + \theta_\gamma^2) \mathbf{T}_\gamma = \mathbf{I} - \boldsymbol{\theta}_\gamma \times \mathbf{I} + \boldsymbol{\theta}_\gamma \boldsymbol{\theta}_\gamma, \tag{80}$$

with $\boldsymbol{\theta}_\gamma = \frac{1}{2}(1 + \gamma) \boldsymbol{\Omega} \, \Delta t$.

Following the procedure described in Section II.A, the field corrector equation is found to be similar to that obtained previously. The susceptibility tensor is now an operator containing the gyroaveraged particle shape. The susceptibility is found from Eq. (79) to be the same as that given by Eq. (35) with $\mathbf{T}$ replaced by $\mathbf{T}_\gamma$ and the addition of spatial smoothing. Thus, the methods described earlier may be used to solve the field corrector associated with a single, small-gyroradius gyroaveraged species, even for $\Omega \, \Delta t \gg 1$.

In a warm, magnetized plasma, spatial scale lengths are such that the electron gyroradius may be neglected for low-frequency electrostatic phenomena. The ion gyroradius is not negligible, however, since fluctuations with scale size comparable to the thermal ion gyroradius are both predicted theoretically and observed experimentally. If both the electron and ion species are treated by gyroaveraged equations similar to Eqs. (73) and (74), with the explicit acceleration $\mathbf{a}_n$ replaced by the implicit acceleration $\bar{\mathbf{a}}_n$, a low-frequency algorithm results in which neither the plasma frequencies nor the cyclotron frequencies constrain Δt.

11. Direct Implicit Plasma Simulation

In this case, however, the field corrector equation will contain the gyroaveraged ion response as part of the susceptibility. The solution of such a nonlocal, velocity-dependent operator equation is cumbersome. This complication is avoided by treating only the electron species implicitly according to Eqs. (73) and (74) with the zero gyroradius acceleration $\bar{\mathbf{a}}_n$. The ion species may be advanced by the explicit equations (73) and (74) as written. Just as there are no ion plasma modes in a neutral plasma, there is no stability constraint on Δt associated with the time scale ω_{pi}^{-1}.

To see this, consider the case of a cold, uniform, unmagnetized plasma in which the electrons are treated implicitly and the ions explicitly. The dispersion relation is a superposition of the implicit equation (13) and the explicit equation (8):

$$1 + \omega_{pe}^2 \Delta t^2 [z/(z-1)^2][z^2/(2z-1) + m_e/m_i] = 0. \tag{81}$$

Note that for $m_e/m_i = 0$ and $\omega_e \Delta t \gg 1$, the destabilizing explicit ion response [represented by the last term on the left of Eq. (81)] is not sufficient to destabilize the time-advancement scheme. In fact, for $m_e/m_i < \frac{1}{3}$ (which is trivially satisfied), there are no unstable roots of Eq. (81) irrespective of the size of Δt.

This observation applies to all implicit plasma simulation algorithms. It is sufficient to advance the electrons implicitly and, therefore, include only their susceptibility in the field corrector equation to ensure numerical stability of all plasma oscillations.

The direct electrostatic GAPS algorithm advances the electrons with the predictor–corrector method outlined in Section II above. The field corrector equation is identical to Eqs. (36) and (35), except that $\mathbf{T}$ is replaced by $\mathbf{T}_y$ defined by Eq. (80) and only the electron susceptibility enters. The ions are advanced explicitly according to Eqs. (73) and (74) with the gyroaveraged acceleration $\mathbf{a}_{\mu n}$ computed from a single gyroperiod of the ion motion in the specified $\mathbf{B}$ field and the self-consistent $\mathbf{E}$ field.

The calculation of the gyroaverages of $\mathbf{a}_n$ and S may be computed by substepping each ion orbit around a single gyroperiod using a fractional time step δt such that a gyroperiod is completed in 4–10 fractional steps.

A more complete development of the properties of GAPS is beyond the scope of this chapter. We remark here only that it can be shown that the guiding centers of GAPS move according to a gyroaveraged Hamiltonian. Thus, a kinetic equation may be obtained for the collisionless GAPS plasma. It may be shown that the linear dispersion relation for drift waves is recovered from this kinetic equation.

Extensions of the GAPS algorithm to inhomogeneous $\mathbf{B}$ and electromagnetic $\mathbf{E}$ are possible. In these cases, the electron gyroradius must be included to first order in the gyroradius, and the gyroaveraged ion equation must include the gyroaveraged $\dot{\mathbf{g}} \times \mathbf{B}$ force in addition to the electric force of Eq. (74).

IV. ELECTROMAGNETIC DIRECT IMPLICIT METHOD

Until recently, plasma simulation including the full electromagnetic field was done with explicit differencing of both the particles and fields. The latter adds a Courant–Levy–Friedrichs time-step limitation $\Delta t < c/\lambda$, where λ is approximately the mesh spacing used for the fields. The limitation has been removed in two ways. One is to alter the field equations so that they no longer support wave propagation. A proven approach here is the Darwin or magnetoinductive model. Another is to use implicit differencing of the field equations. Recently, codes have used implicit differencing of both fields and particles.

Darwin codes eliminate the Courant restriction $\Delta t < c/\lambda$ by dropping Maxwell's transverse displacement current term. These "pre-Maxwell equations" eliminate electromagnetic wave propagation while retaining electrostatic, magnetostatic, and inductive electric fields. The equivalence of this nonradiative approximation to the Darwin Lagrangian, which retains as much of the electromagnetic interaction as possible without including retardation, was shown by Kaufman and Rostler (1971). Nielson and Lewis (1976) provide many references for the historical development of these codes. Although these codes have used explicit differencing for the particles, it is also possible to make an implicit Darwin code.

For applications not requiring a kinetic description of the electrons, codes using a hybrid of particle ions and fluid electrons are indicated. With Darwin and quasi-static approximations, long time scales are accessible, as in a fully implicit code, but with less noise (Hewett, 1980, and references therein).

Implicit fields reproduce electromagnetic wave propagation at long wavelengths ($\gg c \, \Delta t$). At short wavelengths, the electrostatic, magnetostatic, and inductive electric fields are retained, as in a Darwin code. Implicit fields can be used with explicit particles. With implicit particles, Langmuir waves are stabilized at all wavelengths, as in an implicit electrostatic code. The electrostatic fields are accurate for wavelengths longer than the electron transit distance ($v_{te} \, \Delta t$). These properties make an implicit electromagnetic code attractive, for example, to modeling of intense electron flow, which is subject to pinching, to Weibel instability (Brackbill and Forslund, 1982), and to other processes generating magnetic fields that alter the electron flow (Forslund and Brackbill, 1982).

Here we outline the Darwin algorithm and a fully implicit electromagnetic algorithm. The latter is an extension of Langdon (1983a,b) and has been implemented by Hewett and Langdon. Barnes and Kamimura have preliminary results with their version of this algorithm. Other chapters in his volume, by Brackbill and Forslund and by Mason, describe moment method electromagnetic algorithms.

A. Darwin or Magnetoinductive Fields

The Darwin approximation neglects the transverse displacement current $\partial E_T/\partial t$, leaving

$$c \nabla \times \mathbf{B} = \mathbf{J}_T = \mathbf{J} + \partial \mathbf{E}_L/\partial t, \tag{82}$$

$$c \nabla \times \mathbf{E} = -\partial \mathbf{B}/\partial t. \tag{83}$$

Given $\mathbf{E}_n$, $\mathbf{B}_n$, the particles are integrated by explicit differencing to $\mathbf{v}_{n+1/2}$ and $\mathbf{x}_{n+1}$. Extrapolation of $\mathbf{v}_{n+1/2}$ to $\mathbf{v}_{n+1}$ permits collection of $\mathbf{J}_{n+1}$, from which $\mathbf{B}_{n+1}$ is obtained, for example, by solution of $c \nabla^2 \mathbf{B} = -\nabla \times \mathbf{J}$. Unlike usual electromagnetic codes, the Ampere equation (82) cannot be used to advance $\mathbf{E}_T$ in time. Instead,

$$c^2 \nabla^2 \mathbf{E}_T = \partial \mathbf{J}_T/\partial t \tag{84}$$

is used at time level $n + 1$. This creates a time-centering problem. To preserve second-order accuracy in time, Eq. (84) needs a time-advanced expression for $\partial \mathbf{J}_T/\partial t$. To ensure stability, $\partial \mathbf{J}/\partial t$ is expressed in terms of the advanced E, using moments accumulated from the particles:

$$\partial \mathbf{J}/\partial t = -\nabla \cdot \rho \langle \mathbf{vv} \rangle + (q\rho/m)\mathbf{E} + (q\rho \langle \mathbf{v} \rangle/mc) \times \mathbf{B} \tag{85}$$

summed over species. This leads to an elliptic equation for the advanced fields of form

$$c^2 \nabla^2 \mathbf{E}_T - \omega_p^2(\mathbf{x})\mathbf{E}_T = -\sum \nabla \cdot \rho \langle \mathbf{vv} \rangle + (\sum q\rho \langle \mathbf{v} \rangle/mc) \times \mathbf{B} + \omega_p^2 \mathbf{E}_L - \nabla \psi \tag{86}$$

at time level $n + 1$. The divergence of this equation, together with $\nabla \cdot \mathbf{E}_T = 0$, determines ψ. This elliptic equation provides instantaneous propagation of B and $\mathbf{E}_T$, as is necessary for stability. Although this description uses moment equations, it seems possible to make a direct-method Darwin code.

After presenting the implicit algorithm, we make comparisons between the Darwin and the direct and moment implicit algorithms.

B. Implicit Electromagnetic Fields

Here we must select a time-differencing scheme for the fields, and find a method for solving the coupled field and particle equations. Desired features of the implicit differencing of the Maxwell equations include:

(a) accuracy in dispersion Re $\omega(\mathbf{k})$ and weak damping (e.g., Im $\omega(\mathbf{k})/ck = O(ck\,\Delta t)^3$; $\mathbf{k}$ is the wave vector) at *long* wavelengths;

(b) stability (preferably damping) at *short* wavelengths $\cong 2\,\Delta x$; stability despite $c\,\Delta t \gtrsim \Delta x$ (violation of the Courant condition for *explicit* differencing), and dissipation of inaccurately calculated short wavelengths;

(c) compatibility with implicit particles;

(d) adaptability to general boundary conditions;

(e) simplicity and economy in storage;

(f) optional ability to recover the centred second-order scheme now commonly used for the fields;

(g) optional ability to recover the (nearly) centered Darwin scheme (Nielson and Lewis, 1976).

1. D_1 Time Differencing

For the time differencing of the fields, we adapt the D_1 implicit scheme developed for the particle equations of motion. For example, in the particle equations (9) and (10), drop the $\mathbf{v} \times \mathbf{B}$ term, replace $\mathbf{x}$ by $\mathbf{E}$, $\mathbf{v}$ by $c\,\nabla \times \mathbf{B}$, and $\bar{\mathbf{a}}$ by $-c^2\,\nabla \times \nabla \times \bar{\mathbf{E}}$ to obtain the Maxwell equations in rationalized cgs units (Panofsky and Phillips, 1962; Jackson, 1962):

$$c\,\nabla \times \mathbf{B}_{n+1/2} = \mathbf{J}_{n+1/2} + (\mathbf{E}_{n+1} - \mathbf{E}_n)/\Delta t, \qquad (87a)$$

$$-c\,\nabla \times \bar{\mathbf{E}}_n = (\mathbf{B}_{n+1/2} - \mathbf{B}_{n-1/2})/\Delta t, \qquad (87b)$$

where

$$\bar{\mathbf{E}}_n = \tfrac{1}{2}(\bar{\mathbf{E}}_{n-1} + \mathbf{E}_{n+1}) \qquad (87c)$$

is the result of a recursive low-pass filter with phase error $O(\Delta t^3)$. This phase error is an advance, not a lag that one gets if $\mathbf{E}_{n+1}$ is not used, so it provides stability when $ck\,\Delta t \gg 1$; in this limit the fields are close to those in the Darwin approximation. Although equations (87) are heuristically motivated, their desirable properties may be verified rigorously.

To advance the field values implicitly, eliminate $\mathbf{E}_{n+1}$ or $\mathbf{B}_{n+1/2}$ from the coupled equations (87) to obtain a single elliptic equation. Eliminating $\mathbf{B}_{n+1/2}$ to form an equation for $\mathbf{E}_{n+1}$ yields

$$\mathbf{E}_{n+1} + \tfrac{1}{2}c^2\,\Delta t^2\,\nabla \times \nabla \times \mathbf{E}_{n+1} = \mathbf{E}_n - \mathbf{J}_{n+1/2}\,\Delta t + c\,\Delta t\,\nabla$$
$$\times [\mathbf{B}_{n-1/2} - \tfrac{1}{2}c\,\Delta t\,\nabla \times \bar{\mathbf{E}}_{n-1}]; \qquad (88)$$

or eliminate $\mathbf{E}_{n+1}$ to form an equation for $\mathbf{B}_{n+1/2}$:

$$\mathbf{B}_{n+1/2} - \tfrac{1}{2}c^2\,\Delta t^2\,\nabla^2 \mathbf{B}_{n+1/2} = \mathbf{B}_{n-1/2} + \tfrac{1}{2}c\,\Delta t\,\nabla \times [\mathbf{J}_{n+1/2} - \bar{\mathbf{E}}_{n-1} - \mathbf{E}_n]. \qquad (89)$$

11. Direct Implicit Plasma Simulation

In either case the right-hand side is composed of known fields. The left-hand sides have well-behaved elliptic operators. In two dimensions (x, y), it is convenient to solve the z components of Eqs. (88) and (89) for E_z and B_z, respectively. These are two *scalar* uncoupled elliptic equations to solve. Then use Eqs. (87a, b) to find the other fields.

To form a $\mathbf{B}_n$ for use in the particle mover, we use, for example,

$$\mathbf{B}_n = \mathbf{B}_{n-1/2} - \tfrac{1}{2} \Delta t \, c \, \nabla \times \mathbf{E}_n. \tag{90}$$

We use $\mathbf{E}_n$, rather than $\bar{\mathbf{E}}_n$ as in Eq. (87b), to simplify the linearized particle equation.

Equations (87a, b) differ from the usual centered leapfrog scheme only in that the electric field in Faraday's law here is $\bar{\mathbf{E}}_n$ instead of $\mathbf{E}_n$. If we replace $\bar{\mathbf{E}}_n$ in Eq. (87b) with the linear combination $\alpha \bar{\mathbf{E}}_n + (1 - \alpha)\mathbf{E}_n$, then with $\alpha = 1$ we obtain the D_1 scheme above and the leapfrog scheme with $\alpha = 0$. For intermediate values the upper bound on Δt increases as $\alpha \to 1$. In problems where most cells are large and the undamped leapfrog scheme is preferred, but some cells are much smaller (e.g., near a boundary, or for $r \to 0$ in cylindrical or spherical coordinates), one might use $\alpha = 0$ for the large cells and increase α to maintain stability where cells are smaller. B. Godfrey (private communication) observes that electromagnetic wave dispersion is improved by operating close to the stability boundary.

C. The Direct Method for Implicit Particles and Fields

As in Section II.A.4 for the electrostatic case, the source terms in Maxwell's equations (87) are the densities $\{\rho_{n+1}^{(0)}, \mathbf{J}_{n+1/2}^{(0)}\}$ and $\{\delta\rho, \delta\mathbf{J}\}$ corresponding to the extrapolations $\{\mathbf{x}_{n+1}^{(0)}, \mathbf{v}_{n+1/2}^{(0)}\}$, plus the increments $\{\delta\mathbf{x}, \delta\mathbf{v}\}$ that depend on $\mathbf{E}_{n+1}$.

1. *Evaluation of the Extrapolated Densities*

The extrapolated current and charge densities $\{\rho_{n+1}^{(0)}, \mathbf{J}_{n+1/2}^{(0)}\}$ are evaluated as in explicit codes, such as ZOHAR (Langdon and Lasinski, 1976) and the Los Alamos WAVE code, from $\mathbf{x}_{n+1}^{(0)}$, $\mathbf{v}_{n+1/2}^{(0)}$, and $\mathbf{x}_n$. At the grid point located at $\mathbf{X}_j$,

$$\rho_{n+1}^{(0)} = \sum qS(\mathbf{X}_j - \mathbf{x}_{n+1}^{(0)}), \tag{91}$$

and

$$\mathbf{J}_{n+1/2}^{(0)} = \sum q\mathbf{v}_{n+1/2}^{(0)} S(\mathbf{X}_j - \tfrac{1}{2}[\mathbf{x}_n + \mathbf{x}_{n+1}^{(0)}]), \tag{92a}$$

or

$$\mathbf{J}_{n+1/2}^{(0)} = \sum q\mathbf{v}_{n+1/2}^{(0)} \tfrac{1}{2}[S(\mathbf{X}_j - \mathbf{x}_n) + S(\mathbf{X}_j - \mathbf{x}_{n+1}^{(0)})]. \tag{92b}$$

Neither expression for $\mathbf{J}_{n+1/2}^{(0)}$ is quite consistent with $\rho_{n+1}^{(0)}$ and the continuity equation in that

$$\nabla \cdot \mathbf{J}_{n+1/2}^{(0)} + (\rho_{n+1}^{(0)} - \rho_n)/\Delta t \neq 0. \tag{93}$$

Although it is possible to construct a $\mathbf{J}$ that does conserve charge, it has been preferable in most explicit codes to use the above $\mathbf{J}$ and then to apply a correction to the longitudinal component of $\mathbf{J}$ or of $\mathbf{E}_{n+1}$ (Boris, 1970; Langdon and Lasinski, 1976). Here, we replace $\mathbf{J}_{n+1/2}^{(0)}$ with

$$\mathbf{J}'_{n+1/2} = \mathbf{J}_{n+1/2}^{(0)} + (\nabla \psi)/\Delta t, \tag{94a}$$

where

$$-\nabla^2 \psi = \rho_{n+1}^{(0)} + \nabla \cdot [\Delta t \, \mathbf{J}_{n+1/2}^{(0)} - \mathbf{E}_n]. \tag{94b}$$

Note that the source term for ψ would vanish if $\mathbf{J}_{n+1/2}^{(0)}$ were consistent with $\rho_{n+1}^{(0)}$ and ρ_n. Derivation of Eq. (94b) and an alternate correction to $\mathbf{E}_{n+1}$ are discussed below.

2. Evaluation of the Increments due to Future Fields

The care with which we express the increments $\{\delta\rho, \delta\mathbf{J}\}$ is a compromise between complexity and strong convergence. If necessary, they may be evaluated rigorously as derivatives of Eqs. (91) and (92) (strict differencing; see Section II.A.2 and Langdon et al., 1983, Section 4) or as simplified difference representations (Section II.A.2) of

$$\delta\rho = -\nabla \cdot [\rho \, \delta\mathbf{x}], \tag{95}$$

$$\delta\mathbf{J} = \rho \, \delta\mathbf{v} - \tfrac{1}{2}\nabla \times (\mathbf{J} \times \delta\mathbf{x}), \tag{96}$$

for each species, where superscript (0) is understood for ρ and J on the right side. This form for $\delta\mathbf{J}$ trivially conserves charge: $\delta\rho + \Delta t \, \nabla \cdot \delta\mathbf{J} = 0$. This property can easily be preserved in the spatial differencing of $\delta\mathbf{J}$.

The terms in Eq. (96) have both analytic and pictorial justifications; see Fig. 5. A heuristic derivation of δJ uses an analogy to magnetization current. The magnetic moment of the current loop in the last diagram is

$$\frac{I}{2c} \int \mathbf{x} \times d\mathbf{x} = \frac{q}{2c \, \Delta t} \, \delta\mathbf{x} \times (\mathbf{v}_{n+1/2}^{(0)} \Delta t).$$

The current due to a density $n(\mathbf{x})$ of these loops is

$$\delta\mathbf{J} = c \, \nabla \times \mathbf{M} = c \, \nabla \times [n(q/2c \, \Delta t) \, \delta\mathbf{x} \times \mathbf{v}_{n+1/2}^{(0)} \Delta t] = \tfrac{1}{2}\nabla \times [\delta\mathbf{x} \times \rho\mathbf{v}_{n+1/2}^{(0)}],$$

which leads to the last term in Eq. (96). A related term arises in linearized particle codes (Cohen et al., 1980).

11. Direct Implicit Plasma Simulation

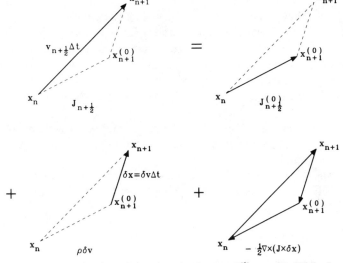

Fig. 5. Geometric interpretation of the terms in $J_{n+1/2} = J^{(0)}_{n+1/2} + \delta J$. While $J_{n+1/2}$ corresponds to moving the particle *directly* from x_n to x_{n+1}, it can be regarded as the sum of three motions: (1) motion from x_n to $x^{(0)}_{n+1}$, giving $J^{(0)}_{n+1/2}$; (2) then motion from $x^{(0)}_{n+1}$ to x_{n+1}, giving $\rho\,\delta v$ plus (3) a circulation term $-\tfrac{1}{2}\nabla \times (J \times \delta x)$ to cancel the effect of the "detour" to $x^{(0)}_{n+1}$. This is not needed to get ρ_{n+1} but does affect B and E_T.

A careful examination of the locations at which E and δJ are evaluated in Eq. (96) shows a $O(kv\,\Delta t)^2$ error. Because its phase is neutral, this error is expected to be innocuous.

3. Construction of the Implicit-Field Equation

We now have everything needed to write an equation for E_{n+1}. On substituting our expressions for $J^{(0)}_{n+1/2}$ and δJ into the field equations (87) and (88), we have

$$c\nabla \times [B_{n+1/2} + (J^{(0)}_{n+1/2} \times \delta x)/2c] = J'_{n+1/2} + \rho^{(0)}_{n+1}\,\delta v \\ + (E_{n+1} - E_n)/\Delta t, \quad (97a)$$

$$-\tfrac{1}{2}c\nabla \times [\bar{E}_{n-1} + E_{n+1}] = (B_{n+1/2} - B_{n-1/2})/\Delta t. \quad (97b)$$

With Eqs. (33), the implicit terms $\rho\,\delta v$ and $J \times \delta x$ are seen to be

$$\rho^{(0)}_{n+1}\,\delta v = \left[\sum_s \frac{\rho_s q_s \Delta t}{2m_s}\tfrac{1}{2}(I + R_s)\right]\cdot E_{n+1} = \frac{\chi \cdot E_{n+1}}{\Delta t}, \quad (98a)$$

$$J^{(0)}_{n+1/2} \times \delta x = \left[\sum_s \frac{J_s q_s \Delta t^2}{2m_s} \times \tfrac{1}{2}(I + R_s)\right]\cdot E_{n+1} = c\zeta \cdot E_{n+1}. \quad (98b)$$

The $[\cdots]$ are sums over *species*, not each particle. If only the electrons are implicit, only they appear in Eqs. (98). In this case, the terms $[\cdots]$ require only a knowledge of the electrons' ρ and $\mathbf{J}$ [in addition to the *net* ρ and $\mathbf{J}$ used on the right side of Eq. (97a)]. In general, it is sufficient to accumulate $\rho_{n+1,s}^{(0)}$ and $\mathbf{J}_{n+1/2,s}^{(0)}$ separately from species with differing q/m. This requires more storage, but *no more computation* than for an explicit code.

Spatial differencing follows in spirit the *simplified differencing* of Section II.A.

Owing to the correction Eq. (94a), the divergence of the Ampere–Maxwell equation (97a) recovers *exactly* our *electrostatic* implicit field equation (36) with $\mathbf{E}_{n+1}^{(0)} = 0$ and $\mathbf{E}_{n+1} = -\nabla \delta \varphi_{n+1}$.

Equations (97) and (98) are the simplest yet proposed for implicit field prediction, both in themselves and in what one must accumulate from the particles.

4. Longitudinal Field Correction

As mentioned above, the solution of Eq. (97a) would fail slightly to satisfy the longitudinal equation corresponding to Eq. (36),

$$\nabla \cdot (1 + \chi) \cdot \mathbf{E}_{n+1} = \rho_{n+1}^{(0)} \qquad (99)$$

if $\mathbf{J}'_{n+1/2}$ were not used in place of $\mathbf{J}_{n+1/2}^{(0)}$. Equation (94b) for the correction is derived by substituting Eqs. (94a) and (98a) into (97a), taking the divergence and comparing with Eq. (99).

Alternatively, one could apply a longitudinal correction to the solution of Eq. (97a) with $\mathbf{J}_{n+1/2}^{(0)}$ used for $\mathbf{J}'_{n+1/2}$. That is, rewrite Eq. (97a) with (98a) as

$$c \, \Delta t \, \nabla \times [\cdots] = \mathbf{J}_{n+1}^{(0)} \Delta t + \chi \cdot \mathbf{E}' + (\mathbf{E}' - \mathbf{E}_n) \qquad (100)$$

and solve for $\mathbf{E}'$, from which $\mathbf{E}_{n+1} = \mathbf{E}' - \nabla \psi$. Comparing the divergence of Eq. (100) to Eq. (99) gives

$$-\nabla \cdot (1 + \chi) \cdot \nabla \psi = \rho_{n+1}^{(0)} - \nabla \cdot (1 + \chi) \cdot \mathbf{E}'. \qquad (101)$$

With $\chi \ll 1$, this step corresponds to Boris's correction to $\nabla \cdot \mathbf{E}_{n+1}$ and has a result identical to the correction to $\mathbf{J}$ (Birdsall and Langdon, 1985). Here the transverse part of $\mathbf{E}_{n+1}$ differs between the two methods; the correction (101) corresponds nearly to using $(1 + \chi) \cdot \nabla \psi$ instead of $\nabla \psi$ in (94a).

D. Comparison of Darwin and Implicit Codes

The need for time-advanced currents and the expression of their dependence on the advanced fields are central to both Darwin and implicit codes. Indeed, the direct implicit codes were conceived with Darwin codes in mind.

The similarity to moment implicit methods, in which Eq. (85) is a key equation, is obvious. In the Darwin code of Nielson and Lewis (1976), although the particle integration remained explicit, most of the price of being fully implicit was paid.

With explicit time differencing in a Darwin code, the time step is still limited by the electron-plasma and Alfvén-wave periods. The field solution in Darwin codes is complicated by the separation of longitudinal and transverse components of **E**. Implicit approaches may subsume the Darwin model in many applications. On the other hand, when electromagnetic wave propagation is not required, nonphysical dissipation due to implicit differencing of the fields could make the Darwin model more accurate.

V. CONCLUDING REMARKS

In this chapter we apply the direct method to derive implicit PIC algorithms that are simpler and less restrictive in some respects than the moment algorithms published to date. A strict application of the direct viewpoint provides tools for analysis of the convergence and stability of both direct and moment algorithms, and shows how to enforce properties such as momentum conservation. Analytic insights and experience in applications gained with both methods have begun to advance the development of each. As moment and direct codes are borrowing features from each other, the distinction becomes more that of viewpoint in deriving algorithms and less in the resulting codes themselves.

We have not discussed boundary conditions. In implicit codes, particle boundary conditions can be complex. Particle deletion or emission at a surface depends on E_{n+1}; therefore, the particle boundary conditions enter into the implicit field equations. For electromagnetic fields it appears that methods used in explicit codes can be adapted. At Palaiseau for example, Adam and Gourdin-Serveniere have adapted the outgoing-wave boundary conditions of Lindman (1975) by using implicit differencing of his boundary wave equations.

Many generalizations are possible. To include relativity, one would linearize the relativistic particle equation of motion (Langdon and Lasinki, 1976; Boris, 1970). In practice, this adds complexity because the susceptibility χ becomes velocity dependent. Electron–ion collisions ($v \lesssim \Delta t^{-1}$) may be described as an addition to the rotation **R** in the equation of motion. If a component of the plasma is modeled by fluid equations, then those equations are linearized to find $\{\delta\rho, \delta\mathbf{J}\}$ (Denavit, 1983, 1984). Combining fluid and particle descriptions is difficult, but not more so in the direct than in the moment method.

ACKNOWLEDGMENTS

This work is an outgrowth of our stimulating and productive collaborations with B. I. Cohen, A. Friedman, D. W. Hewett, T. Kamimura, J.-N. Leboeuf, and T. Tajima. It is a pleasure to acknowledge valuable conversations with J. C. Adam, J. U. Brackbill, J. Denavit, D. Forslund, B. Godfrey, A. Gourdin Heron, R. J. Mason, and D. Nielson. This work was performed under the auspices of the U.S. Department of Energy by the Lawrence Livermore Laboratory under Contract W-7405-Eng-48, and the Institute for Fusion Studies under Contract DE-FGO5-80ET-53088.

REFERENCES

Adam, J. C., Gourdin-Serveniere, A., and Langdon, A. B. (1982). *J. Comput. Phys.* **47**, 229.
Barnes, D. C., and Kamimura, T. (1982). *Res. Rep.—Nagoya Univ., Inst. Plasma Phys.*, **IPPJ-570**.
Barnes, D. C., Kamimura, T., Leboeuf, J.-N., and Tajima, T. (1983a). *Proc. Conf. Number. Simul. Plasmas, 10th, 1983* Paper 2A4.
Barnes, D. C., Kamimura, T., Leboeuf, J.-N., and Tajima, T. (1983b). *J. Comput. Phys.* **52**, 480.
Birdsall, C. K., and Fuss, D. (1969). *J. Comput. Phys.* **3**, 494.
Birdsall, C. K., and Langdon, A. B. (1985). "Plasma Physics Via Computer Simulation." McGraw-Hill, New York.
Boris, J. P. (1970). *Proc. Conf. Numer. Simul. Plasmas, 4th, 1970* p. 3.
Boris, J. P., and Roberts, K. V. (1969). *J. Comput. Phys.* **4**, 552.
Brackbill, J. U., and Forslund, D. W. (1982). *J. Comput. Phys.* **46**, 271.
Brandt, A. (1977). *Math. Comput.* **31**, 333.
Buneman, O. (1967). *J. Comput. Phys.* **1**, 517.
Burger, P., Dunn, D. A., and Halstead, A. S. (1965). *Phys. Fluids* **8**, 2263.
Busnardo-Neto, J., Pritchett, P. L., Lin, A. T., and Dawson, J. M. (1977). *J. Comput. Phys.* **23**, 300.
Cohen, B. I., Auerbach, S. P., and Byers, J. A. (1980). *Phys. Fluids* **23**, 2529.
Cohen, B. I., Freis, R. P., and Thomas, V. (1982a). *J. Comput. Phys.* **45**, 345.
Cohen, B. I., Langdon, A. B., and Friedman, A. (1982b). *J. Comput. Phys.* **46**, 15.
Cohen, B. I., Langdon, A. B., and Friedman, A. (1984). *J. Comput. Phys.* **56**, 51.
Concus, P., and Golub, G. H. (1973). *SIAM J. Numer. Anal.* **10**, 1103.
Denavit, J. (1981). *J. Comput. Phys.* **42**, 337.
Denavit, J. D. (1983). *Proc. Conf. Numer. Simul. Plasmas, 10th, 1983* Paper 1B15.
Denavit, J. D. (1984). To be published.
Doss, S., and Miller, K. (1979). *SIAM J. Numer. Anal.* **16**, 837.
Dubin, D. H. E., Krommes, J. A., Oberman, C., and Lee, W. W. (1983). *Phys. Fluids* **26**, 3524.
Forslund, D. W., and Brackbill, J. U. (1982). *Phys. Rev. Lett.* **48**, 1614.
Friedman, A., Langdon, A. B., and Cohen, B. I. (1981). *Comments Plasma Phys. Controlled Fusion* **6**, 225.
Friedman, A., Langdon, A. B., and Cohen, B. I. (1983). *In* "Laser Program Annual Report-1982," UCRL-80021-82, pp. 3–56. Lawrence Livermore Nat. Lab., Livermore, California.
Hewett, D. W. (1980). *J. Comput. Phys.* **38**, 378.
Hockney, R. W. (1965). *J. Assoc. Comput. Mach.* **12**, 95.
Jackson, J. D. (1962). "Classical Electrodynamics," Wiley, New York. pp. 616ff.
Kaufman, A. N., and Rostler, P. S. (1971). *Phys. Fluids* **14**, 446.
Kershaw, D. S. (1978). *J. Comput. Phys.* **26**, 43.

Kershaw, D. S. (1980). *J. Comput. Phys.* **38**, 114.
Langdon, A. B. (1979a). *J. Comput. Phys.* **30**, 202.
Langdon, A. B. (1979b). *Phys. Fluids* **22**, 163.
Langdon, A. B. (1983a). "Electromagnetic Direct Implicit PIC Simulation," UCRL-88979. Lawrence Livermore National Laboratory, Livermore, California.
Langdon, A. B. (1983b). *In* "Laser Program Annual Report-1982," UCRL-80021-82, pp. 3–53. Lawrence Livermore National Laboratory, Livermore, California.
Langdon, A. B., and Lasinski, B. F. (1976). *In* "Methods in Computational Physics" (J. Killeen, ed.), Vol. 16, p. 327. Academic Press, New York.
Langdon, A. B., Cohen, B. I., and Friedman, A. (1983). *J. Comput. Phys.* **51**, 107.
Langdon, A. B., Hewett, D. W., and Friedman, A. (1984). *In* "Laser Program Annual Report-1983," UCRL-80021-83. Lawrence Livermore National Laboratory, Livermore, California.
Lee, W. W. (1983). *Phys. Fluids* **26**, 556.
Lindman, E. L. (1975). *J. Comput. Phys.* **18**, 66.
Mason, R. J. (1981). *J. Comput. Phys.* **41**, 233.
Morse, R. L., and Nielson, C. W. (1969). *Phys. Fluids* **12**, 2418.
Nielson, C. W., and Lewis, H. R. (1976). *In* "Methods in Computational Physics" (J. Killeen, ed.), Vol. 16, p. 367. Academic Press, New York.
Northrup, T. G. (1961). *Ann. Phys. (N. Y.)* **15**, 19.
Orens, J. H., Boris, J. P., and Haber, I. (1970). *Proc. Conf. Number. Simul. Plasmas, 4th, 1970* p. 526.
Panofsky, W. K. H., and Phillips, M. (1962). "Classical Electricity and Magnetism," p. 461. Addison-Wesley, Reading, Massachusetts.
Petravic, M., and Kuo-Petravic, G. (1979). *J. Comput. Phys.* **32**, 263.
Sakagami, H., Nishihara, K., and Colombant, D. (1981), "Stability of Time-Filtering Particle Code Simulation," Rep. ILE8117P. Inst. Laser Eng., Osaka University.
Tajima, T., and Leboeuf, J. N. (1981). *Bull. Am. Phys. Soc.* [2] **26**, 986.
Yu, S. P., Kooyers, G. P., and Buneman, O. (1965). *J. Appl. Phys.* **36**, 2550.

12

Direct Methods for N-Body Simulations

SVERRE J. AARSETH

Institute of Astronomy
University of Cambridge
Cambridge, England

I. Introduction	378
II. Basic Formulation	381
A. Difference Scheme	381
B. Individual Time-Step Algorithm	383
III. Ahmad–Cohen Scheme	385
A. Principles and Procedures	385
B. Standard Algorithm	386
C. Parameters and Timing Tests	388
IV. Comoving Coordinates	389
A. Basic Principles	389
B. Implementation of the AC Method	391
C. Mergers and Inelastic Effects	393
V. Planetary Perturbations and Collisions	394
A. Grid Method Formulation	394
B. Collision Algorithm and Model Parameters	396
VI. Two-Body Regularization	398
A. KS Transformations and Equations of Motion	398
B. Standard N-Body Treatment	400
C. Algorithm and Parameters	402
D. Regularized AC Procedures	406
VII. Three-Body Regularization	409
A. Isolated Triple Systems	409
B. General N-Body Algorithm	411
VIII. Star-Cluster Simulations	413
A. Models of Open Clusters	413
B. Postcollapse Evolution	415
References	417

I. INTRODUCTION

Newton's law of gravity imprints its signature on a wide range of scales in the universe, giving rise to a variety of characteristic systems in which the boundary effects are small. Thus, to a good approximation, the planetary orbits are not affected by the neighboring stars. Likewise, the internal motions of open star clusters (typically containing 10^3 members in bound orbits) are essentially governed by the individual interactions, with the Galaxy providing a small external perturbation. Other examples of well-defined gravitational systems are found on scales of increasing size; e.g., globular clusters ($\simeq 10^5$ stars), galaxies ($\simeq 10^{11}$ stars), as well as clusters of galaxies ($\simeq 10 - 10^3$ members) and even superclusters. This *hierarchical* structure permits a study of nearly isolated systems on different scales, making dynamical investigations more tractable. Even so, a consistent description of the internal dynamical evolution requires the application of numerical techniques. In addition to modeling real systems, computer simulations may be used for investigating detailed dynamical processes, as well as for checking theoretical predictions. Thus, *numerical experiments* represent an important tool for the modern astronomer.

The gravitational N-body problem poses a formidable challenge to the numerical analyst. Its nonlinear nature places heavy demands on accuracy considerations, and the need to study large systems over significant times pushes even the biggest computers to their limits. The *direct* approach is based on the Lagrangian formulation of following the particle motion in detail, as prescribed by Newton's law. Although this approach is essentially a brute force method, the numerical solutions can be obtained by a variety of techniques. Thus, the design of efficient algorithms presupposes some understanding of the problem under consideration, creating a boot-strapping operation. It is the purpose of this chapter to describe the main computational methods for obtaining direct solutions of the N-body problem and also to be as complete and self-contained as possible. Results of simulations have been reviewed elsewhere (Aarseth and Lecar, 1975).

The main requirements for an efficient integration of large N-body systems are (i) high-order force polynomials, (ii) individual time steps, (iii) neighbor scheme, and (iv) close encounter treatments. These features are considered in the following sections. The methods to be discussed divide naturally into three categories:

1. standard integration,
2. perturbation of dominant motion,
3. regularization of close encounters.

In the simplest case, described in Section II, the force on each particle is represented by a fourth-order polynomial, and the solutions are obtained by stepwise integration. This *basic* formulation is only suitable for small systems, but its simplicity lends itself to experimentation with different force laws and other complications (e.g., mergers by inelastic collisions).

Section III deals with a more efficient scheme of *standard* integration, in which the total force acting on a particle is represented by a sum of two polynomials. Thus, only the local interactions are obtained by summation at each step, whereas the more numerous distant contributions are added by prediction, using a polynomial that is updated less frequently. The *neighbor* scheme enables systems with several thousand particles to be considered. However, except for an appropriate shortening of time steps, standard integration has no provision for dealing with large force terms during close encounters; hence a softened potential is usually adopted. The device of suppressing large-angle scattering can only be justified when the two-body relaxation time is long in comparison with the orbital time (e.g., galaxy simulations).

In the second class of problems, all the orbits are subject to a dominant effect which maintains a high degree of ordered motion. One such special method, described in Section IV, has been developed for expanding systems (e.g., cosmological simulations). By transforming to *comoving* coordinates the problem reduces to integration of the *perturbed* motions, corresponding to small deviations from the expansion. Although localized clustering is an important feature in such models, the comoving representation is still beneficial. This formulation also contains a softening parameter which may be associated with the size of individual galaxies.

Other examples of *dominant* motion are found in the solar system. Section V deals with a new perturbation method for planetary orbits, which is used to study the growth of planetesimals by inelastic collisions. This requires accurate solutions over long times, as well as determination of rare collision events of short duration. In this case the dominant solar term is used to stabilize the equation of motion for each planetesimal when the perturbations are small. It can be seen that the solution method is again of a direct type, although here the *total* motion is integrated.

The dynamical evolution of point-mass systems is to a large extent controlled by encounters that generate chaotic motions. Any attempt at modeling discrete systems must, therefore, include an accurate and economical treatment of the close encounters, in particular, two-body and triple interactions. In *two-body* regularization the relative motion of two particles is cast in a regular form, enabling critical encounters to be studied with confidence. This powerful technique is advantageous whenever the two-body motion is dominant (e.g., close binaries) and can be extended to quite large perturbations. The basic framework for two-body regularization is given in Section VI, together with a scheme for automatic decision making in N-body simulations. Also discussed are a number of special procedures pertinent to an efficient interfacing with the standard integration methods.

Small stellar systems usually develop at least one central energetic binary on a relatively short time scale. The subsequent evolution, therefore, involves interactions between binaries and single particles, which lead to loss of accuracy when only the dominant two-body term is regularized. A *three-body* regularization method for such configurations is outlined in Section VII. Only relatively isolated systems are considered here, since for reasons of programming complexity the triple regularization does not yet include external perturbations.

By now the principle of regularization has proved itself the most powerful tool for studying the long-term evolution of stellar systems. Alternatives, such as time smoothing or classical perturbation methods, are simpler to implement but less effective in the general case. Hence, the success of the regularization schemes can be seen as yet another example of using the natural variables of a problem.

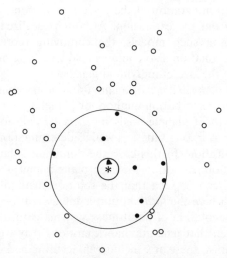

Fig. 1. Schematic illustration of the different length scales in a stellar system. The particle denoted by ∗ has a regularized companion ▲ inside a small separation and several neighbors ● within a second spherical region. Large systems may be studied more efficiently if the slowly varying force due to numerous distant members ○ is recalculated on a longer time scale than required to include the irregular neighbor contributions. Each particle, in turn, is advanced on its own time scale, governed by the local density and velocity dispersion.

12. Direct Methods for N-Body Simulations

All the above methods have been applied to astronomical modeling, such as galaxy clustering, planetary formation, and star-cluster evolution. In these direct methods the concept of *multiple time scales* forms a common theme. Figure 1 shows an example of three spatial regions centered on a typical point-mass particle, giving rise to treatments on different time scales. This feature is especially pronounced in star-cluster simulations. Section VIII contains two recent applications; open clusters with external perturbations and post-collapse evolution of an isolated system, which illustrate the two-body and triple regularization, respectively.

The wide range of methods covered shows the versatility of the direct approach, which is limited only by the particle number and attainable accuracy. Astronomical applications of N-body techniques are often very time consuming. However, it is desirable to extend such calculations as far as possible before adopting more approximate treatments, which are not discussed here. Although there is still a need for even faster methods for some problems, the standard formulation is also sufficiently general to be applicable to other subjects where the interaction law can be specified.

II. BASIC FORMULATION

A. Difference Scheme

Each particle is completely specified by its mass m, position $\mathbf{r}$, and velocity $\mathbf{v}$. Denoting time derivatives by dots, the equation of motion for particle i is given by

$$\ddot{\mathbf{r}}_i = -G \sum_j \frac{m_j(\mathbf{r}_i - \mathbf{r}_j)}{|\mathbf{r}_i - \mathbf{r}_j|^3}, \qquad j \neq i, \tag{1}$$

where G is the gravitational constant and the summation extends over the other $N - 1$ members. The solution of this initial-value problem involves $3N$ coupled differential equations of second order. In the following we employ scaled units with $G = 1$ and introduce the force *per unit mass* $\mathbf{F}$, omitting particle subscripts for convenience. The difference formulation used by Ahmad and Cohen (1973, hereafter AC) has been adopted.

From the values of $\mathbf{F}$ at four previous times t_0, t_1, t_2, t_3, with t_0 being the most recent, we write a fourth-order fitting polynomial at time t as

$$\mathbf{F}(t) = \mathbf{F}(t_0) + \mathbf{D}^1(t - t_0) + \mathbf{D}^2(t - t_0)(t - t_1) \\ + \mathbf{D}^3(t - t_0)(t - t_1)(t - t_2) + \mathbf{D}^4(t - t_0)(t - t_1)(t - t_2)(t - t_3). \tag{2}$$

The first three divided differences are defined by

$$\mathbf{D}^k[t_0, t_k] = (\mathbf{D}^{k-1}[t_0, t_{k-1}] - \mathbf{D}^{k-1}[t_1, t_k])/t'_k, \quad k = 1, 2, 3, \quad (3)$$

$$t'_k = t_0 - t_k,$$

where $\mathbf{D}^0$ symbolically represents $\mathbf{F}$. Equation (2) also includes $\mathbf{D}^4$, which is defined similarly in terms of $\mathbf{D}^3[t, t_2]$ and $\mathbf{D}^3[t_0, t_3]$ over the interval $t - t_3$ and is therefore evaluated at time t. It is convenient to represent the force-fitting polynomial by a Taylor series, which may then be integrated twice to give the new solutions for $\mathbf{r}$ and $\mathbf{v}$ as a set of six first-order equations. Equating terms in Eq. (2) and its successive derivatives with an equivalent Taylor series yields the coefficients

$$\begin{aligned}
\mathbf{F}^{(1)} &= \mathbf{D}^1 + \mathbf{D}^2 t'_1 + \mathbf{D}^3 t'_1 t'_2 + \mathbf{D}^4 t'_1 t'_2 t'_3, \\
\mathbf{F}^{(2)} &= 2![\mathbf{D}^2 + \mathbf{D}^3(t'_1 + t'_2) + \mathbf{D}^4(t'_1 t'_2 + t'_2 t'_3 + t'_1 t'_3)], \\
\mathbf{F}^{(3)} &= 3![\mathbf{D}^3 + \mathbf{D}^4(t'_1 + t'_2 + t'_3)], \\
\mathbf{F}^{(4)} &= 4!\mathbf{D}^4.
\end{aligned} \quad (4)$$

For each term the contribution from $\mathbf{D}^4$ is only added at the end of the step. This so-called *semi-iteration* gives increased accuracy at little cost and no extra memory. Note that these Taylor series coefficients are evaluated at $t = t_0$, whereas the general case needed for combining two polynomials contains additional terms.

The initialization procedure is described first. Given the initial conditions $m_j, \mathbf{r}_j, \mathbf{v}_j$ for $j = 1, \ldots, N$, the required Taylor series coefficients are generated by successive differentiations of Eq. (1). Each term of the interaction is considered separately by introducing the *relative* coordinates $\mathbf{R} = \mathbf{r}_i - \mathbf{r}_j$. We adopt the explicit procedure of Findlay (1983a) instead of the more cumbersome expressions used previously (cf. Aarseth, 1972). All individual terms in $\mathbf{F}$ and $\mathbf{F}^{(1)}$ are first calculated by

$$\begin{aligned}
\mathbf{F}_{ij} &= -m_j \mathbf{R}/R^3, \\
\mathbf{F}^{(1)}_{ij} &= -m_j \mathbf{V}/R^3 - 3a\mathbf{F}_{ij},
\end{aligned} \quad (5)$$

with $\mathbf{V} = \mathbf{v}_i - \mathbf{v}_j$ and $a = \mathbf{R} \cdot \mathbf{V}/R^2$. The total contributions are obtained by $N - 1$ summations for each particle in turn. At the next stage, second- and third-order terms are formed from

$$\begin{aligned}
\mathbf{F}^{(2)}_{ij} &= -m_j(\mathbf{F}_i - \mathbf{F}_j)/R^3 - 6a\mathbf{F}^{(1)}_{ij} - 3b\mathbf{F}_{ij}, \\
\mathbf{F}^{(3)}_{ij} &= -m_j(\mathbf{F}^{(1)}_i - \mathbf{F}^{(1)}_j)/R^3 - 9a\mathbf{F}^{(2)}_{ij} - 9b\mathbf{F}^{(1)}_{ij} - 3c\mathbf{F}_{ij},
\end{aligned} \quad (6)$$

where

$$b = (V/R)^2 + \mathbf{R} \cdot (\mathbf{F}_i - \mathbf{F}_j)/R^2 + a^2,$$
$$c = 3\mathbf{V} \cdot (\mathbf{F}_i - \mathbf{F}_j)/R^2 + \mathbf{R} \cdot (\mathbf{F}_i^{(1)} - \mathbf{F}_j^{(1)})/R^2 + a(3b - 4a^2).$$

A second double summation then gives the corresponding values of $\mathbf{F}^{(2)}$ and $\mathbf{F}^{(3)}$ for all particles. Note that the pairwise quantities of Eqs. (5) which appear in Eqs. (6) are employed in this *boot-strapping* procedure.

We now assume that appropriate initial time steps δt_i have been determined (cf. Section II.B). Setting $t_0 = 0$, the backward times are initialized by $t_k = -k\delta t_i$ ($k = 1, 2, 3$). Starting values of the divided differences are then obtained by inverting Eqs. (4) to order $\mathbf{F}^{(3)}$, giving

$$\begin{aligned}
\mathbf{D}^1 &= \mathbf{F}^{(1)} - (\mathbf{F}^{(2)}/2 - \mathbf{F}^{(3)}t'_1/6)t'_1, \\
\mathbf{D}^2 &= \mathbf{F}^{(2)}/2 - \mathbf{F}^{(3)}(t'_1 + t'_2)/6, \\
\mathbf{D}^3 &= \mathbf{F}^{(3)}/6.
\end{aligned} \quad (7)$$

The choice of a fourth-order polynomial representation is a convenient compromise between efficiency and programming effort, considering the subsequent complexities of combining two force polynomials. The adopted difference scheme possesses inherent stability characteristics (Findlay, 1983a) and is also quite robust to deliberate discontinuous effects (e.g., setting $\mathbf{D}^2 = 0$). Note that a soft potential of the form $m/(R^2 + \varepsilon^2)^{1/2}$ may readily be introduced by modifying all R^3 terms in the denominators of Eqs. (5) and (6).

B. Individual Time-Step Algorithm

We now describe a simple integration algorithm based on one force polynomial. All the subsequent methods are characterized by *individual* time steps. In this way, consistent solutions are obtained over a wide range of time scales while preserving the relative accuracy. Thus, each particle position is recalculated with the largest step for which the force polynomial converges well. An additional requirement for this scheme is a full coordinate prediction at intermediate times.

The main integration cycle begins by determining the next particle to be advanced in time; i.e., $i = \min_j(t_j + \delta t_j)$. The present epoch is defined by $t = t_i + \delta t_i$. In order to avoid a full N search at each step, a list M holds all particles satisfying $t_j + \delta t_j < t_M$. Initially $t_M = \delta t_M$, where δt_M is a small time interval that is modified to stabilize the list membership on $N^{1/2}$ at each updating of t_M and M. A redetermination of the time-step list M is made if $t > t_M$, followed by a second search to include new members, since otherwise the time t might decrease.

The individual time-step scheme requires two sets of coordinates denoted *primary* and *secondary*, defined by $\mathbf{r}_j(t_j)$ and $\mathbf{r}_j(t)$, where the latter are derived from the former by the predictor. All coordinates are first predicted to order $\mathbf{F}^{(1)}$ using the fast expression

$$\mathbf{r}_j(t) = [(\hat{\mathbf{F}}_j^{(1)}\,\delta t'_j + \hat{\mathbf{F}}_j)\,\delta t'_j + \mathbf{v}_j]\,\delta t'_j + \mathbf{r}_j(t_j), \qquad j = 1, \ldots, N, \tag{8}$$

where $\hat{\mathbf{F}}^{(1)} = \mathbf{F}^{(1)}/6$, $\hat{\mathbf{F}} = \mathbf{F}/2$, and $\delta t'_j = t - t_j$. Coordinates and velocities of particle i are then advanced to order $\mathbf{F}^{(3)}$ by standard Taylor series integration. This permits the current total force to be obtained by summation [Eq. (1)]. At this stage the times t_k are updated by replacing t_k with t_{k-1} ($k = 3, 2, 1$) and setting $t_0 = t$. New differences are now formed [Eqs. (3)], including $\mathbf{D}^4$, whose contribution to the first three force derivatives is evaluated [Eqs. (4)]. Together with the equivalent $\mathbf{F}^{(4)}$ term, these corrrections are used to improve the current coordinates and velocities to order $\mathbf{F}^{(4)}$, whereupon the primary coordinates are initialized by $\mathbf{r}_i(t_i) = \mathbf{r}_i(t)$. Note that in the present formulation this entails a second pass through all the Taylor series derivatives [Eqs. (4)].

The integration cycle is completed by prescribing a new time step. After some experimentation, we have adopted the composite expression

$$\delta t_i = \left[\frac{\eta(|\mathbf{F}|\,|\mathbf{F}^{(2)}| + |\mathbf{F}^{(1)}|^2)}{(|\mathbf{F}^{(1)}|\,|\mathbf{F}^{(3)}| + |\mathbf{F}^{(2)}|^2)}\right]^{1/2}, \tag{9}$$

where η is a dimensionless accuracy parameter. This relative criterion ensures that all the force derivatives play a role in determining the time step, which is independent of mass for dominant two-body motion. It is also well defined for special cases; e.g., $\mathbf{F} = 0$, which may occur when a tidal force is included or $\mathbf{F}^{(1)} = \mathbf{F}^{(3)} = 0$ with all particles at rest. Also note that the full value of the time step is assigned during polynomial initializations since all orders are included.

Successive time steps normally change in a smooth manner. However, it is prudent to limit the growth by a stability factor of 1.4 for standard integration. The overall accuracy is controlled by the parameter η, which plays the role of a *relative* tolerance. A typical value $\eta = 0.03$ usually conserves the total energy to better than one part in 10^4 over a characteristic orbital time scale in the absence of close encounters.

The individual time-step method requires a total of 30 variables per particle as follows: m, $\mathbf{r}(t_0)$, $\mathbf{r}(t)$, $\mathbf{v}$, $\mathbf{F}$, $\mathbf{F}^{(1)}$, $\mathbf{D}^1$, $\mathbf{D}^2$, $\mathbf{D}^3$, δt, t_0, t_1, t_2, t_3. On computers with less than about 10-figure standard accuracy, it is recommended that $\mathbf{r}(t_0)$, $\mathbf{r}(t)$, $\mathbf{v}(t_0)$, t_0, as well as the current time t, be defined in double precision at little extra cost. Most of the subsequent algorithms also employ a *secondary* velocity variable $\mathbf{v}(t)$ to facilitate subsidiary calculations.

Equivalent formulations of the basic polynomial method have been employed for some time (Aarseth, 1966; Wielen, 1967). Also of historical interest is the pioneering study of von Hoerner (1960), which demonstrated the feasibility of N-body simulations.

III. AHMAD–COHEN SCHEME

A. Principles and Procedures

The basic idea introduced by AC is to represent the total force on a particle by a sum of two polynomials which are evaluated separately. Neighboring particles provide an *irregular* component which must be considered on an appropriate time scale, as discussed in Section II. The relative change of force due to the more numerous distant particles takes place on a longer time scale, and the *regular* contribution of these particles at intermediate times is obtained by prediction. Thus, an optimal scheme requires the two time scales to be well separated in order to minimize force summations.

Following AC we introduce the irregular and regular times by lowercase and uppercase symbols, respectively, using the formulation of Section II.A. In addition to the two force polynomials, each particle is assigned a neighbor list L containing all the particles inside a sphere of radius R_s. Also included are any approaching particles $[\mathbf{R} \cdot \mathbf{V} < 0]$ within a surrounding shell of outer radius $2^{1/3} R_s$ for which the impact parameter is $< R_s$. This buffer zone serves to identify fast particles before penetrating too far inside the standard sphere.

The neighbor sphere is redetermined at the end of each regular time step when a total force summation is performed. Consideration of the irregular force field suggests a choice of neighbors based on the local density contrast (here number density), approximately determined by

$$C = \frac{2L_1}{N} \left[\frac{\langle R \rangle}{R_s} \right]^3, \tag{10}$$

where L_1 is the membership and $\langle R \rangle$ is the half-mass radius (readily obtained from the potential energy). In order to limit the range of permissible values, we adopt a predicted membership

$$L_p = 0.75 L_{max} (C/20)^{1/2}, \tag{11}$$

subject to L_p being within $(2, 0.75 L_{max})$, with L_{max} a suitably chosen upper limit. The neighbor sphere radius is then modified according to

$$R_s^{new} = R_s^{old} (L_p/L_1)^{1/3}. \tag{12}$$

For stability reasons the volume factor $f = L_p/L_1$ is only allowed to change by 25%. An additional refinement is included for small L_1 to increase the probability of retaining at least one neighbor at the next search.

The number of particles entering or leaving the neighbor sphere depends on the local velocity dispersion and density. Such particles are identified when comparing the old and new neighbor list after the calculation of each regular force $\mathbf{F}_d$. Regular force differences are first evaluated, assuming there has been no change of neighbors. Denoting the irregular force from the old and new neighbors by $\mathbf{F}_n^{old}$ and $\mathbf{F}_n^{new}$, respectively, then gives a new regular force difference

$$\mathbf{D}^1 = \{\mathbf{F}_d(t) - [\mathbf{F}_n^{old}(t) - \mathbf{F}_n^{new}(t)] - \mathbf{F}_d(T_0)\}/(t - T_0), \qquad (13)$$

where the net change of neighbor force is included in the middle brackets. Note that all current force components are evaluated using the *predicted* coordinates, rather than the values based on the irregular semi-iteration, since otherwise Eq. (13) would contain a spurious neighbor force difference. The higher differences are formed in the usual way [Eqs. (3)].

Any changes of the neighbor field necessitates appropriate corrections of both force polynomials. The relevant Taylor series derivatives are accumulated, using Eqs. (5) and (6), to yield the net change. Each polynomial is modified by first adding or subtracting the respective correction terms to the corresponding Taylor series derivatives [Eqs. (4)], followed by a conversion back to differences [Eqs. (7)]. Note that neighbor changes involving large force derivatives also tend to reduce the regular time step, as would occur during an encounter with a remaining member near the boundary of the neighbor sphere. However, this effect is reduced for a soft potential.

B. Standard Algorithm

Given the initial conditions m, $\mathbf{r}$, $\mathbf{v}$, the starting procedure is essentially as described in Section II.A, except that *two* force polynomials are now initialized. A list of neighbors inside a prescribed separation R_s is assigned to each particle when evaluating Eqs. (5), whose contributions are then added to the appropriate component. We also form the *total* force $[\mathbf{F}_n + \mathbf{F}_d]$ and first derivative required for coordinate predictions and neighbor corrections. In the second stage, the next two force derivatives are calculated for each component, as identified from the neighbor list. The initialization of variables is completed by specifying individual time steps δt_j and δT_j [Eq. (9)], followed by conversion to divided differences [Eqs. (7)]. Finally, the time-step list M is initialized to contain all particles satisfying $\delta t_j < \delta t_M$, where δt_M is a suitably small interval.

The subsequent integration cycle consists of the following main steps:

1. Determine the next particle to be advanced, $i = \min_j(t_j + \delta t_j)$, and define the current epoch by $t = t_i + \delta t_i$.
2. Update the time-step list M if $t > t_M$ and modify δt_M.
3. Decide whether the regular force should be predicted [case (i)] or recalculated [case (ii)] by comparing $t + \delta t_i$ with $T_0 + \delta T_i$.
4. Perform coordinate prediction of neighbors [case (i)] or all particles [case (ii)] to order $\mathbf{F}^{(1)}$ [Eq. (8)].
5. Combine polynomials and predict $\mathbf{r}_i(t)$ and $\mathbf{v}_i(t)$ to order $\mathbf{F}^{(3)}$.
6. Obtain the irregular force $\mathbf{F}_n^{\text{old}}$ and update the times t_k ($k = 0, 1, 2, 3$).
7. Form new irregular differences and include the semi-iteration.
8. [Case (i) only.] Extrapolate the regular force and first derivative, $\mathbf{F}_d(t)$ and $\mathbf{F}_d^{(1)}(t)$, and form $\mathbf{F}(t)$ and $\mathbf{F}^{(1)}(t)$. Proceed to step 14.
9. Obtain the new irregular and regular force, $\mathbf{F}_n^{\text{new}}$ and $\mathbf{F}_d$, and form the neighbor list. Include rare cases of $L_1 = 0$ or $L_1 > L_{\max}$.
10. Modify the neighbor sphere R_s and update the times T_k ($k = 0, 1, 2, 3$).
11. Form new regular differences [Eq. (13)] and include the semi-iteration to order $\mathbf{F}^{(4)}$. Also set $\mathbf{F}(t)$ and $\mathbf{F}^{(1)}(t)$ [Eq. (4)].
12. Identify the loss or gain of neighbors and compute derivative corrections for the net change; then convert to differences.
13. Evaluate the new regular time step δT_i [Eq. (9)].
14. Evaluate the new irregular time step δt_i [Eq. (9)].
15. Repeat the cycle at step 1. (To obtain current values for data analysis, predict all coordinates and velocities to order $\mathbf{F}^{(3)}$).

As can be seen from this algorithm, both force polynomials are integrated to the highest order and all derivative corrections are included consistently. The decision of whether to recalculate the regular force is based on comparing the next *estimated* irregular time, $t + \delta t_i$, with $T_0 + \delta T_i$. This leads to an occasional unwarranted reduction of the regular interval if δt_i is shrinking, but is compensated by the case of an increasing irregular step.

A strategy of low-order coordinate prediction has been adopted [Eq. (8)]. We estimate that the median *relative* error of the force from one dominant particle is about $2 \times 10^{-6} \eta$ due to neglecting the $\mathbf{F}^{(2)}$ term. Higher accuracy may therefore be achieved by decreasing the time steps δt_i and δT_i, subject to the available precision. Alternatively, one extra order may be included in the neighbor prediction for dominant contributions only (e.g., small δt_j), provided this is also carried out before each regular force calculation.

Experiments have shown that the amount of coordinate prediction may, in fact, be reduced without affecting the overall accuracy unduly. This is achieved by omitting the coordinate predictions of nonneighbors at regular steps for particles with significant neighbor membership, e.g., above about half the average value. The rationale here is that such particles are typically predicted L_1 times within one *irregular* step, which also tends to be small. It is not advisable to use this procedure for nonequilibrium systems or when massive bodies are present.

Force polynomials must be combined with due care. Thus, at steps 5 and 15, the second regular difference is obtained from the general form of the corresponding Eq. (4), which contains an extra time-step term in $\mathbf{D}^3$. Likewise, step 8 requires the regular force derivative at an intermediate time, evaluated to the highest order. Conversely, the derivatives in Eq. (9) are based on at most one extra order in Eqs. (4).

In keeping with the spirit of the method, the procedure for selecting the next particle uses the irregular times. It is, nevertheless, prudent to impose the condition $\delta t_i < \delta T_i$ in case derivative corrections should temporarily reduce the regular step unfavorably.

C. Parameters and Timing Tests

In the AC method, both types of time steps are determined by the same expression [Eq. (9)]. However, two accuracy parameters are now used, with standard values $\eta_n = 0.03$ and $\eta_d = 0.08$. One reason for adopting a larger convergence ratio for the regular force polynomial is that the corresponding step is truncated to coincide with the nearest irregular step. The average ratio of regular to irregular steps depends on several factors, including softening parameter, N and L_{max}, but may approach 10:1 in favorable cases. The scheme gains in efficiency as the ratio $\delta T_i/\delta t_i$ grows. On the other hand, the optimal choice of neighbor sphere is not too sensitive because of compensating effects; including more neighbors also increases the regular step.

Implementation of the AC integration scheme requires the following basic variables: m, $\mathbf{r}(t_0)$, $\mathbf{r}(t)$, $\mathbf{v}$, $\mathbf{F}$, $\mathbf{F}^{(1)}$, two sets of polynomial variables, $\mathbf{F}_n$ (or $\mathbf{F}_d$), $\mathbf{D}^1$, $\mathbf{D}^2$, $\mathbf{D}^3$, δt, t_0, t_1, t_2, t_3, as well as the neighbor sphere R_s and neighbor list L (of minimum size $L_{max} + 1$). Hence the main memory requirement is now $52 + L_{max}$ variables per particle. For most problems it is adequate to choose $L_{max} = N^{1/2}$, although a slightly more generous allocation reduces the number of special cases when the neighbor membership is outside the permitted range.

We now evaluate the relative performance of the standard methods by some simple comparison tests. Several homogeneous equilibrium systems are integrated over one dynamical time scale (so-called *crossing time*), using a consistent softening parameter $\varepsilon \simeq 4\langle R\rangle/N$, which avoids close-encounter effects. Table I gives the number of integration steps and corresponding computing times.

12. Direct Methods for N-Body Simulations

Table I
Comparison Tests for Standard Integration Methods[a]

N	One polynomial		Two polynomials		
25	2451	0.25	3006	1002	0.26
50	5806	1.03	7219	1932	0.76
100	14627	4.82	17891	3755	2.33
200	—	—	42527	7048	7.52

[a] Integration steps and CPU minutes on VAX 11/780.

These tests employ standard values of the accuracy parameters, $\eta = \eta_n = 0.03$, and $\eta_d = 0.08$, and include all regular coordinate predictions. Reduction of the regular steps by using $\eta_d = 0.03$ is still marginally faster than the basic method for $N = 50$ (7241 irregular and 3401 regular steps). Provided that no binaries are present, a smaller softening parameter requires a modest increase of irregular steps, still conserving the total energy to better than one part in 10^4.

On the basis of the data in Table I, the computing time *per crossing time* for the two methods fits well the power law relations $T_c = 4.7(N/100)^{2.1}$ and $T_c = 2.4(N/100)^{1.6}$. The computational effort involved in the integration itself is even more favorable for the AC method, since both the initialization and total energy calculation scales as N^2 (about 0.43 CPU minutes for $N = 200$). To reduce the time-consuming initialization for large N, we limit Eqs. (6) to separations $< 3R_s$ without significant loss of accuracy. Although the overall accuracy (but *not* the relative energy error) is slightly better for the basic polynomial method when using the preceding parameters, it is noteworthy that the crossover point in efficiency is achieved for quite small systems.

IV. COMOVING COORDINATES

A. Basic Principles

Cosmological N-body simulations have recently become very popular. The main aim of such studies is to account for the observed galaxy distribution which exhibits a high degree of clustering. In order to exploit the dominant motions arising from the universal expansion, we introduce *comoving* coordinates. Only deviations from the overall expansion are now integrated, thereby reducing the computational effort significantly. We first give a basic derivation, followed by an improved formulation of the original comoving AC scheme (Aarseth, 1979).

In Newtonian cosmology the equation of motion for a spherical region of radius S and total mass M is given by

$$\ddot{S} = -GM/S^2. \tag{14}$$

The comoving coordinates ρ are then related to the physical coordinates by

$$\rho_i = \mathbf{r}_i/S. \tag{15}$$

We adopt a standard soft potential of the form $m/(r^2 + \varepsilon_0^2)^{1/2}$, where ε_0 may be associated with the half-mass radius of a typical galaxy. When we use the softened form of Eq. (1), the corresponding comoving equation of motion is derived by twice differentiation of Eq. (15) and substitution for Eq. (14), giving

$$\ddot{\rho}_i = -\frac{2\dot{S}}{S}\dot{\rho}_i - \frac{G}{S^3}\left[\sum_j \frac{m_j(\rho_i - \rho_j)}{[|\rho_i - \rho_j|^2 + \varepsilon^2]^{3/2}} - M\rho_i\right], \quad j \neq i, \tag{16}$$

where $\varepsilon = \varepsilon_0/S$. If desired, ε^2 may be replaced by $\varepsilon_i^2 + \varepsilon_j^2$, where the latter are related to the individual half-mass radii.

Equation (16) may be integrated directly, using a low-order method. However, the presence of S^3 in the denominator is inconvenient in the AC scheme because of the explicit derivative corrections in the neighbor procedure. In order to remove this feature, we introduce a time smoothing, defined by the differential relation

$$dt = S^{3/2}\, d\tau. \tag{17}$$

This gives rise to the velocity transformation

$$\rho_i' = S^{3/2}\dot{\rho}_i, \tag{18}$$

where primes denote differentiation with respect to the fictitious time τ. Conversely, the definition (17) may be used to recover the physical velocities from

$$\mathbf{v}_i = S'\rho_i/S^{3/2} + \rho_i'/S^{1/2}. \tag{19}$$

A second differentiation of Eq. (18) yields the new equation of motion

$$\rho_i'' = -\frac{S'}{2S}\rho_i' - G\sum_j \frac{m_j(\rho_i - \rho_j)}{[|\rho_i - \rho_j|^2 + \varepsilon^2]^{3/2}} + GM\rho_i, \quad (j \neq i). \tag{20}$$

We note that an S^2 time transformation instead of Eq. (17) would remove the damping force while leaving a product term containing S. However, the presence of an explicit viscous term has a desirable numerical effect, especially in hyperbolic models where S'/S increases with time.

The corresponding equations of motion for the boundary radius is readily derived by applying the rule of differentiation [Eq. (17)] twice to Eq. (14).

Setting $G = 1$, this gives

$$S'' = 3S'^2/2S - MS. \tag{21}$$

Because of its simple form, it is convenient to solve Eq. (21) by the method of explicit derivatives. Two successive differentiations give rise to the Taylor series coefficients

$$S''' = \left[\frac{3S'^2}{S^2} - 4M\right]S', \quad S^{(4)} = 15\left[\frac{S'^2}{2S^2} - M\right]\frac{S'^2}{S} + 4SM^2. \tag{22}$$

Knowledge of the physical time is normally not required since S itself is a fundamental parameter. However, the principle of successive differentiation may also be applied to $t' = S^{3/2}$.

When integrating an expanding system where the boundary is governed by the Newtonian force law, it may be desirable to conserve the comoving density. To simulate the balance between loss and gain across the boundary, we perform a mirror reflection. Any particle crossing the boundary (i.e., $\rho_i > 1$, $\rho_i' > 0$) is assigned a negative comoving radial velocity of equal magnitude. New polynomials are initialized to avoid discontinuity effects in the force differences. This simple boundary treatment still permits an energy integral, provided the loss of kinetic energy is accumulated. Conversion to physical scaled units yields

$$\Delta E_i = 2m_i S'(S'\rho_i + S\rho_i' - S')/S^3, \tag{23}$$

where ρ_i' denotes the *old* radial velocity. Note that because of the finite integration interval, this correction is applied at a point just outside the boundary.

B. Implementation of the AC Method

The comoving formulation lends itself readily to the AC scheme discussed in Section III. Inspection of Eq. (20) suggests that the term $M\rho_i$ can be absorbed in the regular force component, since it would tend to cancel the contribution from a homogeneous distribution. Although this procedure is satisfactory (Aarseth, 1979), it was noted by Findlay (1983a) that keeping the smoothing term separate improves the performance further. We therefore write the total force per unit mass as a sum of *three* parts which are treated separately:

$$\mathbf{F} = \hat{\mathbf{F}}_n + \mathbf{F}_d + M\rho_i. \tag{24}$$

As before, we combine the explicit velocity term of Eq. (20) with the actual neighbor force $\mathbf{F}_n$.

Initialization of the comoving force polynomials proceeds as in the standard AC scheme. Differentiation of the softening parameter gives rise to an additional contribution; i.e., $\varepsilon' = -\varepsilon_0 S'/S^2$. This effect was omitted in the original formulation, but a small improvement is noted when the complete first force derivative is evaluated ($\mathbf{F}^{(2)}$ and $\mathbf{F}^{(3)}$ need not be modified). The next two orders are again obtained by the boot-strapping procedure, where Eq. (24) itself is substituted when forming ρ_i'' on the right-hand side, etc.

The comoving integration scheme is also implemented to order $\mathbf{F}^{(3)}$ with semi-iteration added. Irregular and regular time steps are first obtained by Eq. (9), written with prime derivatives. In comoving coordinates, the regular force tends to remain constant, whereas F_n grows in magnitude due to the clustering. A procedure for increasing the regular steps has been introduced. The interval $\delta\tau_d$ is incremented iteratively by a small factor (e.g., 1.05), provided that the predicted change of the regular force does not exceed a specified fraction of the *irregular* component:

$$[(|\mathbf{F}_d'''|\delta\tau_d/6 + |\mathbf{F}_d''|/2) \delta\tau_d + |\mathbf{F}_d'|] \delta\tau_d < \eta_d |\hat{\mathbf{F}}_n|. \quad (25)$$

The adopted regular step is subject to the condition $\delta\tau_d < \delta\tau_{max}$, which is related to a characteristic expansion time corresponding to a relative increment $\Delta S/S = 2^{1/2} - 1$ ($\delta\tau_{max}$ is obtained by iteration as above).

Modification of the neighbor sphere follows the standard procedure, as does the selection of neighbors. Note that Eq. (10) should now be replaced by the comoving density contrast; i.e., $C = (L_1/N)R_s^{-3}$. Simulations of expanding cluster models have the advantage of requiring fewer neighbor corrections, since only the so-called *peculiar* motions (i.e., departures from the local expansion) are integrated. Nevertheless, the rich clusters that are eventually formed create similar chaotic velocity distributions to those found in single bound systems. Such clusters normally contain too many members to be included in the irregular force field.

All the main steps of the standard AC algorithm are retained, with one important exception. Because of the small peculiar motions of distant particles relative to a given particle, there is no need for a full coordinate prediction when obtaining the total force. Moreover, the neighbor prediction [Eq. (8)] is now performed to order $\mathbf{F}$ only. The scale parameter S plays a fundamental role and is, therefore, integrated to high accuracy, which also permits frequent updating of the softening length by $\varepsilon = \varepsilon_0/S$. Finally, the optional feature of boundary reflection is included at each new irregular step.

Except for the smoothing terms in the equation of motion and the omission of complete coordinate predictions, the comoving AC scheme is virtually identical to the standard version. For analysis of results, the velocities ρ' must be transformed to physical variables [Eq. (19)], whereas it is often convenient to work with comoving coordinates. In this formulation it is also possible to achieve a substantially increased ratio of irregular to regular steps, occasionally exceeding 20:1. The original comoving version resulted in a significant saving for expanding systems with 4000 members (Aarseth et al., 1979), and the present formulation is even more efficient.

A supercluster simulation with 10^4 interacting particles was performed recently (Dekel and Aarseth, 1984). Conservative accuracy parameters were used; $\eta_n = 0.02$ and $\eta_d = 0.06$, together with an initial softening $\varepsilon = 0.02$. Although parabolic universe models are less efficient with a comoving formulation than hyperbolic models, the computational effort to cover an expansion factor of 8 does not seem excessive (<90 CPU hr on VAX 11/780). In view of the N^2 nature of the total energy calculation, as opposed to the faster integration speed, it is adequate to monitor the conservation of angular momentum and center-of-mass integrals for large N once the method has been fully tested.

C. Mergers and Inelastic Effects

The present scheme can readily be modified to include the effect of inelastic galaxy collisions resulting in mergers (Aarseth and Fall, 1980). Separate N-body experiments of colliding model galaxies were used to form an empirical cross section for mergers, neglecting a small mass loss and inelastic effects. Let the masses of two interacting particles be denoted by m_k and m_l, and adopt a sequential ordering with $k < l$. If a merger is defined to take place (e.g., overlapping mass distributions and barely hyperbolic velocities), the old particle mass m_k is replaced by the combined mass $m_k + m_l$. New coordinates and velocities are obtained from conservation of center-of-mass motion. Particle m_l is then removed by updating all relevant variables, including a renaming of all the neighbor lists containing members m_j, where $j \geq l$. The polynomial initialization proceeds as in the standard AC method. An energy conservation scheme may be maintained by making appropriate corrections due to the loss of two-body binding energy, as well as changes in the external interactions.

The effect of more weakly interacting galaxies can also be taken into account. Findlay (1983a,b) has formulated an elegant method to calculate the tidal deformations due to galaxy encounters. Such inelastic encounters lead to increased internal energy at the expense of orbital motion, and the corresponding tidal torque may account for the observed rotation and flattening (enhanced by collapse from larger systems). In addition, the effect of a significant fraction of nonluminous matter has been modeled by including further particles with large softening lengths that are not subject to tidal interactions or mergers. This first attempt at consistent galaxy cluster simulations is a good demonstration of what can be achieved by the direct method.

V. PLANETARY PERTURBATIONS AND COLLISIONS

A. Grid Method Formulation

It is commonly accepted that collisions between planetesimals played a crucial role in the formation of the planetary system. Theoretical treatments of such processes are based on assumptions that cannot easily be substantiated, and the final state is subject, therefore, to considerable uncertainties (e.g., runaway accretion). In this scenario, the planetesimals grow by coalescence from intersecting orbits around the Sun. Orbital eccentricities, which tend to increase by gravitational interactions, are damped by collisions, thereby creating the conditions for a competitive process. The main aim of performing computer simulations of such systems is to address the question of the final state. The computational requirements involve long time intervals, as well as high accuracy, necessitating the use of idealized models. In the following we describe a new method for studying a highly flattened system. The results that have been obtained demonstrate that the direct approach is feasible (Lecar and Aarseth, 1985).

It is convenient to carry out the orbit calculations in a *heliocentric* coordinate system. The equation of motion for a particle of mass m_i and relative coordinates $\mathbf{r}_i$ with respect to the Sun is given by

$$\ddot{\mathbf{r}}_i = -\frac{G(M_s + m_i)}{r_i^3}\mathbf{r}_i - G\sum_j m_j\left[\frac{\mathbf{r}_i - \mathbf{r}_j}{|\mathbf{r}_i - \mathbf{r}_j|^3} + \frac{\mathbf{r}_j}{r_j^3}\right], \quad j \neq i, \quad (26)$$

where M_s is the solar mass and the summation is over the $N - 1$ other mass points m_j. In our case, $m_j \ll M_s$ and the particle orbits are subject to a large number of weak encounters interspersed by an occasional dominant close approach. The essential features of the evolution are therefore preserved if we exclude relatively distant interactions. However, an increasing proportion of the perturbers should be included as the system becomes more discrete by inelastic collisions.

The present method has been designed for a flattened system and will be applied to a two-dimensional calculation. We divide the orbital plane into N_a azimuthal zones centered on the Sun. The active perturbers of m_i are then selected within an angle $2\pi(2n_a + 1)/N_a$ centered on the associated zone, where n_a is the number of perturber zones on *either* side. To minimize systematic errors caused by nonsymmetrical edge effects, the choice of N_a is a compromise between resolution and computational effort. Computation of the early stage is also speeded up by omitting *all* indirect terms in Eq. (26). For a uniform angular distribution, $\sum_j m_j \mathbf{r}_j/r_j^3 \simeq 0$ in each annular right, and, since the central distances are comparable, only the small fluctuating part of the indirect accelerations is thereby neglected.

The first few models studied were characterized by a narrow initial region, and no special perturber treatment was included in the *radial* direction. However, it is desirable to minimize the boundary effects by increasing the radial extent (and thereby the particle number), while reducing the computational requirements in a consistent way. The particles are initially distributed within an annular ring defined by R_1 and R_2, where the ring width $R_2 - R_1$ is related to the particle number. Radial zones at $r_k = R_2/k^{1/2}$, with $k = 1, 2, \ldots$, are chosen from accuracy considerations. Let n_r denote the number of such zones to be included on either side of the current particle zone. Each particle j is assigned a radial zone index $k_j = [R_2^2/r_j^2]$ with $k = 0$ for $r_j > R_2$ and an azimuthal index $l_j = 1 + [N_a \varphi_j / 2\pi]$, where φ is the phase angle. Active perturbers for particle i must then satisfy the conditions

$$|l_j - l_i| \leq n_a, \qquad |k_j - k_i| \leq n_r. \tag{27}$$

The storage requirement for this two-dimensional scheme is quite modest; a matrix L_{aj} contains the list of all particles in each azimuthal zone, whereas k_j and l_j are obtained from arrays L_φ and L_r of size N. These lists are updated as necessary at the end of each integration cycle.

We use the difference scheme of Section II, with the active perturbers predicted to order $\mathbf{F}^{(1)}$. The energy *stabilization* of Baumgarte (1973) has been adopted in order to eliminate the usual systematic integration errors. In this formulation the equation of motion takes the form

$$\ddot{\mathbf{r}}_i = -(M_s + m_i)\mathbf{r}_i/r_i^3 + \mathbf{P}_i - \alpha(\hat{h}_i - h_i)\mathbf{v}_i/v_i^2, \tag{28}$$

where $G = 1$ and $\mathbf{P}_i$ is the perturbation (with or without indirect terms). The terms in parentheses represent the difference between the explicitly calculated and *integrated* two-body binding energy per unit mass, respectively. Initially, the latter is given by

$$h_i = v_i^2/2 - (M_s + m_i)/r_i, \tag{29}$$

and subsequent values are obtained by integrating the equation of motion

$$\dot{h}_i = \mathbf{v}_i \cdot \mathbf{P}_i. \tag{30}$$

The quantity $\hat{h}_i$ is calculated from Eq. (29) at the beginning of each integration cycle, when the osculating orbit is known to the highest order. The correction term is then included in the third-order coordinate and velocity prediction (as for the semi-iteration) but is *not* added explicitly to the force polynomial itself. After some experimentation we choose $\alpha = 0.4/\delta t_i$, where δt_i is the integration step (cf. Baumgarte and Stiefel, 1974).

The effectiveness of the stabilization procedure depends on the relative two-body perturbation defined by

$$\gamma = |\mathbf{P}_i|r_i^2/M_s. \tag{31}$$

It is prudent to omit the correction term from Eq. (28) for large perturbations, $\gamma > \gamma_{cr}$, when the solar reference orbit is not a good approximation. The solution is continued with a new osculating energy after the perturbation becomes small again. However, this switching is relatively infrequent for any given particle.

Individual time steps are again determined by the relative convergence criterion of Eq. (9). In addition, the time steps are reduced preferentially by a factor $1 + 4(\gamma/\gamma_{cr})^2$ if $\gamma \leq \gamma_{cr}$ or by 4 if $\gamma > \gamma_{cr}$. The former reduction is included to facilitate detection of collisions between minor particles, whereas the latter factor improves further the integration of close encounters.

B. Collision Algorithm and Model Parameters

Apart from calculating accurate orbits over long times, our main purpose is to determine collisions between finite-size particles. The initial size of a particle, assumed to be spherical (circular in two dimensions), is denoted by s_0. We adopt a simple collision criterion of overlapping spheres (circles) at pericenter. During the integration of particle i, a neighbor search is made within $3s_i$ in the corresponding cell if the perturbation satisfies the condition

$$P_i m_i > P_0 m_0, \qquad (32)$$

where $P_0 = m_0/9s_0^2$ is the perturbation of an initial particle at separation $3s_0$. Since $P_i \simeq m_j/9s_i^2$ for dominant perturbations due to m_j, the adopted criterion is symmetrical with respect to the two particles. However, a positive identification usually occurs when considering the more massive particle, which also has a larger size, and this results in greater accuracy. Any such encounter is defined as a *collision* if $a(1 - e) < s_i + s_j$, where a and e are the osculating semimajor axis and eccentricity, respectively. At separations $3s_i$, the relative *solar* perturbation is small; i.e., $\gamma_s \simeq 54 M_s s_i^3/r_i^3(m_i + m_j) \simeq 10^{-6}/r_i^3$. It has also been established by direct integrations that the actual pericenter distance is in good agreement with the predicted value $a(1 - e)$.

At present, for simplicity, we assume completely inelastic collisions. The collision procedure replaces two strongly interacting particles by their total mass and conserves all dynamical quantities in the local center-of-mass frame. Rotational spin is also conserved such that the new internal spin is the sum of the original spins combined with the orbital angular momentum. The new particle is assumed to grow with constant density; i.e., $s' = (m'/m_i)^{1/3} s_i$, where $m' = m_i + m_j$. After all relevant tables have been updated to the reduced particle number $N' = N - 1$, the new force polynomial is initialized. Because of the predominantly low eccentricities and small perturber masses, only the most dominant (circular orbit) solar terms are included in Eqs. (6), but a conservative time step is adopted.

12. Direct Methods for N-Body Simulations

Table II
Comparison of Systematic Integration Errors[a]

Model	n_a	n_r	$\Delta E/E$	$\Delta J/J$	N_c
IV	2	1	0.010	−0.0050	36
IV	5	1	0.0027	−0.0014	38
IV	10	1	0.0008	−0.0004	31
V	5	2	0.0033	−0.0018	76
VI	10	2	0.0006	−0.0003	17

[a] $N = 100$ (Models IV and VI); $N = 200$ (Model V); $R_2 - R_1 = 0.41$ a.u. (Model IV); $R_2 - R_1 = 1.0$ a.u. (Models V and VI).

The overheads arising from collision determinations are negligible with the present algorithm. Although an extension to three dimensions would require some modifications, the method is suitable for studying a flattened system, provided that the thickness does not exceed too many grid cells.

The simulations usually start with model parameters appropriate to lunar-size planetesimals in circular orbits. We adopt initial masses $m_0 = 3.7 \times 10^{-8} M_s$ and particle sizes $s_0 = 1.1 \times 10^{-5}$ a.u. The particles are distributed with constant surface density inside a ring of maximum width $R_2 - R_1 = 1$ a.u., centered near 1 a.u. About 200 such planetesimals are therefore required to represent the terrestrial planets. So far, we have studied models with $N = 100$ and $N = 200$.

A variety of numerical tests have been performed to determine the most appropriate integration parameters for the method. The neglect of distant interactions introduces small but systematic errors. Accumulated relative errors of energy ($\Delta E/E$) and angular momentum ($\Delta J/J$) for a typical model, evaluated after 1000 yr, are shown in Table II for three values of the azimuthal perturber index n_a (keeping $n_r = 1$). The corresponding number of collisions is shown in Column 6. Also included are the actual errors for Models V and VI at the same time.

To perform more rigorous accuracy tests, it is necessary to write the energy and angular momentum integrals of the motion in heliocentric form (cf. Lecar and Aarseth, 1985). The present collision procedure conserves the total angular momentum, whereas an appropriate energy correction is included for the two-body energy and the difference in solar interaction terms. It can be seen from Table II that there is a systematic loss of angular momentum. This is due to the neglect of distant interactions; however, the shrinkage occurs at comparable rates for different orbits. The corresponding relative decrease of the semimajor axis for the five remaining bodies in Model IV (at

34,000 yr) was $\Delta a/a \simeq 4 \times 10^{-7}$ *per revolution* (ignoring small dynamical effects). It is desirable to include all interactions when the particle number has decreased significantly, as was done for Model V onward, since apparent evolution might otherwise be due to such numerical effects, however small. Typical relative errors in total energy are then reduced to $\simeq 10^{-10}$ per revolution, the small shrinkage being due to nonstabilized motions.

All the model calculations are carried out using double precision in order to include accurately the small individual perturbations and reduce rounding errors. Standard values of the integration parameters are $N_a = 100$, $\eta = 0.005$, $n_a = 10$, $n_r = 2$, $\gamma_{cr} = 0.001$. Note that the width of *one* azimuthal perturber zone at 1 a.u. is about $60\times$ the size of the sphere of influence of an original particle. The computational effort for a complete model is substantial ($\simeq 200$ CPU hr on VAX 11/780), but it is important to continue the integrations until a stable configuration has been reached, usually after $\simeq 10^5$ yr. Typically, about five massive planetesimals settle into stable orbits. Although encouraging, it is expected that even smaller systems may form in 3-D models. However, the final eccentricities are likely to be rather large, requiring further refinements of the model.

VI. TWO-BODY REGULARIZATION

A. KS Transformations and Equations of Motion

The Kustaanheimo–Stiefel (1965, hereafter KS) regularization transforms the equations of the *relative* two-body motion into a form that is well behaved for small separations, including collisions. This involves introducing a set of *four* coordinates u_j, together with a fictitious time τ, obeying the fundamental relations

$$u_1^2 + u_2^2 + u_3^2 + u_4^2 = R, \tag{33}$$

$$dt = R \, d\tau, \tag{34}$$

where R is the separation of the two mass points, subsequently denoted by m_k and m_l. Writing $\mathbf{R} = (X, Y, Z)$, the transformations to *regularized* coordinates take two alternative forms depending on the sign of X:

$$\begin{aligned}
u_1 &= [(R+X)/2]^{1/2}, \quad X > 0, & u_2 &= [(R-X)/2]^{1/2}, \quad X \le 0, \\
u_2 &= Y/2u_1, & u_1 &= Y/2u_2, \\
u_3 &= Z/2u_1, & u_3 &= 0, \\
u_4 &= 0, & u_4 &= Z/2u_2.
\end{aligned} \tag{35}$$

Equally fundamental is the generalized Levi-Civita matrix defined by

$$L = \begin{bmatrix} u_1 & -u_2 & -u_3 & u_4 \\ u_2 & u_1 & -u_4 & -u_3 \\ u_3 & u_4 & u_1 & u_2 \end{bmatrix}. \tag{36}$$

The regularized velocities are obtained by the transformation

$$\mathbf{u}' = L^T \dot{\mathbf{R}}/2, \tag{37}$$

where L^T denotes the transpose matrix. Conversely, the physical coordinates and velocities are recovered from

$$\mathbf{R} = L\mathbf{u}, \tag{38}$$

$$\dot{\mathbf{R}} = 2L\mathbf{u}'/R. \tag{39}$$

The explicit form of Eq. (38) is then given by

$$\begin{aligned} X &= u_1^2 - u_2^2 - u_3^2 + u_4^2, \\ Y &= 2(u_1 u_2 - u_3 u_4), \\ Z &= 2(u_1 u_3 + u_2 u_4). \end{aligned} \tag{40}$$

It is readily verified that Eqs. (40) satisfy Eq. (33).

The regularized equation of motion takes the form

$$\mathbf{u}'' = h\mathbf{u}/2 + R L^T (\mathbf{F}_k - \mathbf{F}_l)/2, \tag{41}$$

where the terms in parentheses represent the external perturbation evaluated in physical units. When expressed in terms of the regularized velocities, the binding energy per unit mass h becomes

$$h = [2\mathbf{u}' \cdot \mathbf{u}' - (m_k + m_l)]/R. \tag{42}$$

It is advantageous to integrate the binding energy explicitly. Together with the time transformation itself, this gives the additional pair of equations

$$h' = 2\mathbf{u}' \cdot L^T (\mathbf{F}_k - \mathbf{F}_l), \tag{43}$$

$$t' = R. \tag{44}$$

Using the definition of R [Eq. (33)], it is possible to express the Taylor series coefficients for the physical time by known quantities, with the first three terms

$$t' = \mathbf{u} \cdot \mathbf{u}, \quad t'' = 2\mathbf{u}' \cdot \mathbf{u}, \quad t''' = 2\mathbf{u}'' \cdot \mathbf{u} + 2\mathbf{u}' \cdot \mathbf{u}'. \tag{45}$$

This concludes the essential transformations and equations of motion for the relative orbit, including external perturbations. All these equations are well behaved for $R \to 0$ and therefore of *regular* form. An alternative method, based on physical variables and the time transformation (34), is also available (Burdet, 1967; Heggie, 1973).

B. Standard N-Body Treatment

The adaptation of KS regularization to standard N-body integration follows closely an earlier formulation (Aarseth, 1972) with some additional features. We begin by selecting two particles k, l with relative coordinates

$$\mathbf{R} = \mathbf{r}_k - \mathbf{r}_l. \tag{46}$$

All transformations concerning the relative motion are given in the previous subsection. The external perturbation required by Eqs. (41) and (43) takes the explicit form

$$\mathbf{F}_k - \mathbf{F}_l = -\sum_j m_j \left[\frac{\mathbf{r}_k - \mathbf{r}_j}{|\mathbf{r}_k - \mathbf{r}_j|^3} - \frac{\mathbf{r}_l - \mathbf{r}_j}{|\mathbf{r}_l - \mathbf{r}_j|^3} \right], \qquad j \neq k, l. \tag{47}$$

The complete solution of the two-body motion is obtained by introducing a set of Jacobian coordinates

$$\mathbf{q} = (m_k \mathbf{r}_k + m_l \mathbf{r}_l)/(m_k + m_l). \tag{48}$$

The corresponding equation of motion for the center of mass simplifies to

$$\ddot{\mathbf{q}} = (m_k \mathbf{F}_k + m_l \mathbf{F}_l)/(m_k + m_l), \tag{49}$$

where again $\mathbf{F}_k$ and $\mathbf{F}_l$ denote the individual perturbations, since the dominant terms cancel analytically. Equations (46) and (48) may be inverted to yield the respective physical coordinates

$$\mathbf{r}_k = \mathbf{q} + m_l \mathbf{R}/(m_k + m_l), \qquad \mathbf{r}_l = \mathbf{q} - m_k \mathbf{R}/(m_k + m_l). \tag{50}$$

The corresponding velocities are then obtained by similar expressions, using Eq. (39) and the differentiated form of Eq. (48) which defines $\dot{\mathbf{q}}$.

To convert a regularized time interval $\delta\tau$ to physical units δt at time t_0, Eq. (44) is integrated by the Taylor series expansion

$$\delta t = t'_0 \, \delta\tau + t''_0 \, \delta\tau^2/2 + t'''_0 \, \delta\tau^3/6 + t_0^{(4)} \, \delta\tau^4/24 + t_0^{(5)} \, \delta\tau^5/120. \tag{51}$$

Additional terms are readily available by differentiation of Eq. (45). Conversely, interpolation within a regularized interval to find $\mathbf{u}$ or $\mathbf{u}'$ at a general time may be achieved by iteration of Eq. (51). However, it is usually adequate to obtain the desired interval by the low-order inversion

$$\delta\hat{\tau} = \dot{\tau}_0 \, \delta t + \ddot{\tau}_0 \, \delta t^2/2 + \dddot{\tau}_0 \, \delta t^3/6. \tag{52}$$

The definition (34) and two differentiations yield the coefficients

$$\dot{\tau}_0 = \frac{1}{R}, \quad \ddot{\tau}_0 = -\frac{t_0''}{R^3}, \quad \dddot{\tau}_0 = \frac{3t_0''^2 - Rt_0'''}{R^5}. \tag{53}$$

Because of the division by R, these equations are not well behaved for $R \to 0$. This difficulty is overcome by employing the center-of-mass assumption instead of resolving the two components for small separations.

Initialization of a given regularization requires an additional procedure for the relative motion. Thus, it is not practical to use the explicit starting scheme of Section II to the highest order.[†] Instead, a fitting procedure is employed for obtaining Taylor series derivatives. Denoting the regularized acceleration at time τ_0 by $\mathbf{G}_0$ instead of $\mathbf{u}''$, we write a third-order force polynomial over the interval $\tau - \tau_0$ as

$$\mathbf{G} = \mathbf{G}_0 + \mathbf{G}_0'(\tau - \tau_0) + \mathbf{G}_0''(\tau - \tau_0)^2/2 + \mathbf{G}_0'''(\tau - \tau_0)^3/6. \tag{54}$$

Successive accelerations $\mathbf{G}_k$ ($k = 1, 2, 3$) are evaluated by Eq. (41) at three equal intervals $\delta\tau_0$ by advancing all coordinates consistently, including the new center of mass. A conservative fitting interval $\delta\tau_0 = \delta\tau/12$ is chosen, where $\delta\tau$ is the standard integration step. The resulting fitting coefficients are given by

$$\begin{aligned}
\mathbf{G}_0' &= (-11\mathbf{G}_0/6 + 3\mathbf{G}_1 - 3\mathbf{G}_2/2 + \mathbf{G}_3/3)/\delta\tau_0, \\
\mathbf{G}_0'' &= (2\mathbf{G}_0 - 5\mathbf{G}_1 + 4\mathbf{G}_2 - \mathbf{G}_3)/\delta\tau_0^2, \\
\mathbf{G}_0''' &= (-\mathbf{G}_0 + 3\mathbf{G}_1 - 3\mathbf{G}_2 + \mathbf{G}_3)/\delta\tau_0^3.
\end{aligned} \tag{55}$$

Conversion to divided differences then proceeds as before [Eqs. (7)].

An expansion for the companion equation h' is determined in a similar manner. Since Eq. (43) is of first order, the corresponding binding energy is obtained by integrating the fitting polynomial *once*.

In the present scheme, the center of mass is treated as a new *composite* particle which must also be predicted when the components are required. This is usually done to the highest order when integrating the relative motion itself. The polynomial initialization follows the standard procedure [Eqs. (5)–(7) and (9)], except that the force is now obtained by a mass-weighted summation over both components [Eq. (49)]. Termination of a regularization is more straightforward. After transforming to physical variables [Eqs. (50)], each of the old components is now initialized by the standard procedure. Although the two-body term usually dominates, the full contributions are included to all orders. Moreover, in both types of initializations the Cartesian coordinates and velocities of any other regularized pairs are obtained by

---

[†] Explicit derivatives of Eqs. (41) and (43) are now obtained directly by the usual bootstrapping method, neglecting second-order changes in the combined perturbation term.

transformations [Eqs. (38)–(39) and (50)], unless the center-of-mass approximation applies. Likewise, the evaluation of Eq. (49) may be speeded up.

C. Algorithm and Parameters

Close encounters between attracting particles lead to shortening of the respective time steps. It is therefore convenient to specify a characteristic time interval δt_{cl} indicating *dominant* two-body motion. In homogeneous systems near equilibrium, the impact parameter for 90° deflection by a typical particle is

$$R_{cl} = 4\langle R\rangle/N, \tag{56}$$

where again $\langle R\rangle$ is the half-mass radius. Conversion to an equivalent relation for the corresponding time step yields

$$\delta t_{cl} = \kappa[\eta/0.03]^{1/2}[R_{cl}^3/\langle m\rangle]^{1/2}, \tag{57}$$

where κ is an empirically determined constant (weakly dependent on h) and $\langle m\rangle$ the average particle mass.

If a particle k satisfies the condition $\delta t_k < \delta t_{cl}$ and the time step is decreasing ($\delta t_k < t_0 - t_1$), a search is made for the particle l forming the dominant relative motion, at the same time noting any other particles j inside a separation $2R_{cl}$. The pair k, l is accepted for regularization, provided that $R < R_{cl}$ and (in the case of other close particles) the relative force is dominant, requiring

$$\frac{m_k + m_l}{|\mathbf{r}_k - \mathbf{r}_l|^2} > \frac{m_l + m_j}{|\mathbf{r}_l - \mathbf{r}_j|^2}, \qquad j \neq k, l. \tag{58}$$

This condition allows for the possibility of particle l being close to another regularized pair. It is also prudent to accept approaching particles only ($\dot{R} < 0$) to minimize excessive regularizations. In order to include circular orbits, the radial velocity test is modified to

$$\mathbf{R} \cdot [\mathbf{v}_k(t_k) - \mathbf{v}_l(t_l)] < 0.1[R(m_k + m_l)]^{1/2}, \tag{59}$$

since the current value of $\mathbf{v}_l$ is not used. Both accepted particles are integrated to order $\mathbf{F}^{(3)}$ before the relative motion is introduced [Eq. (46)], followed by initialization of the KS treatment [Eqs. (35), etc.].

A regularization may be terminated in several ways, depending on the type of orbit. We introduce the dimensionless two-body perturbation

$$\gamma = |\mathbf{F}_k - \mathbf{F}_l|R^2/(m_k + m_l). \tag{60}$$

This quantity is especially useful for deciding when to end the regularization of *hard* binaries, with semimajor axis $a < R_{cl}/2$ for components of mass

$\langle m \rangle$. If $\gamma > 0.5$, a search is made for a new dominant perturber [cf. Eq. (58)], in which case the current binary is terminated.

Soft binaries are normally terminated by the combined criterion

$$R > R_{cl}, \quad \gamma > \gamma_{er}, \qquad (61)$$

where γ_{cr} is a prescribed parameter.[†] The second condition makes it possible to continue the two-body treatment for highly eccentric orbits, whereas hyperbolic motion is now terminated if $R > R_{cl}$ without a perturbation test. Thus, the regularization decision making is essentially controlled by two parameters, δt_{cl} and γ, since Eq. (57) applies to large mass ratios.

Although the regularized equation of motion [Eq. (41)] is numerically well behaved for dominant two-body motion, it is advantageous to include stabilization when integrating over many revolutions. In analogy with the treament in Section V, we write the regularized equation of motion as

$$\mathbf{u}'' = h\mathbf{u}/2 + RL^T(\mathbf{F}_k - \mathbf{F}_l)/2 - \alpha R(\hat{h} - h)\mathbf{u}'/(m_k + m_l). \qquad (62)$$

Again $\hat{h}$ is evaluated explicitly at the beginning of each step [Eq. (42)], whereas h is the *integrated* value [Eq. (43)]. The adopted form of the correction term is regular because of the division by $(m_k + m_l)/R$. As before [Eq. (28)], we take $\alpha = 0.4/\delta\tau$. However, this procedure is only implemented for hard binaries, where the effect is significant [Eq. (41) is still used for differencing].

The integration step for regularized motion is determined from the orbital period. We adopt a constant value modified by the perturbation,

$$\delta\tau = \frac{2\pi}{N_r}\left[\frac{1}{2|h|}\right]^{1/2}\frac{1}{(1 + 1000\gamma)^{1/3}}, \qquad (63)$$

where N_r denotes the number of steps during one *unperturbed* revolution. The additional term is usually small but allows for significant reduction in the presence of strong perturbations, thereby maintaining orbital accuracy. Equation (63) is equally suitable for hyperbolic motion. Since h appears in the denominator, the unperturbed part is subject to the condition $\delta\tau < \beta(2\langle m \rangle/(m_k + m_l))^{1/2}$, where β is a constant parameter appropriate to small binding energies of either sign. Alternatively, a maximum unperturbed step of the type $2\pi[R_{cl}/(m_k + m_l)]^{1/2}/N_r$ may be used. In either case, it is recommended to include the empirical factor $(R_{cl}/R)^{1/2}$ if $R > R_{cl}$. A conservative initial step of half the standard value is chosen, and the increase of subsequent steps is restricted by a factor of 1.2. Note that the period of the regularized equation of motion corresponds to *twice* the Keplerian value. This fundamental property demonstrates the effectiveness of the KS transformations.

Even regularized integration of hard binaries is time consuming, since the number of revolutions involved can be substantial. The problem is alleviated

[†] Alternatively, the initial separation may be used as a termination criterion.

by limiting the perturbation calculation [Eq. (47)] to relatively nearby particles. This procedure is justified because the relative motion is subject to a *tidal* effect [cf. Eq. (43)], whereas direct integration of each component would include all force contributions. Active perturbers are selected from the tidal force approximation

$$|\mathbf{q} - \mathbf{r}_j| < [2m_j/(m_k + m_l)\gamma_{\min}]^{1/3} R, \quad j \neq k, l, \tag{64}$$

where $\gamma_{\min}$ is a parameter denoting significant perturbation. The list of new perturbers is determined initially and at every subsequent binary apocenter $[R = a(1 + e)]$, defined by the turning point conditions

$$t_0'' t'' < 0, \quad R > -(m_k + m_l)/2h, \tag{65}$$

where t_0'' is the radial velocity at the beginning of the step. Conversely, the center-of-mass approximation is used when calculating the force due to a regularized pair for distances exceeding λR, where λ is related to $\gamma_{\min}$ [cf. Eq. (64)].

Unperturbed motion is assumed if there are no particles satisfying Eq. (64), in which case integration of the next period is suppressed. Subsequently, the unperturbed motion may be extended by an integer number of periods, provided that the nearest approaching particles do not enter the sphere of significant perturbation in the meantime. Several conservative criteria (i.e., based on travel times as well as accelerations) are used to estimate the maximum extent of unperturbed motion. A simple restart is performed when reverting to perturbed motion. In this case, it is adequate to determine the new differences by explicit differentiation of the first term of Eq. (41), with $h' = 0$, and by halving the usual interval.

The removal of two-body singularities also permits a well-behaved total energy to be defined. Thus, the explicit potential energy summation over regularized components, together with the relative kinetic energy of each pair, is replaced by the *predicted* binding energy [Eq. (43)] and the kinetic energy of the associated center-of-mass particle. This procedure is only recommended in combination with the stabilized equation of motion, since the small but systematic drift in binding energy, as calculated explicitly, may otherwise lead to unacceptable relative energy errors (thus, $\hat{h}$ now tracks h closely).

The standard integration scheme copes quite well with the additional requirements imposed by the regularization procedures. Occasionally for small N, however, the time steps δt_i are not reduced sufficiently for strong interactions between single particles and hard binaries. Instead of decreasing the overall accuracy parameter η, we introduce a criterion

$$|\mathbf{F}^{(1)}| \delta t_i + |\mathbf{F}^{(2)}| \delta t_i^2/2 = \eta |\mathbf{F}|. \tag{66}$$

12. Direct Methods for N-Body Simulations

This alternative procedure is only adopted for interactions satisfying

$$m_i|\mathbf{F}| > \langle m \rangle^2 / R_{cl}^2, \tag{67}$$

[cf. Eq. (32)] and is, therefore, relatively inexpensive. Moreover, the increase of all new steps is now restricted by a stability factor 1.2 to reduce the force fluctuations of eccentric binaries.

It is desirable to organize the variables in a systematic way to facilitate the simultaneous treatment of *single* particles and regularized pairs, including the associated centers of mass. We distinguish between *global* quantites g_i, such as $\mathbf{r}_i$, $\mathbf{v}_i$, and regularized variables ρ_j. The sequential arrays $\{g_i\}$ are modified to include the regularized components first. For new regularizations this involves exchanging the global arrays of the particles k, l with those of the first two particles. The last member of a global particle array is then g_N, as usual, but the center of mass corresponding to the first regularized pair is added as g_{N+1}. The extension to an arbitrary number of pairs follows readily. Thus, an alteration of the existing configuration is performed by moving all relevant quantities up or down in the tables and deleting or adding the corresponding center of mass.

Consider a general configuration with n_p separate pairs. The particle arrays $\{g_i\}$ where $i \leq 2n_p$ then represent the transformed components with corresponding relative variables $\{\rho_j\}, j \leq n_p$. Subsequent locations $2n_p + 1, \ldots, N$ are assigned to single particles, followed by the center-of-mass arrays $\{g_{N+j}\}$ with $j \leq n_p$. It is then quite simple to distinguish between the different procedures required by the three cases $i \leq 2n_p$, $2n_p < i \leq N$, $i > N$, where i is the next particle to be advanced. Coordinate predictions and force calculations are also more efficient when similar quantities are arranged sequentially. This requires the updating of tables at each new regularization or termination, together with the renaming of any other perturber lists and the time-step list M. However, the polynomial initializations are relatively more expensive.

The automatic decision making is an attractive feature of the present scheme, which is based on a small number of parameters. Most of these are dimensionless or have a dynamical significance. The regularization parameters discussed above are normally assigned the standard values: $\kappa = 0.02$, $\gamma_{cr} = 0.01$, $\gamma_{min} = 10^{-6}$, $\lambda = 70$, $N_r = 50$, $\beta = 5\,\delta t_{cl}/R_{cl}$. In addition, the definition of a hard binary is employed to avoid excessive initializations. A consistent modification of the basic regularization parameters has also been included to take into account significant evolution. This is achieved by introducing the cube root of the *central* density contrast C_{max} [cf. Eq. (10)] in the denominator of Eq. (56). The conservative choice $R_{cl} = \langle R \rangle / N$ is often used in small N simulations, together with $\lambda = 100$, $N_r = 60$, and $\eta = 0.02$.

The regularization scheme employs the following set of variables for each pair: $\mathbf{u}$, $\mathbf{u}_0$, $\mathbf{u}'$, $\mathbf{u}''$, $\mathbf{u}'''$, $\mathbf{D}_u^1$, $\mathbf{D}_u^2$, $\mathbf{D}_u^3$, $\delta\tau$, τ_0, τ_1, τ_2, τ_3, h, h', D_h^1, D_h^2, D_h^3, R, γ,

t'', t'''. Here D_u and D_h denote appropriate differences for the regularized equations of motion [Eqs. (41) and (43)]. When mixed precision is employed, nearly all these variables are defined in double precision, whereas the perturbation [Eq. (47)] need not be evaluated so accurately. Note that the fictitious time τ itself is only used as a reference time. In addition, each center of mass requires the variables of the standard method.

D. Regularized AC Procedures

The introduction of a neighbor scheme adds considerably to the programming complexities of the KS regularization treatment. No new parameters or variables are required when combining the AC scheme with the treatment in the previous subsections. It is an attractive feature of such a formulation that the regularization search and perturber selections are carried out with respect to neighbor particles only. The additional effort is concerned with the renaming of particles, preserving the sequential organization of variables. In the following we outline the main new procedures for the combined scheme.

The close-encounter search is now performed for small *irregular* time steps $[\delta t_k < \delta t_{cl}]$. Having selected a particle pair k, l for regularization, we assign the other neighbors of k and the associated length scale R_s to the new composite particle, whereupon the updating of all tables follows. The neighbor lists are renamed consistently with the new sequence. Although all the particles must be considered, this is a fast procedure involving at most four renamings each time due to exchanging the pair k, l with the first two single particles. We employ a neighbor list convention where both components of a new pair are replaced by the corresponding center of mass (rare cases of only one component being included are allowed). This composite particle is resolved into individual components if the center-of-mass assumption does not apply [cf. λR]. To facilitate neighbor selection, we retain a list of the most recently terminated pairs. This list is updated by removing any pair containing the new components k or l, and by renaming any exchanged components. All other initialization procedures carry over, except that the selection of perturbers [Eq. (64)] is made from the associated center-of-mass neighbor list.

Termination of a regularized pair starts by adopting the corresponding center-of-mass neighbors and the neighbor radius R_s for the standard integration of the components. Following the usual table updating and the deletion of the center of mass, all neighbor lists are modified consistently with the new sequence. This involves renaming any subsequent regularized pair index or corresponding center-of-mass index and replacing the deleted center of mass by the components. The list of terminated pairs is also updated, first by removing any pairs for which the separation is no longer small [e.g., $> 4R_{cl}$] and second by adding the current components.

12. Direct Methods for N-Body Simulations

Both types of polynomial initializations employ the same procedure, with small modifications. Thus, in the case of a new pair, the components are removed from each other's neighbor list since the dominant term is not included in the center-of-mass force [Eq. (49)]. Conversely, the components *are* included in the case of termination. To save time, only the current neighbor velocities are set before the usual force polynomials are evaluated [Eqs. (5)–(7) and (9)]. This is still the most time-consuming part of the initialization procedure, although contributions outside $2R_s$ ($5R_s$ with full double precision) are omitted from the second and third derivatives. The time-step list M must also be consistent with the new particle sequence, with appropriate deletions or additions required by the changing configuration. Finally, the secondary neighbor velocities $v(t)$, which are usually defined in single precision, are restored to their former double-precision values $v(t_0)$. This procedure permits fast coordinate predictions while retaining accuracy for each integration.

The integration procedure itself is very similar to the KS and AC treatments discussed above, although many new features have been added. Compared with the standard N-body regularization, the start of an integration cycle must now include the case of a regular force calculation. Neighbor prediction for a new regularized or irregular step is of the same form as before [Eq. (8)], except that the prediction of one single regularized component, which may occasionally be present in the neighbor list, is suppressed, using the most recent value.[†] The integration method for regularized motion is unchanged; however, the new perturbers are now selected from the corresponding center-of-mass *neighbor* list.

The strategy for total force calculation is based on the need for accuracy combined with speed. It is necessary to consider the force on a single particle or composite particle i due to other single particles or pairs. The center-of-mass approximation is assumed for distances

$$|\mathbf{r}_i - \mathbf{r}_j| > \max(2^{1/3} R_s, \lambda R_{cl}), \qquad j = 2n_p + 1, \ldots, N + n_p, \qquad (68)$$

where the first term corresponds to the outer shell employed in the standard AC scheme. Inside this separation, the center of mass of any composite particle i or j is replaced by the individual components for distances not satisfying the center-of-mass approximation [i.e., $< \lambda R$]. Thus, the force between two neighboring composite particles contains four interaction terms. To avoid spurious force differences, the same criteria are employed when evaluating the irregular force. Although of complicated form, the total force evaluation is still relatively fast, since most interactions comprise only one term.

The selection of perturbers from the corresponding center-of-mass neighbor list assumes that the associated length scale is sufficiently large [cf. Eq. (64)]. The neighbor sphere of binary center-of-mass particles is therefore modified if it contains too few perturbers. We define a volume ratio $(r_p/R_s)^3$

[†] For increased accuracy, include one extra order when predicting dominant neighbors.

corresponding to a typical large perturber mass [e.g., $5\langle m \rangle$] and the maximum binary separation $2a$. If $r_p > R_s$ and $L_p < 0.75 L_{\max}$, the standard neighbor sphere modification [Eq. (12)] is replaced by

$$R_s^{\text{new}} = R_s^{\text{old}}(1 + \delta)^{1/2}, \quad r_p > R_s, \quad i > N, \tag{69}$$

where δ is a stabilization factor defined by

$$\delta = 1 - 1.33 L_1 / L_{\max}. \tag{70}$$

We also impose a time-step-dependent maximum change of R_s for regular steps below 0.01 crossing times in order to reduce neighbor corrections. Using the definition of the crossing time, $t_{\text{cr}} = (8\langle R \rangle^3 / N \langle m \rangle)^{1/2}$, we now replace the permitted relative change of the volume factor by

$$f = \min(0.25, 25 \delta T_i / t_{\text{cr}}). \tag{71}$$

A final set of procedures is included to deal with neighbors undergoing close encounters with particles just outside the boundary. These *optimizing* procedures consist of four separate stages following the determination of new neighbors. First, the current neighbor list is examined for members with small irregular integration steps. A special parameter is introduced for this purpose, with standard value $\delta t_{\min} = 2 \delta t_{\text{cl}}$. If $\delta t_j < \delta t_{\min}$ for any neighbor j, a further search checks whether this particle is within a small separation $4 R_{\text{cl}}$ of another neighbor of particle i. A negative test results in the new particle being added to the neighbor list, with appropriate modifications of the new irregular and regular force.

Provided that $L_1 < L_{\max}$, a second stage involves checking that both components of a recently terminated pair are included. The search is restricted to two possible pairs $[2 n_p < j \le 2 n_p + 4]$ and is therefore fast. A distance test $[4 R_{\text{cl}}]$ is included after eliminating the case of two consecutive neighbor list members belonging to the same pair.

A third procedure is concerned with the missing component of a close pair originally assigned to locations $2 n_p + 1$ and $2 n_p + 2$, but subsequently exchanged with the components of a new pair. The list of recently terminated pairs is examined, and this time the previously considered particle locations [thus, $j > 2 n_p + 4$ now] are excluded. If identified, a missing component is added to the neighbor list, subject to the usual close distance test $[4 R_{\text{cl}}]$, whereupon the force components are modified. Provided that $L_1 < L_{\max}$, other terminated pairs are examined in turn.

A further refinement is introduced to avoid deleting a neighbor experiencing a close encounter. In this case, particle j is retained, provided that $\delta t_j < \delta t_{\min}$ and $|\mathbf{r}_i - \mathbf{r}_j| < 2 R_s$. Temporary retention of a neighbor just outside the standard sphere is beneficial, since it avoids large correction terms in the derivatives, which would otherwise reduce both the irregular and regular time steps [cf. Eqs. (7)].

All of the above special procedures have proved useful in large-scale simulations. Although most such tests are negative, the additional effort is quite modest. The problem of large correction terms in the force derivatives cannot be completely avoided in the AC scheme, however. An attempt to minimize any undesirable effects still present consists of ignoring the contribution to $\mathbf{F}^{(3)}$ in critical cases when the correction term exceeds the third regular derivative by a large factor [e.g., 10] and $\delta t_j < \delta t_{\min}$. A further refinement, implemented recently, avoids premature neighbor sphere penetration by extremely fast particles which are occasionally ejected during multiple encounters. In spite of such complications, the regularized AC scheme has nevertheless proved itself highly efficient and versatile for N-body simulations.

VII. THREE-BODY REGULARIZATION

A. Isolated Triple Systems

Triple systems often display irregular and complex dynamics. Such configurations may nevertheless reveal characteristic motions, and the general three-body problem has, therefore, received much attention. For some time, numerical simulations were best performed using KS regularization (Szebehely and Peters, 1967; Standish, 1972) or the equivalent Burdet method (Valtonen, 1976). A global regularization of the *planar* three-body problem (Waldvogel, 1972) was followed by new treatments of the 3-D case (Aarseth and Zare, 1974, hereafter AZ; Heggie, 1974). The latter formulations are closely related to KS regularization, being generalizations to two and three *simultaneous* KS transformations, respectively. The two methods are comparable for realistic examples (cf. Heggie, 1974), and we therefore discuss the AZ version, which is simpler to implement. However, the formulation of Heggie is *global* and has been generalized to the N-body problem. Although of complicated form, it has been used together with the variation-of-parameters method in simulations with two colliding binaries (Mikkola, 1983).

Consider a three-body system with masses $m_i > 0$ ($i = 1, 2, 3$) and define the two relative distances R_k ($k = 1, 2$) between m_k and m_3. The AZ formulation introduces *eight* coordinates Q_j satisfying two simultaneous KS transformations and employs a new time transformation [cf. Eqs. (33)–(34)]:

$$Q_1^2 + Q_2^2 + Q_3^2 + Q_4^2 = R_1, \qquad Q_5^2 + Q_6^2 + Q_7^2 + Q_8^2 = R_2, \qquad (72)$$

$$dt = R_1 R_2 \, d\tau. \qquad (73)$$

The corresponding Hamiltonian function takes the final form

$$\Gamma^* = R_2\mathbf{P}_1^2/8\mu_{13} + R_1\mathbf{P}_2^2/8\mu_{23} + \mathbf{P}_1^T\mathbf{A}_1\mathbf{A}_2^T\mathbf{P}_2/16m_3$$
$$- m_1m_3R_2 - m_2m_3R_1 - m_1m_2R_1R_2/|\mathbf{R}_1 - \mathbf{R}_2| - R_1R_2H, \quad (74)$$

where $\mathbf{A}_k$ is identical to twice the *transpose* of the generalized Levi-Civita matrix [Eq. (36)], and H is the total energy. The regularized momenta $\mathbf{P}_k$ are related to the physical momenta of m_k by

$$\mathbf{P}_k = \mathbf{A}_k\mathbf{p}_k, \qquad k = 1, 2, \quad (75)$$

and $\mu_{k3} = m_k m_3/(m_k + m_3)$. The equations of motion are then given by

$$\mathbf{Q}'_k = \partial\Gamma^*/\partial\mathbf{P}_k, \qquad \mathbf{P}'_k = -\partial\Gamma^*/\partial\mathbf{Q}_k, \qquad k = 1, 2. \quad (76)$$

These equations are regular for $R_1 \to 0$ or $R_2 \to 0$, since the new momenta are well behaved for small distances. The KS formulation is recovered by setting $m_2 = 0$ and omitting R_2 from the equations (cf. Peters, 1968).

Stabilization has been tried, but with little success (Aarseth, 1976). Instead we recommend a modified time transformation, which affects Eq. (74),

$$dt = R_1R_2 d\tau/(R_1 + R_2)^{1/2}. \quad (77)$$

The basic algorithm is very simple. First, specify initial conditions in the center-of-mass frame and select a reference body m_3 such that

$$|\mathbf{R}_1 - \mathbf{R}_2| > \min(R_1, R_2). \quad (78)$$

New coordinates and momenta $[\mathbf{Q}_k, \mathbf{P}_k]$ are introduced by appropriate transformations of the physical relative coordinates and absolute momenta (cf. AZ). Integration of the equations of motion may proceed until condition (78) is violated. This necessitates transforming back to physical variables, followed by the selection of a new reference body. The loss of accuracy and efficiency from successive relabeling is usually not important. Thus, except for the special case of all three distances approaching zero simultaneously (triple collision), the switching condition (78) implies that, when differentiated, the *singular* terms of Eq. (74) are smaller than the corresponding regular terms.

The AZ method has proved itself in extensive calculations of binary formation from hyperbolic impact velocities (Aarseth and Heggie, 1976). However, the Runge–Kutta 7(8)-order integrator has now been replaced by the even more efficient Bulirsch–Stoer (1966) method. If desired, a relative energy accuracy of 10^{-12} may readily be maintained.

B. General N-Body Algorithm

Close encounters between single particles and hard binaries are a central feature of star-cluster dynamics. Such interactions are best studied by three-body regularization, since frequent changes in the dominant two-body configuration may otherwise lead to loss of accuracy and efficiency. Although the theory for perturbed three-body motion is available (cf. AZ), we only consider the AC implementation for *isolated* systems, which will be adopted for small perturbations. A similar treatment has also been introduced for standard N-body regularization.

The new procedures are added to the regularized AC version without affecting the main decision making. We discuss algorithms for three separate stages: (i) Selection of a critical triple configuration; (ii) termination criteria for unperturbed three-body integration; (iii) initialization of the binary and single particle for the AC scheme. Only the first part is included in the general AC algorithm, whereas (ii) and (iii) are introduced as independent procedures, together with the three-body regularization itself.

Consider a binary at apocenter [cf. Eqs. (65)]. A strong interaction is indicated if the irregular step of the corresponding center-of-mass particle m_j is small; e.g., $\delta t_j < \delta t_{cl}$. In this case a search is made for the first and second most dominant perturber, here denoted by m_3 and m_4. The perturbation γ_4 of m_4 on the *relative* motion of m_3 and m_j is estimated from the tidal force [Eq. (64)]. If $\gamma_4 < 10^{-4}$ (say) and m_3 is approaching the binary, we also compare the impact parameter with the binary apocenter. The critical configuration is selected for special treatment if

$$a_3(1 - e_3) < a(1 + e), \tag{79}$$

where a_3 and e_3 are the orbital parameters of m_3 with respect to m_j. This condition prevents repeated switching of methods for stable hierarchical systems.

The selected three-body configuration is initialized in the local center-of-mass frame using the current coordinates and velocities, whereupon the most massive binary component is chosen as reference body. Unless escape occurs, the new system is integrated until a specified size or time scale is exceeded. We adopt the dual termination conditions

$$|\mathbf{R}_1 - \mathbf{R}_2| > \min(r_\gamma, 2R_*), \tag{80}$$

$$\delta t > 100(m_1 + m_2 + m_3)^{5/2}/|2H|^{3/2}, \tag{81}$$

where r_γ corresponds to the distance for which $\gamma_4 = 10^{-4}$ and

$$R_* = (m_1 m_2 + m_2 m_3 + m_1 m_3)/|H|. \tag{82}$$

If $R_1 + R_2 + |\mathbf{R}_1 - \mathbf{R}_2| > 3R_*$, we use the Standish (1971) escape criterion to check whether particle m_1 or m_2 has reached an asymptotic solution; thus, for m_1

$$\dot{\rho}^2 > 2(m_1 + m_2 + m_3)[1/\rho + M_2 M_3 R_*^2/\rho^2(\rho - R_*)], \qquad \rho > R_*; \quad \dot{\rho} > 0, \tag{83}$$

where ρ is the distance to the binary center of mass and $M_k = m_k/(m_2 + m_3)$.

When terminated, the isolated triple system is initialized for the AC scheme. Because of large derivatives the global time t is only advanced to the smallest value of $t_j + \delta t_j$ for other members j of the time-step list. However, the local interaction time is usually short compared with the maximum duration [Eq. (81)]. Before transforming to global coordinates and velocities, we include the effect of the total force and first derivative on the center of mass during the *partial* interval. New irregular and regular force polynomials are evaluated for the most distant triple component m_k, where $k = \max_k(R_k)$. However, the second and third derivatives can be set to zero here. Irregular and regular steps are specified by relative criteria of the type $\delta t_k = \eta^{1/2}|\mathbf{F}|/|\mathbf{F}^{(1)}|$. An alternative criterion based on the distance ρ and the binary mass ensures a conservative value for the irregular step [cf. Eq. (57)]. After the total force and first derivative are defined, the single particle can be advanced by the general scheme. Note that the ejected particle may be different from the original intruder; hence the initial identities must be preserved. The remaining binary components are accepted for KS regularization using the standard initialization procedure. Usually the termination conditions (80) or (83) ensure a small perturbation, in which case the regularized polynomials [Eqs. (55)] are well behaved. Finally, the new center-of-mass particle is initialized in the standard way; the irregular step may still be below δt_{cl}, but $\dot{\rho} > 0$ in most practical cases.

An equivalent procedure for studying two colliding binaries has recently been implemented. This method is based on Mikkola's (1985) formulation of four-body regularization (Heggie, 1974). In this case, the selection criterion for strong interaction [Eq. (79)] is modified to include the additional term $1.5a_2$ on the right-hand side, where a_2 is the semimajor axis of the second binary, with an assumed eccentricity of 0.5. Termination is again controlled by a distance criterion [cf. Eq. (80)], with R_* of Eq. (82) generalized to four bodies. As before, the two closest particles, which also here form a bound pair, are accepted for KS regularization, whereas the two other particles are initialized for direct integration. Some care is exercised at termination to avoid troublesome configurations for the subsequent solutions; e.g., a compact triple system or eccentric binary orbits near the pericenter, in which case the four-body treatment is extended.

A new scheme has also been adopted for hierarchical triples. Thus, provided the configuration is relatively stable, the outer component is regularized with respect to the inner binary, which is subsequently restored.

VIII. STAR-CLUSTER SIMULATIONS

A. Models of Open Clusters

The simulation of star-cluster dynamics represents an ideal application of N-body techniques. Currently the direct approach is limited to a few thousand members and is therefore suitable for typical *open* clusters. Such systems differ from the classical N-body problem in several important respects that complicate a consistent theoretical treatment. Realistic effects may readily be included in numerical cluster models. Moreover, the computational requirements are often reduced by the dissipative nature of the additional features. The following dynamical effects have been incorporated:

1. the galactic tidal force,
2. interstellar clouds,
3. mass loss by evolving stars,
4. escape from the cluster.

We now discuss the regularized AC implementation of these modifications.

The external perturbations on open clusters may be modeled by a smooth *tidal* force due to the distant mass distribution, combined with an irregular component arising from passing clouds and field stars. We express the tidal field as a first-order expansion about the cluster center. This yields equations of motion in local rotating coordinates with the x-axis pointing toward the galactic center (Aarseth, 1967):

$$\ddot{x}_i = F_x + 4A(A - B)x_i + 2\omega_z \dot{y}_i,$$
$$\ddot{y}_i = F_y - 2\omega_z \dot{x}_i, \qquad (84)$$
$$\ddot{z}_i = F_z - Cz_i, \qquad i = 1, \ldots, N.$$

Here the first terms represent the internal force components [Eq. (1)], A and B the Oort constants, and C the vertical force gradient. For circular cluster motion in the galactic plane, the angular velocity ω_z is constant and Eqs. (84) admit an approximate energy integral for orbits in low-density regions, as well as a conserved total energy.

The tidal force varies smoothly and is therefore included in the regular force polynomial, whereas the small but more rapidly fluctuating Coriolis terms are added to the irregular component. Higher derivatives of the additional terms are included in all force polynomial initializations by the usual boot-strapping [Eqs. (5), (6)]. However, only the tidal force itself is added to the regularized two-body perturbation [Eq. (47)], neglecting the Coriolis terms which do not contribute to h'. It is also convenient to use dimensionless units here; e.g., $\sum_i m_i = N$ and total energy $E = -N^2/4$. The galactic terms are then scaled by $\langle R \rangle^3/\langle m \rangle$ (square root for ω_z) in physical units. Thus, each model is characterized by the half-mass radius and mean mass, which can be chosen from observations.

Discrete clouds of different types form an important constituent of the interstellar medium. Their relevance for star-cluster dynamics depends on the mass spectrum and velocity dispersion. Consider a simplified model of N_c *standard* clouds with isotropic velocities, selected inside a spherical region of size R_b. The stochastic effects of each cloud, represented as a polytrope of index 5, are included in the equations of motion, providing an additional regular force component

$$\mathbf{F}_c = -Gm_c \left[\frac{\mathbf{r}_i - \mathbf{r}_c}{(|\mathbf{r}_i - \mathbf{r}_c|^2 + R_c^2)^{3/2}} - \frac{\mathbf{r}_i - \mathbf{r}_d}{R_b^3} \right]. \tag{85}$$

Here m_c, $\mathbf{r}_c$ and R_c are the corresponding mass, coordinates, and half-mass radius, and $\mathbf{r}_d$ is the current cluster density center. The repulsive force term is added to cancel a spurious net inward attraction, which would otherwise reduce the escape rate. The coordinates $\mathbf{r}_c$ are advanced frequently in the rotating frame, with the cluster force omitted from Eqs. (84). Clouds crossing the boundary are replaced at random position angles with isotropically distributed velocities. This discontinuity is alleviated by introducing a "sunset" and "sunrise" mass smoothing near R_b. Typical simulation parameters are $N_c = 5$ clouds of 50–500 solar masses and half-mass radii 3.6–7.7 pc inside a region $R_b = 28$ pc, with a Maxwellian velocity distribution of mean 10 km/sec and dispersion 6 km/sec (Terlevich, 1983). This compares with $N = 1000$, $\langle m \rangle = 0.5$ m$_\odot$, $\langle R \rangle = 2$–3 pc for realistic cluster models. These standard clouds do not affect the cluster lifetimes appreciably.

Massive stars in young clusters evolve rapidly to the red giant phase, which is characterized by significant mass loss. Although there is still some uncertainty about the relevant time scales, we adopt a simple scheme of *instantaneous* mass loss. Stars above 6 m$_\odot$ are assumed to turn into high-velocity neutron stars of 1.5 m$_\odot$; otherwise they form quiescent white dwarfs of 0.7 m$_\odot$, on time scales given by stellar evolution models. Typically the first supernova event occurs after 2×10^7 yr, compared to cluster half-lives of 5×10^8 yr (Terlevich, 1983). Moreover, the most massive stars are usually concentrated toward the center, thereby maximizing the disruptive effect. The AC scheme is quite robust to the instantaneous mass loss procedure.

However, it is prudent to reduce the time steps of neighbors and terminate affected regularized pairs.

The stability of star clusters is mainly controlled by the smooth galactic field, which defines the classical tidal radius:

$$R_t = [GN\langle m\rangle/4A(A - B)]^{1/3}. \tag{86}$$

This is the distance for which a particle on the x-axis is likely to escape, giving $R_t \simeq 11$ pc for $N\langle m\rangle = 500$ m$_\odot$. However, even if the energy is sufficiently large, a considerable delay may occur until the orbit is deflected toward the Lagrangian points. This trapping is caused by a significant restoring force in the z direction; thus, $4A(A - B)/C \simeq 0.2$ near the Sun. Even so, the escape rate per crossing time is considerably higher than for isolated systems (Wielen, 1972; Aarseth, 1973).

Escaping stars outside *twice* the tidal radius are removed from the calculations. Reduction of the particle number N also speeds up the simulation and focuses attention on the bound cluster members. In the AC scheme this entails updating all the global variables, as well as the relevant lists. A special procedure is also included for escaping regularized pairs. In order to monitor the overall integration errors, consistent corrections are made to the total energy, including the mutual interaction terms of the escaper.

In conclusion, the *realistic* simulations provide detailed results for observational comparisons. It is especially encouraging that the whole life history of typical clusters can now be studied by the direct method.

B. Postcollapse Evolution

A considerable theoretical effort has been devoted to the study of core collapse in isolated systems. These investigations are primarily motivated by the globular cluster problem, but more modest N-body simulations may still be instructive. However, technical problems due to multiple encounters affect progress. Several lines of attack have been tried in order to learn more about the *postcollapse* phase.

Energetic binaries are likely to play a crucial role during the late stages of collapse and subsequent evolution. The effects would be even greater for a significant initial binary population, which cannot be ruled out observationally. Exploratory calculations with initial binaries have been made, using the regularized AC scheme (Aarseth, 1980; Giannone, 1983). Hard binaries are destroyed or ejected from the core during this process, and it is still uncertain whether new binaries can form at a sufficient rate to delay the core collapse indefinitely. Interesting results on postcollapse evolution have also been obtained by a new hybrid scheme that combines standard N-body regularization in the core with a Fokker–Planck method in regions of low density (McMillan and Lightman, 1984).

Modifications of the regularized AC scheme to include critical triple encounters enable advanced stages to be considered. New calculations that illustrate this method have been made for an isolated system of $N = 250$ equal-mass particles. To maintain accuracy, all variables are now converted to full double precision. The simulation covers more than 300 *initial* crossing times and provides a fascinating preview of long-term cluster evolution. In the following we highlight some aspects of multiple encounters in the core. A parameter ε_b is used to describe the fraction of the *current* total energy contained in binaries; thus, $\varepsilon_b = -\sum_{kl} m_k m_l / 2 a_{kl} E$. Conversely, $\varepsilon_f(1 - \varepsilon_b)$ is a measure of the core expansion, where ε_f denotes the ratio of final to initial *total* energy.

In the numerical model, permanent binaries do not develop during the first 40 crossing times. At this stage the first hard binary is formed, with $\varepsilon_b \simeq 1/N$. Further significant binary evolution takes place over a short time scale. This epoch may be identified with the end of the runaway phase when a dominant binary is formed, causing termination of previous simulations. The most critical stage is reached after 50 crossing times, with a nearly isolated *hierarchical* triple configuration in which the inner binary accounts for an energy fraction $\varepsilon_b \simeq 0.22$. This compact system is quite stable because of the small outer eccentricity, and the inner binary is therefore replaced by the corresponding center-of-mass body to permit further progress after regularization of the outer component. The new binary configuration is also strongly bound, giving $\varepsilon_b = 0.08$ with respect to the modified total energy. Further energetic triple encounters increase the binding energy fraction, reaching $\varepsilon_b = 0.67$ after about 100 crossing times. This energy is now shared by two binaries, whereas only one dominant pair is present for much of the earlier phase. The most energetic pair is ejected from the cluster by a strong recoil effect, thereby reducing the current total energy by 59%.

Binary activity in hierarchical configurations also plays an important part in the subsequent development. During the next 200 crossing times the binary energy increased from 15 to 70% of the total energy, the latter having been reduced to 40% of the initial value. At this stage 95 distant particles had been removed from the cluster, using a nominal escape criterion of 20 initial length units, and the simulation was terminated.

The cluster structure is significantly affected by the binary evolution. Thus, subtracting the energy stored in binaries only leaves a final energy fraction $\varepsilon_f(1 - \varepsilon_b) \simeq 0.12$ binding the cluster itself. This corresponds to a four-fold increase of the half-mass radius $\langle R \rangle$, as well as a considerable core *expansion*.

A total of 50 separate triple interactions took place during the whole interval, often involving highly energetic transitions. It may also be noted that the final cluster configuration contains four hard binaries; this clearly shows the need for the new three-body treatment, which will be adopted in future simulations.

ACKNOWLEDGMENTS

I thank Dr. Douglas Heggie for critical comments. FORTRAN programs using the N-body algorithms described above are available upon request.

REFERENCES

Aarseth, S. J. (1966). *Mon. Not. R. Astron. Soc.* **132**, 35–65.
Aarseth, S. J. (1967). *Bull. Astron.* **2**, 47–57.
Aarseth, S. J. (1972). *In* "Gravitational N-Body Problem" (M. Lecar, ed.), pp. 373–387. Reidel Publ., Dordrecht, Netherlands.
Aarseth, S. J. (1973). *Vistas Astron.* **15**, 13–37.
Aarseth, S. J. (1976). *In* "Long-Time Predictions in Dynamics" (V. Szebehely and B. D. Tapley, eds.), pp. 173–177. Reidel Publ., Dordrecht, Netherlands.
Aarseth, S. J. (1979). *In* "Instabilities in Dynamical Systems" (V. Szebehely, ed.), pp. 69–80. Reidel Publ., Dordrecht, Netherlands.
Aarseth, S. J. (1980). *In* "Star Clusters" (J. E. Hesser, ed.), pp. 325–326. Reidel Publ., Dordrecht, Netherlands.
Aarseth, S. J., and Fall, S. M. (1980). *Astrophys. J.* **236**, 43–57.
Aarseth, S. J., and Heggie, D. C. (1976). *Astron. Astrophys.* **53**, 259–265.
Aarseth, S. J., and Lecar, M. (1975). *Annu. Rev. Astron. Astrophys.* **13**, 1–21.
Aarseth, S. J. and Zare, K. (1974). *Celestial Mech.* **10**, 185–205.
Aarseth, S. J., Gott, J. R., and Turner, E. L. (1979). *Astrophys. J.* **228**, 664–683.
Ahmad, A., and Cohen, L. (1973). *J. Comput. Phys.* **12**, 389–402.
Baumgarte, J. (1973). *Celestial Mech.* **8**, 223–228.
Baumgarte, J., and Stiefel, E. L. (1974). *Lect. Notes Math.* **362**, 207–236.
Bulirsch, R., and Stoer, J. (1966). *Numer. Math.* **8**, 1–13.
Burdet, C. A. (1967). *Z. Angew. Math. Phys.* **18**, 434–438.
Dekel, A., and Aarseth, S. J. (1984). *Astrophys. J.* **283**, 1–23.
Findlay, D. A. (1983a). Ph.D. Dissertation, University of Cambridge.
Findlay, D. A. (1983b). *In* "Clustering in the Universe" (D. Gerbal and A. Mazure, eds.), pp. 35–50. Editions Frontières, Gif sur Yvette.
Giannone, G. (1983). Dissertation, University of Palermo.
Heggie, D. C. (1973). *In* "Recent Advances in Dynamical Astronomy" (B. D. Tapley and V. Szebehely, eds.), pp. 34–37. Reidel Publ., Dordrecht, Netherlands.
Heggie, D. C. (1974). *Celestial Mech.* **10**, 217–241.
Kustaanheimo, P., and Stiefel, E. L. (1965). *J. Reine Angew. Math.* **218**, 204–219.
Lecar, M., and Aarseth, S. J. (1985). *Icarcus* (to be published).
McMillan, S. L. W., and Lightman, A. P. (1984). *Astrophys. J.* **283**, 801–812.
Mikkola, S. (1983). *Mon. Not. R. Astron. Soc.* **203**, 1107–1121.

Mikkola, S. (1985). *Mon. Not. R. Astron. Soc.* (to be published).
Peters, C. F. (1968). *Bull. Astron.* **3**, 167–175.
Standish, E. M. (1971). *Celestial Mech.* **4**, 44–48.
Standish, E. M. (1972). *Astron. Astrophys.* **21**, 185–191.
Szebehely, V., and Peters, C. F. (1967). *Astron. J.* **72**, 876–883.
Terlevich, E. (1983). Ph.D. Dissertation, University of Cambridge.
Valtonen, M. J. (1976). *Astrophys. Space Sci.* **42**, 331–347.
von Hoerner, S. (1960). *Z. Astrophys.* **50**, 184–214.
Waldvogel, J. (1972). *Celestial Mech.* **6**, 221–231.
Wielen, R. (1967). *Veroeff. Astron. Rechen Inst. Heidelberg* No. 19, pp. 1–43.
Wielen, R. (1972). *In* "Gravitational N-Body Problem" (M. Lecar, ed.), pp. 62–70. Reidel Publ., Dordrecht, Netherlands.

13

Molecular Dynamics and Monte Carlo Simulations of Rare Events

BRUCE J. BERNE

Columbia University
New York, New York

    I. Introduction. 419
    II. Activated Barrier Crossing: Theory and Methodology . . . 421
   III. Some Methods for Accelerating Simulations 430
       A. Mass Tensor Dynamics. 431
       B. Force Bias and Smart Monte Carlo 432
   IV. Summary 435
       References. 435

I. INTRODUCTION

In this chapter we focus on methods for simulating systems with strong energy barriers. In such systems barrier crossing is an infrequent event. The barriers divide phase space into regions such that most trajectories starting in one region stay in that region for very long times before making a transition to another region. Two problems arise in connection with such systems. First, they take a very long time to come to thermal equilibrium, so that it is difficult to determine equilibrium properties. Second, since the crossings are infrequent, it is very difficult to determine the rate constants for barrier

---

* Work supported by funds from NSF and NIH.

crossing. In this chapter we focus on a rigorous method for accelerating barrier crossing and thereby determining rate constants for barrier crossing. This method is based on a time-correlation-function description of the barrier crossing. It is shown that, using time-correlation-functions, one can calculate the rate constant by studying trajectories originating at the barrier maximum. This permits one to calculate only initially "activated" trajectories and thus to avoid calculating trajectories that require a long time before being activated. The method springs from the pioneering work of Keck (1962, 1967), Anderson (1973), and Bennett (1977). The original time-correlation-function approach to rate constants (Yamamoto, 1960) was generalized and extended by Chandler (1978). In systems where the time required for energy activation is long compared with the time required for an activated trajectory to cross the barrier, Chandler (1978) showed that the reactive flux rapidly decays to a plateau value and very slowly decays thereafter. In this case the decay constant can be determined from the plateau value of the reactive flux. The simulation need only be carried out for a time sufficiently long compared with the time required to reach the plateau value—a short time compared with the time it would take the system to cross the barrier.

In a set of papers, Montgomery *et al.* (1979, 1980), Rosenberg *et al.* (1980, 1982), DeLeon and Berne (1981, 1982), and Adams and Doll (1982) have used this method to compute rate constants in a wide variety of systems.

To determine the rate constant from the reactive flux, one must also determine the probability of finding the system at the barrier maximum. This, unfortunately, requires the equilibrium structure of a system with large energy barriers. Given the long time required to sample the full phase space in such systems, this would itself require very long simulations. Umbrella sampling techniques (see, e.g., Valleau and Whittington, 1977) allow us to solve this problem.

The general approach advocated in this review depends on our ability to define good dividing surfaces in phase space. Bennett (1977) discusses this problem for general systems. The simple systems described here present no problem with respect to either defining the dividing surface or devising umbrella sampling techniques to determine the configurational distribution functions. Bennett advocates using a dividing region in phase space rather than a single dividing surface. Recently, Connick and Alder (1983) followed Bennett's prescription in studying solvent exchange around an ion.

In this chapter we focus only on the simplest aspects of activated barrier crossing. In complicated systems there may well be other factors leading to inefficient or slow sampling of phase space. In water, for example, protons move much more rapidly than oxygen. The time step for stable numerical integration of the equations of motion is governed by the very fast protonic motions. Thus, to observe a large-scale reorientation of the water molecules or hydration of an ion requires a very large number of small integration steps. These problems exist even when one uses the methods discussed here for studying barrier crossing. The problem of simulating dynamics in systems with wide separations of time scales has yet to be solved. In classical equilibrium systems the configurational distribution function is independent of the particle masses so that one can perform a molecular dynamics study with rescaled masses to determine these functions. Bennett (1975) has devised a mass-tensor, molecular-dynamics method for slowing down the high-frequency vibrations and rotations, thereby allowing a more rapid determination of the configurational distribution functions. The basic idea is to replace Newton's equations for the real system ($M\, dq_i^2/dt^2 = F_i$) by ($M_{ij}\, d^2q_j/dt^2 = F_i$), where M_{ij} is a mass tensor suitably chosen to make all frequencies roughly comparable. Another method is to freeze out high-frequency vibrations by fixing the offending bond lengths using the shake algorithm (Ryckaert *et al.*, 1977), but this can change both the dynamics and equilibrium structure of the system. The shake algorithm provides a method for solving the equations of motion for molecular systems in which prescribed bond lengths and/or bond angles are constrained. Other related and interesting methods for performing more efficient simulations are force-biased Monte Carlo (Pangali *et al.*, 1978, Rao *et al.*, 1979) or smart Monte Carlo (Rossky *et al.*, 1978) techniques for increasing the rate at which configuration space is sampled. In the usual Monte Carlo algorithm (Metropolis) particle displacements are uniformly made and either accepted or rejected. In these biased algorithms the displacements in the direction of the instantaneous force are sampled more frequently than displacements in the opposite direction and then either accepted or rejected.

In Section II we provide a simple account of a method for efficiently studying activated barrier crossing. In Section III we provide a short account of Bennett's mass-tensor dynamics (Bennett, 1975) and force bias (Pangali *et al.*, 1978, Rao *et al.*, 1979) and smart Monte Carlo (Rossky *et al.*, 1978).

II. ACTIVATED BARRIER CROSSING: THEORY AND METHODOLOGY

A simple model for activated barrier crossing consists of a particle moving in the one-dimensional bistable potential of Fig. 1 and experiencing a fluctuating force originating from its interactions with the solvent. The molecule will vibrate in either the A or B well until it receives enough energy ($E_A^\ddagger = V^\dagger - V_A$ or $E_B^\ddagger = V^\dagger - V_B$, respectively) from the bath to allow it to pass over the barrier. If the activation energies ($E_A^\ddagger, E_B^\ddagger$) are large compared with the thermal energy kT of the bath, the particle will cross the barrier infrequently.

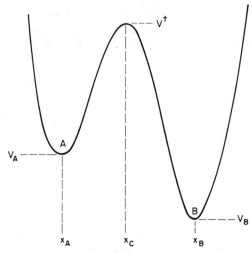

Fig. 1. A schematic potential energy curve for system requiring activated barrier crossing.

Let us consider a solution containing N particles each moving in its own double well and each interacting with the solvent. Let $x_j(t)$ be the configuration of particle j at the time t. The surface $x_j = x_c$ divides the configuration space into two subspaces. For $x_j < x_c$ molecule j is in the reactant state A (see Fig. 1) and for $x_j > x_c$ it is in the product state B. The step function

$$\theta_B(x) \equiv \theta(x - x_c) = \begin{cases} 0, & x < x_c, \\ 1, & x > x_c, \end{cases} \tag{1}$$

defines the instantaneous chemical state of molecule j, and

$$\hat{N}_B(t) = \sum_{j=1}^{N} \theta_B[x_j(t)] \tag{2}$$

gives the number of molecules in state B at time t. The average of $\hat{N}_B$ in an equilibrium ensemble is

$$N_B^0 = \langle \hat{N}_B(t) \rangle = N X_B, \tag{3}$$

where X_B is the equilibrium mole fraction $[X_B = N_B^0/(N_A^0 + N_B^0)]$ and $\langle \cdots \rangle$ denotes an average over $\rho_{eq}(\Gamma)$, the equilibrium ensemble distribution function.

According to the fluctuation–dissipation theorem of linear response theory (see, for example, Forster, 1975), a small deviation from equilibrium of the population in state B decays to equilibrium in precisely the same way as the autocorrelation function of the spontaneous thermal fluctuations of the population; that is,

$$\frac{\bar{N}_B(t) - N_B^0}{\bar{N}_B(0) - N_B^0} = \frac{\langle \delta N_B(t)\, \delta N_B(0)\rangle}{\langle \delta N_B(0)^2\rangle} \equiv C_B(t), \qquad (4)$$

where $\bar{N}_B(t)$ is the ensemble-averaged number of molecules in state B at time t, N_B^0 the equilibrium number, and δN_B the spontaneous fluctuation:

$$\delta N_B(t) = \sum_{j=1}^{N} \{\theta_B[x_j(t)] - \bar{\theta}_B\}. \qquad (5)$$

According to the phenomenology of chemical kinetics, the reaction is described by

$$A \underset{k_b}{\overset{k_f}{\rightleftarrows}} B, \qquad (6)$$

where k_f and k_b are forward and backward rate constants in the equilibrium ensemble. Solving the first-order rate equations for $\bar{N}_B(t)$, subject to the initial condition $\bar{N}_B(0)$, and taking into account that $\bar{N}_A(t) + \bar{N}_B(t) = N$, where N is the total number of molecules in the closed system, one finds that

$$[\bar{N}_B(t) - N_B^0]/[\bar{N}_B(0) - N_B^0] = \exp(-t/\tau_R), \qquad (7)$$

where

$$\tau_R^{-1} = k_f + k_b \qquad (8)$$

is called the kinetic rate constant.

Comparing Eqs. (4) and (7), we see that if phenomenology is to describe the barrier crossing, that is, if a unimolecular rate constants exist, then after some short time transients

$$C_B(t) \to \exp(-t/\tau_R) \qquad (9)$$

decays exponentially. For high-activation barriers $C_B(t)$ decays slowly; that is, molecules spend long periods in one isomeric state before passing over the barrier to another. Thus, τ_R is expected to be a long relaxation time.

Two fundamental questions immediately arise. In a particular case is the above phenomenology valid; that is, do the rate constants k_f and k_b exist? If the rate constants exist, how can we determine them? One possible strategy for answering these question is to evaluate $C_B(t)$ [cf. Eq. (4)] by molecular dynamics. For simplicity, consider a system consisting of only one reacting molecule in a nonreactive solvent bath. Then

$$C_B(t) = \langle \delta\theta_B(t)\, \delta\theta_B(0)\rangle/\langle \delta\theta_B(0)^2\rangle, \qquad (10)$$

where

$$\delta\theta_B(t) = \theta_B[x(t)] - \bar{\theta}_B. \qquad (11)$$

In molecular dynamics one generates a very long trajectory at a sequence of times $0, t_1, \ldots, t_M$ where $t_j = j\,\Delta t$. At each time one determines θ_B for the molecules, and from the sequence of M values of θ_B, one computes

$$\bar{\theta}_B = \frac{1}{M} \sum_{l=1}^{M} \theta_B(t_l) \tag{12}$$

$$C_B(t_n) = \frac{1}{M-n} \sum_{l=1}^{M-n} \{\theta_B[x(t_l)] - \bar{\theta}_B\}\{\theta_B[x(t_{l+n})] - \bar{\theta}_B\}. \tag{13}$$

The time dependence of $C_B(t)$ would provide an answer to both questions. If $C_B(t)$ decays exponentially at long times, then one could determine the rate constant $1/\tau_R$. Unfortunately, the trajectory would so rarely cross the barrier that even in incredibly long computations one could not gather the proper statistics.

An alternative to the direct determination of $C_B(t)$ is the determination of

$$k_B(t) \equiv -dC_B(t)/dt. \tag{14}$$

From

$$C_B(t) = \int d\Gamma\, \rho_{\text{eq}}(\Gamma)\, \delta\theta_B(x)\, e^{iLt}\, \delta\theta_B(x)/\langle \delta\theta_B^2 \rangle, \tag{15}$$

where $\rho_{\text{eq}}(\Gamma)$ is the equilibrium distribution function, Γ a point in phase space, and L the Liouville operator, it is possible to derive an explicit formula for $k_B(t)$.

First we note that $\langle \delta\theta_B^2 \rangle = \langle \theta_B^2 \rangle - \langle \theta_B \rangle^2$. Since $\theta_B^2 = \theta_B$ and $\langle \theta_B \rangle = X_B$, it follows that

$$\langle \delta\theta_B^2 \rangle = \langle \theta_B \rangle - \langle \theta_B \rangle^2 = X_A X_B. \tag{16}$$

Substitution of Eq. (15) into Eq. (16) gives

$$k_B(t) = -\frac{1}{X_A X_B} \int d\Gamma\, \rho_{\text{eq}}(\Gamma)\, \delta\theta_B(x)\, iLe^{iLt}\, \delta\theta_B(x). \tag{17}$$

Since L is a Hermitian operator and

$$iL\theta_B(x) = \dot{x}\, \delta(x - x_C), \tag{18}$$

it follows that

$$k_B(t) = \langle \dot{x}(0)\, \delta[x(0) - x_C] \theta_B[x(t)] \rangle / X_A X_B. \tag{19}$$

The first term $\dot{x}(0)\, \delta[x(0) - x_C]$ gives the initial microscopic flux across the transition state, and $\theta_B[x(t)]$ is zero unless the molecule is in the product state at time t; $k_B(t)$ is thus called the *reactive flux*. This quantity provides us with a rigorous method for studying rare barrier crossings efficiently (Chandler, 1978; Montgomery et al., 1979).

Equation (19) can be expressed as

$$k_B(t) = \frac{1}{QX_AX_B} \int d\Gamma \, e^{-\beta H(\Gamma)} \dot{x} \, \delta(x - x_C) \, \theta_B[x(t)], \qquad (20)$$

where Q is the canonical partition function, $Q = \int d\Gamma \, e^{-\beta H(\Gamma)}$. Only those trajectories contribute nonzero values to Eq. (20) for which the reacting particle is initially at the transition state x_C (because of the delta function) and which at time t are found in the B well.

The reactive flux has several interesting properties (Chandler, 1978). First we note that $k_B(t)$ is discontinuous at $t = 0$.

$$\lim_{t \to 0+} k_B(t) = k_{\text{TST}}, \quad \lim_{t \to 0-} k_B(t) = -k_{\text{TST}}, \qquad (21a)$$

where

$$k_{\text{TST}} = \langle \dot{x} \, \delta(x - x_C) \, \theta(\dot{x}) \rangle / X_A X_B \qquad (21b)$$

and

$$\theta(\dot{x}) = \begin{cases} 1, & \dot{x} > 0, \\ 0, & \dot{x} < 0. \end{cases} \qquad (22)$$

In the classical canonical ensemble,

$$k_{\text{TST}} = \langle \dot{x} \theta(\dot{x}) \rangle \langle \delta(x - x_C) \rangle / X_A X_B, \qquad (23)$$

where we have separated the momentum and configurational averages. Furthermore, since the configurational distribution function of the reacting particle is

$$S(x_1) = \langle \delta(x - x_1) \rangle, \qquad (24)$$

it follows that

$$k_{\text{TST}} = (2X_A X_B)^{-1} \langle |\dot{x}| \rangle S(x_C). \qquad (25)$$

The first factor is the mean speed of the reaction coordinate, and the second factor the probability distribution of finding the molecule at the top of the bistable barrier. We shall return to a discussion of how $S(x_C)$ is determined. Suffice it to say here that this is a nontrivial problem.

As we have seen, the phenomenology of chemical kinetics predicts Eq. (9). Suppose we push the phenomenology to the extreme and assert that $C_B(t)$ is a single exponential for all time; that is $C_B(t) = \exp(-|t|/\tau_R)$. Then the reactive flux given by Eq. (14) will also decay as a single exponential for $t > 0$:

$$k_B(t) = \tau_R^{-1} \exp(-t/\tau_R). \qquad (26)$$

We then obtain an approximation for τ_R^{-1}:

$$\tau_R^{-1} = \lim_{t \to 0_+} k_B(t) = k_{TST} \equiv \tau_{TST}^{-1}. \tag{27}$$

The kinetic relaxation time is then approximated by

$$\tau_{TST}^{-1} \equiv k_{TST} = (\langle |\dot{x}| \rangle / 2X_A X_B) S(x_C). \tag{28}$$

This is the famous transition state formula (Pechukas, 1976) for the kinetic rate constant. It is equivalent to the assertion that all trajectories originating at the transition state and moving toward B will get trapped in the B well for random times before recrossing the transition state. In this approximation there are no rapid recrossings of the potential. In the real system there will be rapid recrossings, which give rise to nonexponential short time decay in $k_B(t)$. Often, τ_{TST}^{-1} is a very good approximation to the real rate constant. In the following we present an algorithm for determining the time dependence of $k_B(t)$. The initial value of $k_B(t)$ can be used to determine the transition state rate constants. This provides a reference point for an understanding of the actual rate constant. By determining the time dependence of $k_B(t)$, one can observe to what degree transition state theory is valid.

Returning now to a method for determining the reactive flux (Montgomery et al., 1979; Rosenberg et al., 1980), insertion of $\theta(\dot{x}) + \theta(-\dot{x}) = 1$ into Eq. (20) allows the resulting equation to be expressed as

$$k_B(t) = \left\{ \Delta^{(+)} \int d\Gamma \, P^{(+)}(\Gamma) \theta_B[x(t)] + \Delta^{(-)} \int d\Gamma \, P^{(-)}(\Gamma) \theta_B[x(t)] \right\}, \tag{29}$$

where

$$\Delta^{(\pm)} \equiv \frac{1}{X_A X_B} \frac{\int d\Gamma \, \dot{x} \theta(\pm \dot{x}) \delta(x - x_C) e^{-\beta H(\Gamma)}}{\int d\Gamma \, e^{-\beta H(\Gamma)}} \tag{30}$$

and

$$P^{(\pm)}(\Gamma) = \frac{\dot{x} \theta(\pm \dot{x}) \delta(x - x_C) e^{-\beta H(\Gamma)}}{\int d\Gamma \, \dot{x} \theta(\pm \dot{x}) \delta(x - x_C) e^{-\beta H(\Gamma)}} \tag{31}$$

are normalized distribution functions.

Since the Hamiltonian is even in the momenta, it follows from the simple transformation $\{\vec{p}\} \to \{-\vec{p}\}$ that

$$\Delta^{(-)} = -\Delta^{(+)}. \tag{32}$$

Furthermore, comparison of $\Delta^{(+)}$ in Eq. (30) with Eq. (21b) shows that

$$\Delta^{(+)} = k_{TST}. \tag{33}$$

Substitution of Eqs. (32) and (33) into Eq. (29) then leads to the result

$$\hat{k}(t) \equiv \frac{k_B(t)}{k_{\text{TST}}} = \int d\,\Gamma[P^{(+)}(\Gamma) - P^{(-)}(\Gamma)]\,\theta[x(t) - x_c]. \tag{34}$$

To compute $\hat{k}(t)$, one samples initial states from the distribution functions $P^{(\pm)}(\Gamma)$; from the definition $P^{(\pm)}$ it is clear that the initial configurations are sampled from $\delta(x - x_c)\,e^{-\beta V(\mathbf{R})}$ and the initial velocities from $\dot{x}\theta(\pm\dot{x})e^{-\beta T(P)}$. Since $T(P)$ is an even function of $\tilde{p}$, one can sample $\dot{x}$ from $\dot{x}\theta(\dot{x})e^{-\beta T(P)}$ giving values $\{\dot{x} > 0\}$ and then simply reverse all momenta instead of resampling from $\dot{x}\theta(-\dot{x})e^{-\beta T(P)}$. Thus, for every trajectory starting from x_c and moving forward toward well B there is also a trajectory moving backward from x_C into well A. For each such pair of trajectories one computes the difference in $\theta[x(t) - x_c]$. Averaging this difference uniformly over all such sampled initial states then gives $\hat{k}(t)$.

As already mentioned, $\hat{k}(t)$ decays from an initial value $\hat{k}(t \to 0_+) = 1$. In a purely one-dimensional double well the barrier-crossing dynamics are periodic with a period dependent on the energy. Therefore, $\hat{k}(t)$ will consist of a superposition of periodic functions with different periods. Although such a function may decay to zero, it will not exhibit the long-time exponential decay required for the existence of a rate constant. In a complex but isolated molecule the reaction coordinate will be coupled to other intramolecular modes, and the decay of $\hat{k}(t)$ may be exponential at long times (Berne et al., 1982; DeLeon and Berne, 1981, 1982). If the energy rapidly equilibrates, the rate constant is expected to be given by TST. In liquids we expect that the rate constant will exist, but because of rapid collisions, it is not expected that a simple transition state theory will apply. Simulations of complicated systems using Eq. (34) allow one to determine $\hat{k}(t)$. It is expected that $\hat{k}(t)$ will rapidly decay to a plateau value, which gives $(k_{\text{TST}}\tau_{\text{Rxn}})^{-1}$.

To determine the absolute reaction rate $(\tau_{\text{Rxn}})^{-1}$, it is necessary to evaluate k_{TST}. From Eq. (28), we see that the equilibrium conformational distribution function $S(x)$ must be determined for $x = x_C$. This is not an easy task since at x_C the potential may be very high. Ordinary Monte Carlo or molecular dynamics simulations will exhibit very few configurations with the reacting molecule at the top of the barrier (these are very improbable). It is thus necessary to develop a method for studying improbable configurations in an equilibrium system. In the following we present one possible method based on umbrella sampling (Valleau and Whittington, 1977; Rebertus et al., 1979; Rosenberg et al., 1982).

Let us note that by definition

$$S(x) = \langle \delta(x - x_1) \rangle, \tag{35}$$

where x_1 is the reaction coordinate of a molecule and $\langle \ \rangle_T$ denotes a canonical ensemble average. Thus,

$$S(x) = \int dq\, e^{-\beta V(q)}\, \delta(x - x_1) \bigg/ \int dq\, e^{-\beta V(q)}, \tag{36}$$

where q stands for all coordinates in the system and x_1 is the coordinate describing the reactions. The total potential $V(q) = V(x, R)$ is a function of the reaction coordinate x and all other coordinates of the system. If R is held fixed, the x dependence of $V(x, R)$ should exhibit a barrier for each possible value of R. For purposes of illustration let us assume that we can write

$$V(x, R) = U_0(x) + \Delta V(x, R), \tag{37}$$

where $U_0(x)$ exhibits a strong barrier, whereas $\Delta V(x, R)$ has weak dependence on x. There are a variety of ways of finding such a subdivision. Substituting Eq. (37) into Eq. (36), applying the properties of the delta function, and recognizing that $S(x)$ is normalized allow one to write

$$S(x) = c \exp(-\beta U_0)\, s^{(0)}(x), \tag{38}$$

where

$$S^{(0)}(x) = \int dq\, e^{-\beta \Delta V(x_1, R)}\, \delta(x - x_1) \bigg/ \int dq\, e^{-\beta \Delta V(x, R)} \tag{39}$$

is the distribution function in a system in which $u_0(x)$ is turned off (the cavity distribution). In this latter system, by construction there are no strong barriers so that an ordinary Monte Carlo simulation with $\Delta V(x_1, R)$ instead of with $V(x_1, R)$ will sample all regions of x_1 more favorably. The procedure is then obvious; evaluate $S^{(0)}(x)$ in the model system and multiply the result by $\exp{-\beta U_0(x)}$ normalization then gives $S(x)$, and substituting $S(x_c)$ into Eq. (28) yields k_{TST}.

The procedure outlined above has proved very useful in the computer simulation of systems with activated barrier crossing. It suffices to give a brief description of some of the recent publications using this method:

Montgomery et al. (1979) have studied model systems in which a particle moves in a double-well potential (see Fig. 1) and suffers random instantaneous impulsive collisions. After each collision the particle is given a new velocity sampled from the Maxwell distribution function at a temperature T. The reactive flux is evaluated by initially placing the particle at the barrier maximum, and then by sampling initial velocities from the Maxwell distribution function. For each sampled initial state $(x(0) = x_c, \dot{x})$ a set of collision times is sampled from the Poisson distribution $(e^{-\alpha t})$. Numerical solution of the Newtonian equations for the particle moving in the potential then gives $[x(t_1), \dot{x}(t_1)]$ at the time of its first collision. At this instant a new velocity is sampled from the Maxwell distribution, and the trajectory is continued until the time of the next collision, and so on. The reactive flux given by Eq. (34) is computed on the basis of an ensemble of such trajectories, and the rate constant is determined. This method provides a numerical solution to the master equation

$$\left[\frac{\partial}{\partial t} + \dot{x}\frac{\partial}{\partial x} - \frac{1}{m}\frac{\partial V(x)}{\partial x}\frac{\partial}{\partial \dot{x}}\right]\rho(x,\dot{x},t)$$

$$= \int_{-\infty}^{+\infty} dx' \int_{-\infty}^{+\infty} d\dot{x}'[W(x,\dot{x}|x',\dot{x}')\rho(x',\dot{x}',t) - W(x',\dot{x}'|x,\dot{x})\rho(x,\dot{x},t)], \quad (40)$$

where $V(x)$ is the double-well potential, $\rho(x,\dot{x},t)$ the probability distribution function of the position and velocity at time t, and $W(x,\dot{x}|x',\dot{x}') \, dx' \, d\dot{x}'$ the transition rate out of states in the neighborhood of $dx' \, d\dot{x}'$ of the state $(x',\dot{x}')$ and into the state $(x,\dot{x})$. If we take

$$W(x,\dot{x}|x',\dot{x}') = \alpha\varphi_M(\dot{x})\,\delta(x-x'),$$

where

$$\varphi_M(\dot{x}) = \left(\frac{m}{2\pi kT}\right)^{1/2} \exp(-m\dot{x}^2/2kT), \quad (41)$$

then the kinetic equation is completely equivalent to the dynamic model specified above. This is the well-known BGK kinetic equation (Bohm, and Gross, 1949; Bhatnagar et al., 1954; Skinner and Wolynes, 1980). Thus it is possible to define dynamic models that provide a simulation technique for solving master equations (Berne, 1979; Berne et al., 1980). Other choices of the transition rate have been discussed in the literature. A detailed discussion of the equivalence between the master equation and the trajectory model is given by Montgomery et al. (1979). Moreover, this paper provides a detailed description of the methodology and results for the reactive flux.

Rosenberg et al. (1980) have evaluated the reactive flux in a full molecular dynamics simulation of a "butane" molecule dissolved in a Lennard-Jones liquid. In this case the reaction coordinate is the dihedral angle φ, defining the conformational state of butane. This coordinate moves in a tristable potential with the three wells corresponding, respectively, to the gauche($-$), trans, and gauche($+$) states. In the simulation φ is fixed at the barrier maximum separating the gauche($-$) and trans wells. A molecular dynamics simulation (with fixed φ) is performed using the shake algorithm (Ryckaert et al., 1977). The trajectory is aged at a given temperature; that is, it is run for a long time with velocity rescaling to ensure that the states generated are typical of an equilibrium system. Widely separated states on this trajectory are taken as initial states for the calculation of the reactive flux. For each of these initial states φ is "unfrozen", $\dot{\varphi}$ is sampled from its equilibrium distribution, and a trajectory is run. Equation (34) is then evaluated, thereby giving the reactive flux. Rebertus *et al.* (1979) have calculated $s(\varphi_c)$ using the umbrella sampling technique. This allows for the determination of the kinetic rate constant and the transition state rate constant.

DeLeon and Berne (1981) have used the reactive flux method to determine rate constants in isolated systems consisting of two degrees of freedom. This activated barrier crossing was discussed in connection with the transition to chaos in Hamiltonian flows. DeLeon and Berne (1982) also studied barrier crossing in very idealized systems consisting of stadium billiards. Berne *et al.* (1982) have applied reactive flux ideas to the determination of the gauche to trans isomerization in isolated butane. Montgomery *et al.* (1980) have studied a BGK model of conformational dynamics and barrier crossing in longer chained hydrocarbons.

In a closely related study Adams and Doll (1982) have applied the reactive flux formalism to the study of thermal desorption from solid surfaces. Grummelmann *et al.* (1981) have studied a Langevin model for the thermal desorption of xenon from a platinum surface. Finally McGammon and Karplus (1980) have applied these methods to studies of protein dynamics.

The reactive flux formalism provides a method for determining infrequent barrier-crossing events. It provides a well-defined method for sampling initial states so that all trajectories are activated. Thus, one does not have to wait for infrequent events to take place. Other methods have been used to study activated barrier crossings, but these do not place the system initially in an activated state (see, for example, Helfand *et al.*, 1979, 1980).

III. SOME METHODS FOR ACCELERATING SIMULATIONS

Systems in which there are high and low frequencies are difficult to simulate. The time steps used in molecular dynamics must be sufficiently small to guarantee the stability of the integration algorithm. To follow the low-frequency motion, one must run trajectories over a very large number of these small time steps. In this section we summarize several different approaches that accelerate convergence in systems with widely separated frequencies. The first method discussed involves replacing the equations of motion by fictitious set equations of motion in which the correct masses of the particles are replaced by mass tensors. The resulting trajectories can be used to determine the equilibrium properties of the system but are useless for the study of the time dependence, that is, for the determination of dynamic properties. The second and third method avoids the problem of multiple time scales by using efficient Monte Carlo methods based on biasing

A. Mass-Tensor Dynamics

Bennett (1975) has devised an interesting strategy for accelerating convergence in such systems. This method is based on the fact that in the classical canonical ensemble the positions and momenta are statistically independent. Thus, if the true Hamiltonian for the system of f degrees of freedom,

$$H = \frac{1}{2}\sum_{i=1}^{f} m_i \dot{q}_i^2 + U(q_1 \cdots q_f), \tag{42}$$

is replaced by the Hamiltonian

$$H' = \frac{1}{2}\sum_{i,j=1}^{N} \dot{q}_i M_{ij} \dot{q}_j + U(q_1 \cdots q_f), \tag{43}$$

the configurational distribution functions will be unchanged. Here M_{ij} is a symmetric mass tensor. The Hamiltonian H' can be written more compactly as

$$H' = \tfrac{1}{2}\dot{\mathbf{q}}^T \cdot \mathbf{M} \cdot \dot{\mathbf{q}} + U(q), \tag{44}$$

where $\mathbf{q}, \dot{\mathbf{q}}$ are column vectors. Newton's equations are then given by

$$\ddot{\mathbf{q}} = -\mathbf{M}^{-1} \cdot \nabla U(\mathbf{q}). \tag{45}$$

By a judicious choice of **M**, it is possible to slow down the original system's high-frequency motion, so that in the same number of iterations the new system will evolve further in configuration space than does the original system. Bennett (1975) provides a heuristic method for finding a reasonable **M** for this purpose. Recognizing that the local oscillation frequencies can be obtained from the instantaneous matrix of force constants, we obtain

$$A_{ij}(\mathbf{q}) = \partial^2 U(\mathbf{q})/\partial q_i \, \partial q_j. \tag{46}$$

At low temperatures the system is nearly harmonic, and **A** will be approximately constant throughout the accessible region of phase space. Then a first approximation of **M** is

$$\mathbf{M} = \text{const } \mathbf{A}. \tag{47}$$

If this assumption is not valid, Bennett suggests several other possible choices for **M**. When this method is applied to a flexible chain with nearest neighbor harmonic forces and Lennard-Jones forces between nonnearest neighbors, the acceleration of convergence is between a factor of 5 and 10. It is our belief that this will be very useful in simulations of quantum systems.

B. Force Bias and Smart Monte Carlo

The Monte Carlo procedure of Metropolis *et al.* (1953) is widely used to determine the equilibrium structural and thermodynamic properties of gases, liquids, solids, and mesophases. Several years ago Pangali *et al.* (1978) introduced a modification of the usual Metropolis procedure that gives more rapid convergence and thereby more efficient Monte Carlo runs. In this new procedure each particle displacement is chosen with greater probability in the direction of the instantaneous force on the particle than in other directions. The particle displacements, therefore, usually lead to a lowering of the overall potential energy and thereby to a higher acceptance probability than in the usual Metropolis procedure. The new procedure, appropriately called the force bias method, was then applied to a study of ST-2 water (Rao *et al.*, 1979) leading to a fourfold increase in efficiency.

It was clear at the outset that there were no objective criteria for comparing the efficacy of two different Monte Carlo procedures or for optimizing any given procedure. Thus, in an ensuing publication (Rao *et al.*, 1979) we argued that a good measure is the diffusion in configuration space. It is clear that if the Monte Carlo procedure does not involve the exchange of particles, then the step sizes, etc., for the procedure should be chosen in such a way as to optimize the diffusion coefficient.

Moreover, the best procedure is that which gives the largest diffusion coefficients for optimized step sizes. This will guarantee that more independent configurations are sampled in a given number of particle advances. Rao *et al.* (1979) made an exhaustive study of diffusion in the Metropolis scheme and in the force bias scheme and showed that for ST-2 water the force bias scheme gave a substantial improvement in convergence over the Monte Carlo scheme.

An equivalent Monte Carlo procedure has been suggested (Rossky *et al.*, 1978). This method is based on Brownian dynamics and is referred to by its inventors as the smart Monte Carlo (SMC) method. This method also generates displacements biased in the direction of the force and is in many ways similar to the force bias (FB) procedure.

In both the FB and Brownian dynamic SMC methods, a new configuration of the N-particle system $\mathbf{R}' = (R'_1, \ldots, R'_N)$ is generated from an old configuration $\mathbf{R} = (R_1, \ldots, R_N)$ by sampling R' from a transition probability $T(\mathbf{R}'|\mathbf{R})$. The transition probability is chosen such that it is normalized and such that if a state **R′** is accessible from **R**, then the state **R** is accessible from **R′**. The new state **R′** is then accepted with probability

$$p = \min[1, q(\mathbf{R}'|\mathbf{R})], \tag{48}$$

or the old state **R** is kept with probability $(1 - p)$, where

$$q(\mathbf{R}'|\mathbf{R}) \equiv T(\mathbf{R}|\mathbf{R}')P(\mathbf{R}')/T(\mathbf{R}'|\mathbf{R})P(\mathbf{R}) \tag{49}$$

and

$$P(\mathbf{R}) = Z^{-1} e^{-\beta V(\mathbf{R})} \tag{50}$$

is the Boltzmann distribution. The sequence of configurations so generated will then be distributed according to the Boltzmann distribution, Eq. (50).

The choice of the transition probability is what distinguishes the various methods. In all of the methods discussed here, only one particle is moved at a time, so that **R′** differs from **R** with respect to the position of a particular particle.

In the FB method the transition probability is taken as (Pangali *et al.*, 1978; Rao *et al.*, 1979; Rao and Berne, 1979).

$$T_{\mathrm{FB}}(\mathbf{R}'|\mathbf{R}) = \begin{cases} C(\mathbf{F}(\mathbf{R}), \Delta) \exp[\lambda \beta \mathbf{F}(\mathbf{R}) \cdot (\mathbf{r}' - \mathbf{r})], & \mathbf{r}' - \mathbf{r} \in D, \\ 0, & \mathbf{r}' - \mathbf{r} \notin D, \end{cases} \tag{51}$$

where $\mathbf{r}' - \mathbf{r}$ stands for the displacement of the particle being moved, $\mathbf{F}(\mathbf{R})$ the force acting on the particle before it is moved (when the whole system has the old configuration **R**), λ a parameter to be discussed below, $\beta = (kBT)^{-1}$, and $C(\mathbf{F}(\mathbf{R}), \Delta)$ a normalization constant. From Eq. (51) we note that the transition probability is zero if the particle displacement $\Delta \mathbf{r} = \mathbf{r}' - \mathbf{r}$ falls outside a certain domain D defined such that $-\Delta/2 \leq \Delta x$, $\Delta y, \Delta z \leq \Delta/2$. This means that the maximum step size allowed is $\frac{3}{2}\Delta$. Clearly the normalization constant C depends both on the instantaneous force on the particle and on the value of Δ. As we have shown elsewhere, for each system there is an optimum choice for Δ.

It is clear that Eq. (51) gives rise to a biased sampling of Δr in the direction of the force F. If $\lambda = 0$, the transition probability gives uniform sampling in the domain D and reduces to the usual Metropolis scheme.

In the Brownian dynamic SMC scheme (Rossky *et al.*, 1978)

$$T_{\mathrm{SMC}}(\mathbf{R}'|\mathbf{R}) = \frac{1}{(4\pi A)^{3/2}} \exp\left\{ -\frac{[(\mathbf{r}' - \mathbf{r}) - \beta A \mathbf{F}(\mathbf{R})]^2}{4A} \right\}, \tag{52}$$

where A is a parameter that must be chosen. Here, too, the sampled particle displacement is biased in the direction of the instantaneous force.

Explicit evaluation of the exponent allows us to write Eq. (52) in the form

$$T_{\text{SMC}}(\mathbf{R}'|\mathbf{R}) = \frac{B(F, A)}{(4\pi A)^{3/2}} \exp\left(-\frac{(\mathbf{r}' - \mathbf{r})^2}{4A}\right) \cdot \exp\left[+\frac{1}{2}\beta F(\mathbf{R}) \cdot (\mathbf{r}' - \mathbf{r})\right], \tag{53}$$

where $B = \exp[-\beta^2 A F^2(\mathbf{R})/4]$. Thus, in reality the T_{SMC} is very similar to T_{FB} with one major and one minor difference. The major difference is that in the FB method there is an upper bound on the displacement, whereas in the SMC there is no upper bound. For example, if in a certain initial configuration $F(R) = 0$, the T_{FB} would sample Δr uniformly, whereas T_{SMC} would sample Δr from a normalized Gaussian whose width is specified by the parameter A. From the properties of the Gaussian we note that in this case the mean square displacement on a move would be $\langle \Delta r^2 \rangle = 6A$. Clearly, $\sqrt{A}$ then gives a measure of the kind of displacement generated in this method and is thus analogous to the parameter Δ in the FB method. In applying the SMC method one should choose a value of A that optimizes the method.

The minor difference is the parameter λ. If this parameter λ in Eq. (51) is adjusted to be $\lambda = \frac{1}{2}$, then the force bias part of the sampling is the same for the two methods. Although the FB allows freedom in the choice of λ it gives the best results and is equivalent to the SMC for λ fixed at $\frac{1}{2}$. One advantage FB has over SMC is that it is readily generalized to rigid molecules that can both rotate and translate. In this case one biases the displacements with respect to both forces and torques. Pangali *et al.* (1978) have applied this to simulation of liquid H_2O and find a fourfold acceleration in convergence over uniform sampling techniques. This method has been adopted by Beveridge and co-workers for studies of aqueous solutions. These workers find that, when FB is combined with preferential sampling, there is a very considerable acceleration of convergence. In preferential sampling, solvent molecules close to the solute molecule are moved more frequently than those far away. Unfortunately, it is more difficult to generalize SMC to systems conforming to rigid rotors.

In unpublished work, M. Lee and B. J. Berne (unpublished work, 1980) have generalized FB Monte Carlo to include the second derivatives of the potential. Then

$$T(R'|R) = \begin{cases} \exp[\lambda\beta F(\mathbf{R}) \cdot (\mathbf{R}' - \mathbf{R}) - \alpha(\mathbf{R} - \mathbf{R}') \cdot \mathbf{A}(R) \cdot (\mathbf{R}' - \mathbf{R})], & \mathbf{R} - \mathbf{R}' \in D, \\ 0, & \mathbf{R} - \mathbf{R}' \notin D \end{cases} \tag{54}$$

where

$$\mathbf{A}(\mathbf{R}) = (\partial^2 V(\mathbf{R})/\partial \mathbf{R}\, \partial \mathbf{R})_\mathbf{R}. \tag{55}$$

13. Monte Carlo Simulations of Rare Events

This modification has been tested in one-dimensional systems but has yet to be tested in many-body systems. For highly anharmonic potentials like the Morse oscillator, this method gives much more rapid convergence. In the multidimensional case, if one diagonalizes **A** and expresses Eq. (54) in terms of normal modes, it can be seen that modes with small force constants will suffer larger accepted displacements than modes with larger force constants. This should lead to larger displacements in configuration space. It has yet to be determined whether this sampling function for single-particle moves using local coordinates will lead to more rapid convergence. Given the high overhead in the computation of forces and second derivatives, it is not yet clear whether one gains by using this latter approach.

IV. SUMMARY

In this chapter we have discussed two distinct problems. The first concerns systems in which energy barriers form bottlenecks in phase space. Barrier crossings are sufficiently infrequent that one must invent new methods for the computer simulation of such systems. The reactive-flux methodology (Section II) allows us to simulate such systems and to study thereby chemical reaction dynamics in condensed systems.

The second problem involves systems without strong barriers but, nevertheless, with multiple time scales. The major goal is to develop methods for efficiently simulating the dynamics of such systems. This goal has not yet been met. In Section III several methods have been discussed for determining the equilibrium properties of such systems. One method involves introducing anisotropic masses. This is Bennett's mass-tensor dynamics. The other methods are based on a biased sampling of displacements in Monte Carlo simulations. These are the force bias and smart Monte Carlo techniques of Section III.B. These methods have already proved useful in the study of aqueous systems.

REFERENCES

Adams, J. E., and Doll, J. D. (1982). *J. Chem. Phys.* **77**, 2964.
Anderson, J. B. (1973). *J. Chem. Phys.* **58**, 4684.
Bennett, C. H. (1975). *J. Comput. Phys.* **19**, 297.
Bennett, C. H. (1977). *In* "Algorithms for Chemical Computations" (R. E. Christofferson, ed.), p. 63. Am. Chem. Soc., Washington, D.C.
Berne, B. J. (1979). NRCC Workshop on Stochastic Molecular Dynamics, Woods Hole, Massachusetts, 1979, p. 4.
Berne, B. J., Skinner, J. L., and Wolynes, P. G. (1980). *J. Chem. Phys.* **74**, 5300.
Berne, B. J., DeLeon, N., and Rosenberg, R. O. (1982). *J. Phys. Chem.* **86**, 2166.

Bhatnagar, P. L., Gross, E. P., and Krook, M. (1954). *Phys. Rev.* **94**, 511.
Bohm, D., and Gross, E. P. (1949). *Phys. Rev.* **75**, 1864.
Chandler, D. (1978). *J. Chem. Phys.* **68**, 2959.
Connick, R. E., and Alder, B. J. (1983). *J. Phys. Chem.* 87, 2764.
DeLeon, N., and Berne, B. J. (1981). *J. Chem. Phys.* **75**, 3495.
DeLeon, N., and Berne, B. J. (1982). *Chem. Phys. Lett.* **93**, 162, 169.
Forster, D. (1975). "Hydrodynamic Fluctuations, Broken Symmetries, and Correlation Functions." Benjamin, Reading, Massachusetts.
Grummelmann, E. K., Tully, J. C., and Helfand, E., (1981). *J. Chem. Phys.* **74**, 5300.
Helfand, E., Wasserman, Z. R., and Weber, T. A. (1979). *Physica (Amsterdam)* **70**, 2016.
Helfand, E., Wasserman, Z. R., and Weber, T. A. (1980). *Macromolecules* **13**, 526.
Keck, J. C. (1962). *Discuss. Faraday Soc.* **33**, 173.
Keck, J. C. (1967). *Adv. Chem. Phys.* **13**, 85.
McGammon, J. A., and Karplus M. (1980). *Annu. Rev. Phys. Chem.* **31**, 29.
Metropolis, N., Metropolis, A. W., Rosenbluth, M. N., Teller, A. H., and Teller, E. (1953). *J. Chem. Phys.* **21**, 1087.
Montgomery, J. A., Chandler D., and Berne, B. J. (1979). *J. Chem. Phys.* **70**, 4056.
Montgomery, J. A., Holmgren, L., and Chandler, D. (1980). *J. Chem. Phys.* **73**, 3688.
Pangali, C., Rao, M., and Berne, B. J. (1978) *Chem. Phys. Lett.* **55**, 413.
Pechukas, P. (1976). *In* "Dynamics of Molecular Collisions" (W. H. Miller, ed.), Part B, p. 269. Plenum, New York.
Rao, M., and Berne, B. J. (1979). *J. Chem. Phys.* **71**, 129.
Rao, M., Pangali, C., and Berne, B. J. (1979). *Mol. Phys.* **37**, 1773.
Rebertus, D., Berne, B. J., and Chandler, D. (1979). *J. Chem. Phys.* **70**, 3395.
Rosenberg, R. O., Berne, B. J., and Chandler, D. (1980). *Chem. Phys. Lett.* **75**, 162.
Rosenberg, R. O., Rao, M., and Berne, B. J. (1982). *J. Am. Chem. Soc.* **104**, 7647.
Rossky, P. J., Doll, J. D., and Friedman, H. L. (1978). *J. Chem. Phys.* **69**, 4628.
Ryckaert, R. Y., Ciccotti, G. and Berendsen, H. J. C. (1977). *J. Comput. Phys.* **23**, 327.
Skinner, J. L., and Wolynes, P. G. (1980). *J. Chem. Phys.* **72**, 4913.
Valleau, J. P., and Whittington, S. G. (1977). *In* "Statistical Mechanics" (B. J. Berne, ed.), Part A, p. 137. Plenum, New York.
Yamamoto, T. (1960). *J. Chem. Phys.* **33**, 281.

Index

A

Activated barrier crossing, 421, 428
Adaptive grid, 15, 127 ff., 148, 153, 157
  storage, 135
  tensor product, 138
Advection, 164
Advective remapping, 260
Ahmad–Cohen difference scheme, 381, 385–388
  boot-strapping procedure, 383, 392
  irregular component, 385–388, 392, 406
  regular component, 385–388
  semi-iteration, 382
  standard algorithm, 386
Aliasing, 243
Ampere's law, 274
Anisotropic wave motion, 192
Asymptotic expansion, 31, 139
Asymptotic solution, implicit methods, 13, 296–299
Atmospheric motions, 49

B

Backward differentiation formula, 38, 124
Balance equations, nonlinear initialization, 60–61
Barrier crossing, 419
Baye's theorem, 61
BDF, *see* Backward differentiation formula
BGK kinetic equation, *see* Master equation
Binary stars
  hard, 402
  soft, 403

Binding energy, 399, 404
Boltzmann transport equation, 74
Boundary layers, 7, 9, 30
Bounded derivative method, 14, 30
Braginskii equations, 236, 253

C

Cell Reynolds number, 18
Charge dilation method, 234
Chemical kinetics
  phenomenology, 425
  rate constant, 423
Cherenkov numerical instability, 314
Christoffel symbols, 196
Classical tidal theory, 60
CO_2 lasers, 234
Collisionless plasma, 273
  shocks, 305–308
  transport, 188
Collisions, elastic 346
Comoving coordinates, 389
  in Ahmad–Cohen scheme, 391
  mergers and inelastic effects, 393
  peculiar motions, 392
Compressible flow, 203
Convection–diffusion equation, 3, 7–9
Convergence acceleration, 430
  mass tensor, 430
Coordinate transformation method, 128–130, 192
Coriolis force, 63, 66
Courant condition, electron, 235, 245
Current-driven instability, 305

D

DAE, *see* differential algebraic equation
Darwin model, 272, 314, 367
Debye length, 277, 282
Defect correction iteration, 139–140
Detonation front, 2
Diamond differencing, 95
Differential-algebraic equation, 123–127
Differential constraint equation, 210
Diffusion, artificial mass and heat, 162
Diffusion equation
  discretized, 94
  frequency-dependent, 80
  multi-group, 103
Diffusion-synthetic acceleration
  method, 77–82
  numerical considerations, 83
Direct implicit method
  comparison with moment method, 348
  electromagnetic, 366–373
  electrostatic, 341
  extrapolated orbits, 341
  multidimensional, 344–346
  one-dimensional, 342–343
  outline, 341–342
  simplified and strict spatial differencing,
    248, 343, 344, 360
  in particle simulation of plasmas, 248, 313
Direct implicit plasma simulation, *see* Direct
  implicit method
Discrete Fourier transform, 19
Dispersion, numerical plasma, 275–278, 340
Dissipation, numerical, 319
Drift-kinetic formulation, 315
DSA, *see* Diffusion-synthetic acceleration
  method
Dynamic grid method, 194, 204

E

Eddington factor, variable, 151
Electron Debye length, 349
Electrostatic plasma, 337–338
Elliptic equation, variable coefficient, 347–348
Energy conservation
  direct implicit method, 357
  implicit moment method, 299–300
Energy stabilization scheme, 395
Ensemble average, 286

Equidistribution, 15, 132
  of arc length, 127
Euclidean norm, 32
Euler differencing, *see also* Backward
  differentiation formula
  backward, 74
  explicit, 119
Euler method, 36
  backward, 36, 120, 275–278, 339
  implicit, 31
Explicit time differencing
  centered leapfrog, 339
  stability, 339
Explosion, 117
Extrapolation, unperturbed orbits, 369–370
Extrapolation methods, 20

F

Faraday's law, 274
FB, *see* Monte Carlo, force bias
Field reversing ion rings, 328
Filtering, 31
  asymmetric temporal, 148, 157
  digital, 321
Finite difference method
  explicit, 203
  implicit, 203
Finite-grid instability, 235, 277
Finite-sized particles, 272
  and collisions, 280, 282
  and interpolation, 282
Fixup, 96
Fluctuation–dissipation theorem,
  autocorrelation function, 422
Flux-corrected transport, 260
Fokker–Planck method, 415
Free expansion, of plasma, 348–349
Functional iteration, 120

G

GAPS, *see* Gyroaveraged motion
Gaussian elimination, 142
Gauss's law, 274, 342
Geostrophic balance, 49
Global iteration
  method of Concus–Golub, 294–295, 347
  variable-coefficient elliptic equations, 347
Gravity wave, 60, 61–62, 63

Index

Gravitational N-body problem, 378–417
  neighbor scheme, 378
  relative coordinates, 382
  standard integration of, 378
Green's function method, 201
  infinite medium toroidal function, 201
Guiding center equations, 273
Gyro-averaged motion, 10, 361–364
Gyro-kinetic formulation, 315

H

Hilbert program, 148, 179–183
Hot electrons, 249
  drag, 250
  emission, 250
  scatter, 250
  thermal conduction, 254, 262
Hough function, 64
Hybrid finite-element method, 22
Hybrid plasma simulation, 233
  implicit, 234, 250
Hydrostatic assumption, 53
Hyperbolic equations, 189

I

Implicit differencing in time, 273, 275 ff. *see also* Euler method
  plasma dispersion, 276
Implicit moment method, 234, 274–280, 313
  collisional velocity coupling, 257
  comparison with direct method, 248, 289–291, 348
  convergence, 247
  emission, drag, and thermal coupling, 256
  implicit pressure, 244, 289
  Lagrangian advancement, 259
  particle-field iterations, 246
  stability analysis, 241, 296–300
  time-step control, 255, 280
Incompressible flow, 203
Individual time-step algorithm, 383
Infrequent event, 419
Initial data, homogeneous, 32
Initialization, 30
Initial-value problem, 29, 139
Interchange instability in magnetized plasma, 351
Interpolating approximation, 281

Interpolation
  area weighting, 338
  convolution, 338
  Gaussian, 350
  nearest-grid-point, 338
  particle-grid, 338
  smoothing, 338
Ion acceleration, 301–304
Ion acoustic wave, 327, 349
Isolated triple systems, 409

J

Jacobian matrix
  computation, 140–142
  in chemical kinetics, 140

K

Kinetic rate constant, 423

L

Lamb's parameter, 63
Langmuir frequency, 338
Large-mass method, 209
Laser-heated plasma, 301–304
Latent heat, 70
Lattice
  normal modes, 4–5
  of point masses, 4–8
Leapfrog differencing, 39, 237, 275
Lenard–Balescu collision operator, 357
Levi–Civita matrix, generalized, 399, 410
Lewis number, 134
Linear lattice, 4
Linear-multistep method, method of averaging, 21–22, *see also* Orbit averaging; Subcycling
Local thermodynamic equilibrium (LTE), 84, 147
Lorentz force, gyroaveraged, 362
LU (lower–upper) decomposition, 166

M

Magnetic flux coordinates, 194–196
Magnetic moment, 283, 362
Magnetofluid dynamics, *see also* Magnetohydrodynamics

Magnetohydrodynamics, 55, 239, 272, 273
  waves in Tokamaks, 189
Master equation, 429
Maxwell's equations, 274
  implicit differencing, 291
Maxwell–Vlasov equations, 236
MC, *see* Monte Carlo
Method of averaging, 9–11
Method of lines, 16, 122–127
Metric tensor, 196
Metropolis, 11, 432
MHD, *see* Magnetohydrodynamics
Modified equation, 17
Molecular dynamics
  force-biased Monte Carlo, 421, 431–432
  mass tensor, 421
  shake algorithm, 421, 429
Moment equations, 236–239, 275
Momentum conservation, direct method, 355
Monte Carlo, 11, 281
  force bias, 432
  smart, 432
  transition probability, 433
MP, *see* Metropolis
Multifrequency-gray (MFG) acceleration, 82, 102, 152
Multifrequency (multigroup) method, 89
Multiple time scales, 2

N

Natural boundary conditions, 213
Newton's method, 127, 139, 166
  bandwidth, 127
  Jacobian matrix, 127
NGP, *see* Interpolation, nearest-grid-point
Noise
  in numerical weather prediction, 59–60
  gravity wave, 60
Nonlinear constraint, balanced state, 65–66
Nonlinear equations, solution, 138–142
Nonlinear partial differential equations, Godunov's method, 18
Numerical algorithms, 2
Numerical methods for multiple-time-scale problems, 11–22
Nyquist oscillation, 322

O

ODE, *see* Ordinary differential equation

One-sided differencing, 141
Opacity
  absorption-mean, 150
  flux-mean, 151
  Planck mean, 150
  Rosseland-mean, 152
Operator splitting, 139
Orbit averaging, 245, 315, 320–331
  direct implicit method with, 326
  implicit, 323
  linear dispersion for magneto inductive model, 322
  linear dispersion relation for direct implicit method with, 326
  magnetoinductive model, 320
  stability analysis for implicit moment method with, 325
Ordinary differential equation, 32, 123

P

Parabolic equations, 189
Partial differential equations, 40
  numerical methods, 16–19
Particle-in-cell, method, 234, 273, 336
  cloud, 247, 337
Particle orbits
  in collisional plasma, 284
  guiding center approximation, 282
  in magnetic field, 9–11, 282
PDE, *see* Partial differential equation
Perturbation analysis, 19–20
Phenomenological modeling, 273
Photon-mean-free path, 147
PIC, *see* Particle-in-cell method
Pipe flow, 124–127
  comparison of ODE and DAE, 126–127
  Von Mise coordinate transformation, 125
Planetary perturbations and collisions, 393–395
  collision algorithm, 396
  grid method, 394
  helio-centric coordinates, 394
Plasma dispersion
  with explicit equations, 273
  with implicit equations, 273, 296–299
Plasma oscillations, 233
Plasma skin time, 190
Plasma turbulence, 305–308
Poisson's equation, 343
Poloidal magnetic field, 188
Protostellar cloud, 174

Index

Q

Quasi-equilibrium, effect of fast-time-scale variation, 9–11
Quasi-geostrophic theory, 67
Quasi-neutral condition, 234

R

Radiation
 coupled flows, 146
 energy equation, 150
 equilibrium, 86
 momentum equation, 151
 radiative transfer, 75
 space and time scales, 146
 specific intensity of, 85
 thermal, 74
 transport equation, 149
Radiation hydrodynamics, 74, 84, 146, 179
 implicit adaptive-grid method for, 145–183
Rate constant, 419
Ray-mode representation, 6–7
Reaction-diffusion equations, 134
Reactive flux, 420, 424, 425, 426
Reduced equations, 2, 23, 313
Regularization, 378
 Aarseth and Zare (AZ) method, 409
 in Ahmad–Cohen scheme, 406
 Kustaanheimo–Stiefel transform, 398, 400, 409
 N-body algorithm, 411–412
 regularized companion, 380
 three body, 380, 409–410
 two body, 378, 396–399
 two body algorithm, 402–405
Resistive magnetohydrodynamics, 209
Return currents, 235
Richardson's equation, 53
Rotational waves, 61–62

S

Safety factor, 189
Self-consistent solution, of field equations, 272
Shallow water theory, 44, 62–65
 gravity waves, 44
 Rossby waves, 44
 on rotating earth, 63

Shock, *see also* Collisionless plasma
 supercritical, 173
 tube, 169
SI, *see* Source-interation
SLIC technique, 261
Slow manifold, first order, 66
SMC, *see* Monte Carlo, smart
Source-iteration method, 75
 iteration eigenvalue, 76
Spectral compression, 273
Spectral radius, 79, 95
Stability
 Doppler limit, 354
 trapped particle modes, 354
Star clusters
 core collapse, 415
 open, 378, 413
 post collapse evolution, 415
 simulations, 413–416
 super clusters, 378
Stiff equations
 chemical kinetics, 12
 definition, 13, 116
 and implicit methods, 115–122
Stiffness operator, 148
Stochastic fields, 285
Stoichiometric condition, 117
Subcycling, 245, 314, 316
 linear dispersion and stability, 317
Subspace projection, 3, 5, 7
Susceptibility
 implicit, 344
 in magnetized plasma, 294

T

Thermal equilibrium in uniform plasma, 331
Tidal effect, 404, 413
Tidal waves, operator, 64
Time scales, plasma physics, 311
Toroidal magnetic field, 188
Transition state formula, kinetic rate constant, 426
Transport
 laser-generated electron, 234
 neutron, 75
Transport equations, in one-dimension, 210
Trapezoidal rule, 38
TST, *see* Transition state formula

U

Umbrella sampling, 427
  activated barrier crossing, 428
Upwind differencing, 164

V

Vacuum equation, 201
van Leer corrected donor-cell differencing, 262
Variable mesh method, 130–138
Velocity moments, 275
  asymptotics, 280–281
  calculation by extrapolation, 289

Viscosity,
  artificial, 156
  artificial tensor, 148, 160
  kinematic, 161
Volume of fluid approach, 261

W

Wave phenomena, 189
Weibel instability, 235

Z

Zero-mass method, 209